Cost Engineering f
Prevention and

Cost Engineering for Pollution Prevention and Control

Paul Mac Berthouex and Linfield C. Brown

CRC Press
Taylor & Francis Group
Boca Raton London New York

CRC Press is an imprint of the
Taylor & Francis Group, an **informa** business

First edition published 2021 by
CRC Press
6000 Broken Sound Parkway NW, Suite 300, Boca Raton, FL 33487-2742

and by

CRC Press
2 Park Square, Milton Park, Abingdon, Oxon, OX14 4RN

Library of Congress Cataloging-in-Publication Data

Names: Berthouex, P. Mac (Paul Mac), 1940- author. | Brown, Linfield C., author.
Title: Cost engineering for pollution prevention and control / Paul Mac Berthouex and Linfield C. Brown.
Description: First edition. | Boca Raton : CRC Press, 2021. | Includes index.
Identifiers: LCCN 2020046726 (print) | LCCN 2020046727 (ebook) | ISBN 9780367710606 (hbk) | ISBN 9781003154693 (ebk)
Subjects: LCSH: Sanitary engineering--Cost effectiveness. | Pollution--Economic aspects.
Classification: LCC TD152 .B47 2021 (print) | LCC TD152 (ebook) | DDC 628.5068/1--dc23
LC record available at https://lccn.loc.gov/2020046726
LC ebook record available at https://lccn.loc.gov/2020046727

ISBN: 978-0-367-71060-6 (hbk)
ISBN: 978-0-367-72415-3 (pbk)
ISBN: 978-1-003-15469-3 (ebk)

Typeset in Times
by Deanta Global Publishing Services, Chennai, India

Contents

Preface

Environmental engineering projects compete for resources with projects in medicine, transportation, education, and other fields that serve the public welfare. The goal is to efficiently use available resources. This book is about cost-estimating, identifying and evaluating alternatives, financial management, and some mathematical and statistical methods that facilitate making the necessary evaluations.

The methods for doing this are presented in 20 chapters that are divided into six sections.

- *Planning and Estimating Capital and Operating Costs* (*Chapters 2–5*)
- *Economic Evaluation of Projects* (*Chapters 6–10*)
- *Financial Management of Operations and Design Projects* (*Chapters 11–13*)
- *Linear and Nonlinear Optimization* (*Chapters 14 and 15*)
- *Model Building Optimization by Experimentation* (*Chapters 16 and 17*)
- *Uncertainty, Monte Carlo Simulation, and Reliability* (*Chapters 18–20*)

Most real problems start with an imprecise problem statement, e.g., this operation costs too much; the emissions are too high. The design engineer has the key role in formulating a clear problem statement, devising a strategy for solving the problem, and calculating the solution. Solving problems involves writing a set of possible prescriptions and doing analysis to select the best design. *Best* is judged with respect to measures of performance, such as total capital cost, annual cost, annual net profit, return on investment, cost-benefit ratio, net present worth, minimum production time, maximum production rate, minimum energy utilization, minimum weight, and so on.

These methods are illustrated with examples about wastewater, air pollution, sludge and solid waste management, remediation projects, and public health. They can be understood without an advanced knowledge of process design. An introductory course in environmental engineering will be helpful, more for the vocabulary than for process design. The book is suitable for engineers, managers, and environmental scientists.

Mathematical complexities are minimized to focus on the essence of the engineering problem. Emphasis is more on formulating the problem statement than on the mathematical details of the calculations. The book can be used with a knowledge of algebra and the most basic concepts of calculus.

The boundary between process design and cost engineering is fuzzy. Many examples involve some aspects of process design, such using material and energy balances and sizing treatment processes and equipment. The design-related ideas are basic and an introductory course in environmental engineering should provide sufficient prerequisite knowledge.

This is not a design manual or a cost-estimating manual. It is about concepts and methods of cost engineering that are universal. This is most evident in the cost data, which come from different countries and from different times. We wish they could all be made current and brought to the same basis, but this is not possible. Nor should it be necessary because the examples and the data are meant only to support and make more realistic explanations of the concepts.

The clumsy US units of measurement (gallons, feet, etc.) have been used in examples when the data came from case studies. We felt it was proper to keep them in the original context. The concepts will be understood without converting units from one system to the other.

These good friends deserve our gratitude and thanks for improving the book.

- James L. Nemke, formerly Director and Chief Engineer, Madison Metropolitan Sewerage District, reviewed the manuscript and contributed material for Chapters 10 and 11.

- Don M. Marske, formerly Vice President, CH2M Hill, reviewed the manuscript and contributed material for Chapters 12 and 13.
- David K. Stevens, Professor of Civil and Environmental Engineering, Utah State University, reviewed the book and provided technical advice.
- Tom Foltz, Strand & Associates, Madison, WI, collected and organized the wastewater equipment cost data and cost-estimating methods in Chapter 4, with the generous assistance of Bill Buckles from LW Allen, Josh Gable from Centrisys, Sue Canney and Larry Henderson from Energenecs, Rich Hussey from Ley Associates, Dean Wiebenga from Peterson Matz, Rich Knoelke from Mulcahy Shaw, Todd Steinbach from Aeromod, and Andrew Synhorst from Vulcan Industries.

Paul Mac Berthouex
Madison, Wisconsin

Linfield C. Brown
Braintree, Vermont
August 2020

Access the Solutions Manual at www.routledge.com/9780367710606

Author Bios

Paul Mac Berthouex is Emeritus Professor of civil and environmental engineering at the University of Wisconsin-Madison, where he worked for 29 years. Awards and honors include the Rudolph Hering Medal (1975 and 1995) from ASCE, and Harrison Prescott Eddy Medal (1971) from the WPCF, and the Radebaugh Award (1989 and 1991) from the Central States WPCA. He is a member of the University of Iowa Distinguished Alumni Academy. Work experience includes Chief Research Engineer for GKW Consult, (Mannheim. Germany), consultant to UNDP, UNESCO, WHO, and TVA, project manager of Asian Development Bank Projects in Korea and Indonesia, treatment plant design in Nigeria and Samoa, visiting professor in England, Denmark, New Zealand, Taiwan, Singapore, India, and Indonesia. Prof. Berthouex is co-author of eight books, 200 papers and technical reports, and was graduate advisor to almost 100 graduate students.

Linfield C. Brown is Emeritus Professor and former chairman of civil and environmental engineering at Tufts University. He taught courses on water quality modeling, water and wastewater chemistry, industrial waste treatment, and engineering statistics, and was the recipient of the prestigious Lillian Liebner and Seymour Simches Awards for excellence in teaching and advising. He has served as consultant to the US Environmental Protection Agency and the National Council for Air and Stream Improvement. Prof. Brown is the author or co-author of over 60 technical papers and reports, has offered over two dozen workshops in the United States, Spain, Poland, England, and Hungary on water quality modeling with QUAL2E, and is the co-author of five books on statistics and pollution prevention and control.

1 An Introduction to Cost Engineering

Cost binds a variety of components and considerations into an understandable whole. What all engineering projects have in common is the competition for resources and the challenge to efficiently invest the resources that are available. *Cost engineering* is about methods that are used to evaluate investments so that available monies are used wisely.

The fundamentals are estimating costs, making economic comparisons of alternative solutions, financing construction and facility operations, and financial management of projects. Some special methods are formulating and solving problems to minimize costs, optimizing processes by experimentation, building statistical models, and engineering to accommodate uncertainty, variability, and reliability.

These methods and concepts apply to all engineering disciplines and specialties, to private and public works projects, and to decisions in business and personal finance. This book shows how they are used to solve pollution prevention and control problems.

1.1 HEALTH AND HAPPINESS

Albert Einstein said that the environment was "everything that is not me." For "me" to be healthy and happy, everything that is "not me" must be healthy and happy. We disturb the environment as we draw from it food, water, shelter, clothing, and energy for our material needs, and as we dispose into it our wastes. Our disturbances must be managed carefully to avoid damaging our health and happiness.

The goal is to promote health and happiness. The difficulty for engineers, who like to measure and quantify outputs, is that there is no metric for measuring happiness, and not very precise ones for measuring healthiness. The things we can count and measure are at the bottom of the hierarchy of goals shown in Figure 1.1.

1.2 THE COST OF ACTION AND NO ACTION

The cost of *no action* toward environmental protection is greater than the cost of action. Action delayed will increase future costs, endanger lives, and it may cause irreversible environmental damage. Action is required by law in all parts of the world, and the required actions improve environmental quality and protect public health.

Table 1.1 lists a few of the impacts of investing money to improve health and environmental quality.

1.3 ABOUT THIS BOOK

This book has 20 chapters that are organized into six sections.

Chapters 2–7: Planning and Cost Estimating

Capital costs buy fixed assets, such as equipment, land, and structures. The capital cost estimates in this book are the kind that is made early in the design process to evaluate project feasibility and budgets. The accuracy of cost estimates improves as design details are fixed and uncertainties are removed. Operating costs for labor, electricity, natural gas, and various chemicals keep personnel, equipment, and processes working.

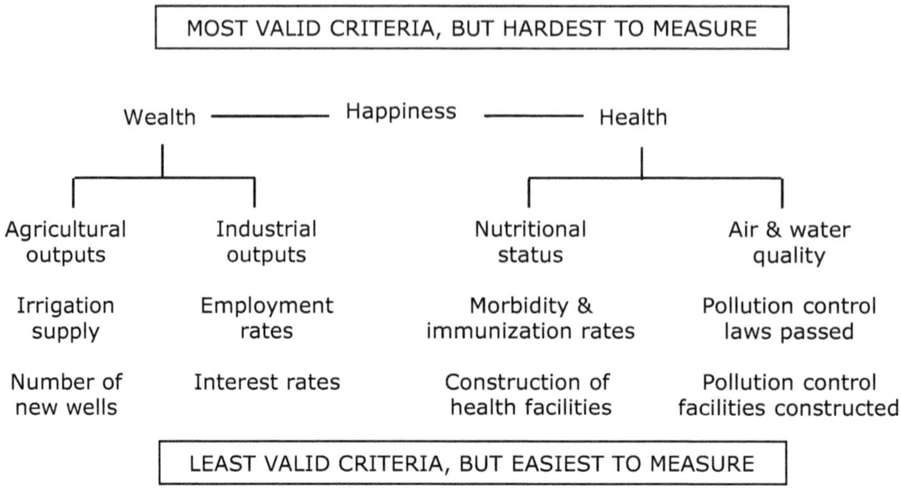

FIGURE 1.1 The hierarchy of goals for environmental and human health.

TABLE 1.1 Examples of *No-Action* Impacts on Health, Environmental Quality, and Productivity

Impacts on	Examples of Impacts
Health	Increased burden of disease due to reduced drinking water quality
	Increased burden of disease due to reduced bathing water quality
	Increased burden of disease due to unsafe food (contaminated vegetables, fish, and other farm products)
	Increased risk of disease when working or playing in wastewater irrigated areas
	Increased financial burden on health care
Environment	Diminished recreational opportunities
	Decreased biodiversity
	Degraded ecosystems (e.g., eutrophication and dead zones)
	Increased greenhouse gas emissions
	Increased odors and unsightly conditions
Production	Reduced industrial productivity
	Reduced agricultural productivity
	Reduced market value of harvested crops due to unsafe irrigation
	Reduced tourism and the willingness to pay for recreational activities
	Reduced fish and shellfish catches, or reduced market value of fish and shellfish

Chapters 8–10: Economic Evaluation of Projects

Projects have a life cycle cost that consists of the capital costs and the operating costs. Capital costs often are a lump sum that is invested at the beginning of the project. Operating costs go on year after year. These two kinds of costs need to be combined into a single measure of the economic attractiveness of a project, or for several projects that are to be compared. One approach is to convert an initial cost into an equivalent series of annual costs; this is *amortization* of the capital cost. A second approach is to convert a stream of annual costs into an amount that is equivalent to a lump sum at time zero that is called the *present value*. Cash flow over the project lifetime and the value of money over time must be known to make these calculations.

Chapters 11–13: Financial Management of Construction, Operations, and Design Projects

Every organization or project, whether private or public, profit-making or non-profit, has a budget for planned spending. Fees are established to collect the required revenue. A public utility relies mostly on user charges that are paid by the customers that benefit from the services provided. An engineering firm gets its revenue from fees that are paid by clients. Spending is monitored and tracked against revenues to monitor and protect the financial health of the project or organization.

Chapters 14 and 15: Linear and Nonlinear Optimization

The complete mathematical description of a problem is a statement of effectiveness, known as the objective function, and a set of equations that constrain the solution to fall within specific bounds. The objective is usually to maximize profits or revenues or to minimize costs. The constraints are limitations on money, personnel, raw materials, or operating conditions. A feasible solution cannot violate any of the constraints.

Chapters 16 and 17: Optimization by Experimentation

Many pollution control processes must be understood by doing site-specific or process-specific experiments. A common problem is to build, fit, and evaluate mathematical models. The experimental strategy depends on the form of the models. Mechanistic models are based on fundamental engineering principles. Empirical models are polynomial equations that can describe the test results, but they do not necessarily explain the underlying scientific mechanisms.

Efficient strategies are needed to maximize the rate of learning with the least investment of time and money in experiments. An iterative approach usually is the best experimental strategy. The first iteration may be to identify the factors that are most important in the process performance. Optimum-seeking investigations should start with a modest experiment to discover how factor levels should be changed in order to move toward the region of optimal performance. Factorial experimental designs are recommended for their efficiency.

Chapters 18–20: Uncertainty, Variability, and Reliability

Uncertainty is caused by a lack of information and data. Variability is inherent in physical phenomena and processes, and data can be used to identify patterns. Reliability is the probability that a process or system will function as intended when it is needed. Engineering to accommodate variability, and evaluating uncertainty and reliability, makes processes safer and more stable. Monte Carlo simulation is a method for understanding system behavior when inputs are variable, process data is uncertain, or system components may fail.

1.4 CONCLUSION

This book introduces a variety of methods that are used to evaluate and optimize engineering designs. The methods and concepts can be useful in all areas of pollution prevention and control.

The goal is to raise the level of health and happiness through rational investments in pollution prevention and control, where *rational* means that decisions are based on data and financial analysis and not on emotions or unsubstantiated opinions.

Most real problems have more than one technically feasible solution, and the alternatives need to be compared so that the best can be selected. The basis of comparison in engineering design is almost always the cost, usually the life cycle cost, which combines the cost of building the process with the annual cost of operation and maintenance. The most fundamental parts of cost engineering deal with evaluating and comparing alternatives.

This book is about the application of cost engineering methods to well-defined problems in pollution prevention and control, with an occasional mention of drinking water. Well-defined means that the criteria for evaluating projects can be precisely defined and quantified. Part of the definition is often the level of treatment required by a regulation, in which case it is assumed that all solutions that satisfy the regulation produce equal benefits, so the comparison can be made entirely in terms of cost. There are some problems where non-monetary factors may need to be considered. In that case, the subjective factors can be listed and evaluated separately from cost.

Many important problems are not like this, in particular problems that have benefits that are diffuse and difficult to evaluate. Climate change, large water resource projects, and evaluating the net benefits of cleaner air are difficult in this way. These broad policy issues and the methods used to evaluate them are outside the scope of this book.

1.5 PROBLEMS

1.1 SUSTAINABILITY

Figure P1.1 shows a sustainability region at the intersection of three sectors that bear on the importance and effectiveness of pollution prevention and control. Explain.

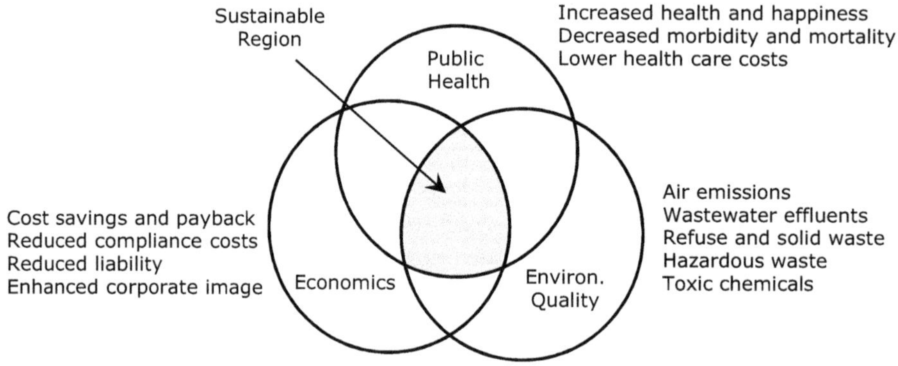

FIGURE P1.1 Possible measures of health and wealth.

1.2 INCOMPATIBLE GOALS

Minimum cost and minimum environmental impacts are competing goals in most pollution prevention and control projects. Explain why.

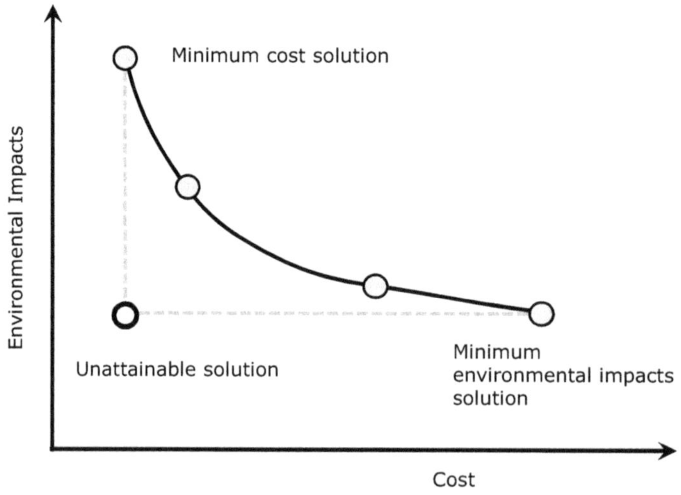

FIGURE P1.2 Competing project goals.

1.3 HEALTH AND HAPPINESS

Engineers like to measure and quantify inputs and outputs, but there is no metric for measuring happiness and not very precise ones for measuring healthiness. Figure 1.1 shows some possible measures of health and wealth, which we presume are some measures of happiness. The things we can count and measure are shown at the bottom of the hierarchy of goals.

(a) Define, in your own words, health and wealth.
(b) Rank the factors listed under the category of health in order of importance.
(c) Rank them in order of our ability to measure or quantify them.
(d) How can the factors be compared quantitatively?
(e) Add more factors to the list.
(f) Compare our ability to measure and quantify the factors under wealth with those under health.
(g) Are the health and wealth factors independent? That is, does nutritional status stand as an independent factor, or is it linked to or dependent on wealth factors?

1.4 COST OF INACTION 1

A UK company was fined £15,000 for polluting a river. The total cost, including remediation costs, is given in Table P1.4. This does not include internal costs and loss of production. Discuss the company's failure to act.

TABLE P1.4

Activity	Cost (£)
Fine	15,000
Hauling away the chemical	32,000
Plugging the leak	8,300
Initial fish survey	6,000
Installing new equipment to detect a leak	60,000
Payment to government officers responding to the incident	2,270
Further fish surveys	20,000
Restocking stretch of river	63,500
Prosecution costs	3,493
Total	211,164

1.5 COST OF INACTION 2

A fire at a chemical company polluted a river. This accident cost the company £6,000,000 in damages and lost business. In addition, they were billed £300,000 for the environmental agency's investigation of the incident. Discuss the company's failure to act.

1.6 COST OF INACTION 3

During 1975 and 1976, a company legally disposed of 104 drums of hazardous waste in a landfill. The landfill later leaked, the landfill operator went bankrupt, and the company was legally liable for $12,000,000 in cleanup costs under the joint and several liability provisions of the US EPA Superfund legislation. When the wastes were generated, at the rate of one drum per week, disposal costs were $10 per drum. In 1974, the company evaluated a process modification that would have eliminated this waste stream at a capital cost of $2,000,000 and operating cost of $2,500 per drum of waste avoided. How much would the pollution prevention project have saved, assuming that the capital equipment was purchased on October 1, 1974, and operation began on January 1, 1976?

1.7 HIDDEN AND AVOIDED COSTS

Managers often have difficulty justifying investments because the costs of hazardous chemicals are hidden as part of overhead. Consider the hidden and direct costs of hazardous materials use and disposal, and the indirect benefits and cost avoidance gained through using pollution prevention. Facilities that separate the costs of using, managing, and disposing of toxic substances have an average of three times as many pollution prevention projects as facilities that don't.

(a) Two hidden costs are permits and spill reporting. Name others.
(b) List at least five benefits of pollution prevention.

2 Defining the Engineering Problem

Design is most often used to mean the conception and preparation of detailed written and graphical descriptions of all the system components. It is more than that. Important design decisions are made long before detailed drawings are prepared. In some projects there are no drawings at all. The design might be a better way of financing bonds, a new way of scheduling waste hauling, or improved instructions for operating and maintaining pumps or motors.

Every design problem will involve answering these questions:

- What problem needs to be solved?
- What alternative designs can reasonably be expected to solve the problem?
- What is the cost of each technically feasible alternative?
- Which is the best solution for implementation?

Sometimes non-technical and non-monetary factors are identified and evaluated for each alternative. Such factors are system reliability, worker safety, esthetics and public relations, flexibility and ease of operation, constructability, compatibility with existing systems, and ease of expansion.

2.1 THE STAGES OF DESIGN

Figures 2.1 and 2.2 show that creative decisions are made early in the design process. As the work moves toward detailed drawings and specifications, the opportunity for conceptual innovation is diminished. The accumulating volume of detailed information consumes the design effort.

2.2 IDENTIFYING THE PROBLEM

The systematic solution of an engineering problem follows a logical process that is no different from that which everyone should follow in conscious everyday reasoning. It does differ, however, from casual judgments that lack a thorough and impartial pursuit of fact and data. Casual decisions are distinguished from engineering decisions largely by the clarity with which hypotheses and assumptions are stated, and by the careful collection and analysis of factual information.

A problem is a difficulty that needs to be resolved. It is an opportunity to make someone happier, healthier, wealthier, or wiser; an opportunity to do something in a more efficient and less expensive way.

The first step is to identify a problem worth studying. Some things are important but cannot be managed. Some things can be managed but are not important. The problems of interest to engineers are at the intersection of these two categories.

A problem:

- Begins with a suspicion that things could be improved, and that decisions for change are needed.
- Can arise because the system has always been wrong; for example, people have always polluted a river.
- Can emerge slowly, because the system is slowly getting worse or the community's expectations are increasing.
- Can arise because of a sudden change. For example, a new factory has started operations or a new law has been passed.

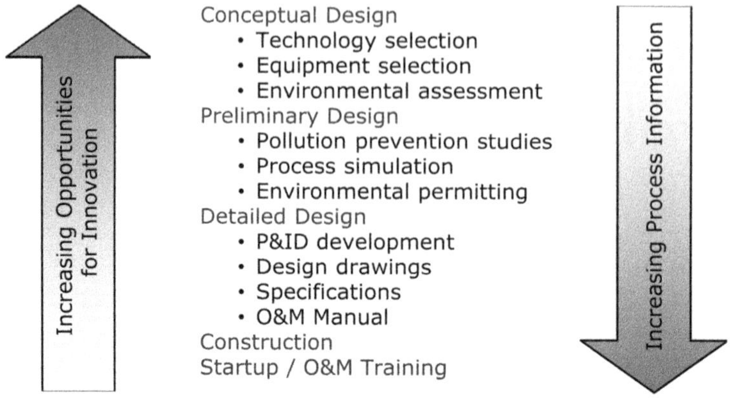

FIGURE 2.1 The early stages of design are when innovation occurs.

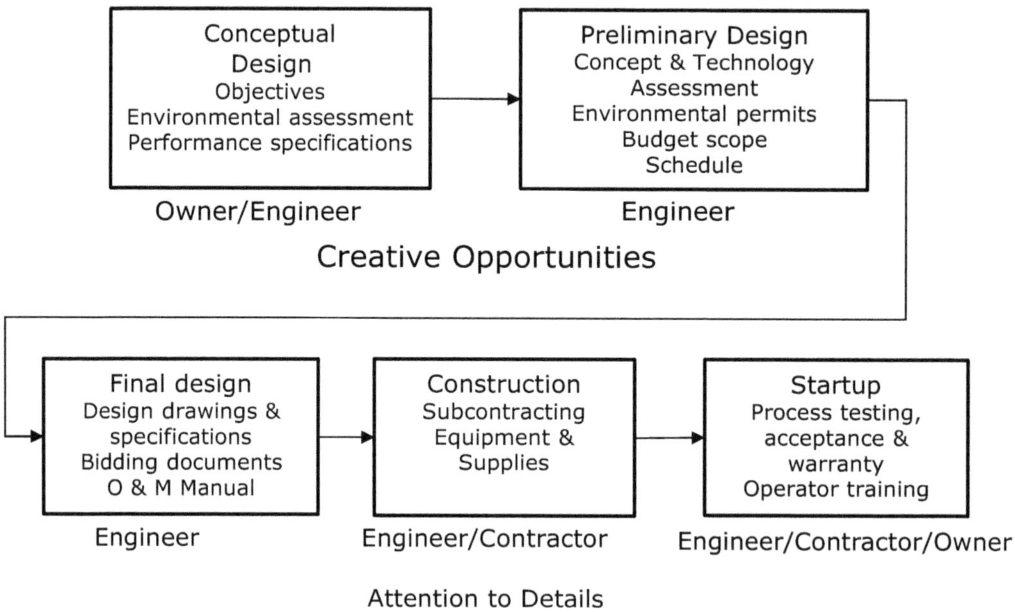

FIGURE 2.2 Stages of design and construction.

2.3 DEFINING THE PROBLEM

A clear definition of the problem is essential. Sometimes the problem is defined for you.

The state regulatory agency orders that effluent phosphorus cannot exceed 0.5 mg/L and the current average is 5 mg/L. The dominating problem is to remove phosphorus. There are many ways to do this, which means there are sub-problems to be identified and solved.

Most opportunities to make things better do not appear as fully formed problem statements. Clarification and precise definition are required.

A storage lagoon contains sludge that was produced over a period of 40 years, including the era when PCBs were in use. The treatment plant is shifting from sludge storage to land application. This raises questions about managing the PCB-contaminated sludge. Can it be applied to corn fields? In what amounts and at what PCB concentrations? What are the risks? How will contaminated sludge be identified and managed separately from the sludge that does not contain PCBs? Can the contaminated sludge be kept in the lagoon or will other treatment and disposal be required?

Clients and managers usually have excellent ideas but no time to develop or try them out. Talking to people who are involved in the system is clearly sensible. If nothing else, we may spot inconsistencies. Divergent views are often the source of the best ideas for change. Differences of opinion indicate that there is a good chance that the current system is not fully satisfactory.

Above all, be attentive and absorbent, but critical. Ask questions in a purposive way. Do not prejudge, but neither accept what you are told without question. Do not accept confused responses or unsubstantiated assertions unless the respondent admits to guessing or expressing opinion rather than fact.

Some vital follow-up questions are:

Why do you think that?
What evidence is there to support your view?
Has that ever been tested?

Imprecise words, like *small*, *feasible*, *increase*, and *expensive*, can cause confusion and misunderstanding. These are examples:

- The level of health in the village needs to be improved.
- The ammonia concentration is too high.
- Make the process easy to operate.
- Use the smallest number of units that is feasible.

Small and *feasible* may start a useful conversation, but more precise language is needed to avoid confusion and misunderstanding. This *pseudo-technical language* may lead to *pseudo-solutions*. A pseudo-solution answers a question that is of no interest to the person who has the problem.

An acceptable problem statement explains why the identified problem seems important and why we think it is capable of being solved. The manager or client needs to see this statement and accept its validity. Without this acceptance, the recommended solutions may not be implemented. If the design is of a structure or physical process, this statement comes easily. If the problem is to design a new way of operating a process, or an organization, it is more difficult.

Problem formulation begins with developing a model of the system. This could be a diagram, a flowchart, or a set of equations. Whatever its form, the model must help us understand how the real system works. In particular, it should show us how the system reacts to changes.

One decision is how wide a view needs to be taken, especially if there are shortages of data or no precise model structure for part of the system. It can be a serious mistake to restrict the analysis to only those parts of the system that are amenable to precise mathematical analysis. It is better to have an imprecise but valid model of the whole problem than a precise solution to the wrong problem.

2.4 A HYPOTHETICAL CONVERSATION ABOUT A PROBLEM

Here is a hypothetical conversation between a plant manager (M) and a pollution control engineer (E) when a printing plant receives notice that the volatile organic compounds (VOCs) emissions exceed the allowable limits.

M We are in violation of the Clean Air Act for emission of VOCs. We must become compliant within 18 months or else the company will be penalized with fines and unfavorable publicity. We want to be in compliance. If the cost is reasonable, we prefer to be well under the statutory limits for emissions.

E A variety of technologies exist to solve the problem. There will be several workable technical solutions. The VOCs are solvents that we use in manufacturing. The net cost of becoming compliant will depend upon the kind and quantity of the solvents being emitted. This may be an opportunity to save money. Some captured solvents can be reused and some can be used as fuel. Another approach is to substitute non-volatile chemicals for VOCs.

M We use a variety of solvents, and we know the amounts that are purchased. The largest volume is toluene. We can tell you which printing processes and cleanup operations use solvents and in what quantities. We can tell you which solvents are emitted from the printing presses and when the presses are operated. Is that the information you need?

E It is. With that data we can estimate the average and maximum emissions for each solvent type and calculate the reduction needed to meet the new standards. Then we can identify useful technologies and make preliminary cost estimates. With some luck – I should say good engineering – the project will save money after the first year or two.

M I understand how that could happen. We use a lot of solvents, and lose a lot, and the price per kilogram is quite high. The net effect of solvent recovery is like using less and losing less.

E If the polluted air stream contains a single solvent, we may be able to capture it, concentrate it, and recycle it to the printing process. We can do this with adsorption onto resins or activated carbon. Or, we might do it with a membrane process, for example, pervaporation.

M What if adsorption won't work? Or if the solvents are emitted as a mixture, which would make recycle and reuse problematic?

E If the exhaust air contains a mixture of solvents, we might separate them at the source. If the concentrations are too low for economic separation and recovery, we may be able to use incineration and recover heat instead of solvent.

M The technology – adsorption, membranes, and incineration – must be expensive.

E The pollution control equipment may cost a lot of money. Whatever it costs, non-compliance with the Clean Air Act will cost more. Non-compliance is not an option.

M True. It's necessary, but, still, it will be expensive.

E "Expensive" is a pseudo-technical word that carries a lot of emotion. The person who says "expensive" and the person who hears "expensive" may have quite different ideas about what that means. We will have alternative technical solutions. Some will cost more than others to install, and the same is true for operating costs. Let's get the costs. What is the net annual cost? What is the net cash flow, year by year? What is the payback period?

M I agree. That's how we evaluate manufacturing operations. If I say, "We had a good day," my boss will want to know what "good" means in terms of product shipped, product rejected, materials used, personnel, and quality control data. Let's get the data.

They set to work accounting for all material used in the plant. The first steps produced the material flow diagram shown in Figure 2.3. This is just a preliminary model of the screen-printing operation. Some inputs and outputs may be missing, and the quantities will be revised as the design progresses.

Alternative ways of meeting the emission standards included modifying processes to reduce losses and installing new pollution control equipment to recover solvents. The capital cost, operating cost, and net life cycle cost of the alternative solutions were compared.

Management may have been told something like this: VOC emissions need to be reduced by 80 percent. We propose a mixed strategy.

1. Half of the solvents used for cleaning can be replaced by aqueous cleaners. This reduces emissions by 5 percent.
2. Some solvents can be reused after purification by distillation. There will be a cost for disposal of the residue that is not recovered as distillate. This reduces emissions by 40 percent. The payback period is 18 months.
3. Air from ventilation hoods carries a solvent mixture that can be incinerated to produce useful heat energy. This reduces emissions by 40 percent. The payback period is two to three years.

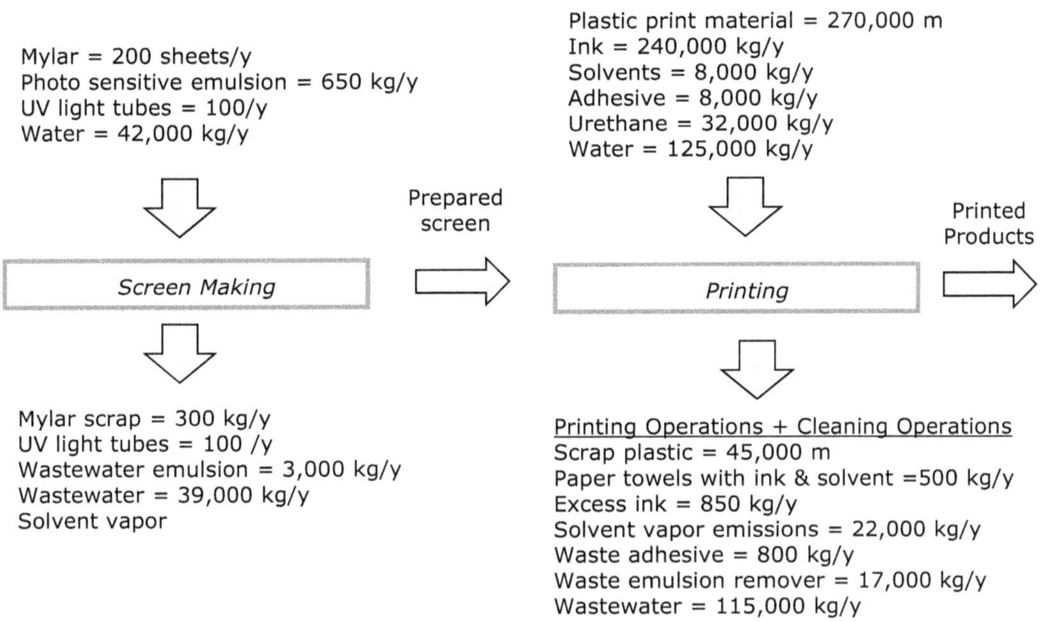

FIGURE 2.3 A preliminary schematic for the screen-printing operation. More details will be needed on the solvent composition, point of use, and point of ventilation or emission.

Each feasible solution will have pros and cons. Management may prefer a solution with a high operating cost and a low construction cost over one that costs more to build and less to operate. These financial decisions depend on interest rates, company debt, and competition for money among possible company investments.

2.5 EVALUATING ALTERNATIVE SOLUTIONS

Identifying alternatives combines imagination with engineering savvy. Constraints imposed at the formative stage are often imaginary, and this creates a risk of prematurely rejecting promising solutions. Avoid judgments like "it is too expensive" or "it is too complex" and let subsequent analysis select the alternatives that need to be studied in more detail.

A solution is a prescribed intervention that will produce better information, apply better physical technology, improve analytical techniques, modify management styles, reduce the cost, or accomplish more than one of these objectives. Solutions can be stimulated by the emergence of new knowledge, techniques, technologies, materials, or methods.

Most real problems have several technical solutions that will satisfy the design constraints (effluent quality, production capacity, etc.). They will differ in construction cost, operating cost, ease of maintenance, the flexibility of operation, robustness to shifts in ambient conditions, changes in loading rates, the amount of chemicals used, and the amount of solid waste and sludge that must be hauled away. The engineer narrows the options. Weak solutions are discarded, and good solutions are polished to get very good solutions. Details are refined. Flaws are identified and eliminated.

Proposed solutions are judged with respect to some measure of system performance. There is considerable choice in defining such a performance criterion. It could be, for example, total capital cost, annual cost, annual net profit, return on investment, cost-benefit ratio, or net present value. Or the measure might be stated in terms of technological factors, like minimum production time, maximum production rate, minimum energy utilization, minimum weight, and so on.

Environmental criteria could be reduced algal blooms in lakes, fewer beach closings due to high bacterial counts, prevention of premature deaths by reducing fine particulates in the air, or more miles of streams available for fishing and boating. In practical situations, it may be desirable to find a solution that is good with respect to more than one criterion. Evaluating multiple competing objectives is possible when judgment and experience complement mathematical solutions.

2.6 CONCLUSION

A problem is a condition that needs to be improved by providing more effort, more knowledge, more personnel, more money, or new technology. It may arise from a new law or regulation, changes in a wastewater discharge or air emissions permit, new demands from a community, or an internal recognition of an opportunity to reduce costs.

A problem statement should be precise and clear, and all parties should agree that it identifies and defines the problem that needs to be solved.

2.7 PROBLEMS

2.1 GOAL STATEMENTS

Rewrite these goal statements.

The wastewater treatment plant will work on odors.
The water system will provide sufficient pressure.

2.2 PROBLEM SELECTION

There are innumerable pollution prevention and control problems. Some must be tackled to comply with environmental regulations. Others are worthy but less demanding. Explain how Figure P2.2 helps to direct attention to worthwhile problems.

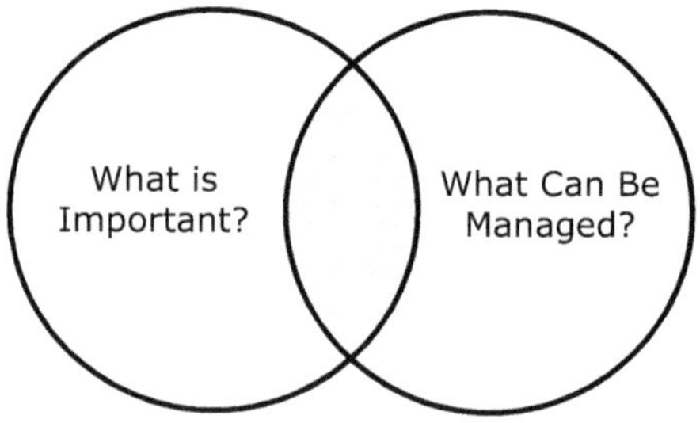

FIGURE P2.2 Tackling worthwhile problems.

2.3 CONFUCIOUS SAYS

"Man has three ways of acting wisely. First, on meditation, this is the noblest. Second, on imitation, this is the easiest. Third, on experience, this is the bitterest." Find examples of acting with good intentions that resulted in the bitter experience of creating a problem.

2.4 DEFINING THE PROBLEM 1

Degreasing metal parts before they go to plating or painting has traditionally used organic solvents. Solvent emissions are regulated under the Clean Air Act, so industries are looking for ways to reduce the use of solvents or to reduce solvent losses to air or water. Figure P2.4 shows that dirty

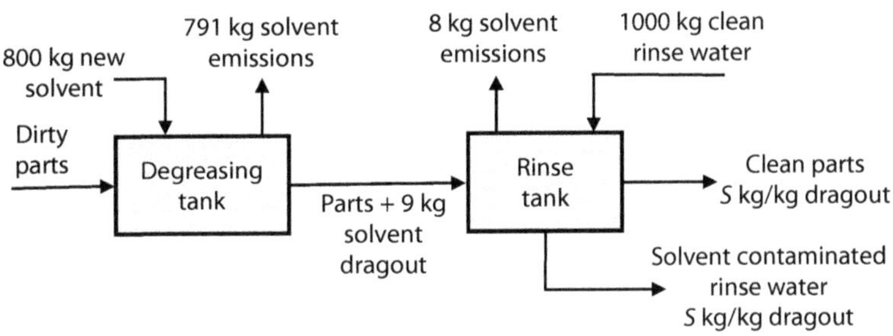

FIGURE P2.4 Degreasing metal parts.

parts carry some solvent when they leave the degreasing tank. This is dragout. The parts are washed and they carry dragout from the cleaning tank, but the dragout concentration is much less going out than coming in. Major solvent emissions (to the air) are from the degreasing tank and there are minor losses from the cleaning tank, as well as some solvent carried out with the rise water. Define the pollution prevention and control problems created by the system.

2.5 DEFINING THE PROBLEM 2

Executives of an industry asked for a summary of material and energy flow for a production unit. Engineers McBride and Murphy used Figure P2.5 in their presentation. Do you like one better than the other? Is one diagram more informative than the other? More complete? What is not revealed that could be important? Does either diagram sufficiently define the problem?

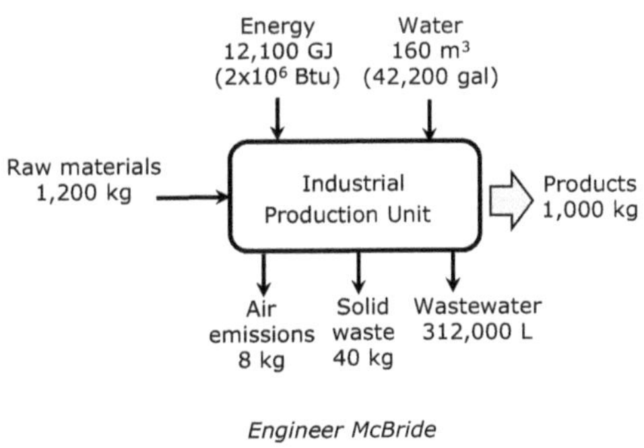

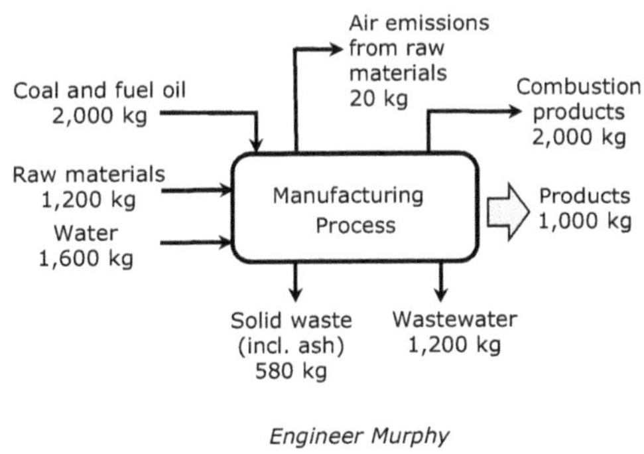

FIGURE P2.5 Material and energy balance presentations.

3 Planning for the Future

Planning requires forecasting future demands for drinking water, wastewater flows, amounts of pollutants that must be removed, amounts of sludge and solids to be treated and managed for disposal, and tons of solid wastes to be landfilled. It also includes adapting to new technology, new regulations, and changes in the composition of raw water, wastewater, solid waste, or contaminated air.

The historical practice for municipal water and wastewater utilities has been to forecast and plan for 20–30 years with an expectation of steady growth in population and the demand for services. Wastewater flow estimates are linked to water demands. Wastewater flows from residential and commercial areas are less than the purchased water. Industrial flows are added individually. Inflow and infiltration or stormwater and groundwater are added separately.

Recent history shows that contrary to historical patterns, populations of cities are increasing, but water demands and wastewater flows are not. The mass of pollutants in wastewater are increasing in proportion to population.

A grassroots project – the design and construction of an entirely new facility on a new site – is unusual in municipal engineering. Most municipal projects are expansions and modification of older facilities, usually with the incorporation of new technologies to reduce costs or to comply with new regulations. These projects are more complicated than simply adding a module to the existing process.

Industry uses shorter forecasting periods and accelerates the design and construction. The motivation is "time is money." Having money tied up over a long planning and construction period is undesirable.

3.1 FORECASTING AND UTILITY DECISIONS

Planning is done with different forecast horizons. Less than one year is short term, one to ten years is medium term, and more than ten years is long term. These can also be thought of as operational, tactical, and strategic planning levels. The forecasting frequency is greater than the decision-making frequency, as shown in Table 3.1.

A tactical planning decision relates to projecting revenues that are needed to operate the facilities and for managing the replacement of major equipment and sinking funds. The most important strategic decisions are planning for capacity expansions and changes in technology.

TABLE 3.1 Relationship among Planning Level, Water Utility Decision Problems, and Forecast Attributes

Planning level	Decision problem	Forecast horizon	Forecast periodicity
Operational	System operation management and optimization Short-term forecasts for optimizing pumping schedules.	Short term ($\leq$ 1 year)	Hourly, daily, weekly, monthly
Tactical	Revenue forecast; investment planning; staging system improvement	Medium term (1–10 years)	Monthly. annual
Strategic	Capacity expansion	Long term (> 10 years)	Annual

3.2 POPULATION FORECASTS

Population forecasts and land use plans are the basis for projections of demand for services. Forecasts made by cities and regional planning commissions are used in many engineering projects.

Many methods of population forecasting have been used:

- *Component methods* base forecasts on birth, death, and migration rates.
- The *cohort survival method* uses birth, death, and migration rates for age-classes (cohorts) of the population and forecasts how the size of each age class will change.
- *Multiple correlation methods* use symptomatic data, such as vehicle registrations, births, school enrollments, and sometimes ratios for different years to derive regression equations for population growth.
- The *ratio method* uses ratios of city/state or city/county populations, assuming that forecasts for larger areas are more accurate than forecasts for smaller entities.
- *Graphical extrapolation* is simply drawing a smooth curve through the relevant historical data and using the projected trend as the forecast, assuming the pattern of future growth will be similar to the past.

3.3 FORECASTS OF DEMAND FOR SERVICES

A long-used method of forecasting demand is to multiply the forecasted population by historical per capita values for water use, wastewater discharges and pollutant loads, and the volume and mass of solid wastes. This kind of forecasting has become unreliable because the historical per capita values have been changing. Over the last 30 years, populations have continued to grow, but water use and wastewater flows have decreased due to the success of water conservation programs. Modern forecasts use data on housing density, family size, family income, geographical patterns of growth, weather, and forecasts of climate change.

3.4 DEMAND FORECASTING BY PROJECT TYPE

Municipal drinking water systems are designed for long life, typically 20 years for treatment plants and processes and 40–50 years for storage and distribution systems. There are seasonal and daily fluctuations in demand. Daily fluctuations are mitigated by temporary storage of water until it is needed by consumers.

Wastewater collection and treatment plant expansion plans are usually made by estimating average flows and loads for some time in the future and applying peaking factors. The maximum and minimum flow into a wastewater treatment plant can vary by a factor of 2 or 3 over a 24-hour cycle. If the sewer system collects stormwater along with domestic sewage, the peaking factors may be tenfold. The concentrations of pollutants (biochemical oxygen demand; total suspended solids, Nitrogen, and Phosphorus) are also variable, but the mass loadings are more constant.

Interceptor sewers have a 50-year design life, due to a great economy of scale and the difficulty of increasing the capacity once the pipes are in the ground. In densely populated areas, the design for new sewers tends to be for full future capacity because the cost of future expansion would require great investments to dig up streets and move underground utilities. This expense is the reason that so many older cities, such as Boston, have combined sewers for stormwater and domestic sewage when separate sewers are preferable.

Municipal solid waste landfills have long lifetimes. One reason is the difficulty of finding a site and getting the necessary approvals. Landfill operation develops in stages or cells. The time between opening and closing each cell is short relative to the total lifetime. Other solid waste handling facilities, like transfer stations and centers for sorting and recycling, have shorter planning periods because they are more easily expanded (or abandoned).

Air pollution projects at power plants have a constant design capacity over a long period of time. The megawatt power capacity does not change, but process loadings shift as electricity production is adjusted to daily and seasonal patterns of energy use. Changes in air pollution regulations may force changes in control technology.

Wastewater treatment plants and landfills have gas treatment processes that are built to purify biogas so it can be used for heating or power generation. The volume of gas produced will increase as the landfill matures, and then decrease after waste deposits stop. Gas yields in wastewater treatment plants gradually increase over time as solid loadings increase.

3.5 WATER DEMAND

Forecasts of water demand are made by applying per capita factors to the estimated population. The specific water demand for a modern urban household might be 0.5 m^3/cap-d for water consumption and 0.4 m^3/cap-d for wastewater.

$$\text{Water demand of households}\left(\text{m}^3/\text{y}\right) = \left(\text{Population}\right)\left(\text{Specific water demand, m}^3/\text{cap-y}\right)$$

$$\text{Water demand of commercial firms} = \left(\text{Employees}\right)\left(\text{Specific water demand, m}^3/\text{emp-y}\right)$$

More complex methods incorporate detailed local information, including changes in water use habits and climate change. Water conservation programs have been effective due to:

- Rainwater reuse
- Introduction of water meters for apartments
- Modernization of sanitation facilities (dual flush toilets)
- More efficient washing machines

3.6 UNCERTAINTY IN FORECASTS

A 2015 survey asked water utility managers, "What do you consider to be the three main drivers of uncertainty about water demands in the next 20–30 years?" The five most frequent replies were (Keizer 2015) as follows:

- Future population or number of customers
- Future climate
- Future economic conditions
- Irrigation and outdoor water use
- Efficiency of future water technologies
- Characteristics of individual large customers

Engineering forecasts are often made by extrapolation of historical data. The historical record for wastewater flow will reflect changes in the pattern of water use (for example, toilets and plumbing fixtures that use less water) and sewer rehabilitation projects to reduce inflow and infiltration. The problem with extrapolation is that things change after the forecast is made, and dramatic changes are possible over a 10- or 20-year forecast period.

Figure 3.1 shows a series of water demand forecasts for Seattle, WA. Forecasts before 1980 were extrapolations of the historical demands, a method that had been used with success for many years. There was an obvious reluctance to acknowledge the flat demand curve from 1970 to 1990. After 1990, the forecasters started using a different approach but still could not completely accept the new trend. By the year 2000, the declining demands were incorporated in the forecasts, as shown by Figure 3.2. This pattern of decreasing demand has been observed in other large cities.

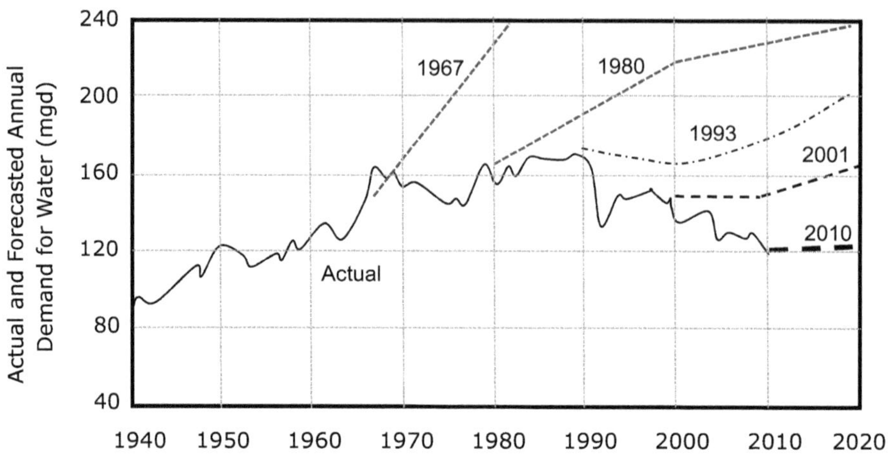

FIGURE 3.1 Actual water consumption and demand forecasts for Seattle, Washington (*Source:* Seattle Public Utilities, 2012). *Note:* In the United States, water and wastewater demands and flows are usually reported as million gallons per day (mgd). 1 mgd = 3,785 m^3/d = 158 m^3/h = 43.8 L/s.

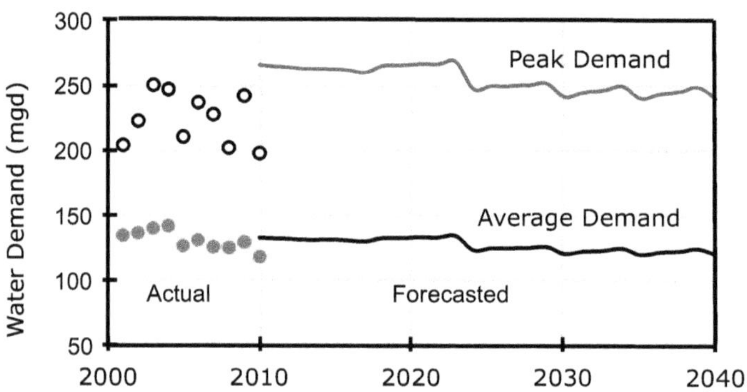

FIGURE 3.2 Water demand forecasts for Seattle (1 mgd = 3,785 m^3/d).

Demographers typically define forecasting error as the percentage by which a forecasted population deviates from the population actually achieved. Suppose that a population is estimated to grow from 800,000 to 1,000,000 over a 20-year period, but the actual population turns out to be 900,000. Demographers would define this error as 10% error.

An engineer is more interested in the discrepancy relative to the projected incremental growth. Suppose the forecast is for population growth from 800,000 to 1,000,000, an increase of 200,000, over a 20-year period and with a proportional increase in the demand for services. If the actual population after 20 years is 900,000, the project is overdesigned by 50% and the forecast error is 50%. The service population pays for the excess capacity.

Water demands have decreased, and the growth in wastewater flows has also slowed. For the moment, forecasting for wastewater is more about changes in regulations and technology than growth. In time, demands must increase again because water conservation measures will have a limit. There will be some minimum amount of water that households need. When that limit is approached, the rate of decrease will slow, but the service area populations will increase and demand will start to increase.

There are some demands that are growing. One is the demand for refuse disposal and landfill capacity. Another is the need for industrial wastewater treatment in expanding industrial parks, which are popular in developing countries.

3.7 CASE STUDY: NINE SPRINGS WASTEWATER TREATMENT PLANT, MADISON, WI

The Nine Springs Wastewater Treatment Plant is a regional plant that serves Madison, WI, and adjacent areas. Table 3.2 and Figure 3.3 show the average wastewater flow rates (mgd) for 1966–1992. These data were used to plan a plant expansion.

- A linear projection using all the data gives an annual increase of 1.05 mgd/y. The capacity to be added for a 2012 design life was 21 mgd, and the new capacity was 41 mgd + 21 mgd = 62 mgd.
- A linear projection using the 1982–1992 data gives an annual increase of 1.5 mgd/y. The capacity to be added for a 2012 design life was 30 mgd, and the estimated average design flow was 41 mgd + 30 mgd = 71 mgd.

It is interesting to examine the average flows from 1996 to 2017, which are given in Table 3.3, along with wastewater concentrations and mass pollutant loads. Concentrations are milligrams per liter (mg/L) and mass loads are pounds per day (lb/d). (Figure 3.4).

The population of this region has been growing rapidly, but the wastewater flowing to the Nine Springs plant has not. The steady growth in flow rate that seemed so obvious before 1990 did not continue. The same is true for the BOD, SS, and TP loads, which did not increase, but TKN did (Figure 3.5).

TABLE 3.2 Average Annual Flow Rates (mgd) at the Nine Springs Plant for 1966–1992

Year	Flow	Year	Flow	Year	Flow
1966	14.0	1975	22.5	1984	31.0
1967	14.1	1976	21.5	1985	30.2
1968	16.6	1977	22.0	1986	32.5
1969	17.1	1978	22.5	1987	33.5
1970	17.5	1979	22.5	1988	37.5
1971	17.5	1980	24.0	1989	38.5
1972	17.4	1981	25.5	1990	40.2
1973	17.5	1982	26.0	1991	39.7
1974	21.0	1983	28.5	1992	40.9

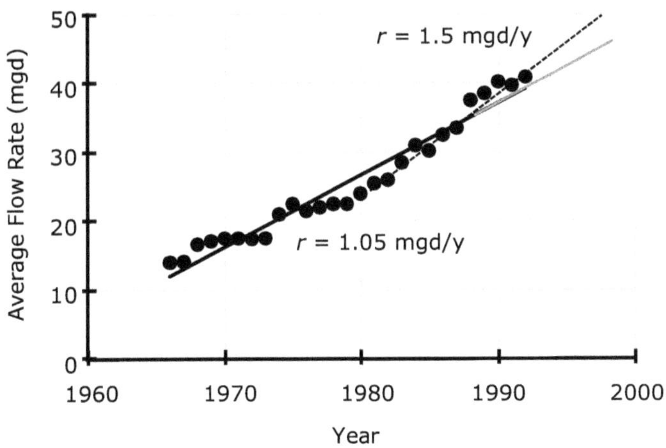

FIGURE 3.3 Plot of the data at the Nine Springs Plant for 1966–1992.

TABLE 3.3 Average Annual Flow Rates (mgd) and Pollution Mass Loads (lb/d) at the Nine Springs Plant for 1996–2017 (Note: 1 mgd = 3,785 m^3/d; 1 lb/d = 0.4536 kg/d)

Year	Flow (mgd)	TSS (lb/d)	BOD_5 (lb/d)	TKN (lb/d)	TP (lb/d)
1996	38.18	78,127	75,424	11,045	2,102
1997	36.92	76,269	74,933	11,162	2,045
1998	41.12	81,509	75,107	11,204	2,039
1999	41.59	83,769	78,115	11,342	2,087
2000	42.10	86,915	80,860	11,915	2,186
2001	41.76	80,197	81,648	12,439	2,132
2002	40.14	78,214	83,722	13,185	2,165
2003	38.56	75,592	84,396	12,955	2,125
2004	41.93	76,712	76,796	11,462	2,111
2005	39.37	78,127	75,424	11,045	2,102
2006	41.22	76,269	74,933	11,162	2,045
2007	42.88	81,509	75,107	11,204	2,039
2008	47.26	83,336	81,983	13,480	2,207
2009	42.63	80,090	77,506	13,333	2,062
2010	42.96	76,462	75,957	13,436	1,935
2011	40.45	73,994	76,242	13,562	1,687
2012	36.64	78,498	74,867	13,598	1,833
2013	40.94	80,626	76,824	14,272	1,844
2014	38.6	79,516	77,262	14,454	1,899
2015	38.23	75,919	73,333	14,571	1,945
2016	40.62	84,347	76,223	14,906	1,965
2017	42.08	85,298	74,050	15,758	1,930

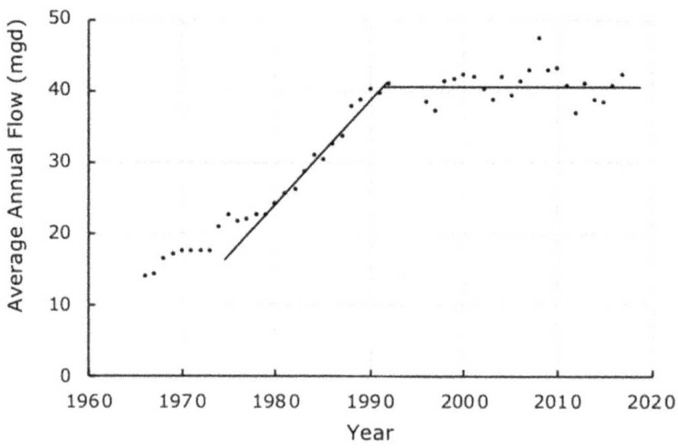

FIGURE 3.4 Plot of the average annual flow rate at the Nine Springs Plant for 1966–2017.

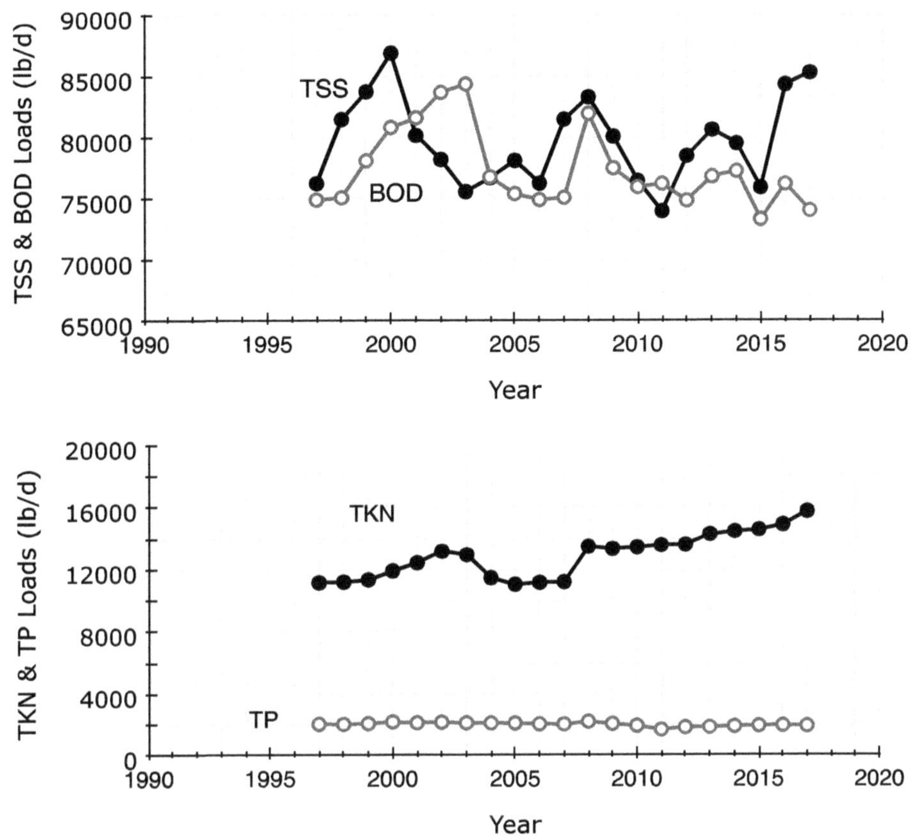

FIGURE 3.5 TSS, BOD, TKN, and TP loads at the Nine Springs Treatment Plant (1 lb/d = 0.4536 kg/d).

TABLE 3.4 Regional Planning Commission Projections for Population and Flows for the Nine Springs Wastewater Treatment Plant

Year	Population Served	Average Flow (mgd)	Average Flow gal/cap-d
2030	406,000	47	116
2040	431,000	50	116
2050	491,000	53	108
2060	560,000	60	107

In 2010, the Regional Planning Commission for the Madison area made the forecasts in Table 3.4. Notice the substantial population growth and the shrinking per-capita loads.

3.8 CASE STUDY: TAMPA BAY, FLORIDA

Figure 3.6 shows the historical water demand for Tampa Bay, Florida. The solid line is a 12-month moving average that distinguishes the trend from the seasonal variations. From 2007, the trend is downward and then flat from about 2009 (Hazen and Sawyer 2013).

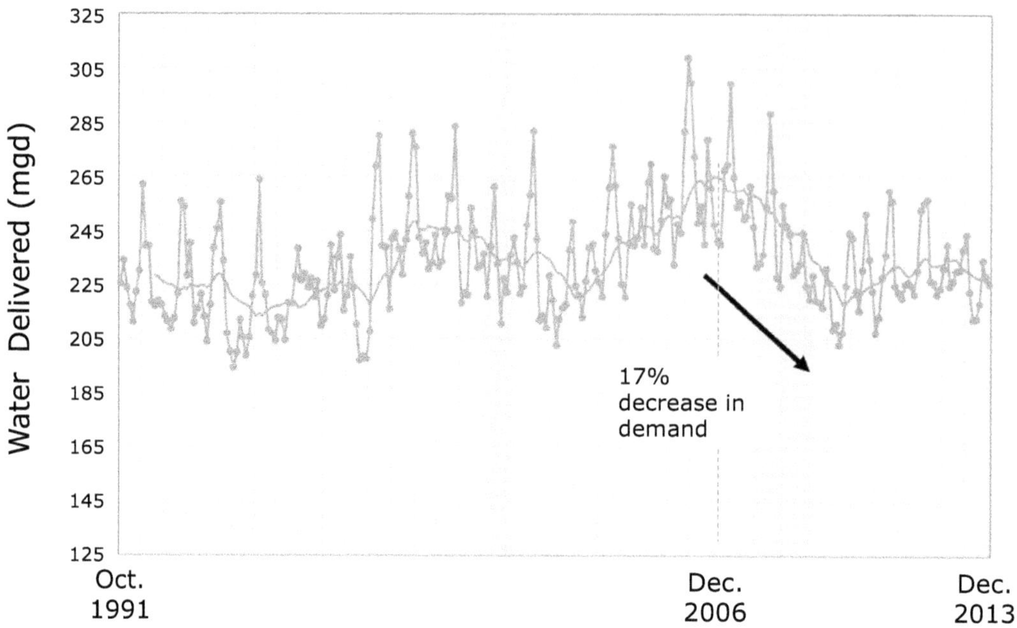

FIGURE 3.6 Tampa Bay Regional Water deliveries from 1991 to 2013. The smoothed line is a 12-month moving average (Keifer 2015).

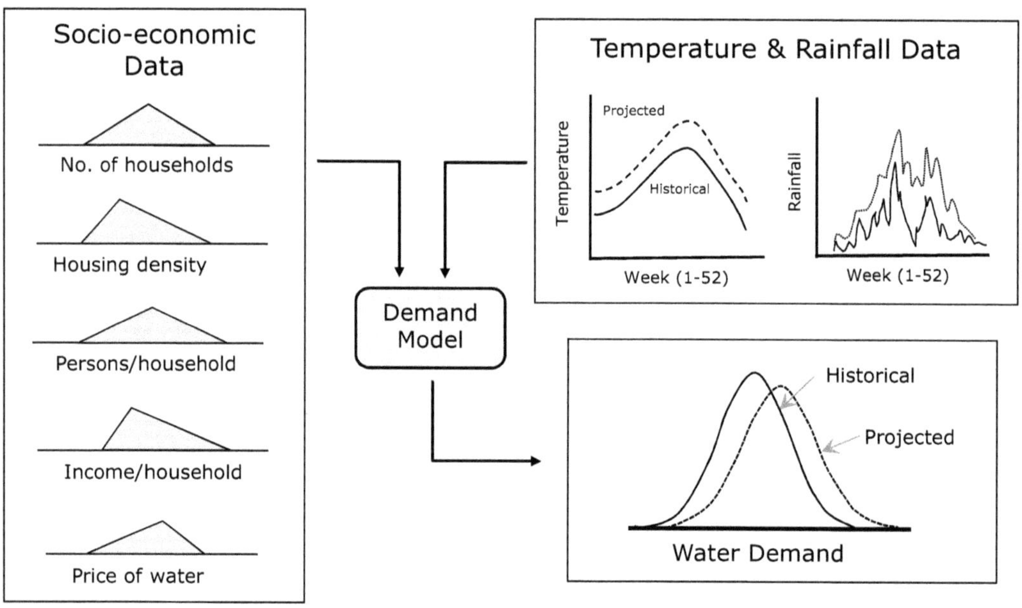

FIGURE 3.7 Tampa Bay, FL, water demand forecasting uses detailed information about the water consumer and local weather (*Source:* Adapted from Keifer, JC 2015; Hazen and Sawyer 2013).

The Tampa Bay water demand forecasting model, shown in Figure 3.7, is a giant step away from extrapolating historical data. Data on the number of households, housing density, household income, persons per household, and the price of water are inputs to the demand model, not as point values but as distributions of values. Temperature and rainfall are important factors, and they are part of the model. Water conservation is expected to reduce demand significantly (Hazen and Sawyer 2013).

Tampa Bay has used simulations to study the effects of uncertainty due to future household income, price of water, and housing density. Climate change is considered as well. The historical and projected air temperatures and rainfall are graphed in the upper right-hand corner of Figure 3.7. The output of the model is a distribution of water demands, as shown by the bell-shaped curve in the lower right-hand corner of Figure 3.7, which is labeled "Projected 2035 Water Demand." Monte Carlo simulation was used to generate probability distributions of water use based on historical weather patterns and for projected warmer weather. (Simulation is explained in Chapter 19.)

The most likely demand or expected value is located at the peak of the distribution. The height indicates the frequency of a particular demand, but not the demand itself. The spread of the distribution above and below the expected value is the important feature. The spread reflects variability (real changes in temperature, price, etc.) and uncertainty in the input values.

The expected demand and the variation in demand are predicted to increase in the future. The bell-shaped curves are labeled with respect to weather, but the weather is only one factor. The information given does not reveal how much of the future change is due to weather and how much is due to changes in housing, water use patterns, and other factors.

3.9 PLANNING FOR LANDFILL CONSTRUCTION

A landfill has a finite lifetime that depends on the area of land that has been acquired, the total volume and mass of solid waste that is collected from the community, and the amounts that can be diverted from the landfill by source reduction and recycling. These changes need to be projected. After the landfill site has been purchased, there is almost no economy of scale in landfill construction. Typically, large areas are purchased because suitable sites are hard to locate and develop due to reasons that are largely non-technical.

The rate of waste deposits in the landfill also needs to be measured and projected. Landfill performance is measured by airspace utilization density (AUD) or airspace utilization factor (AUF). Air space is landfill volume. AUD is the mass of waste landfilled divided by the total landfill volume utilized. A landfill that receives 100,000 tons/y (45,454 T/y) and consumes 180,000 yd^3 (137,520 m^3) of the landfill (total, including solid waste, cover soil etc.) has an AUD of 0.56 tons/yd^3 of airspace (0.33 T/m^3).

Figure 3.8 shows the projections for Hilo, Hawaii. The 2005 drop in landfilled waste was based on plans to reduce waste production and encourage more recycling. Table 3.5 shows additional projections of waste placement volume, the need for daily cover material, and when landfill cells will be opened and closed. Projections were also made for leachate and biogas production.

The pattern of growth is slightly exponential. The landfill will have an ultimate capacity and closure date. The ultimate capacity is known, but the time of closure is not because the forecasts, like all long-term forecasts, will differ from what was projected years before closure. The actual lifetime will be longer or shorter than the design life because population growth deviates from projections, per capita waste generation changes, and diversion from landfills by recycling becomes more or less a factor.

3.10 ECONOMIC FACTORS THAT INFLUENCE DESIGN LIFE

This section uses concepts from Chapters 7 and 8, specifically how construction cost is related to process capacity, and how present and future amounts of money can be compared. Some readers may want to move ahead and come back when they feel better prepared.

Civil engineering projects, including municipal wastewater treatment plants, sewers, and pumping stations, typically are designed to serve for 20 or more years before expansion or replacement is needed. They are large projects with a planning, design, and construction period that is measured in years, and it is not practical to repeat the design-bid process more often than necessary.

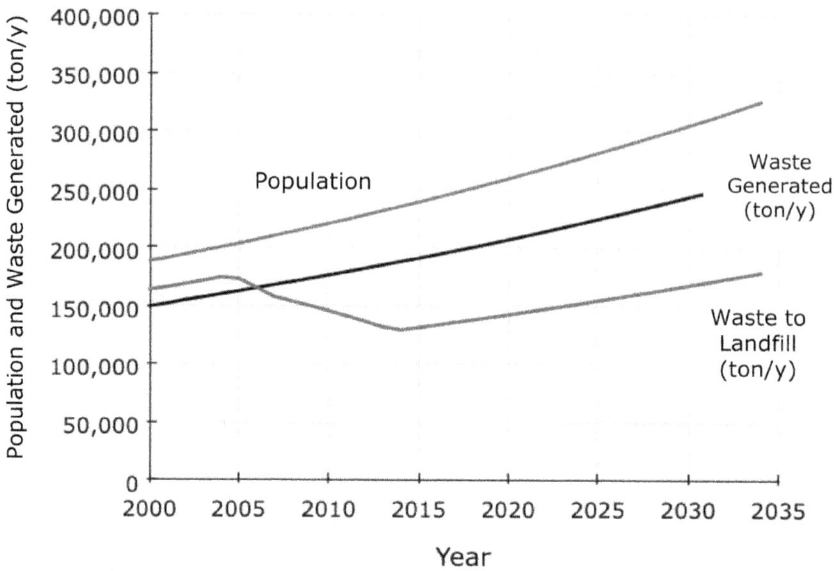

FIGURE 3.8 Landfill Capacity Planning for Hilo, Hawaii.

TABLE 3.5 Projections for Landfill Disposal in Hilo, Hawaii

Year	Disposal Rate (tons/y)	Landfill Volume Required: Refuse Volume (yd³)	Landfill Volume Required: Daily Cover (yd³)	New Cells (acres)	Cells Closed (acres)
2005	74,382	91,000	39,000	8	
2006	71,152	87,000	37,000	0	
2007	67,796	83,000	35,000	0	
2008	66,149	81,000	35,000	5	
2009	64,430	79,000	34,000	0	
2010	62,637	76,000	33,000	0	
2011	60,768	74,000	32,000	3	8
2012	58,821	72,000	31,000	0	0
2013	56,794	69,000	30,000	0	0
2014	55,698	68,000	29,000	0	0
2015	56,609	69,000	30,000	3	5
2016	57,535	71,000	30,000	0	0
2017	58,477	71,000	31,000	0	0
2018	59,433	73,000	31,000	0	0
2019	60,406	74,000	32,000	3	3
2020	61,394	75,000	32,000	0	0
⋮	⋮	⋮	⋮	⋮	⋮
2030	72,211	88,000	38,000	4	4
2031	73,392	90,000	38,000	0	0
2032	74,593	91,000	39,000	0	0
2033	75,813	92,000	40,000	4	0
2034	77,053	94,000	41,000	0	0
2035	0	0	0	0	11
Totals	1,965,319	2,401,000	1,031,000	37	37

Source: Friesen and Associates, date not reported.

If there is a consistent trend in growth, the capacity to be constructed, Q, is related to the design life, t, and the rate of growth in demand, r.

$$\text{Design capacity, units} = (\text{Rate of growth, units/y})(\text{Design life, y})$$

$$Q = rt$$

Generally, a larger addition is planned if rapid growth is expected. If there is no growth in demand, the design period will be determined by the need of new technology and the service life of that technology.

The interest rate is a key economic factor. The ability to borrow money at a low interest rate is an incentive to build a large project and finance it over a long period of time.

Another factor is capital cost and economy of scale, as reflected in the cost-capacity model:

$$\text{Cost} = K(\text{Capacity})^M$$

where K is the capital cost of one unit of capacity and the exponent M measures the economy of scale. Economy of scale means that the cost per unit of capacity decreases as the capacity increases. Doubling the size of a project with $M = 0.3$ will increase the cost by only 23%. The favorable result is that building excess capacity to accommodate future demand costs very little. Therefore, low values of M encourage a longer design life. Pipelines and interceptor sewers have M of 0.3–0.4 and the typical design life is 40–50 years.

Wastewater treatment plants have M values between 0.6 and 0.8, so there is less incentive to build excess capacity and a typical design life is 20 years. For $M = 0.6$, doubling the capacity will increase the cost by 52%; for $M = 0.8$, the increase is 74%.

Another consideration is the ease of expansion. Urban pipelines and sewers are difficult to dig up and replace because streets and underground utilities are also disturbed. Therefore, one large project is preferable to two smaller projects that would serve for the same period of time. Treatment plants, by comparison, are easier to expand or modify, and modifications may be attractive because of changes in technology, or they may be required by changes in environmental regulations. Also, wastewater treatment plants have pumps, blowers, and other machines that have a physical lifetime of 15–20 years.

These generalities lead to some useful rules of thumb:

- A 20-year design life is a good number for treatment plants.
- A 40- to 50-year design life is a good number for large hydraulic projects.
- Adjust the project size to account for high or low interest rates.
- Larger projects are favored when the growth rate is high. This does not always correspond to a longer design lifetime.
- Take advantage of economy of scale.
- Consider ease of expansion and the periodic need to replace machinery or change technology.

We can show that this advice is correct with a simple economic model that relates economy of scale, interest rate, growth rate, and design capacity in order to determine the optimal design capacity or design life. *Optimal* is used in the mathematical sense to mean the solution of an equation that minimizes some measure of the construction cost of present and future expansions. The solution is not optimal in an absolute sense because the model is too simple. The purpose of the model is to show how the "optimal" design life is shifted by changes in the growth rate, interest rate, and economy of scale. This information can also be used to show that the economic consequences of picking a "non-optimal" solution are small, probably within the margin of error for the interest rate and growth rate that are used.

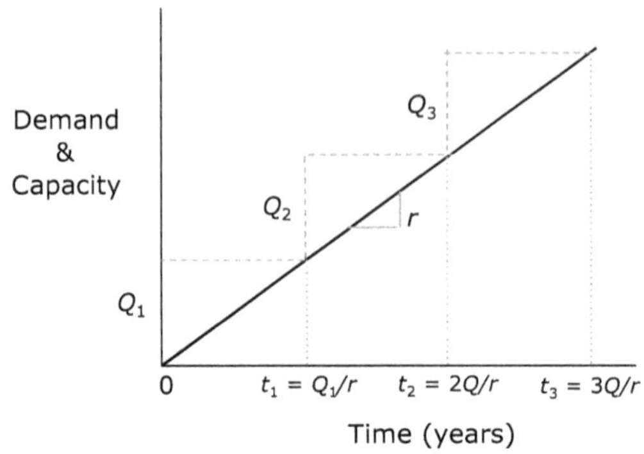

FIGURE 3.9 The first three stages to satisfy a linear growth in demand with a growth rate of *r*.

Several assumptions are made in order to simplify the model. Assume that demand increases linearly, all capacity remains useful into the future, costs and interest rates do not change in time, and maintenance costs do not influence the selection of the optimal expansion capacity. The relevant "future" is 40–60 years.

Figure 3.9 defines the capacity expansion problem for the three stages of construction. The linear growth rate is r (units/y). (The assumption of linear growth is unrealistic, but a forecast is needed and not knowing what the future growth will be, it is convenient to use the simplest possible model.) The assumption of linear growth means that the future looks the same everywhere on the straight-line demand curve. When it is time for a second expansion, the future appears the same as when the first-stage decision was made. As a result, the same capacity Q should be built at each stage of the expansion plan, and the design life of each stage will be $t = Q/r$.

For linear growth rate r, define the optimal design life as $t^* = Q^*/r$, where Q^* is the optimal design capacity (Rudd & Watson 1966, Manne 1967). The simple economic model is

$$M(1-e^{-it^*})-it^*e^{-it^*}=0$$

where i is the interest rate and M is the economy of scale factor. (The derivation is in Appendix A.)

There is no analytic solution for t^*. The graphical solution is shown in Figure 3.10. The ordinate is the optimal design life, $t^* = Q^*/r$. The growth rate does not change the recommended design life, but it does change the design capacity.

Two examples show how the diagram is used.

Example 3.1 Design Life for a Wastewater Treatment Plant

A conventional wastewater treatment plant has $M = 0.7$. Using this value and $i = 3\%$, the design period from Figure 3.10 is $t^* = 22$ years. For $i = 4\%$, the design period is about $t^* = 17$ years. There is no significant economic penalty for using 20 years instead of 17 or 22 years.

Example 3.2 Design Life for an Interceptor Sewer

For a water main or an interceptor sewer, $M = 0.3$. Using this value and $i = 4\%$, the design period from Figure 3.10 is about $t^* = 50$ years.

A useful reference point is $M = 0.6$ and $i = 0.05$ (5%), which gives a design period of 20 years. A ±20% change in the interest rate (from 0.05 to 0.04 or 0.06) changes the design period by about 10

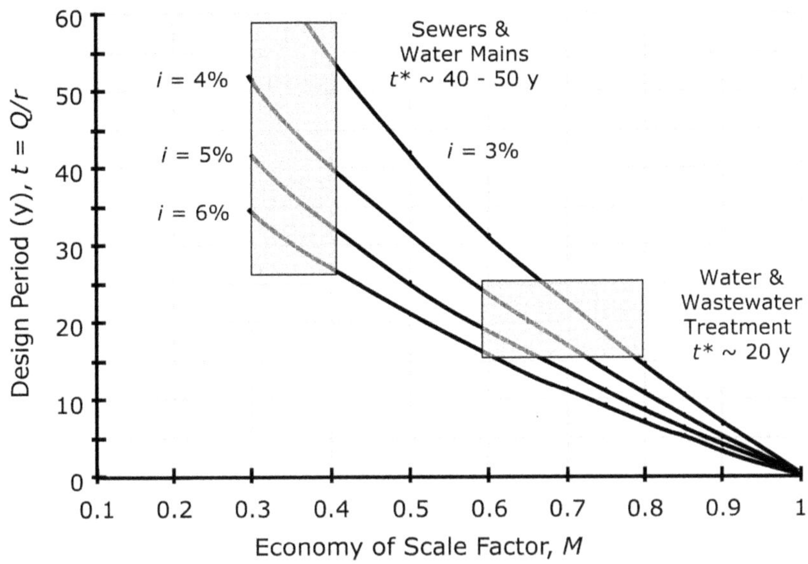

FIGURE 3.10 Optimal design capacity (dimensionless) as a function of interest rate, growth rate, and economy of scale for linear growth in demand.

percent. A 10% change in M, from $M = 0.6$ to $M = 0.54$ or 0.66, changes the design period by about 10%. The shaded rectangle maps a region where the conventional 20-year design life is reasonable. The model confirms what engineers have been doing for many years.

The optimal design period is 20 years for $M = 0.6$ and $i = 5\%$; it is 18 years for $M = 0.7$ and $i = 5\%$. If the correct value of M is 0.6 and $M = 0.7$ is used instead, the economic penalty, measured as the increase in present value, is about 1%.

A similar argument can be made for low values of M, which means a large economy of scale. The shaded rectangle indicates a suitable design life of 40 years or more.

The model also recommends reducing the design capacity when the interest rate is uncertain, but this result is not important because the interest rate for the first design stage will be known within a narrow margin of error.

3.11 CONCLUSIONS

Some technologies are stable. The hydraulics of sewer and water main design has not changed, and changes in materials or construction methods should lead to greater economy of scale. Therefore, traditional design periods of 40–50 years are valid.

The declining demand for water that has occurred in many large cities is a warning against very long design periods. The decline has been due to water conservative toilets, showers, and washing machines, leak prevention, and higher charges for water. Public information programs have been effective in changing habits within households and commercial properties. Industries have a strong financial incentive because using less water, beyond reducing the cost of water, also saves energy and reduces wastewater disposal costs. The savings in treatment plant expansions can be invested in improving water mains and other infrastructure.

The design period for treatment plants is typically 20 years, which is reasonable for several reasons:

- It gives an economic balance between savings due to economy of scale and relative value of money spent today and money spent in the future.

- The plan-design-construct period is 3–5 years, making a short design life unattractive in management terms.
- Treatment technologies change, for example, the growing use of ultrafiltration and other membrane processes.
- Regulations change, usually setting lower limits for traditional pollutants and adding restrictions for new pollutants as more is learned about harmful effects.

An idealized economic model, given without mathematical details, shows how the design period and design capacity can be adjusted to take advantage of economy of scale and interest rates. The rate of growth in demand links the design period and the design capacity.

3.12 PROBLEMS

3.1 FLOW AND LOAD PROJECTIONS

The current population and wastewater data are

Population = 8,920	
Average daily flow = 3,568 m^3/d	Per capita flow = 400 L/d
BOD = 225 mg/L = 0.09 kg/cap-d	TSS = 275 mg/L = 0.11 kg/cap-d
TKN = 35 mg/L = 0.014 kg/cap-d	TP = 8 mg/L = 0.003 kg/cap-d

(a) Project the population for 5, 10, and 20 years into the future using a 1.5% per year growth rate.

(b) Project future wastewater flows, assuming the average per capita flow decreases by 0.2% per year.

3.2 FORECASTING WASTEWATER TREATMENT DEMAND 1

Use the data in Table P3.2 to make a forecast of future wastewater flows. The peak day is the highest daily flow recorded in the given year. Some processes may be designed for peak day. Calculate the peaking factor for peak day.

TABLE P3.2 Historical Flow Data

Year	Annual Average (m^3/min)	Peak Day (m^3/min)	Year	Average Day (m^3/min)	Peak Day (m^3/min)
1993	82.1		2003	76.2	118.1
1994	79.1		2004	79.1	142.1
1995	82.3		2005	75.9	137.5
1996	82.6	103.9	2006	75.6	138.6
1997	90.1	165.2	2007	74.3	107.6
1998	76.4	99.8	2008	78.6	103.3
1999	80.7	128.9	2009	75.3	121.9
2000	79.4	107.1	2010	77.8	105.2
2001	75.1	112.5	2011	67.3	111.9
2002	78.8	150.2	2012	74.5	135.1

3.3 WASTEWATER TREATMENT FACILITY EXPANSION

Table P3.3 gives the annual average daily wastewater flow and the peak day for the years 1993–2011. Make a planning forecast for future wastewater flows. Does it appear that a plant expansion is needed?

TABLE P3.3 Historical flow Data

Year	Annual Average Daily Flow (mgd)	Peak Day Flow (mgd)	Year	Annual Average Daily Flow (mgd)	Peak Day Flow (mgd)
1993	1.25	1.77	2003	1.53	2.00
1994	1.34	1.97	2004	1.52	2.24
1995	1.29	1.71	2005	1.52	2.02
1996	1.32	1.91	2006	1.50	2.30
1997	1.33	1.74	2007	1.50	1.97
1998	1.34	1.85	2008	1.44	2.00
1999	1.41	1.92	2009	1.32	1.83
2000	1.41	2.27	2010	1.34	2.10
2001	1.49	2.03	2011	1.44	2.06
2002	1.44	1.92			

3.4 WASTEWATER PLANNING

Tables P3.4a and P3.4b give projections of population and water consumption. The population growth assumes that currently non-sewered areas will be attached to the collection system at the rate of 20% per year between 2020 and 2060. Assume the average dry weather wastewater flow is 85% of the average daily water use. Estimate the peak daily flow using a peaking factor of 3.0. Assume an additional flow for wet weather flow (inflow and infiltration to the sewer system) of 20% of the average daily flow. Project the anticipated water consumption for 10-year intervals to 2060.

TABLE P3.4a Projected Contributing Population

Year	Population		
	Residential	Commercial	Industrial
2020	710,000	610,000	38,000
2030	750,000	680,000	37,000
2040	800,000	770,000	39,000
2050	860,000	860,000	40,500
2060	730,000	940,000	42,000

TABLE P3.4b Projected Water Consumption Rates

Type of Consumption	Year 2020 (gal/cap-d)	Year 2030 and Beyond (gal/cap-d)
Residential (city)	56	50
Commercial	33	30
Industrial	55	50

3.5 WASTEWATER FLOW FOR WET AND DRY SEASONS

Figure P3.5 defines inflow and infiltration and shows how the extra flow changes the influent load on a treatment plant. The dry season is May–October; the wet season is November–April (Table P3.5). Develop a seasonal forecasting model.

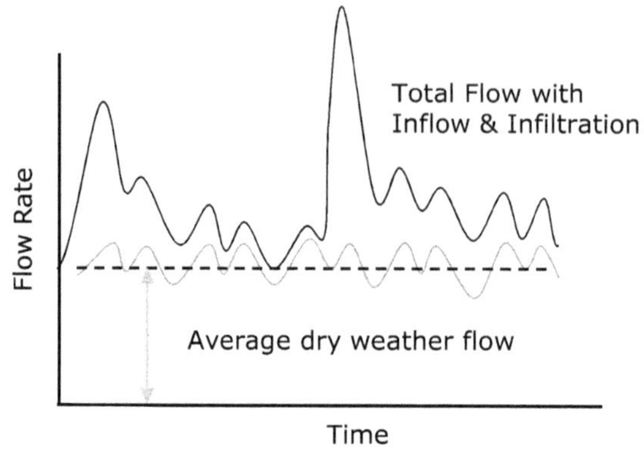

FIGURE P3.5 Effect of infiltration and inflow on total wastewater flow rate.

TABLE P3.5 Historical Dry Weather and Wet Weather Flows

	Dry Season		Wet Season	
Water Year	Rainfall (inch)	Average Flow (mgd)	Rainfall (inch)	Average Flow (mgd)
2009	17.04	3.68	(a)	(a)
2010	16.19	3.79	47.07	8.63
2011	7.39	3.58	34.56	7.96
2012	5.56	3.07	40.09	10.21
2013	6.90	3.14	22.97	7.39
2014	8.35	3.04	32.37	7.48
2015	5.69	2.89	37.27	7.86
2016	5.63	3.06	37.84	7.35
2017	9.51	3.13	28.82	6.46
2018	12.50	3.51	17.99	4.37
2019	5.24	3.12	38.61	8.17
2020	11.1	3.40	36.56	7.42
Averages	8.55	3.25	34.01	7.57

(a) November rainfall and flow data are missing.

3.6 SELECTING THE DESIGN CAPACITY

Table P3.6 gives the historical annual average flows of a treatment plant that operated a 32-mgd activated sludge plant plus a parallel 8-mgd trickling filter plant, giving a total capacity of 40 mgd. It is evident that the plant had to be expanded. There was a complicating factor. The trickling filter plant could fail structurally at any time, and it would then be abandoned. Replacement parts could no longer be purchased. The existing activated sludge plant was in good structural condition and

TABLE P3.6 Historical Flow Data: Annual Average Flow (mgd)

Year	Flow (mgd)	Year	Flow (mgd)	Year	Flow (mgd)
1966	14.0	1975	22.5	1984	31.0
1967	14.1	1976	21.5	1985	30.2
1968	16.6	1977	22.0	1986	32.5
1969	17.1	1978	22.5	1987	33.5
1970	17.5	1979	22.5	1988	37.5
1971	17.5	1980	24.0	1989	38.5
1972	17.4	1981	25.5	1990	40.2
1973	17.5	1982	26.0	1991	39.7
1974	21.0	1983	28.5	1992	40.9

was expected to serve another 20 years with normal maintenance. Develop an expansion plan for the plant.

3.7 DESIGN LIFE FOR AN INTERCEPTOR SEWER

The economy of scale exponent for an interceptor sewer is $M = 0.4$. Assume linear growth and recommend a design period for an interest rate of 4.5%.

3.8 LINEAR PROJECTIONS OF WASTEWATER POLLUTANT LOADS

Figure P3.8 shows 12 years of data for TSS, BOD, total Kjeldahl nitrogen (TKN), and total phosphorus (TP). A straight line seems to be a reasonable description of the increase in loadings. Is a linear projection into the future reasonable?

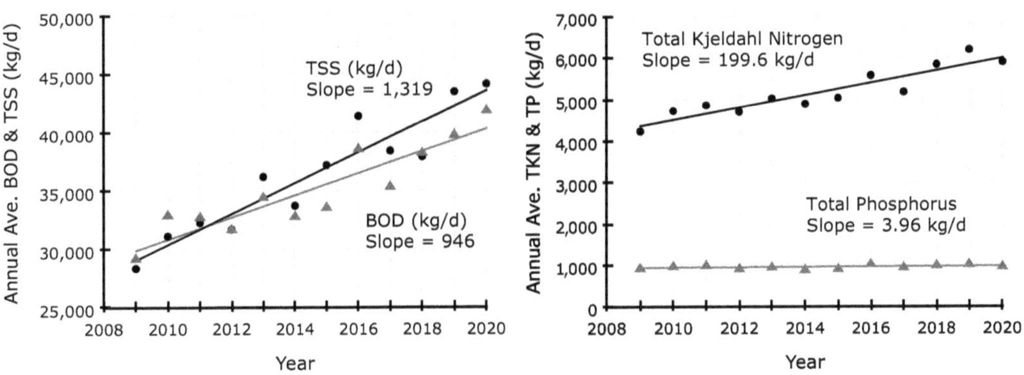

FIGURE P3.8 Historical load data for wastewater pollutants.

3.9 LANDFILL PLANNING

Planning for a landfill requires estimates of the amount of waste that will be delivered in future years. The estimates will be developed by source, as organized in Table P3.9. The percentage of total waste is given for each source category; for example, single-family dwellings account for 33.5% of the total waste. Based on history and demographic forecasts, single-family dwelling waste will increase by 6,047 tons/y and multifamily dwelling waste will increase by 1,317 tons/y. Commercial waste will increase by 6% per year, and other categories will increase by 3%/y. (Other includes government facilities, construction and demolition, and authorized outside haulers.) The percentages of waste diverted from the landfill to recycling or other disposal methods are given for each category. For single-family dwellings, the diversion percentage is 33.5%.

(a) Fill out the table of projected waste quantities, Table P3.9, assuming that the percentages for source categories and waste diversion remain constant in the future.
(b) Determine the growth rate for single-family dwelling waste diverted and waste to landfill.
(c) Determine the growth rate for all residential waste diverted and waste to landfill.

TABLE P3.9 Current and Projected Amounts of Waste (tons/y)

Source	% of Total	Year			
		2020	2030	2040	2050
Single-family dwelling – Diversion rate = 55.7%					
Landfilled					
Diverted					
Total	36.6	680,444			
Multifamily dwelling – Diversion rate = 33.5%					
Landfilled					
Diverted					
Total	24.4	512,175			
Commercial – Growth rate = 6%/y; Diversion rate = 64.1%					
Landfilled					
Diverted					
Total	16.6	319,910			
Other sources – Growth rate = 3%/y; Diversion rate = 90.7%					
Landfilled					
Diverted					
Total	22.4	433,845			
Total waste quantities					
Landfilled		1,138,335			
Diverted		808,039			
Total		1,946,374			

4 Capital Cost Estimates

This chapter is about estimating capital costs early in a project when design concepts are being tested and budgets are being planned. These are *screening* and *preliminary* estimates. At this stage of design, the designer knows the size of major equipment and processes, for example, the capacity and power required for pumps and blowers and the volume of tanks.

The goal is to not to prescribe a method of making cost estimates, because there are differences between companies and organizations in how it is done. What is universal is that the estimate should include all the costs. The early-stage cost estimates do this with a few large categories, such as major equipment, equipment installation, and electrical. The equipment manufacturer will be unknown and so will many physical details. Electrical details are undeveloped, and this is estimated a fraction of the equipment cost.

The categories are configured to the project. A solid waste recycling center, a municipal wastewater treatment plant, an industrial process to removal metals, and a system to remove particulates from an incinerator's gaseous emissions need different categories. Likewise, the fraction of the total cost allocated to each category will be different. Therefore, consider the examples only as examples, and not as templates that apply to all projects.

Capital costs are for tangible assets. Direct capital costs include equipment, tanks, piping, electrical gear and control systems, buildings, land, and site improvements. Indirect capital costs are for assembling the components into a functioning system. These include design engineering, construction services, legal and bond counsel, and contingencies.

Operating costs are paid repeatedly over time, and they are sometimes referred to as *repeating costs* in contrast to the capital cost or *first cost*. The operating cost keeps the enterprise, project, operation, plant, or piece of equipment running and producing and will include operating labor, labor and materials for maintenance, energy, and chemicals.

Bid estimates are provided to the owner to solicit competitive bids for construction, arrange for financing, or fixed to negotiate a price for construction. The bids are based on costs obtained from equipment manufacturers and subcontractors, and on quantity takeoffs from design drawings, unit costs for cubic meters of concrete, meters of pipe of specified quality, and many similar details. They also use subcontractor quotations and site-specific knowledge of construction procedures. These detailed costs are a special business that is outside the scope of this book.

A warning is needed about using cost data in this book. This is not a cost-estimating manual. Much of the data in the book comes from case studies and, therefore, is real cost data, but it comes from different countries and different times. Most often the costs are presented as given in the original reference. The cost basis is usually given. For example, US$ 2018 means US dollars in the year 2018. The dollar sign also denotes Mexican pesos (MXN) and Canadian dollars (CAD). Other currencies, such as the Euro (EUR), Indonesia rupiah (INR), Indian rupee (INR), Singapore dollar (SGD), Middle Eastern rial or riyal, and the Chinese yuan may be cited in case studies or problems. There is no need for currency conversion because the relative value of cost items will stay the same, and only the relative values are important when comparing alternatives.

4.1 CONTRACTING THE COST

Figure 4.1 shows two methods for designing and contracting the cost. The traditional project moves from planning and budget estimates to preliminary design to detailed design, and then to getting bid prices based on the detailed design, and finally to construction and start-up.

The traditional method for public works projects in the United States is Design/Bid/Build. Three parties are involved – the owner, the designer, and the contractor (builder). The owner's design must

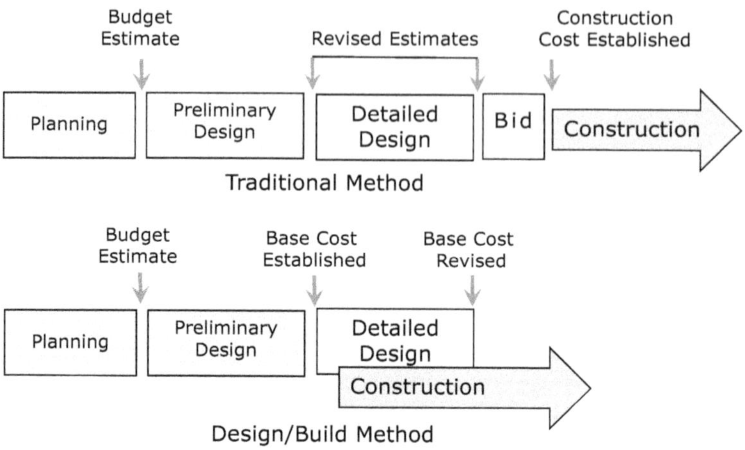

FIGURE 4.1 The traditional design-bid-build method of contracting compared with the design/build method.

be 100% complete before bidding. The contractor is selected by bidding when the design is final. All details are known to all bidders. The designer's and the contractor's interests are independent. This puts most of the responsibility and risk on the owner as the result of multiple contracts and points of contact. Public works projects favor this process because it is open and understandable.

An alternative method – Design/Build – reduces the start-to-finish time of the project. The owner describes a project with preliminary plans, specifications, and sometimes performance requirements (schedule, capacity, safety, etc.). The owner sets a budget for the project based on 30% complete plans. Design/build proposals are solicited with technical and cost components. Selection is usually based on a technical score and a costs score that are combined according to predefined weights. The cost is established before the detailed design is complete, and construction starts while the final design is incomplete. For example, earthwork and foundations may be under construction, while structures are being designed. There is a single contract with one firm for the design and the construction; if conflicts occur there is one point of responsibility. This reduces the owner's risk and shifts it to the contractor.

There are at least 30 different distinct contract types for engineering services. Two are the *cost-plus-fixed-fee* and the *lump-sum-fixed-fee* types with numerous variations of each.

Cost-plus-fixed-fee contracts are used to deliver a specified level of effort over a specified period of time. The contract allows compensation for consultant's labor at agreed billing rates plus a fixed fee (profit) for the term of the contract. There is a low risk for the consultant, so the profit will be relatively low. The designer may have to cover additional costs, but the fee or profit is fixed. Fixed-fee compensation is often used for the consultant's construction-phase services.

Lump-sum-fixed-fee contracts are used when the project is highly defined and performance specifications are available, which is typical of design-build projects. It is preferred by the owner because it places the responsibility and risk on the engineer or contractor. It can also be preferred by the engineer because it can earn a higher level of profit as a result of the higher risk. The lump sum can be amended if the project scope changes. This type of compensation is often used for consultant's design-phase services.

4.2 CLASSES OF COST ESTIMATES

The stages of design and the improvement in cost-estimating accuracy are shown in Figure 4.2 and Table 4.1. The estimated cost and the accuracy tend to increase from conceptual design to

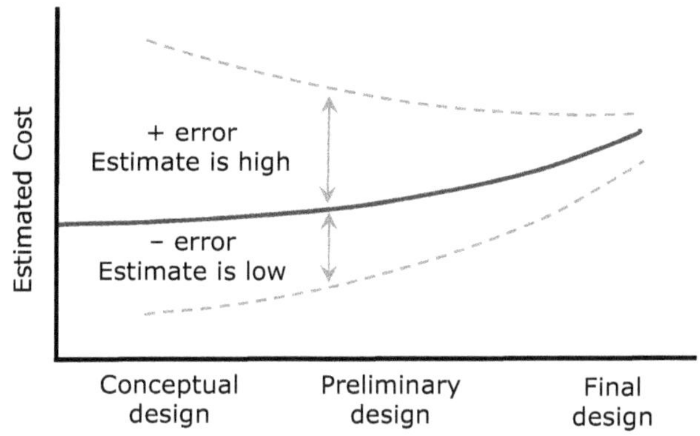

FIGURE 4.2 Estimates tend to increase but become more certain as the design progresses.

TABLE 4.1 Cost Estimation Matrix for Process Industries (AACE 2019)

Estimate Class	Percent Completion	Purpose	Methodology	Expected Accuracy
Class 5	0–5%	Planning	Cost curves	Low −20% to −50% High +30% to +50%
Class 4	5–30%	Feasibility Budget authorization	Equipment factors or parametric models	Low −15% to −30% High +20% to +50%
Class 3	10–40%	Budget authorization and control	Semidetailed unit costs	Low −10% to −20% High +10% to +30%
Class 2	60–90%	Budget control or bid tender	Detailed unit cost with forced detail takeoff	Low −5% to −15% High +5% to +20%
Class 1	90–100%	Bid tender	Detailed unit costs with detailed takeoff	Low −3% to −10% High +3% to +15%

final design and bidding because risk and uncertainty are reduced. The estimating uncertainty may decrease from ±40% for conceptual and preliminary estimates to 5% for the final design estimates.

When plans are forming, details are lacking, and what is called estimating error is a contingency. *Contingency* is an amount added to the base cost estimate to cover uncertainty and risk exposure. There is always uncertainty as to the precise content of all items in the estimate, how work will be performed, what work conditions will be like when the project is executed, and so on. As the design moves from preliminary to final, and as details are added, uncertainties are eliminated, and the contingencies are reduced.

Table 4.1 describes five classes of cost estimates for projects as they develop from planning (Class 5) to bidding (Class 1), and it gives the expected accuracy of the estimates. Class 1, the most accurate and complete estimate, is made when all details are defined in contract documents and drawings and unit prices are applied to quantities (cubic meters of concrete, meters of dimensioned pipe, etc.) that are taken from the final design drawings for the specified equipment and materials of construction. Table 4.2 lists the work that is completed in each stage of design.

TABLE 4.2 Cost Estimation Classes and Percent of Project Completion (AACE 2019)

Project Activity	Status of Activity			
	Class 4 Predesign	Class 3 60% Development	Class 2 90% Development	Class 1 Final Design
Project scope description	Prelim.	Defined	Defined	Defined
Plant production/facility capacity	Prelim.	Defined	Defined	Defined
Plant location	Approx.	Specific	Specific	Specific
Soils and hydrology	Prelim.	Defined	Defined	Defined
Integrated project plan	Prelim.	Defined	Defined	Defined
Escalation strategy	Prelim.	Defined	Defined	Defined
Project code of accounts				
Contracting strategy	Assumed	Prelim.	Defined	Define
Block diagrams	P/C	C	C	C
Plot plans	S	P/C	C	C
Process flow diagrams	S/P	P/C	C	C
Utility flow diagrams	S/P	P/C	C	C
Piping and instrument diagrams	S	P/C	C	C
Heat and material balances	S	P/C	C	C
Process equipment	S/P	P/C	C	C
Utility equipment list	S/P	P/C	C	C
Electrical one-line drawings	S/P	P/C	C	C
Specifications and datasheets	S	P/C	C	C
General equipment arrangement drawings		S/P	P	C
Spare parts listings		S	P	P/C
Mechanical pipeline drawings		S	P	P/C
Electrical discipline drawings		S	P	P/C
Civil/structural drawings		S	P	P/C

S = Started; development is sketches and rough outlines.
P = Preliminary; work is advanced.
C = Complete; deliverable has been reviewed and approved.

4.3 COST-CAPACITY ESTIMATES

A quick estimate of the cost of project A, C_A, which has capacity Q_A, can be made from the cost and capacity of project B, C_B and Q_B. Estimates of this kind are accurate to ±30%.

$$\frac{C_A}{C_B} = \left(\frac{Q_A}{Q_B}\right)^M \Rightarrow C_A = C_B\left(\frac{Q_A}{Q_B}\right)^M$$

The condition is that projects A and B have similar technology. The meaning of *similar technology* is clear when applied to machinery or to a simple process (e.g., neutralization or chlorination), but murky when applied to entire plants or process complexes because technology and equipment are changed and improved.

TABLE 4.3 Economy of Scale Exponents for Some Wastewater Treatment Processes

Treatment Process	Units of Capacity	M
Wastewater treatment processes		
Activated sludge, aeration basin	Volume	0.50–0.70
Primary sedimentation	Area	0.60–0.76
Final clarifier	Area	0.57–0.76
Sludge handling processes		
Digestion (gas mixing)	Volume	0.34
Vacuum filter	Filter area	0.34
Centrifuge	Solids loading	0.81

The economy of scale exponent, M, is near 0.7 for many kinds of equipment and processes. A few values of M are given in Table 4.3. More values are given in Chapter 7 and Appendix B.

Example 4.1 Activated Sludge Process

A conventional activated sludge process (aeration tanks with final clarifiers) was recently constructed at City A at a cost of $8,000,000. A similar design is proposed for a capacity that is four times larger. Construction cost at City A and the planned location are the same (if they were not, an adjustment would be made).

Use $M = 0.7$ to make a quick planning estimate of the cost.

$$C = \$8,000,000(4)^{0.7} = \$8,000,000(2.639) = \$21,100,000 \pm 30\%$$

4.4 EQUIPMENT COST ESTIMATES

Current equipment cost data are not readily available, except to design engineers who have project-specific information to present to an equipment manufacturer. Table 4.4 gives some representative data for recent years.

Table 4.4 gives costs from an online source for order-of-magnitude costs. The costs are US$ 2014, FOB Gulf Coast. *FOB* means *free on board*. The term is used to indicate who is responsible for paying transportation charges. If the seller quotes a price that is *FOB shipping point*, the sale takes place when the seller puts the goods on a common carrier at the seller's dock. Therefore, when the goods are being transported to the buyer, they are owned by the buyer and the buyer is responsible for the shipping, handling, and storage costs. FOB Gulf means the owner owns the goods at the Gulf port and pays the shipping, handling, and storage costs. If a seller quotes a price that is *FOB destination*, the sale takes place when they are unloaded at the buyer's destination. This means that the seller owns the goods until they are delivered and the seller is responsible for the shipping costs.

Matches says, "These costs are helpful during a project's early development and budgeting. The actual cost of a piece of equipment depends upon many factors. You should exercise caution in use of this educational content." The table shows that the cost can depend on the material of construction as much as by size.

TABLE 4.4 Equipment Costs (US$ 2014) FOB Gulf Coast

Pumps and blowers				
Pump, centrifugal (horiz.)	*Cast iron*	*Bronze*	*Stainless 304*	*Hastalloy*
4-inch diameter	$10,300	$9,800	$13,600	$37,700
6-inch diameter	$13,200	$12,600	$17,400	$48,300
8-inch diameter	$15,800	$15,000	$20,700	$57,500
Blower, centrifugal	*Carbon steel*			
1,000 ft^3/min	$6,300			
5,000 ft^3/min	$10,600			
10,000 ft^3/min	$16,900			
50,000 ft^3/min	$88,300			
Process equipment				
Centrifuge	*Carbon steel*	*Stainless 304*	*Monel*	*Hastalloy*
24 inch	$75,800	$125,100	$178,100	$329,000
36 inch	$121,800	$200,900	$288,200	$616,500
48 inch	$170,500	$281,200	$400,600	$739,800
Screen, vibratory, 2-deck	*Carbon steel*	*Stainless*	*Nickel alloy*	
25 ft^2	$17,200	$21,200	$30,100	
50 ft^2	$29,500	$36,400	$51,500	
75 ft^2	$40,500	$49,900	$70,400	
Air pollution control equipment				
Electrostatic precipitator				
500,000 ft^3/min	$1,353,500			
1,000,000 ft^3/min	$2,587,800			
2,500,000 ft^3/min	$6,374,900			
Incinerator, catalytic				
Feed material	*Low hazard*	*Corrosive*	*Hazardous*	
10 million Btu/h	$291,900	$437,800	$583,700	
20 million Btu/h	$494,300	$741,400	$988,500	
30 million Btu/h	$672,700	$1,009,000	$1,345,300	

Source: Matches Equipment Costs (http://www.matche.com/).

4.5 FACTORED ESTIMATES OF CAPITAL COST

One method for making planning-level cost estimates is to start with the cost of the major process equipment and add the cost of installation, piping, instrumentation, and other costs as a percentage of the equipment cost. These are called *factored estimates.* The factors are the multipliers that are applied to the equipment cost to estimate the direct capital costs. Indirect capital costs are a fraction of the direct costs. Contractors' costs for bonds, insurance, etc. are a factor of the total capital cost, as are contingencies and engineering.

Some forms of factored estimates carry the level of detail to the materials of construction (carbon steel, stainless steel, etc.). This is beyond our scope, which is limited to the general structure of the estimating procedures.

Three formats will be illustrated. First is a format from the chemical process industries that can be used on many pollution control projects. Second is a format that is more specific to water and wastewater as practiced by civil and environmental engineering firms. Then, in the next section, is the format that was used for a large air pollution control project at a solid waste processing facility.

The format that is used must capture all of the cost elements in one of the cost categories. Land is not shown as a cost because most pollution control projects are expansions and the required space is available within the boundaries of the existing facility. If this is not the case, the cost of land is added as a separate item.

The magnitude, or relative values, of the factors will change with the character of the project. The equipment used in a solid waste handling process is vastly different than a process that handles only liquids, or an air pollution project. For example, a liquid processing plant has tanks and reactors, chemical feeders, pumps, piping, and dedicated process control instrumentation. A solid waste processing plant has conveyors, grinders, screens, and various kinds of separation equipment. Therefore, the factor values can be adjusted to reflect the kind of processing that is being designed.

Table 4.5 gives the cost factors that can be used to make preliminary estimates for chemical process plants. The categories, and the total capital cost, are estimated as percentages of the purchase price of major equipment. Plants that process solids materials will have a lower multiplier of about 3.1. Equipment installation covers such costs as equipment pads and foundations and support structures and connections of piping and equipment.

Many water and wastewater treatment processes are chemical processes, and these factors can be used. Examples are membrane separation, ion exchange, adsorption, precipitation, filtration, and oxidation processes. These all have short retention times and do not require large concrete tanks. This is in contrast to biological wastewater treatment plants which have long detention times and require more tankage.

Table 4.5 gives single values for the fractions (percentages), but a range of values might be used in different applications. For example, a plant that handles solid materials will have a lower fraction of the cost for piping. If the pipes are installed underground, or if they are insulated, the fraction of the total cost that is spent on piping will change. At the preliminary stage of design, such details may be unknown, thus the inherent inaccuracy of these estimates.

TABLE 4.5 Example of a Factored Estimating Format

Cost Component	Cost
Direct costs	
Major processes and equipment	1.0 E
Purchased equipment installation	0.3 E
Instrumentation and controls	0.1 E
Piping (installed)	0.3 E
Electrical (installed)	0.1 E
Buildings/yard improvements	0.2 E
Service facilities (installed)	0.4 E
Total direct cost =	*2.4 E*
Indirect costs	
Engineering	0.3 E
Construction expenses	0.3 E
Total indirect costs =	*0.6E*
Total direct and indirect costs =	*3.0 E*
Contractor's fee (5% of direct + indirect)	0.05(3.0 E) = 0.15 E
Contingency (10% of direct + indirect)	0.1(3.0 E) = 0.30 E
Fixed capital investment =	*3.45 E*

Cost of land and working capital not included. E = capital cost of major processes and equipment (Humphreys 2005).

The factor for buildings may be larger for public works projects than for commercial projects because they are planned for 40 years or more of use, and appearance and aesthetics are more important. There are other site-specific conditions that can change the factors, which is why the estimates have an error range of ±30%.

4.6 COST ESTIMATING FOR MUNICIPAL WASTEWATER TREATMENT

Biological wastewater treatment plants have 12–24 h of detention time, including the detention time in primary settling tanks and final clarifiers, for the liquid wastewater treatment and a 15-d detention time for sludge digestion. This reduces the fraction of total cost for equipment and increases the costs for tankage and site work. The cost of civil works can equal or exceed the cost of mechanical equipment. A rough breakdown is given in Table 4.6. These percentages of the total cost are not helpful in developing a cost estimate.

A brief diversion on wastewater treatment will be helpful in using the cost data and estimating model in this section. Figure 4.3 shows a typical municipal wastewater treatment plant that may receive industrial wastewater. Grit is inorganic solids, such as sand and glass, that is removed first to protect downstream equipment from abrasion and to prevent the heavy solids from accumulating in downstream tanks where it will be difficult to remove. Screens remove such materials as large solids, cloth, and plastics. The size of the solids to be removed is controlled by the kind of screen and the size of the openings.

TABLE 4.6 Typical Division of Cost Items as a Proportion of Total Investment Cost, Excluding the Cost of Site Acquisition and Infrastructure (van Haandel 2012)

Cost Item	Fraction	Description
Design and management services	10–20%	
Initial studies		Feasibility study, system selection, geotechnical survey
Design and engineering		Basic and detailed design and engineering, bidding, and procurement
Project management		Planning and budget control
Construction management		Site supervision, testing, and commissioning
Site acquisition	Location dependent	Acquisition of building plot, real estate fees
Infrastructure	Location dependent	Access roads, influent and effluent lines, power supply
Site preparation	0.5–2%	Demolition, ground work, rerouting pipes and cables, roads
Construction	70–85%	
Civil	23–29%	Construction of tanks, building, and foundations
Mechanical	21–27%	Equipment costs, including installation, local piping, etc.
Electrical and instrumentation	10–16%	Local instrumentation and electrotechnical equipment
Piping	2–5%	Interconnection piping, utilities, sewers,
Central process control	2–5%	Central processing including software, cable work
Contingency	10–20%	Allowance for unforeseen expenses
Start-up	1–3%	
Equipment		Including maintenance and laboratory equipment
Start-up supplies/spares		Chemicals, first-fills (filters, etc.), fittings, cables
Personnel		Hiring and training

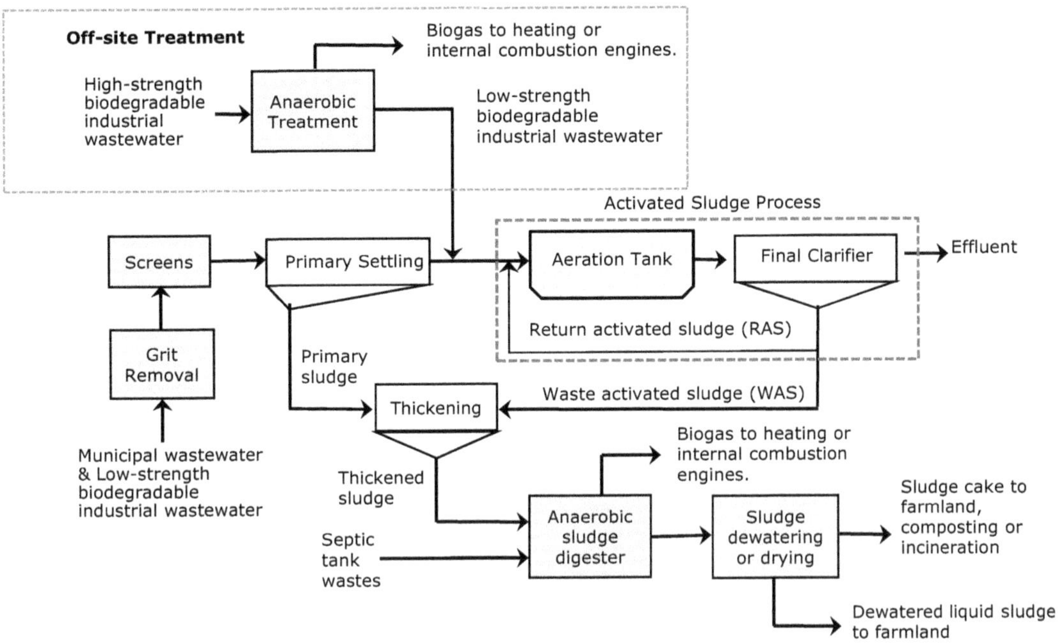

FIGURE 4.3 Conventional wastewater treatment plant that can accept pretreated industrial wastewater. Air flow to the aeration basin is not shown. Recycle of dilute streams from sludge thickening and dewatering are not shown.

The solids remaining after grit removal and screening can be classified as total suspended solids (TSS) or dissolved solids. These terms refer to how solids are measured in the laboratory. Suspended solids can be captured on a laboratory filter; dissolved solids pass through the filter.

Another important characterization is volatile solids (VS) and inert solids (IS), which can be subdivided into suspended and dissolved forms.

Sludge processing and disposal, sometimes called solids handling, includes thickening sludge that is fed to the digester and additional thickening or dewatering of the digested sludge. The diagram shows anaerobic digestion to convert organic material into biogas, which is a mixture of methane gas and carbon dioxide. The biogas is used for heating, electric generation, or fuel for internal combustion engines. Aerobic digestion is an alternative process that is sometimes used in small treatment plants. Useful biogas is not produced, and air must be supplied just as for the activated sludge process.

Wastewater entering the primary settling tank has a TSS concentration of 200–400 mg/L. (Primary refers to the position of the process in the treatment system; it does not mean it is the main or most important process.) About 60% of the TSS in the feed can be removed by gravity settling. The collected solids are removed as a liquid sludge that has a solids concentration of 2–4% total solids on a mass basis (2% solids means 2 kg dry solids in 100 kg of liquid sludge). As a rough guide, primary settling tanks are designed for a surface overflow rate at a peak flow of 24–32 m^3/m^2-d (79–106 ft^3/ft^2-d or 600–800 gal/ft^2-d) with a water depth of 3.6 m (12 ft).

The activated sludge process combines an aeration tank with a final clarifier. Microorganisms, primarily bacteria, in the aeration tank convert organics in the wastewater to cell mass (biosolids) and carbon dioxide. The suspended solids concentration in the aeration tank is 1,500–2,500 mg/L, and the solids are about 70–75% organic. The biosolids are removed in the final clarifier; the clarifier effluent TSS concentration is less than 20 mg/L and may be as low as 5 mg/L. Most of the solids captured in the final clarifier are returned to the aeration tank as return activated sludge (RAS), but a portion is removed as waste activated sludge (WAS). The WAS has a solids concentration of 0.8–1.2% (8,000–12,000 mg/L). WAS will be thickened to 2–4% before being pumped to aerobic or anaerobic digesters. Mixed primary sludge and WAS can be thickened to 5–6% solids. As a rough guide, the aeration tank detention time is 6–8 h. The air supply is designed for 94 m^3/kg BOD (1,500 ft^3/lb BOD) applied at peak flow. The final clarifier is designed for a surface overflow rate of 245 m^3/m^2-d (800 gal/d-ft^2) at an average flow and 370 m^3/m^2-d (1,200 gal/d-ft^2) at peak flow.

A conventional anaerobic sludge digester operates at a temperature of 35°C (95°F), a detention time of 15 days. The reason for thickening the feed sludge is to use the available digester volume efficiently and reduce the cost of heating. A conventional anaerobic sludge digester operates at a temperature of 35°C (95°F) with a detention time of 15 days. Doubling the solids concentration will reduce the sludge volume by half. For example, 1 m^3 of sludge with 1% solids contains (1 m^3)(1,000 kg/m^3)(0.01) = 10 kg of dry solids and 990 kg of water. Increase the solids concentration to 2% and the sludge mixture is 10 kg of solids and 490 kg of water. The volume is 50% less, and the cost of heating the feed sludge is reduced by 50%.

The sludge treatment devices listed in Table 4.7 are intended for thickening and dewatering either stabilized (digested) or unstabilized sludge. Digestion produces sludge that can be thickened more easily and to a higher solids concentration than the undigested sludge. Undigested sludge disposal is a problem because it has a strong and offensive odor, and it is a public health hazard because it contains pathogenic bacteria, virus, and parasites.

Thickened digested sludge is liquid with pumping properties not much different than water. It can be hauled in tanker trucks and spread onto farmland, or incorporated in the soil, as a liquid.

Dewatering removes water from liquid sludge to produce a solid sludge cake with 20% solids and 80% water content (more or less). Solid means that water is not freely released from the product so it can be shoveled, handled by front-end loaders, and hauled in open trucks.

Table 4.7 gives 2019 costs (US$) for equipment that is commonly used for wastewater treatment. Table 4.8 gives some preliminary estimating costs for tanks and buildings. These are average costs. Costs will change with location. Tankage costs include a nominal allowance for excavation/backfill. Costs can be higher due to poor soils, need for sheeting, rock, etc.

The cost estimating worksheet given in Table 4.9 is designed for using the data from Tables 4.7 and 4.8. This methodology is appropriate for Class 3 or 4 estimates.

The total purchased cost of major equipment delivered on site is TPE. Equipment installation is to connect the equipment to the piping plus things such as equipment pads, contractor attending stored equipment on site, equipment commissioning and start-up, simple acceptance testing, and operator and maintenance training. Yard piping is buried piping that connects treatment processes and gets the influent and effluent into and out of the plant. Electrical is the capital cost of bringing power to the point of use and providing the necessary instrumentation and controls. Site work is excavation, grading, landscaping, and the like. Miscellaneous metals are stairs, handrails, and other fittings not included in the cost of equipment.

Engineering services at 20% of TDC are for design plus engineering services during construction such as resident engineer, shop drawing review, O&M manual, record drawings, etc.

The percentages are aggregated over a project, assuming it includes a collection of typical processes. They can be, and often should be, changed to accommodate project or local conditions. For example, plumbing and HVAC are associated with administration buildings, workshops, and laboratories. Process buildings typically have less HVAC and plumbing.

TABLE 4.7 Basic Equipment Costs, Delivered on site, US$ 2019

Equipment	Size or Capacity		Cost ($)
Centrifugal pump – with motor	Flow (gpm)		
1 gpm = 189 L/m	Up to 50		6,800
	50–100		13,100
	100–500		19,300
	500–1,000		36,600
	1,000–5,000		52,300
Submersible pump – solids handling type	Flow (gpm)		
1 hp = 0.7457 kW; 1 gpm = 189 L/m			
7.5 hp/30 ft TDH	300		12,000
10 hp/30 ft TDH	700		12,500
30 hp/40 ft TDH	1,500		30,000
40 hp/40 ft TDH	3,000		35,000
125 hp/45 ft TDH	7,000		80,000
Note: TDH = total dynamic head			
Positive displacement pumps	Flow (gpm)		
(for sludge pumping)	Up to 50		15,000
1 gpm = 189 L/m	50–100		20,000
	100–500		30,000
Vortex grit removal	Peak flow (mgd)		
grit pump, and grit washer	2.5		145,000
	5.0		150,000
1 mgd = 3,780 m^3/d	10		155,000
	20		190,000
	30		200,000
Influent step screens/perforated plate	Channel width (inch)		
including screenings washer and compacter	18		125,000
	24		135,000
	36		150,000
	48		170,000
	60		215,000
	72		228,000
Circular clarifier – includes access bridge, drive mechanism, center influent column, baffles, scum removal, and sludge collection	Scraper mechanism	$/ft tank diameter	2,400
	Dome cover	$/ft² tank area	35
1 ft = 0.08 m			
Aeration – positive displacement blowers	Airflow (cfm)	hp	
hp = horsepower	1,000	60	34,000
1 hp = 0.7457 kW	2,000	100	55,000
1 ft^3 = 0.0283 m^3	3,000	150	78,500
	4,000	200	95,000

(Continued)

TABLE 4.7 (CONTINUED) Basic Equipment Costs, Delivered on site, US$ 2019

Equipment	Size or Capacity		Cost ($)
Aeration – high-speed turbine blower	Airflow (cfm)	hp	
	2,000	100	100,000
	3,000	150	100,000
	4,000	200	135,000
	5,000	250	170,000
	6,000	300	200,000
Fine bubble aeration diffusers	$/ft^2 aeration tank area		21
With tank piping			
(uninstalled)			
Ultraviolet disinfection equipment	$ per mgd peak hourly flow		21,000–26,000
(uninstalled)			
Sludge thickening and dewatering			
Gravity belt thickener	Width (m)	Feed rate (gpm)	
	1.5	130	175,000
	2.0	150	200,000
Disk thickener	Feed rate	Feed rate	
(DS = dry solids)	(gpm)	(lb DS/h)	
	65	400	166,000
	130	800	200,000
Thickening centrifuge (biosolids)	Feed rate with no	Feed rate with	
(DS = dry solids)	polymer (gpm)	polymer (gpm)	
	50–125	100–250	320,000
	150–350	250–450	450,000
	300–750	450–1,000	600,000
Dewatering centrifuge (biosolids)	Feed rate	Feed rate	
(DS = dry solids)	(gpm)	(lb DS/h)	
1 gpm = 189 L/m	50–80	1,485	325,000
1 lb = 0.4545 kg	100–175	3,330	425,000
	200–350	5,310	550,000
Dewatering screw press (biosolids)	Feed rate	Feed rate	
(DS = dry solids)	(gpm)	(lb DS/h)	
	20	125	200,000
	100	700	375,000
Gravity belt press (biosolids)	Feed rate	Feed rate	
(DS = dry solids)	(gpm)	(lb DS/h)	
Width = 1.0 m	50	500	300,000
Width = 1.5 m	75	750	325,000
Width = 2.0 m	100	1,000	350,000
Polymer feed systems	Liquid polymer		
(for biosolids thickening and dewatering)	(gal/h)		
(liquid polymer feed rate is for undiluted	0.4		20,000
solution as purchased)	1.0		22,000
	2.5		25,000
gal/h = 10.76 L/h	5.0		27,000
	10.0		30,000
	Dry polymer		
	(lb/h)		
	4		60,000

(Continued)

TABLE 4.7 (CONTINUED) Basic Equipment Costs, Delivered on site, US$ 2019

Equipment	Size or Capacity		Cost ($)
	16		95,000
	32		120,000
		$/ft length	3,100

Source: Courtesy of Tom Foltz and others noted in Preface, September 2019.

TABLE 4.8 Preliminary Cost Estimates for Tanks and Buildings

Cost Component	Cost	
Buildings		
Administration Building/Lab	$220/ft^2	$2,400/m^2
Process equipment building	$160/ft^2	$1,725/m^2
Tanks (based on volume of tank)		
Circular concrete tank	$0.50–$0.75/gal	
	$3.75–$5.60/ft^3	$132–$192/m^3
Rectangular concrete tank	$0.80–$1.00/gal	$210–$264/m^3
	$6.00–$7.50/ft^3	

Source: Courtesy of Tom Foltz, 2019.

TABLE 4.9 Factored Cost Estimating Worksheet for Conventional Wastewater Treatment

Structure	Cost ($)
Direct costs	
Structures	S
Total purchased equipment	TPE
Equipment installation @ 35%	0.35 TPE
Subtotal	A
Other direct costs	
Yard piping (10%)	0.10 A
Process piping (15%)	0.15 A
Electrical (20%)	0.20 A
Site work (10%)	0.10 A
Ladders, stairways, weirs, etc. (6%)	0.06 A
Plumbing / HVAC (7%)	0.07 A
Other direct costs (ODC)	0.68 A
Total direct costs	TDC
Indirect costs	
Insurance, performance bonds, etc. (8% of TDC)	
Engineering and construction services (20% of TDC)	
Contingencies (20% of TDC)	
Total capital cost	TCC

Source: Courtesy Tom Foltz, personal communication 2019.

Example 4.2 Cost Estimate for a Circular Final Clarifier

This estimate is for a secondary clarifier that is part of an activated sludge process. It is developed in Tables 4.10 and 4.11. The design flow is 2 mgd. The design surface overflow rate of the clarifier is 450 gal/ft²-d, which gives a total area of (2,000,000 gal/d)/(450 gal·ft²-d) =4,400 ft².

TABLE 4.10 Circular Final Clarifier Preliminary Design Summary

Design Information	Quantity
Design flow (gal/d)	2,000,000
Design surface overflow rate (gal/ft²-d)	450
Design surface area (ft²) = (2,000,000 gal/d)/(450 gal/ft²-d)	4,444
Number of clarifiers	2
Clarifier diameter (ft)	55
Area per clarifier (ft²)	2,376
Total clarifier area (ft²)	4,752
Side water depth (ft)	14
Side wall height (ft)	16
Volume per clarifier (ft³) = (16 ft)(2,376 ft²)	38,016
Total volume (ft³) = 2(38,016 ft³)	76,032

TABLE 4.11 Cost Estimate for Circular Final Clarifiers (US$ 2019)

Cost Item	Units	Cost/Unit ($)	Total ($)
Direct capital cost (DC)			
Structure			
Cost per clarifier = 38,016 ft³ @ $5.50/ft³	2	210,000	420,000
Equipment			
Scraper mechanism ($2,300/ft diameter)(55 ft diameter)	2	126,500	253,000
RAS pump (700 gpm)	2	36,600	73,200
Total purchased equipment (TPE)			326,200
Installation @ 35% of TPE			114,200
Total equipment installed			440,400
Total structures and equipment = A			860,400
Other direct costs			
Process piping @ 10% of A			86,000
Yard piping @ 15% of A			129,100
Electrical @ 10% of A			86,000
Site work @ 10% of A			86,000
Ladders, stairways, weirs, etc. @ 6% of A			51,600
Plumbing/HVAC			None
Subtotal other direct costs			438,700
Total direct costs (TDC)			1,299,100
Indirect costs			
Insurance, performance bonds, etc. @ 8% TDC			103,900
Engineering design and construction services @ 20% TDC			259,800
Contingencies @20% TDC			259,800
Subtotal indirect costs			623,500
Total capital cost =			1,922,600

Example 4.3 Cost Estimate for Two Sludge Thickening Options

A treatment plant will install a process to thicken aerobically digested biosolids: either a new gravity belt thickener or a thickening centrifuge that would be located in an existing building. The capital cost estimate is developed in Table 4.12.

The plant generates approximately 16,000 gal/d at a solids concentration of 1.5% (mass percent). This is (16,000 gal/d)(8.34 lb/gal)(0.015) = 2,000 lb/d of solids on a dry weight basis. Sludge thickening equipment will operate 3 d/week for 6 h/d. Sludge is stored in the digester between thickener operating cycles. The thickened sludge will be applied to farmland, but this is allowed only six months of the year, so thickened sludge storage is needed.

Operating costs are the subject of the next chapter, but a few lines have been added to Table 4.12 to provide the estimated values for electricity and polymer that is used to enhance the thickener performance. Laboratory testing indicates that a representative polymer cost for dewatering is $7,500/y. The local power cost is $0.07/kWh, and contract hauling and land application of biosolids cost $0.03/gal.

The thickening centrifuge has higher capital costs, but the overall cost is less due to having to store and haul a smaller volume of thickened sludge.

TABLE 4.12 Cost Estimates for Gravity Belt Thickener and Thickening Centrifuge Systems for Biosolids Processing (US$ 2019)

Parameter	Gravity Belt Thickener	Thickening Centrifuge
Process performance		
Solids production rate (lb dry solids/d)	2,000	2,000
Feed solids concentration (wt.%)	1.5	1.5
Operating schedule (3 d/week; 6 h/d)		
Pumping rate to device (gpm)	103	103
Thickened solids concentration (%)	4.5	6.5
Total solids to land application, (gal/y)	1,945,000	1,347,000
Storage volume required (days)	200	200
Storage volume required (gal)	1,066,000	738,000
Modular tank dimensions, diameter (ft)	95	80
Side wall height (ft)	20	20
Storage volume provided (gal)	1,060,500	752,000
Horsepower (connected/operating)	6 / 4	60 / 50
Equipment	**Costs ($)**	**Costs ($)**
Dewatering unit cost, installed	236,000	432,000
Polymer conditioning system, installed	29,700	---
Storage tank @ $0.65 gal	689,300	488,800
Storage tank dome cover @ $32.50/ft^2	230,400	163,400
Electrical @ 20% installed equipment cost	53,200	86,400
Total capital costs	1,238,600	1,170,600
Direct annual operating costs ($/y)		
Electrical power @ $0.07/kWh	2,790 kWh	34,900 kWh
	$200	$2,500
Polymer	$7,500	---
Biosolids hauling and disposal @ $0.03/gal	$58,400	$40,400
Total operating costs	$66,100	$42,900

There will be other costs in a complete capital cost projection for a project or in a complete operating cost budget. Other local considerations or assumptions could change the total costs presented above.

4.7 CASE STUDY: PALM BEACH RENEWABLE ENERGY FACILITY NO. 2

The Palm Beach Renewable Energy Facility No. 2 is a waste-to-energy (WTE) power plant built for the Solid Waste Authority (SWA) of Palm Beach County in West Palm Beach, Florida. It is the first WTE project of its kind in more than 20 years in the United States. The plant processes 3,000 tons/d (1,000,000 tons/y) of mostly unprocessed municipal solid waste, enough to reduce the amount of waste landfilled by more than 90%. Metals recovery is 27,000 tons/y.

Figure 4.4 shows a moving grate furnace for waste incineration, the steam boiler, and economizer to recover heat and the first two stages of the air pollution control system.

This was a design/build contract with an operation and maintenance (O&M) agreement. The design/build lump sum contract cost was $668,000,000. Construction started in 2012, and commercial operation started in 2015 (Kitto et al. 2016, Schauer 2016).

The heat generated by combustion produces 284,400 lb/h steam in three Babcock & Wilcox boilers that drive turbine generators to produce 95 MW of electric power that is sold to Florida Power & Light. Natural gas is used as an auxiliary fuel.

This is one of the cleanest renewable energy facilities in the United States. Emissions are controlled with a spray dryer/absorber to remove acid gases (primarily sulfuric and hydrochloric), fabric filters to remove particulates, and selective catalytic reduction (SCR) to remove nitrogen oxides. Figure 4.5 shows the selective catalytic reduction (SCR) for removing NO_x. The uncontrolled NO_x emission level is 250 ppmv in an actual flue gas volume of 5,219 m^3/min. The catalytic reduction system will remove 541,182 kg NO_x/y.

Bottom and fly ash are delivered, along with any metallic items, to the ash management system. A rotary magnet removes ferrous metals, and an eddy current separator removes non-ferrous metals to be sold on the scrap metal market.

Three bidders submitted performance guarantees, as given in Table 4.13. Bidder 1 dominated every item except the carbon usage rate. The higher electric generation values, the higher metals

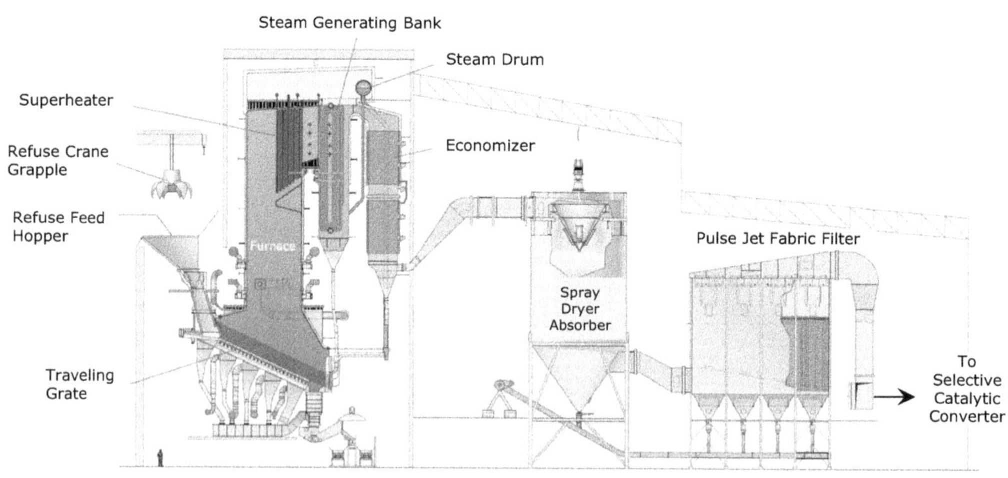

FIGURE 4.4 Palm Beach Renewable Energy Facility No. 2 furnace, boiler, economizer for heat recovery, and fabric filters for particulates removal. Turbine generator is not shown (Courtesy of the Babcock & Wilcox Company; Kitto et al. 2016).

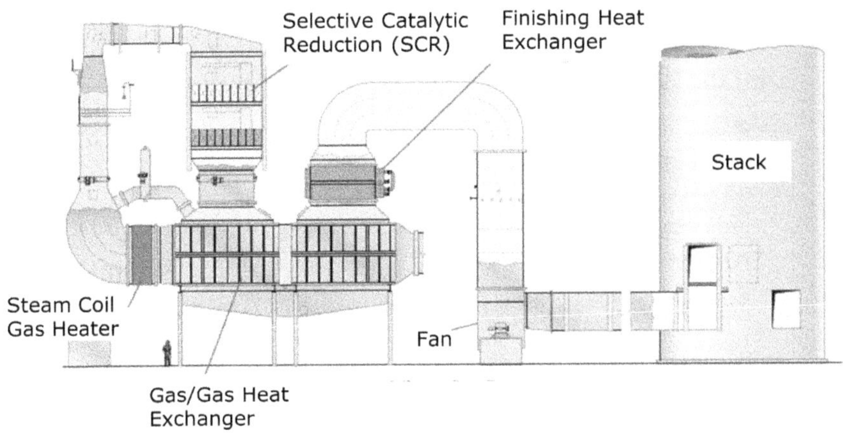

FIGURE 4.5 Selective catalytic reduction for NO_x control for the Palm Beach Renewable Energy Facility No. 2 (Courtesy of the Babcock & Wilcox Company; Kitto et al. 2016).

TABLE 4.13 Performance Guarantees Tendered by Three Bidders for the Palm Beach Renewable Energy Facility No. 2 (Kitto et al. 2016, Schauer 2016)

Performance Guarantee	Bidder 1	Bidder 2	Bidder 3
Electrical generation			
Gross electric generation rate (kWh/ton)	681	636	640
Electric generation rate (kWh/ton)	575	554	550
Electric capacity guarantee (MW/month)	77.3	72	69.5
Metals recovery			
Ferrous recovery rate	90%	80%	85%
Non-ferrous recovery rate	85%	80%	85%
Reagent usage			
Pebble lime usage rate (lb/ton)	11.9	17	14
Ammonium hydroxide usage rate (lb NH_3/ton)	0.95	1.0	1.425
Carbon usage rate (lb/ton)	0.38	0.5	0.2

recovery, and the lower chemical reagent usage add up to a large advantage in the operating costs.

Table 4.14 gives the present value of the lifetime O&M costs and the total net present value. The present value of the lifetime O&M costs converts them to an equivalent lump sum at time zero that can be added to the construction cost to get the *total net present value.* The total net present value is the best measure of the life cycle cost of the projects. The qualifier *Net* is needed because the present value for O&M costs for bidders 1 and 2 are negative, so the negative values are subtracted from the construction cost to get total net value.

Bidder 1 was awarded the design/build contract at the agreed lump sum construction cost of $667,981,128 and a contracted first-year operation and maintenance cost of $20,490,000.

Table 4.15 gives the factored capital cost estimates for the fabric filter dust removal system for the municipal solid waste combustor of the Palm Beach facility.

TABLE 4.14 Summary of Bids for the Design/Build Contract for the Palm Beach Renewable Energy Facility No. 2

Cost	Bidder 1	Bidder 2	Bidder 3
Total lump sum construction cost	672	830	605
Total annual O&M (first year)	20.5	24.5	28.3
Present value O&M costs	–171	–50	24
Total net present value	501	780	629

Costs are rounded to millions US$ 2016 (Kitto et al. 2016, Schauer 2016).

TABLE 4.15 Cost Estimate for the Fabric Filter Dust Removal System for the Palm Beach Renewable Energy Facility No. 2 (Malcolm Pirney 2009)

Cost Item	Factor	Cost ($)
Direct costs		
Purchased equipment and auxiliaries	1.0 E	8,404,700
Instrumentation	0.1 E	840,470
Freight	0.05 E	420,235
Total purchased equipment cost (TPE)	1.15 E	9,665,405
Direct installation costs		
Foundations and supports	0.16 TPE	1,546,465
Handling and erection	0.40 TPE	3,866,162
Electrical	0.05 TPE	483,270
Piping	0.02 TPE	193,308
Insulation for ductwork	0.01 TPE	96,654
Painting	0.01 TPE	96,654
Total direct installation cost (TDIC)	0.65 TPE	6,282,513
Total direct cost (TDC = TPE + TDIC)	1.65 TPE	15,947,918
Indirect costs		
Engineering	0.1 TPE	966,541
Construction and field expenses	0.05 TPE	483,270
Contractor fees	0.1 TPE	966,541
Start-up	0.02 TPE	193,308
Performance test	0.01 TPE	96,654
Contingency	0.15 TPE	1,449,811
Total indirect costs (TIC)	0.43 TPE	4,156,125
Total capital cost (TDC + TIC)	2.08 TPE	20,104,043

4.8 CASE STUDY: WASTEWATER TREATMENT PLANT REHABILITATION

A 70-year-old treatment plant must be replaced. One option is to connect to a nearby local municipality. This would cost $2.44 million with an operating cost of $376,000 per year. A second option is to demolish most of the old plant and replace it with a sequencing batch reactor (SBR). The cost to demolish and remove the unwanted parts of the old plant is $150,000. The total direct cost of the major processes and equipment, given in Table 4.16, is $1,555,000. The SBR buildings and

TABLE 4.16 Estimated Cost for Rehabilitating a Treatment Plant

Cost Item	Estimate
Direct costs	
Major equipment and processes	
Headworks (screens, flow meters, comminutor)	$250,000
SBR digester equipment and controls	$1,000,000
Chemical feed equipment	$125,000
Post equalization pumps	$60,000
Chlorination process	$120,000
Subtotal major equipment and processes	$1,555,000
Mobilization/demobilization	$50,000
Demolition and renovation of existing plant	$150,000
Site work, earthwork, and roadways	$150,000
Yard piping	$250,000
Electrical (installed)	$450,000
New control building	$100,000
New buildings with auxiliary equipment	$2,000,000
Subtotal support systems	$3,150,000
Total direct cost	$4,705,000
Indirect costs	
Bonds and insurance	$50,000
Allowance for construction sequencing	$25,000
Engineering and administration @ 20% of direct cost	$945,000
Total indirect cost	$1,020,000
Miscellaneous and contingency @ 20% of total direct cost	$945,000
Total capital cost	$6,670,000

structures, auxiliary pumps and blowers, and sludge digester cost $2,000,000, some part of which will be for piping and electrical work. The total direct capital cost is $4,705,000. Indirect costs and contingency bring the total to $6,670,000.

Indirect costs include performance bonds and insurance for the contractor. The allowance for construction sequencing is for extra work to keep the plant operating during construction.

4.9 CASE STUDY: WATER SUPPLY

A city evaluated three options for improving the water supply system. The city is divided into two regions (north and south) by a river and a 100-year floodplain which will never be developed. At this time, no water mains cross the river. State law requires 24 h of water storage capacity, and all lines shall be designed for fire flow conditions (minimum of 20 psi at 1,500 gpm). Hydraulic modeling provided the storage tank sizes and pipe diameters.

Three options are shown in Table 4.17. Option 1 purchases water from two wholesalers – one that will serve the northern region and one that will serve the south. Option 2 serves the entire city with water purchased from one wholesaler; this involves connecting the north and south regions. The third is for the city to build two treatment plants, one for each region.

Table 4.18 gives the estimated capital cost and annual O&M costs for Option 1 to meet the 2030 demands. These two kinds of costs can be combined to get a life cycle cost or, alternatively, an annual cost that pays for the project. This calculation is explained in Chapter 8.

TABLE 4.17 Options for Water Supply

Option	Capital Cost	Annual O&M Cost
1. Purchase from two wholesalers	$11,458,800	$2,675,500
2. Purchase all water from one wholesaler	$12,733,400	$2,699,300
3. City builds two treatment plants	$26,318,900	$2,312,000

TABLE 4.18 Costs for Option 1 – Purchase Water from Two Wholesalers

Item Description	Unit	Quantity	Unit Price	Cost
Northern region				
500,000-gal elevated storage tank	EA	1	$525,000	$525,000
750,0000-gal elevated storage tank	EA	1	$750,000	$750,000
12-inch ductile iron pipe (DIP)	LF	102,400	$27/ft	$2,764,800
	Subtotal northern region			$4,039,800
Southern region				
1,000,000-gal elevated storage tank	EA	2	$1,000,000	$2000,000
12-inch ductile iron pipe (DIP)	LF	66,500	$27/ft	$1,795,500
12-inch DIP/replace existing 6-in PVC	LF	35,000	$27/ft	$945,000
	Subtotal southern region =			$4,740,500
	Subtotal northern and southern regions =			8,780,300
Contingencies (15%)				$1,317,000
Engineering, legal, administration, etc. (15%)				$1,317,000
Land and easements				$44,500
	Total capital cost =			$11,458,800
Annual O&M cost				$2,675,500

4.10 CONCLUSION

Capital costs buy land, equipment, buildings, and other physical assets. Preliminary cost estimates are based on information about the major processes and equipment. Piping and electrical details, and many others, are unknown, so they are estimated from factors based on equipment costs. The method must be clear on whether the factors apply to installed or uninstalled equipment costs.

The accuracy, or inherent uncertainty, of the preliminary cost estimates requires adding a contingency cost. Details are added as the design progresses, allowing uncertainties and contingencies to be reduced. The accuracy of estimates based on the final detailed designs is ±10%, compared with ±30% for preliminary designs. The contingency after the bidding is often 3%.

Almost every water or wastewater treatment plant design in the United States is an expansion, remodel, or rehabilitation. This complicates the design, construction, and cost estimating. For example, an anaerobic digestion project may include the construction of two new digestion tanks, draining and cleaning four digesters while removing old mixers and installing new ones, modifying the inlet and outlet piping; adding gas cleaning equipment and gas storage tanks, and rerouting gas piping. All of this must be done while the digestion process continues to operate. The cost thus becomes less dependent on the cost of installed new equipment, and more dependent on the arrangement and physical condition of the existing equipment. All of this is influenced by the sequence of construction.

The time from conception through construction and start-up of a complex project can be five years or more. The designer, owner, and builder therefore have to manage fund raising and spending over the span of the project. Borrowing and cost inflation become factors. These are discussed in Chapter 10.

4.11 PROBLEMS

4.1 CAPITAL COST

The construction cost for a treatment plant rehabilitation is $1,000,000. Other related costs are engineering fees = $85,000; legal fees = $26,000; environmental information document = $10,000; land = $200,000; geotechnical testing = $22,000. What is the total capital cost?

4.2 PIPE SIZE AND PUMPING COSTS

Table P4.2 shows the relationship between pipe size and pipe capacity, and between pipe capacity and pipe cost. Velocities are determined so the pressure difference between the pipe inlet and outlet is the same for all pipe sizes. The flow rate, measured in million gallons per day (mgd), for this fixed pressure difference is given in Table P4.2. Develop a relation between cost, flow rate, and pipe diameter.

TABLE P4.2 Relationship between Pipe Size, Pipe Capacity, and Pipe Cost

Pipe Size (inch)	Capacity (mgd)	Cost ($/lin. ft.)
8	0.6	12.0
12	1.8	16.0
16	3.9	23.0
20	7.0	29.0
24	11.4	33.0
30	20.2	39.0
36	32.6	46.0
42	48.5	62.0
48	70.0	78.0

4.3 FACTORED ESTIMATES

Use the equipment costs in Table P4.3 to make a factored estimate of the capital cost for two 5,000 ft^3/min centrifugal compressors and three centrifugal pumps with a 5-inch discharge. No new buildings, yard improvements, and service facilities are needed.

TABLE P4.3 Blower Cost

Blower Capacity (ft³/min)	Blower Cost FOB, Gulf Coast ($)	Cost with Motor FOB, Gulf Coast ($)	Discharge Diameter (inch)
200	7,400	56,800	2
500	15,300	62,400	3
1,000	26,400	72,200	4
2,000	45,700	92,900	5
5,000	94,200		6
10,000	162,900		8
25,000	336,000		10

Source: Matches website.

4.4 CAPITAL COST OF A FABRIC FILTER

Table P4.4 gives the US EPA format for a factored estimate of a fabric filter system to remove particulates from air. The purchased equipment cost is $340,000. Estimate the total capital investment. The cost of site preparation and buildings will be ignored. (Notice that the indirect costs are factors of the purchased equipment costs. An alternative method is to estimate these costs as fractions of the total direct cost.)

TABLE P4.4 Factored Estimate of Fabric Filter System Costs (US EPA 2002)

Cost Item	Factor
Direct Costs	
Fabric filter with bags and auxiliary equipment	E
Instrumentation	0.10 E
Sales taxes	0.03 E
Freight	0.05 E
Purchased equipment costs = PE	1.18 E
Direct installation costs	
Foundations and supports	0.04 PE
Handling and erection	0.50 PE
Electrical	0.08 PE
Piping	0.01 EE
Insulation for ductwork	0.07 PE
Painting	0.04 PE
Direct installation cost = DIC	0.74 PE
Total direct cost = TDC	1.92 PE
Indirect costs (installation)	
Engineering	0.10 PE
Construction and field expense	0.20 PE
Contractor fees	0.10 PE
Start-up	0.01 PE
Performance test	0.01 PE
Contingencies	0.03 PE
Total indirect cost = TIC	0.45 PE
Total capital investment	2.37 PE

4.5 CAPITAL COST ESTIMATE IN THE CSI FORMAT

The Construction Specifications Institute (CSI) master format organizes costs according to 17 divisions. Table P4.5 lists capital cost estimates for a 500-gpm ion exchange process to remove chromium. The equipment can be installed in an existing building, so some categories have a cost of zero, but they are still shown. The subtotal for direct costs is $1,143,489. Indirect costs, as factors of the total direct cost, are 2.5% for insurance, 2% for bonds, 10% for overhead and profit, and 10% for engineering. Add a contingency of 20% of the total direct and indirect costs. Estimate the total construction cost.

TABLE P4.5 Estimated Capital Costs for the 17 CSI Divisions

CSI Division	Estimated Cost ($)
Direct costs	
Division 1: General conditions	50,000
Division 2: Site construction	81,690
Division 3: Concrete	55,500
Division 4: Masonry	0
Division 5: Metals	15,500
Division 6: Wood and plastics	0
Division 7: Thermal and moisture protection	250
Division 8: Doors and windows	0
Division 9: Finishes	14,750
Division 10: Specialties	500
Division 11: Equipment	604,059
Division 12: Furnishings	0
Division 13: Special construction	21,600
Division 14: Conveying systems	0
Division 15: Mechanical	34,640
Division 16: Electrical	150,000
Division 17: Instrumentation and control	115,000

4.6 AIR SPARGING AND SOIL VAPOR EXTRACTION SYSTEM

A soil remediation project includes air sparging injection wells and soil vapor extraction wells. Quantities and unit costs are given in Table P4.6a. Table P4.6b shows the cost of an air sparging injection wells as bid by a local drilling subcontractor. Health and safety protection are built into the quote. Subcontractor overhead and profit are included. Prime contractor overhead and profit are added. Unit prices from standard cost-estimating guides typically are broken down into labor, equipment, and materials categories, but these were not provided in the quote. Drilling oversight is based on typical labor rates in the area for a geologist and technician. This unit cost can then be rolled into the cost for the individual activity. Table P4.6b gives the construction cost of an air sparging/soil vapor extraction system.

This is a feasibility study estimate with an accuracy of −30% to +50%. Complete the cost estimate.

TABLE P4.6a Cost Estimating Data (Goldstein & Ritterling 2001)

Description	Quantity	Unit	Unit Cost
Direct costs			
Mobilization/demobilization	1	LS	$106,723
Monitoring, sampling, testing and analysis	1	LS	$60,838
Site work	5	acres	$2,588
Air sparging/vapor extraction system			
Mobilization	1	LS	$1,534
Impermeable surface cover	105,000	sq. ft.	$0.84
Vapor extraction wells	8	each	$3,725
Air sparging injection wells (Table P4.6b)	2	each	–
Soil vapor extraction system	1	each	$93,510
Air sparging blower	1	each	$5,712
Soil vapor extraction piping	400	ft	$8.66
Air sparging piping	100	ft	$5.03
Electrical hookup	1	LS	$9,900
Start-up and testing	1	LS	$11,000
Off-site treatment and disposal	1	LS	$1,550
Total direct costs (TDC)			$433,348
Indirect costs and contingency			
Contingency (25% of TDC)			
Project management (5% of TDC)			
Remedial design (8% of TDC)			
Construction management (6% of TDC)			
Institutional controls (1% of TDC)			
Total indirect costs and contingency			
Total capital cost			

TABLE P4.6b Cost Per Air Sparging Injection Well

Description	Quantity	Unit	Unit Cost
Mobilization/demobilization	2	Each	100
Setup and decontamination	2	hours	$125/h
Wellhead completion	15	Feet	$55/ft
IDW handling	1	hours	$950/h
Drilling oversight	7	hours	$110/h
Subtotal = A			A
Prime contractor overhead			B = 15% of A
Subtotal			
Prime contractor profit			10% of B
Total unit cost			

4.7 WATER REUSE PIPELINE PROJECT ESTIMATE

Table P4.7 gives the estimated unit costs for three related pipeline projects that will create a water reuse system for irrigation and other non-potable uses. Calculate the total cost for the projects.

TABLE P4.7 Capital Cost for Combined Water Reuse Pipeline (US$ 2005)

Item	Quantity	Unit	Unit Cost
Direct cost – pipeline A			
12" PVC pipe installed	10,000	LF	$47
14" PVC pipe installed	17,500	LF	$61
16" PVC pipe installed	4,600	LF	$69
Fittings @ 15%	1	LS	$277,309
Jack pipe under two highways	1,600	LF	$284
Jacking pits	4	EA	$12,000
Pumping/dewatering	330	EA	$1,150
Asphalt @ 10% of pipeline length	3,050	LF	$65
Mobilization @ 5%	1	LS	$160,317
Direct cost – pipeline B			
8" PVC pipe installed	10,500	LF	$25
10" PVC pipe installed	8,500	LF	$34
12" PVC pipe installed	46,500	LF	$47
Fittings @ 15%	1	LS	$411,869
Asphalt @ 10%	6,550	LF	$65
Pumping/dewatering	600	EA	$1,150
Mobilization @ 5%	1	LS	$213,687
Direct cost – connect pipelines A and B			
12" PVC pipe installed	8,500	LF	$47
Fittings @ 15%	1	LS	$60,193
Asphalt @ 10% of pipeline length	822	LF	$65
Pumping/dewatering	285	EA	$1,150
Mobilization @ 5%	1	LS	$42,133
Direct cost – treatment plant			
Membrane filtration facilities	3	MGD	$1,300,000
Mobilization @ 5%	1	LS	$268,611
Indirect costs (% of total direct cost = TDC)			
Unlisted items @ 10%			
Contingency @ 10%			
Contract cost @ 2%			
Field costs @ 2%			
Engineering design and construction oversight @ 12%			

(LF = lineal feet, LS = lump sum, EA = each)

4.8 SLUDGE TREATMENT

A gravity belt press will dewater 50 gal/min of digested sludge. A polymer feed system of 0.4 gal/h capacity is needed. The sludge treatment building will be 800 ft^2. A centrifugal pump will feed the belt press, and dewatered sludge will be carried away an 80-ft-long conveyor belt. No site work or yard piping is needed. There is no HVAC. Plumbing is 4% of installed equipment. Estimate the cost of this system.

4.9 CLARIFIER EXPANSION

A wastewater treatment plant needs to install new scraper mechanisms in two 55-ft diameter final clarifiers and build one new clarifier of 100-ft diameter. The existing RAS pumps (2 units of 700 gpm) will be replaced and one new pump of 1,400 gpm will be installed for the new clarifier. Estimate the capital cost.

4.10 TREATMENT PLANT COST ESTIMATE

An activated sludge system will have a 2,000,000-gal aeration tank. Air will be supplied by three 6,000-cfm high-speed turbine blower and a full floor covering of fine bubble diffusers. There will be two (circular final clarifiers with a total area of 12,500 ft^2 and sidewater depth of 15 ft. There will be three 800-gpm return activated sludge (RAS) pumps.

Primary settling will be in rectangular tanks with a volume of 1,000,000 gal, a liquid depth of 12 ft, and a side water depth of 14 ft. Sludge removal will be with positive displacement pumps with a total capacity of 200,000 gal/d. Assume the scraper mechanism cost is the same as final clarifier with the same surface area. Prepare preliminary cost estimate.

4.11 BID COSTS

Table P4.11 shows the bid costs from two companies. Complete the bid tallies and calculate the total bid costs.

TABLE P4.11 Bids from Two Companies

Description	Quantity	Unit	Bid A		Bid B	
			($/unit)	($)	($/unit)	($)
Clearing and grubbing	1	LS				
36-inch diameter pipe	9,684	Foot	107.00		104.00	
Jack and bore under road	200	Foot	555.00		730.00	
Roadway overpass area	520	Foot	143.25		154.00	
Connections @ PS 3 and PS 4	1	LS				
Two air release valves	1	LS				
Foundation stone	3,200	Ton	9.00		13.00	
Geotextile fabric	1,000	Foot	1.50		2.00	
Contaminated soils (off-site)	100	Ton	5.00		5.00	
Contaminated soils (on-site)	500	Ton	3.75		4.00	
Select backfill	2,000	Ton	6.00		8.00	
Pavement restoration	1	LS		50,000		75,800
Site restoration	1	LS		29,000		41,000
Total bid						

LS = lump sum.

5 Operating Costs

Operating costs can be divided into direct costs, indirect costs, and general expenses.

- *Direct operating costs*, sometimes called *variable costs*, tend to be proportional to output and raw material use. Raw materials, utilities, O&M supplies, and waste disposal are direct operating expenses. Energy is a dominant direct operating cost.
- *Indirect operating costs*, sometimes called *fixed costs*, include insurance and depreciation. These tend to be independent of output and have to be paid even if the plant is not operating. Engineering is in this category.
- *General expenses* represent an overhead burden that is necessary to carry out the functions of the organization. They include administration, accounting, and research.

Labor is the term for operating personnel, maintenance personnel, and O&M supervision. *Labor cost* normally has two parts: the gross amount of paychecks issued to employees (excluding personnel included in general expenses), and fringe benefits paid by the employer on behalf of the employee, including FICA tax, workers' compensation, pensions, vacations, holidays, sick leave, overtime premium, company contribution to profit sharing, life insurance, and medical insurance.

5.1 CASE STUDY: PALM BEACH RENEWABLE ENERGY FACILITY NO. 2

This example continues the Palm Beach Renewable Energy Facility No. 2 from Chapter 4. The facility comprises several major processes, including the furnace, boiler, heat recovery economizer, fabric filter for dust removal, and the selective catalytic reduction (SCR) process for removing NO_x. Table 5.1 lists the annual operating costs for the SCR process. The total annual cost is $3,920,042/y, which is $7.24/kg of NO_x removed (Kitto et al. 2016, Schauer 2016).

Total annual operating cost	$1,979,860
Annualized capital cost	$1,940,182
Total annual cost	$3,920,042

Calculation of the annualized capital cost is explained in Chapter 8.

5.2 COST OF CHEMICALS

Estimating current chemical costs is easy because prices are available online. Cost estimates for chemicals are based on current prices. It is tempting to assume that prices will increase each year, but the direction and magnitude of change are uncertain.

Figure 5.1 shows the prices for chlorine in the United States, Germany, China, and the Philippines. Imagine yourself in Germany in 2010 making a budget for future chlorine costs, which have been generally increasing. Three years later, in 2013, the price has dropped by 30%.

Once a process is in operation, actual purchase prices are determined by competitive bidding. The 2012 accepted total bid costs for chemicals for the three water treatment plants of Des Moines,

TABLE 5.1 Operating Costs – Selective Catalytic Reduction (SCR) System for a 1,000 T/D Municipal Waste Combustor (Kitto et al. 2016, Schauer 2016)

Operating Cost Item	Cost Factor	Cost ($/y)
Direct operating costs		
Operating labor (three shifts per day)	1 h/shift ($50/h)	54,750
Supervisory labor	15% of operating labor	8,210
Maintenance labor	1 h/shift ($50/h)	54,750
Maintenance materials	100% of maintenance labor	54,750
Catalyst replacement	$900,000 every 5 years	145,000
Reagent cost (ammonia/urea)[a]	$1.40/gal urea solution	253,900
Electricity[b]	$0.063/kWh	158,680
Lost power generation ($0.063/kWh)		324,230
Total direct operating costs		1,504,270
Indirect operating costs		
Overhead	60% of labor and materials	103,480
Administration and insurance	4% of capital cost	822,110
Total indirect operating cost		925,590
Total operating cost		1,979,860

[a] Reagent cost based on 0.39 kg NH_3 per kg NO_x removed. Delivered cost = $0.32/kg of urea solution.

[b] 184,310 actual ft^3/min at 8-inch pressure drop and 60% combined fan and motor efficiency.

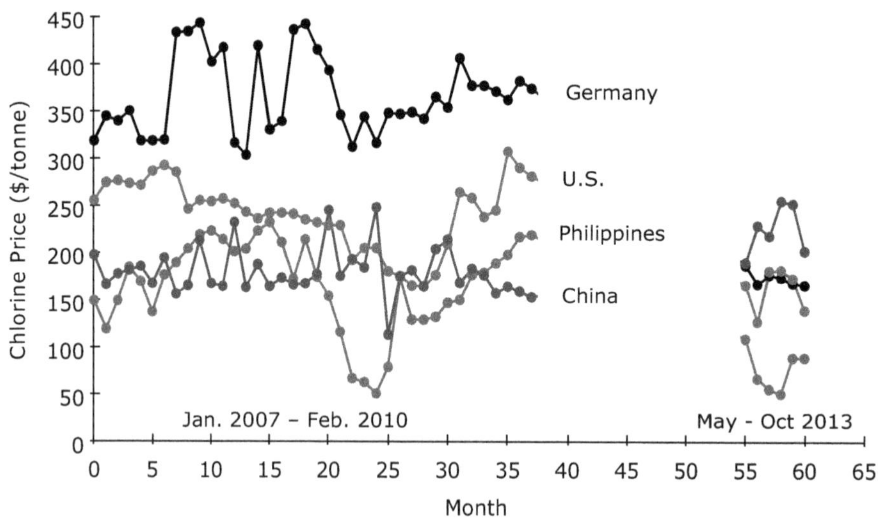

FIGURE 5.1 Prices for chlorine ($/ton).

TABLE 5.2 Bid Prices in 2012 for Water Treatment Plant Chemicals, Des Moines, Iowa

Chemical	Unit Price ($)	Quantity	Total Price ($)
Fluer Drive Plant			
Activated carbon (lb)	0.493	423,722	$208,895
Aluminum sulfate (lb)	0.2225	1,422,616	$316,532
Carbon dioxide (lb)	0.02575	1,621,714	$41,759
Ferric chloride (lb)	0.120	1,418,361	70,203
Hydrofluorosilicic acid(lb)	0.2655	212,711	$56,475
Lime (ton)	127.57	11,316	$1,443,615
Soda ash (ton)	0.1675	253,552	$42,470
Solar salt (ton)	120.0	449	$53,880
Sodium hypochlorite (gal)	0.870	169,744	$147,677
Polyphosphate (lb)	0.470	97,847	$45,988
Fleur Drive subtotal			$2,527,462
McMullen Plant			
Carbon dioxide (lb)	0.02575	831,665	$21,415
Sodium hypochlorite (gal)	0.870	80,564	$70,091
Ferric chloride (lb)	0.120	1,210,468	$145,256
Hydrofluorosilicic acid (lb)	0.2655	126,601	$33,613
Lime (ton)	158.0	5,129	$810,382
Polyphosphate (lb)	0.470	36,529	$17,169
McMullen subtotal			$1,097,924
Saylorville Plant			
Antiscalant (lb)	2.440	26,179	$63,877
Citric acid (lb)	0.600	23,592	$14,155
Hydrofluorosilicic acid (lb)	0.2655	19,254	$5,112
Polyphosphate (lb)	0.470	5,555	$2,611
Sodium bisulfite (lb)	0.127	139,648	$17,735
Sodium hydroxide, 30% (lb)	0.1159	450,527	$52,216
Sodium hypochlorite (gal)	0.870	46,651	$40,586
Sodium permanganate (lb)	0.735	21,309	$15,662
Saylorville Subtotal			$211,953
Des Moines Waterworks Total			$3,737,339

IA, was $3,854,644, as shown in Table 5.2. This was approximately $183,000 less than the estimated cost presented in the 2012 budget.

5.3 ENERGY AND POWER

Power is energy transfer per unit of time. It may be measured at any point in time. Electrical *power is* usually measured in watts (W), kilowatts (kW), megawatts (MW), etc. The power of internal combustion engines and motors are rated in kilowatts, except in the United States where horsepower (hp) is also used. The word "horsepower" gives an intuitive idea that power defines how much

TABLE 5.3 Arithmetic Equivalence of Energy Units

Power Unit		kW	hp	Btu/h
1 kW	=	1.0	1.341	3.412
1 hp	=	0.7457	1.0	2,544
1 Btu/h	=	2.931×10^{-4}	3.930×10^{-4}	1.0
Energy unit		**Btu**	**kWh**	**Joules**
1 kilowatt-hour (kWh)	=	3,412	1.0	3.6×10^{6}
1 Joule (J)	=	9.478×10^{-4}	2.778×10^{-7}	1.0
1 British thermal unit (Btu)	=	1.0	2.931×10^{-4}	1,055
1 horsepower-hour (hp-h)	=	2,544	0.7457	2.685×10^{6}

Notes: 1 Joule (J) = 1 Watt-second.
1 Btu raises the temperature of 1 lb of water by 1°F
4,166 J raises the temperature of 1 kg of water by 1°C

"muscle" a generator or motor has, whereas energy tells you how much "work" a generator or motor performs during a certain period of time. The conversion is 1 kW = 1.359 hp.

Energy is the amount of work a physical system is capable of performing. It has to be measured during a certain period, e.g., a second, an hour, or a year. The usual units are kilowatt hours (kWh), Joules (J), or British thermal units (Btu).

Energy may be converted to different forms: The kinetic energy of moving air molecules may be converted to rotational energy by the rotor of a wind turbine, which in turn may be converted to electrical energy by the wind turbine generator. *Energy loss* is that part of the energy input that cannot be used directly in the next link of the energy conversion system because it is converted to heat. Rotors, gearboxes, and generators are never 100% efficient because part of the energy from the source is converted to heat energy by friction.

The *efficiency* of a machine refers to how well it converts one form of energy into another. The efficiency is 50% if one unit of energy supplied to a machine yields an output of one-half unit.

A wind turbine is a convenient example because the energy is all mechanical/electrical (no boilers or other thermal processes are involved). A wind turbine with a rated power (nameplate power) of 1,000 kW will produce 1,000 kWh of energy per hour of operation when running at its maximum performance (i.e., at high winds above, say, 15 m/s). Wind turbines will usually be running about 75% of the hours of the year, but most of this will not be at rated power.

The arithmetic equivalence of energy units indicates how numerical values are converted from one set of units to another. It provides no useful information about the efficiency of converting energy from one form to another. Table 5.3 lists some frequently needed energy and power equivalence factors.

5.4 ENERGY USE IN WATER AND WASTEWATER TREATMENT

Table 5.4 gives estimates of electric energy use, in kWh/d, for public water supply unit processes. Table 5.5 gives similar cost estimates for wastewater treatment plants.

There is no total given because no plant will use all the listed processes. For example, few water treatment plants will pump raw surface water and raw groundwater. A plant usually has one kind of disinfection. Electricity use is proportional to the flow for most processes.

TABLE 5.4 Estimates of Electricity Use (kWh/d) by Unit Processes in a 5-mgd Water Treatment (Pabi et al. 2013)

Unit Process	Electricity Use (kWh/d)
Pumping	
Raw surface water pumping	725
Raw groundwater pumping	4,600
Finished water pumping	5,325
Filter backwash water pumping	60
Pumping for membrane processes	
Microfiltration (in lieu of sedimentation)	500
Ultrafiltration (contaminant removal)	4,000
Reverse osmosis (brackish water)	29,800
Reverse osmosis (ocean water)	60,000
Chemical processes	
Rapid mixing	175
Flocculation	50
Sedimentation	45
Chemical feed systems	65
Disinfection	
On-site chlorine generation for disinfection	420
Ozone disinfection	560
UV disinfection	310
Non-process loads (HVAC, lighting, etc.)	1,200

Wastewater treatment plants can use many variations of secondary treatment, and not all are listed here. The classical activated sludge process (without nitrification) is aeration with final clarification. The aeration supports microorganisms that remove carbonaceous biochemical oxygen demand (BOD). The energy used is for blowers that supply the air. The process with nitrification removes BOD and also converts ammonia to nitrate. The oxidation of ammonia, which is also done by microorganisms, needs more aeration and more energy. The process can be modified for biological phosphorus removal and nitrogen removal. This will change the energy requirements. Two changes are that energy is needed for internal recycle and mixing. It is possible that bio-phosphorus removal will reduce the aeration requirement.

Anaerobic digestion is almost always part of the system, and there are variations that operate at different temperatures, and variations that have two stages. Usually several options are investigated for sludge treatment, dewatering, and disposal, and for recovering heat energy and using the digester gas. The process is heated and mixed, and sludge must be pumped into and out of the digester and through heat exchangers.

Organic compounds are converted to methane and carbon dioxide gases. This combustible gas mixture is used for heating and in internal combustion engines to generate electricity and drive pumps and compressors. This energy recovery is reflected as a credit of 1,440 kWh/d in Table 5.5.

TABLE 5.5 Estimates of Electricity Use (kWh/d) by Unit Processes in a 5-mgd Wastewater Treatment Plant (Pabi et al. 2013)

Unit Process	Electricity Use (kWh/d)
Wastewater pumping	1,100
Primary treatment	
Odor control	600
Grit removal	180
Primary clarifiers	140
Secondary biological treatment	
Trickling filters	2,540
Biological nutrient removal mixing	550
Aeration without nitrification	3,600
Aeration with nitrification	5,400
Secondary clarifiers	350
Membrane bioreactors	13,530
Solids treatment, dewatering, and disposal	
Anaerobic digestion	550
Gravity belt thickening	140
Dissolved air flotation thickening	N/A
Centrifuge thickening	290
Centrifuge dewatering	1,300
Filtration and disinfection	
UV disinfection	1,170
Depth filtration	350
Surface filtration (e.g., cloth filters)	175
Non-process loads	
Plant utility water	220
Buildings, lighting, computers, etc.)	1,200
Energy recovery from biogas utilization	(1,440)

Example 5.1 Energy Savings at Merrimac, Wisconsin

Pumps, motors, and blowers offer great opportunities for saving electricity. New motors are more efficient than old ones. Variable speed drives (VSDs) allow machinery to operate closer to their peak efficiency. New air diffusers transfer more oxygen than older technology, and dissolved oxygen controllers will regulate and minimize the airflow. Table 5.6 lists some savings from four upgrades at the Merrimac, WI, wastewater treatment plant.

TABLE 5.6 Energy Efficiency Measure Results for Merrimac, Wisconsin, Wastewater Treatment Plant

Energy-Efficiency Measure	Project Cost ($)	Electricity Savings (kWh/y)	Cost Savings ($/y)
1. Replaced two 200-hp influent pump station motors with premium efficient motors and VFDs	170,000	157,000	18,840
2. Replaced two 125-hp process motors with premium efficient motors and VFDs	150,000	79,140	9,497
3. Blower replacement	773,000	358,000	42,960
4. Dissolved oxygen control	128,000	459,000	55,080
Totals =	*1,221,000*	*1,053,140*	*126,377*

Source: ACEEE Case Study: Sheboygan, WI Energy Efficiency in Wastewater, 2011.

5.5 COST OF ELECTRICITY

The cost for an electric utility to supply electricity changes with time of day and by season. Most consumers pay rates based on the seasonal cost of electricity. Changes in prices generally reflect variations in electricity demand, availability of generation sources, fuel costs, and power plant availability. Prices are usually highest in the summer when total demand is high because more expensive generation sources are added to meet the increased demand.

In April 2019, the annual average price of electricity in the United States was \$0.1326/kWh and the median was \$0.1242. Thirty-one states were below the average. Hawaii, Connecticut, and Alaska have the highest average retail electricity prices, \$0.3445/kWh, \$0.2335, and \$0.2293/kWh, respectively, as listed in Table 5.7.

Electricity prices are usually highest for residential and commercial consumers because it costs more to distribute electricity to them. Industrial consumers use more electricity and can receive it at higher voltages, so the supply is more efficient and less expensive. The price of power to industrial customers is generally close to the wholesale price of electricity.

The annual average prices by major types of utility customers were as follows:

Residential	12.55¢ per kWh	Commercial	10.37¢ per kWh
Industrial	6.75¢ per kWh	Transportation	9.48¢ per kWh

5.6 PEAK ENERGY CHARGES

The cost of electricity is based on an *energy charge* and a *demand charge*. The *energy charge* is based on the number of kilowatt hours used during the billing cycle. The total kilowatt hours are multiplied by the unit charge (\$/kWh) for total energy billing. The energy charges can vary with the type of service, voltage, and energy consumption.

The *demand charge* compensates the electric utility for the capital investment required to serve peak loads. Peak demand, or peak load, is measured over a short interval, generally 15 or 30 min. The interval with the highest demand is the one used for billing demand charges.

TABLE 5.7 Average Electricity Price (¢/kWh) Comparison by State for April 2019

State	¢/kWh	State	¢/kWh	State	¢/kWh
Louisiana	9.62	Mississippi	12.01	Illinois	14.12
Washington	9.72	Texas	12.01	Pennsylvania	14.17
Idaho	9.93	Colorado	12.09	Maryland	14.25
Arkansas	10.01	New Mexico	12.26	District of Columbia	14.29
Utah	10.31	N. Carolina	12.34	Wisconsin	14.85
Missouri	10.53	W. Virginia	12.36	Michigan	15.45
N. Dakota	10.63	Nevada	12.42	Vermont	17.19
Kentucky	10.91	Virginia	12.43	New York	17.56
Tennessee	11.01	Ohio	12.49	Maine	17.92
Oregon	11.03	Arizona	12.95	California	18.05
Oklahoma	11.17	Indiana	13.00	New Hampshire	20.65
Wyoming	11.2	S. Carolina	13.07	Rhode Island	22.37
Montana	11.24	Iowa	13.24	Massachusetts	22.61
Nebraska	11.4	Delaware	13.25	Alaska	22.93
Georgia	11.52	Kansas	13.35	Connecticut	23.35
S. Dakota	11.53	Alabama	13.36	Hawaii	34.45
Florida	11.89	Minnesota	13.38	US average	13.26

Source: Choose Energy.

Load factor is the ratio of the average kilowatt load over a billing period to the peak demand. The user will obtain the lowest electric cost by operating at a load factor close to 1.0. The key to a high-load factor, and corresponding lower demand charge, is to even out the peaks and valleys of energy consumption.

Example 5.2 Demand Charge

Suppose that the power demand for a plant is 7,000 kW for 30 days except for a 140,000 kW demand for a 30-min period. The demand charge for that month is based on the peak demand of 140,000 kW, which is the maximum energy demand from the utility company for the month.

Example 5.3 Pumping Costs

A 50-hp pump is operated only 50 h in the month of June. Electricity costs $0.10/kWh plus a peak demand charge of $8/kW. The total electricity cost is

Demand charge = (50 hp)(0.7457 kW/hp)($8/kW) = $298.28
Energy charge = (50 hp)(0.7457 kW/hp)(50 h)($0.10/kWh) = $186.42
Total electric charge = $484.70

If the required service could be provided by running a smaller pump, say 25 hp, for 100 h, the demand cost would be halved, and the total cost would be reduced to $149.14 + $186.42 = $335.56.

Example 5.4 Load Factor

A facility consumed 1,080,000 kWh during 720 h in a 30-d billing period and had a peak demand of 2,000 kW. The load factor is

Average demand = 1,080,000 kWh/720 h = 1,500 kW
Peak demand = 2,000 kW
Load factor = 1,500 kW/2,000 kW = 0.75

5.7 LOAD SHEDDING

Electricity use can be assigned to two categories:

- *Essential loads* that maintain production or safety. Unscheduled shutdowns of these loads cannot be tolerated.
- *Nonessential or sheddable loads* that can be reduced, rescheduled, or shut down temporarily without significantly affecting operations or worker comfort. Examples of such loads are air-conditioning, exhaust and intake fans, chillers and compressors, water heaters, and battery chargers.

Three ways to reduce the use of grid electricity during peak period are:

- *Load shedding* strategies include dimming or turning off lights, changing HVAC temperature set points, and turning off non-critical equipment.

- *Load shifting* of equipment use from on-peak to off-peak time periods (e.g., using off-peak power to pump water).
- *Switch to onsite generation* to meet a portion of the on-peak.

These strategies can be combined.

Example 5.5 Shedding Peak Loads

The potential savings for achieving a given kilowatt reduction is determined by tabulating the size and frequency of peak demands over a representative time period. Seasonal or production variations may also be studied. Table 5.8 lists the 11 highest peak demands for a typical month in descending order. Eliminating the highest ten will reduce the peak demand from 6,320 to 5,990 kW. Based on a peak demand charge of $20/kW, this would save $6,600 per month, or $79,200 annually.

TABLE 5.8 Ranked Peak Demands for Hypothetical Month of May

Date	Time	kW	kW Above 5,900
May 10	10:00 am	6,320	330
May 24	10:30 am	6,220	230
May 14	11:00 am	6,145	155
May 5	1:30 pm	6,095	105
May 20	2:30 pm	6,055	65
May 15	10:30 am	6,025	35
May 15	10:00 am	6,010	20
May 8	2:00 pm	6,000	10
May 9	2:00 pm	5,995	5
May 13	1:30 pm	5,995	5
May 5	2:00 pm	5,990	–

Example 5.6 On-Peak and Off-Peak Charges

Electric companies have on-peak and off-peak rates for certain days of the week and time of day. Table 5.9 shows two companies that are charged for electricity by time-of-use (i.e., on-peak and

TABLE 5.9 On-Peak and Off-Peak Demand Charges

Energy Consumption	Company A	Company B
Total consumption	17,000 kWh	15,000 kWh
On-peak energy charge	9,000 kWh($0.12/kWh) = $1,080	8,000 kWh($0.12/kWh) = $960
Off-peak energy charge	8,000 kWh($0.09/kWh) = $ 720	7,000 kWh($0.09/kWh) = $630
On-peak demand	50 kW	100 kW
Demand charge	50kW($10/kW) = $500	100 kW($10/kW) = $1,000
Total charges	$2,300	$2,590

off-peak time of day) and by peak demand. The total electric consumption (kWh) by company B is less than company A, but company B's total charges are higher because of a higher on-peak demand (kW).

5.8 PUMP EFFICIENCY

Pumping systems account for nearly 20% of the world's energy use by electric motors and 25–50% of the total electrical usage in certain industrial facilities. Energy used in pumping, treating, delivering, and preparing water for end uses accounts for about 13% of the US total annual energy consumption. Every pump provides an opportunity to waste energy or save energy. Efficiency is the key.

Pumping systems involve two extremely simple, yet efficient machines – the centrifugal pump and the AC induction motor. The pump converts mechanical energy into hydraulic energy (flow, velocity, and pressure) and the motor converts electrical energy into mechanical energy. Many medium and larger centrifugal pumps operate at 75–90% efficiency and even smaller ones usually fall into the 50–70% range. Large AC motors can approach an efficiency of 97% and any motor of 4 kW (5 hp) and above can be designed to break the 90% barrier.

Most pumps will operate at peak efficiency only intermittently because the discharge is frequently changing, and the pipe networks change over time either by aging or by intentional modifications. The pumps drift away from optimal performance due to wear and age.

Figure 5.2 shows a pump curve, system head curve, and efficiency curve for a centrifugal pump system. The vertical axis of the pump curve is the discharge pressure of the pump at the indicated flow rate. When measured in units of water depth (m or ft), this is called *head*. This is convenient

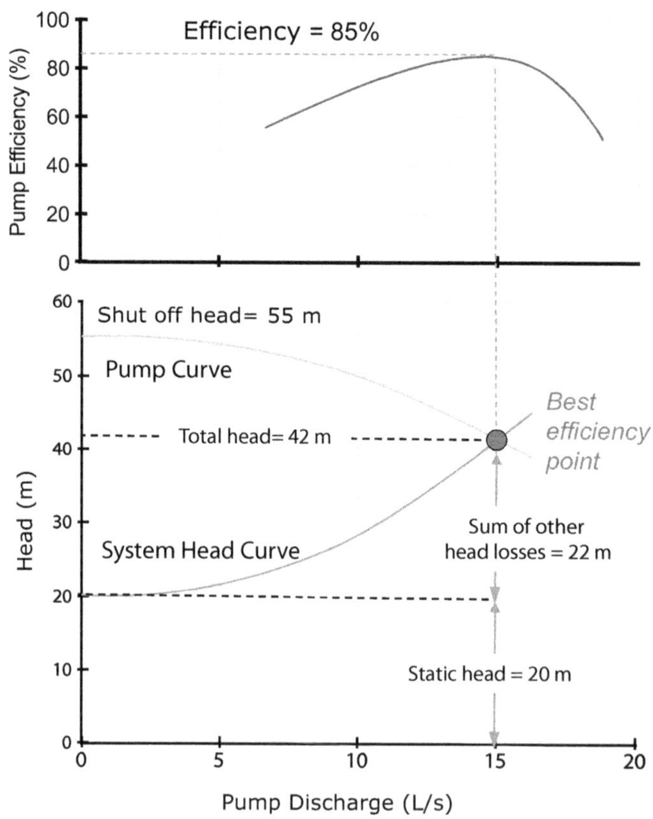

FIGURE 5.2 The pump head-discharge curve with the efficiency curve (top).

when the pump is imagined as a device for lifting fluid from one elevation to another. The static head is the system pressure at zero flow rate. The total head is the static head plus pressure losses due to friction between the pump and the point of discharge. For a fixed-diameter pipe, the fluid velocity and the energy loss due to friction increase as the discharge increases.

The *pump curve* and the *efficiency curve* are determined by the physical characteristics of the pump (the type of pump, impeller diameter, and speed of rotation).

The *system head curve* shows the operating head as a function of the discharge. It is defined by the system of piping, valves, filters, heat exchangers, and other devices through which the fluid must be pumped, the properties of the fluid that is being pumped (density, viscosity), and the velocity of the fluid. It is quadratic because the friction resistance increases as the square of the velocity and the flow.

The *duty point* is at the intersection of the curves. If the duty point is at the peak efficiency, it is called the *best efficiency point*. This curve shows a correspondence of the duty point and best efficiency point (85% efficiency) at flow Q = 15 L/s and total head h = 42 m.

5.9 PUMPING PRESSURE AND POWER

Note: *Some basic knowledge of hydraulics is needed to understand the calculations in the examples that follow. Readers who lack this knowledge may choose to read the examples for the concepts related to energy use, or with slightly more effort they may learn something about hydraulics.*

The power required for pumping depends on the mass of water being moved (lifted) and the pressure difference (resistance) between the pump outlet and the discharge point.

Resistance is measured as *pressure* in units of kg/m^2, lb/ft^2, or kPa. Another measure for pressure, but with different units is, *head*. Head is pressure expressed as an equivalent water depth, measured in meters or feet. This is convenient because the main job of a pump is to lift water, and it is natural to think of the lift as so many meters or feet. *Total head* is the lift plus the resistance from moving the liquid through the system of pipes and valves.

- 1 m of water depth = 1,000 kg/m^2 = 9.807 kPa
- 1 ft of water depth = 62.4 lb/ft^2 = 0.433 lb/inch^2
- 1 atm of pressure = 10,330 kg/m^2 = 101.35 kPa =14.7 lb/inch^2

The power delivered by the pump to the fluid is P_o:

$$\text{Power in kW:} \quad P_o = \rho g H Q / 1{,}000 = 9.81 \rho Q / 1{,}000$$

$$\text{Power in horsepower:} \quad P_o = \rho g H Q / 550 = 32.2 \rho H Q / 550$$

where

ρ = density of water (kg/m^3, lb/ft^3)
g = acceleration of gravity (9.81 m/s^2, 32.2 ft/s^2)
H = total head (m, ft)
Q = flow rate (m^3/s, ft^3/s)

The factors in the denominators have units of 1,000 W/kW and 550 ft-lb/s/hp, and convert the power to kW and hp, respectively.

The power that must be supplied by the drive motor to the pump is P_p, where

$$P_p = P_o / \eta_p$$

where η_p = pump efficiency.

The power input that must be supplied to the motor is

$$P_m = P_p / \eta_m = P_o / (\eta_p \eta_m)$$

where η_m = efficiency of the motor.

P_o is sometimes called the *water power.* P_p is also known as the *shaft* or *brake power.* P_m is the power value used to estimate the pump power costs.

Example 5.7 Pumping Power

The pumping system shown in Figure 5.3 has a static lift of 10 m between two reservoirs. The piping and other hydraulic elements (elbows, etc.) have a friction loss equivalent to a lift of 27.7 m at a flow rate of 0.5 m³/s. This gives an effective total lift of

$10\,\text{m} + 27.7\,\text{m} = 37.7\,\text{m}$ total head, or a pumping pressure of 37.7 m of water.

Pump output power (kW)

$$P_o = \frac{\rho g Q H}{1{,}000 \text{ W/kW}} = \frac{(1{,}000 \text{ kg/m}^3)(9.81 \text{ m/s}^2)(0.5 \text{ m}^3\text{/s})(37.7 \text{ m})}{1{,}000 \text{ W/kW}}$$

$$= 185 \text{ kW}$$

The efficiencies of the pump and motor are

Pump efficiency = η_p = 0.7
Motor efficiency = η_m = 0.9
Overall efficiency $\eta = \eta_p \eta_m = 0.7(0.9) = 0.63$

$$P_m = \frac{P_o}{\eta_p \eta_m} = \frac{184.9\text{KW}}{(0.7)(0.9)} = 293.5\text{KW}$$

The annual electricity use, assuming 90% operation (7,884 h/y) is

$$\text{Electricity} = (7{,}884 \text{ h/y})(294 \text{ kW}) = 2{,}320{,}000 \text{ kWh/y}$$

The annual pumping cost, at $0.14 /kWh, is

$$\text{Cost}(\$/\text{y}) = (2{,}320{,}000 \text{ kWh/y})(\$0.14 \text{ /kWh}) = \$325{,}000/\text{y}$$

The pumping power can be reduced by using a larger pipe, which reduces the friction resistance in the piping system.

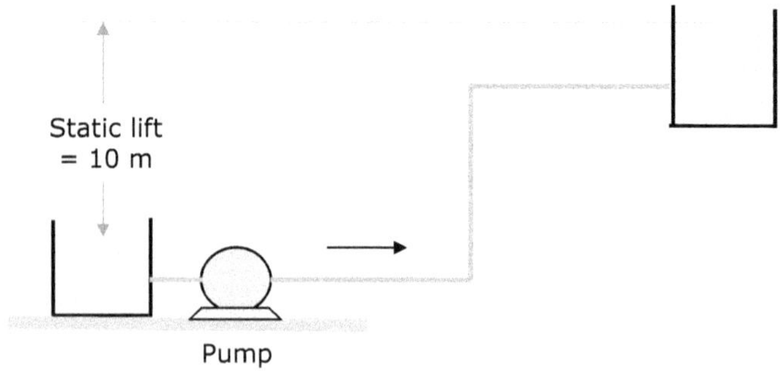

FIGURE 5.3 Geometry of pumping system.

5.10 AIR BLOWERS AND WASTEWATER AERATION

The activated sludge process relies on aerobic (oxygen using) microorganisms to degrade organic pollutants. The organisms can only use oxygen that is dissolved in the wastewater. Insufficient air supply causes an absolute failure to meet performance standards. This poses a design challenge because air requirements fluctuate continually due to changes in flow, organic load, temperature, and atmospheric pressure. Oxygen concentration can be monitored in the aeration basins, and process control computers can adjust variable speed compressors to keep pace with the demand and give the process only the air it needs to maintain a healthy dissolved oxygen concentration of 1 mg/L.

The average amount of electricity used by modern treatment plants is 0.3 kWh/m^3 of wastewater treated (1,200 kWh per million gallons). Aeration accounts for 60–70% of the energy demand of a typical activated sludge treatment plant. Pumping is about 12%. Energy consumption will account for about 75% of the total life cycle cost for an air blower, compared with 15% for capital cost and 10% for maintenance.

The three important parts of an activated sludge wastewater treatment plant are the (a) aeration tanks, (b) blowers to supply air, and (c) diffusers to disperse the air into the aeration tank contents. These are shown schematically in Figure 5.4.

A conventional activated sludge process will consume 0.12–0.20 kWh/m^3 of wastewater treated (465–775 kWh per million gallons) (Henze 2008). If ammonia is oxidized to nitrate (nitrification), the air requirement increases by about 30%. A modern process that uses nitrification–denitrification with biological phosphorus removal needs less air, and less energy, about 0.10–0.15 kWh/m^3.

Centrifugal blowers are a popular means to supply air to an activated sludge process. The centrifugal blower, like a centrifugal pump, has an impeller on a rotating shaft. The output will vary depending on the output pressure. The output of a centrifugal blower can be adjusted with a variable speed drive.

Multiple blowers can be installed to match the demand, or variable speed drives can be used. The installed blower capacity should be sufficient to meet the full load when one blower is out of service. The aeration system is usually designed to supply 200% of the average demand for air.

The most efficient aeration systems use fine bubble diffusers that can dissolve 25–30% of the oxygen in the air that is supplied to the aeration basin. Fine bubbles transfer more oxygen than large bubbles because there is larger gas–liquid interface. Dry air has a density of 1.2754 kg/m^3 at 0°C. Air is 23.2% oxygen, by mass, so one normal cubic meter (Nm^3) of dry air (1 atm, 0°C) contains 0.275 kg O_2. Therefore, 1 Nm^3 can deliver 0.0688–0.082 kg of dissolved oxygen.

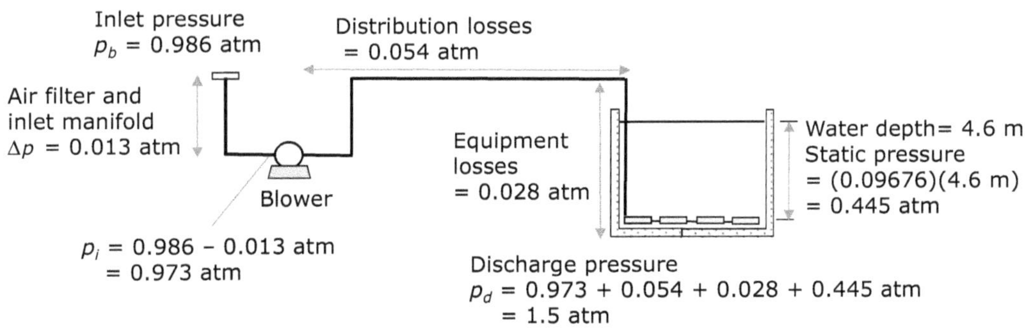

FIGURE 5.4 Schematic of blower piping, aeration tank, and fine bubble diffusers, with pressure losses for the system components.

In US units, the density of air at 70°F = 0.074887 lb/ft^3 and 1 scfm of air contains 0.0173 lb O_2, so 1 scfm of air might transfer 0.0044–0.0052 lb of dissolved oxygen.

The basic concepts of air delivery systems are similar to liquid pumping systems, with the difference that air is compressible, and liquids are not. Compression will raise the temperature of the air, and the required power increases as the temperature increases.

The blower has an operating curve that determines the relation of airflow and discharge pressure. The air delivery piping and equipment (filters, diffusers, etc.) determine the system pressure curve. The intersection of the operating curve and the system curve is the operating point.

Figure 5.4 shows the components that determine the operating pressure:

1. Blower inlet filter and inlet manifold piping
2. Header losses from blower to aeration tanks
3. Drop-pipe and diffuser piping in basin
4. Diffuser losses
5. Static pressure water depth to diffuser
6. Allowance for diffuser clogging
7. Safety factor (usually 0.50–0.75 psia)

Items 1, 2, 5, and 7 are the responsibility of the process designer. Items 3, 4, and 6 are the responsibility of the diffuser equipment supplier (Environ. Dynamics 2005, 2011, Siemens AG/Turblex, Inc. 2009, Mueller et al. 2002).

Example 5.8 Operating Pressure in fine bubble diffuser system

The aeration tank of an activated sludge process is supplied with oxygen by blowing air through diffusers, as shown in Figure 5.4. The barometric pressure at the blower air filter inlet is 0.986 atm. The loss in the filter and the blower inlet manifold is 0.013 atm, so the inlet pressure at the blower is 0.986 – 0.013 = 0.973 atm.

There is a pressure loss of 0.054 atm in the air distribution piping and a loss of 0.027 atm in the aeration tank downpipe and the fine bubble diffusers. The 4.6 m water depth above the diffusers causes a static pressure of 0.445 atm. This gives a diffuser discharge pressure of

$$p_d = 0.973 + 0.054 + 0.445 + 0.027 \text{ atm} = 1.50 \text{ atm} (\text{rounded from } 1.499)$$

The design pressure usually includes an allowance for a diffuser clogging (0.035 atm) and perhaps a safety factor (0.035 atm), which would give a design discharge pressure of

$$p_{\text{design}} = 1.50 \text{ atm} + 0.07 \text{ atm} = 1.57 \text{ atm}$$

The power the blower must impart to the air is

$$P = \rho_{\text{air}} Q_{\text{air}} c_{P,\text{ air}} T_1 \left[\left(\frac{p_2}{p_1} \right)^{0.286} - 1 \right]$$

where

Q_{air} = actual volume flow rate of air at the inlet (m^3/s, ft^3/sρ)
ρ_{air} = density of inlet air (kg/m^3, lb/ft^3)
$c_{P,\text{air}}$ = specific heat of air = 1.006 kJ/kg K (0.24 Btu/lb°R)
T_1 = inlet air temperature (K, °R)
p_1 and p_2 = inlet and outlet air pressures, atm

T_1 is absolute temperature (K = 273 + °C, °R = 460 + °F). The pressures are absolute pressures (i.e., gauge pressure plus atmospheric pressure). (*Note:* In some references the exponent on the pressure ratio is 0.283.)

Example 5.9 Blower Power

Calculate the blower power required for an inlet airflow rate of 40 m^3/s for these conditions:

Blower inlet pressure = p_1 = 0.985 atm
Blower outlet pressure = p_2 = 1.52 atm
Blower inlet temperature = T_1 =10°C = 283 K
Density of dry air at atm and 0°C = 1.2754 kg/m^3
Heat capacity = $c_{P,\,air}$ = 1.006 kJ/kg K

$$P = \rho_{air} Q_{air} c_{P,\,air} T_1 \left[\left(\frac{p_2}{p_1} \right)^{0.286} - 1 \right]$$

$$= \left(1.2754 \frac{kg}{m^3} \right) \left(40 \frac{m^3}{s} \right) \left(1.006 \frac{kJ}{kg\text{-}K} \right) (283\ K) \left[\left(\frac{1.52\ atm}{0.985\ atm} \right)^{0.286} - 1 \right]$$

$$= 1{,}918\ kJ/s = 1{,}918\ kW$$

Blower efficiency = 70%
Input power = 1,918 kW/0.7 = 2,740 kW (3,675 hp)

5.11 CONCLUSION

The price of electricity has been increasing by 4–5% per year in many states. This is a strong incentive to look for ways to conserve energy to reduce the kilowatt-hour charges. Peak load charges can be reduced by load shedding.

Any process that uses energy, in any form, presents an opportunity for savings. Pumping water and blowing air into the activated sludge process are profitable targets for energy reduction. The physical equipment can be changed. Worn or inefficient pumps can be replaced with new equipment that has higher mechanical efficiency. New motors are more efficient. Fine bubble aeration systems use much less air and, therefore, much less energy for air delivery. Variable speed drives for pumps and blowers increase efficiency.

Heat energy can be recovered from thermal devices such as internal combustion engines. Combined heat and power is the ultimate example of this.

5.12 PROBLEMS

5.1 COST OF WATER FOR BEER MAKING

Producing 1 L of beer in a modern brewery (without water recycling) uses 4 L of freshwater and results in 2.5 L of wastewater. The costs for water and wastewater are 1.50 and 2.50 €/m^3, respectively. Calculate the costs for a brewery with an annual output of 10,000,000 L.

5.2 ANNUAL BULK CHEMICAL PURCHASE

Table P5.2 gives the bid tabulation for bulk chemical quotations by a city for all water and wastewater treatment. Four vendors submitted quotations for unit costs. No vendor can supply all the chemicals, so the vendors can be chosen as the city thinks best. Calculate the total annual cost for chemicals. Recommend a purchasing plan.

TABLE P5.2 Chemical Bidding Information (NB = no bid)

Chemical	Unit	Amount	Vendor Bid Price ($/unit)			
			A	B	C	D
Liquid ammonium sulfate	gal	40,000	1.175	0.92	2.07	NB
Anhydrous ammonia	ton	56	NB	1,085	995	NB
Aquamag	lb	7,245	0.584	0.54	0.6787	0.574
Citric acid	gal	5,830	6.43	NB	NB	7.77
Liquid chlorine	ton	250	860	NB	719	719
Copper sulfate	lb	8,000	1.84	1.52	NB	1.93
Liquid ferric chloride	gal	240,000	NB	1.56	1.436	1.67
Liquid ferric sulfate	gal	241,000	NB	1.1198	NB	1.108
Hydrofluorosilicic acid	gal	14,500	NB	NB	2.16	2.06
Lime slurry	gal	214,000	NB	NB	NB	0.48
Liquid oxygen	gal	27,000	NB	0.4928	0.4928	0.45
Calcium oxide	ton	3,000	NB	NB	164.85	197
Potassium permanganate	lb	4,350	1.9345	NB	1.85	2.37
Powdered activated carbon	ton	30	1,393.40	NB	1,780	1,930
Scale inhibitor	gal	17,200	7.2742	NB	8.09	10.47
Liquid sodium bisulfate	gal	2,475	738.37	NB	NB	NB
Liquid sodium hydroxide	gal	14,500	1.128	0.982	2.002	1.062
Liquid sodium hypochlorite	gal	16,080	NB	1.08	NB	4.19
Sulfur dioxide	ton	90	900	749	NB	NB
Liquid sulfuric acid	gal	15,080	NB	2.87	NB	NB

5.3 CHERRY BRINE RECLAMATION

The manufacture of Maraschino cherries involves soaking the fruit in brine to remove the natural color before the distinctive red color is added. The brine, which is colored (but not bright red) and contains dissolved salts, can be reused if the color is removed. The color can be removed by passing the brine through granular activated carbon columns. When the carbon becomes saturated with the adsorbed color-causing molecules the carbon has to be replaced or regenerated. Table P5.3 gives the

TABLE P5.3 Net Costs and Savings for Treating Cherry Brine (Soderquist 1971)

Cost Item	Production level (T/d)	
	1,000	5,000
Capital cost ($/y)	2,240	5,360
O&M cost ($/y)		
• Labor	1,040	1,560
• Power	40	50
• Carbon replacement	2,500	12,400
• Maintenance	1,000	1,500
Annual cost savings ($/y)		
Water purchased	40	200
Sewerage charges	4,500	22,600
Make-up chemicals		
• Sulfur dioxide	870	4,400
• Calcium chloride	810	4,050
• Lime	60	300

annual capital cost and the annual O&M cost for 1,000 T/y and 5,000 T/y of cherry production. It also gives the savings in process chemicals (sulfur dioxide, calcium chloride, and lime), purchased water, and wastewater treatment charges. Calculate the total annual cost ($/y), the total annual savings ($/y), and the total net annual cost ($/y).

5.4 NEUTRALIZATION OF ACID WASTES

Shanahan Valley Chemicals needs an effluent neutralization plant to treat 22.5 m^3/h of acid wastewater that has a specific mass of 1,000 kg/m^3 and contains an average of 3% free acid. Neutralization can be done using lime (CaO) or caustic soda (NaOH). Lime neutralization produces sludge that has a disposal cost of $37.50/T. The caustic process produces no sludge. The plant will operate 2,500 h per year. Use the data in Table P5.4 to evaluate the two alternatives.

TABLE P5.4 Neutralization Costs

Cost Factor	Caustic	Lime
Construction ($)	110,000	150,000
Chemical cost ($/T)	100	25
Chemical required T/T acid	0.8	1.1
Operating labor ($/y)	7,500	10,000
Operating energy ($/y)	500	1,000

5.5 ENERGY FOR WASTEWATER TREATMENT

In Austin, Texas, 60% of municipal electric energy is for water and wastewater treatment, 10% is for streetlights and traffic signals, and 30% is for other city uses. The fraction of total cost going for energy is 34% for the water utility and 28% for the wastewater system. This compares to 35% and 48% of the cost, respectively, going for staffing. Looking for opportunities to reduce energy costs is very worthwhile. Energy audits at 200 US treatment plants suggest that 10–20% energy savings are available through process changes and another 10–20% through equipment modifications. How could these savings change the distribution of municipal energy use of Austin?

5.6 STOP WATER LOSSES TO SAVE ENERGY

Three towns will reduce the volume of water lost from leaking distribution systems by the projected amounts shown in Table P5.6. Also given are the current energy used for pumping and the current price of electricity. Calculate the projected cost savings. Explain how reducing lost water saves energy.

TABLE P5.6 Water Distribution System Losses

Town	Projected Reduction (10^6 gal)	Current Energy Usage (kWh/10^6 gal)	Rate ($/kWh)
Gonzales	12.5	1,604	0.35
Eustis	70	1,013	0.07
Palatka	219	997	0.062

5.7 TREATMENT PLANT ENERGY

A survey by the National Council of Clean Water Agencies (2005) reported that 47 wastewater treatment plants used a combined 2.1 billion kWh of electricity. The breakdown was 38% for in-plant pumping, 26% for aeration, 25% for effluent pumping, and 11% for other uses. Suppose the pumping efficiency can be improved by 15% and aeration efficiency can be improved by 20%. How much electricity can be saved?

5.8 DEMAND CHARGE FOR TWO COMPANIES

Company A runs a 50 MW load continuously for 100 h. Company B runs a 10 MW load for 500 h. The rate for both companies is $0.09/kWh, and the demand charge is $1.75/kW. Calculate the total electricity costs for the two companies and comment on why they are different.

5.9 POSSIBLE LOAD SHIFTING

Company A consistently requires 100 kW over the course of a 720-h month. Company B uses just 50 kW/h for 716 h. However, they use 500 kW for 1 h every week (4 days/month) to start up and bring their machines on line. Can load shifting benefit the companies?

5.10 ELECTRIC RATE SCHEDULES

Most electric service providers offer time of day electric power rates. Table P5.10 is an example rate schedule. A wastewater utility has no more opportunities for load shifting, and it wants to generate electricity on-site during periods of peak demand. Does this seem a rational strategy to reduce power costs?

TABLE P5.10 Electric Rate Schedule

Description	Summer	Winter
On-peak demand charge:		
0–1,000 kW	$12.98/kW	$7.64/kW
2,000–5,000 kW	$11.89/kW	$6.54/kW
Economy demand charge	$1.03/kW	$1.03/kW
Energy charge:		
On-peak energy	$0.042485/kWh	$0.042485/kWh
Off-peak energy	$0.021131/kWh	$0.021131/kWh

5.11 ELECTRIC BILL

The monthly electric bill asks payment for delivery of 10,400 kWh at $0.14/kWh. In addition, the peak demand was 56.2 kW and the peak demand charge is $16.65/kW. There is also a base charge of $65/month per customer. Calculate the total payment due.

5.12 PEAK ELECTRICAL DEMAND CHARGES

The data in Table P5.12 are the highest demands for power during a hypothetical billing period. The peak demand charge is calculated using the single largest peak demand for *power* during the billing period (usually the average over a 15-min period). The normal peak demand at a pollution control

TABLE P5.12 Highest Electrical Energy Demands for a Hypothetical Billing Period in May

Date	Time	Demand (kW)	kW Above 590 kW
May 10	10:00 am	6,320	420
May 24	10:30 am	6,220	320
May 14	11:00 am	6,140	240
May 5	1:30 pm	6,090	190
May 20	2:30 pm	6,050	150
May 15	10:30 am	6,020	120
May 15	10:00 am	6,010	110
May 8	2:00 pm	6,000	100
May 9	2:00 pm	5,990	90
May 13	1:30 pm	5,950	50
May 5	2:00 pm	5,920	20

facility is 5,900 kW, but this has been exceeded several times in this billing period due to special operating demands. Assume the peak demand charge is \$19.40/kW.

(a) Calculate the peak demand charge for a typical month.
(b) How much has the peak demand increased because of the excursions above 5,900 kW?
(c) Assume that peak power demand excursions of this magnitude occur every month. How much could be saved per year if the some of the peak demand could be shifted to one of the low periods of demand?

5.13 LAGOON AERATION COMPARISON

Aerated lagoons treat sewage with less capital expense than activated sludge systems, but most aerated lagoons use surface aerators which have high energy consumption. Currently, a lagoon is aerated by nine 5-hp surface aerators. It is proposed to replace the surface aerators with bottom-laid fine bubble diffusers that would use a 12-hp compressor. The system will operate 24 h/d. For simplicity, assume the aerators always run at the same rate. Calculate the cost of electricity to operate each system, using \$0.08/kWh and a wire-to-motor efficiency of 80%.

5.14 AIR SUPPLY FOR AN INCINERATOR

Calculate the power requirement of fan that will move a volume of $Q = 30{,}000$ acfm through an incinerator system that has a pressure drop of 19 inches of water and efficiency of $\eta = 0.6$. (The total pressure drop depends on the equipment and the details of the installation.) The system operates 8,000 h/y.

5.15 DIFFUSER CLEANING

Figure P5.15 shows the oxygen transfer efficiency of fine bubble diffusers that are cleaned regularly versus those that are not. One treatment plant found that an annual diffuser cleaning accomplished savings of \$50,000/y. The cost of energy was \$0.05/kWh. Explain the economics of diffuser cleaning.

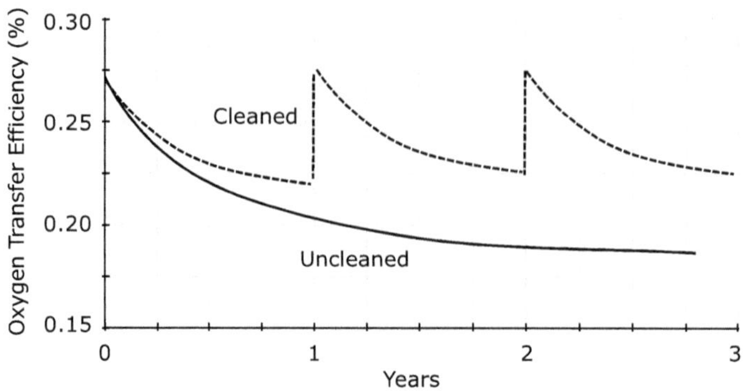

FIGURE P5.15 Oxygen transfer efficiency as a function of fine bubble diffuser cleaning.

5.16 BLOWER POWER

Calculate the blower power required for an inlet airflow rate of 50 m³/s for these conditions:

Blower inlet pressure = p_1 = 0.98 atm
Blower outlet pressure = p_2 = 1.6 atm
Blower inlet temperature = T_1 = 8°C = 281 K
Density of dry air at 1 atm and 0°C = 1.2754 kg/m³
Heat capacity = $c_{P,\text{air}}$ = 1.006 kJ/kg K

5.17 BOILER EFFICIENCY

A boiler has an output of 6,700,000 BTU/h and burns No. 6 oil that yields 150,000 BTU/gal. Show that

(a) A 3% drop in efficiency (85% → 82%) increases fuel costs 3.7%
(b) A 5% drop in efficiency (85% → 80%) increases fuel costs 6.3%

5.18 COMPRESSED AIR COST 1

Compressed air is often taken for granted as a necessary cost and is often squandered and wasted, but it costs real money. A plant has two 50-hp and one 25-hp compressors. Each compressor operates 4,000 h/y and has an efficiency of 90%. Calculate the average cost of compressed air, assuming $0.06/kWh.

5.19 COMPRESSED AIR COST 2

A typical small job-shop manufacturer operates a 100-hp air compressor 4,160 h annually. It runs fully loaded, at 94.5% efficiency, 85% of the time. It runs unloaded, at 25% of full load, at 90% efficiency, 15% of the time. The electric rate is $0.06/kWh, including energy and demand costs. Calculate the annual cost per 1,000 ft³ of compressed air, assuming 3.6 ft³/hp when the compressor is fully loaded.

5.20 AUDIT OF PUMPING ENERGY

Centrifugal pumps are one of our most efficient pieces of equipment and, when properly maintained, they will have a long life. An energy audit of a water pumping installation provides the information listed in Table P5.20 for four pumps. The oldest, and the largest, was installed in 1992 and the newest in 2002. Rated horsepower or nameplate horsepower is the measure of the motor's mechanical output rating. The measured efficiency is the line-to-water value. (a) Calculate the actual power consumption, the energy use per year, and the cost per year if electricity costs $0.18 kWh. (b) Comment on opportunities to upgrade the pumping complex and reduce the cost of operation.

TABLE P5.20 Energy audit for four pumps

Pump	Installed	Rated Power (hp)	Rated Power (kW)	Efficiency (line to water)	Operation (h/y)
1	1992	200	149	0.75	2,000
2	1994	150	112	0.93	4,000
3	1995	80	60	0.85	4,000
4	2002	40	30	0.93	5,000

5.21 METHANE PRODUCTION AND ELECTRICITY GENERATION

An estimate is needed of the digester gas production and electric generation potential for a future treatment plant. Since the plant does not yet exist, there are no measurements of solids loading to the digester. These are estimated to be

Raw wastewater BOD = 3,958 lb/d
Raw wastewater TSS = 3,500 lb/d

Reasonable assumptions are as follows:

Cost of electricity = \$0.12/kWh
Primary clarifiers remove 50% of the TS and 30% of the BOD
Primary solids are 75% volatile
Secondary sludge production = 0.7 lb TS/lb BOD removed
Secondary sludge is 70% volatile
Volatile solids destruction in digester = 50%
Digester gas production = 15 ft^3/lb volatile destroyed
Microturbine will generate 30 kW at a digester gas feed rate of 16 ft^3/min

5.22 NATURAL GAS PRICES AND COGENERATION

Figure P5.22 shows the price of natural gas and electricity over a 20-year period, starting in January 2000. Explain how these trends encourage or discourage the use of cogeneration in pollution prevention projects.

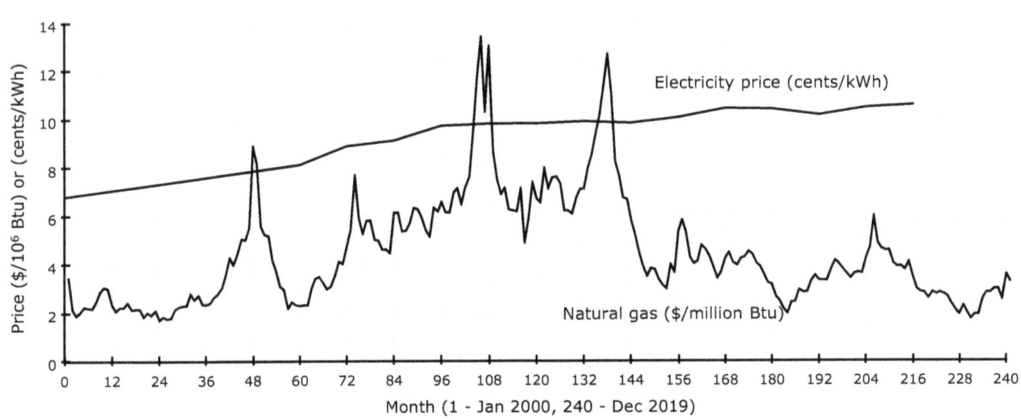

FIGURE P5.22 Price of natural gas ($\$/10^6$ Btu) and electricity (cents/kWh) for January 2000–December 2019).

6 Cost Indexes

A *cost index* is a weighted cost of specified materials in known quantities that are used to track cost changes from year to year. It is used to adjust costs over time by using the ratio of index values at two different times.

The US government's Consumer Price Index (CPI) includes food, housing, transportation, medical care, and a few other categories: The Engineering News Record Construction Cost Index (ENR-CCI) includes steel, cement, lumber, and labor. The Chemical Engineering Plant Cost Index (CEPCI) includes equipment, machinery, fabrication and installation, and labor. There are other specialized indexes, such as the Marshall and Swift (M&S) index for process equipment and the Vatavuk Air Pollution Control Index. RSMeans, an excellent source of detailed unit cost information, also maintains a cost index. European and Asian countries have cost indexes for their economies and economic sectors, for example, the Chemie Teknik Index for Germany and the EU-28 Index for the European Union.

The reason for making an index for a class of industry or an economic sector is to include components (labor, structural steel, pumps, pollution control equipment, etc.) that are most relevant to that industry.

6.1 TRENDS IN COST ESCALATION

Figure 6.1 shows the historical trends of the Consumer Price Index, Engineering News Record Index, and the Chemical Engineering Index up to 2016. It is not possible to make accurate adjustments by reading values from the graph. The index values for 1990–2017 are listed in Table 6.1. These apply to U.S. average costs. The indexes have been refined to include country and regional subindexes.

6.2 ADJUSTING COSTS OVER TIME

The ratio of the index values, and not the absolute values of the index, is used to adjust for cost changes over time. C_1 and C_2 are the costs of similar projects that have been, or will be, constructed at different times. The cost index values at the two times are I_1 and I_2. The cost C_2 can be estimated from C_1 and the ratio of the appropriate cost index values for the two times.

$$C_2 = C_1 \left(I_2 / I_1 \right)$$

Example 6.1 Time Adjustment of Costs

The average construction cost of a project in 2005 was \$900,000. The estimated cost of the same project in 2015, using only the ENR-CCI to make the estimate, is

$$\text{Cost in 2005} \left(\frac{\text{Index for 2015}}{\text{Index for 2005}} \right) = \$900{,}000 \left(\frac{10{,}031}{7{,}446} \right) = \$1{,}212{,}450$$

Example 6.2 Regional Adjustment of Costs

Convert the \$900,000 cost of a building in New York in 2016 to the cost in Toronto in 2016. The 2016 RSMeans Index for Toronto is 109.9 and 131.1 for New York. The adjustment is

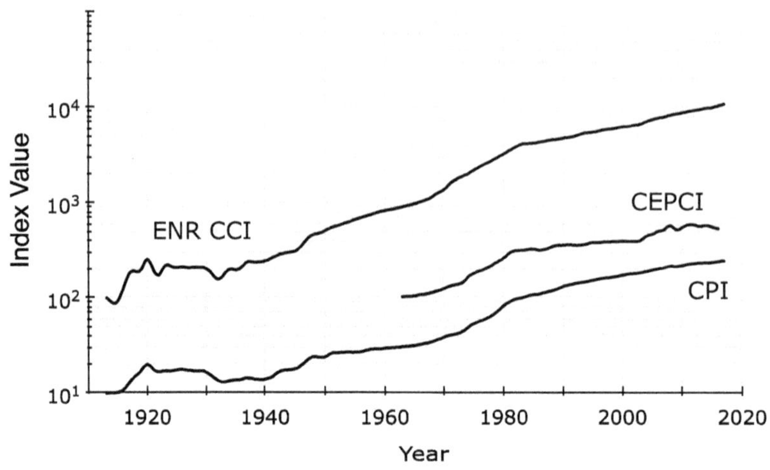

FIGURE 6.1 Historical trends for the Consumer Price Index (CPI), Engineering News Record Construction Cost Index (ENR-CCI), and the Chemical Engineering Plant Cost Index (CEPCI).

TABLE 6.1 Values for the Engineering News Record CCI and the Chemical Engineering PCI

Year	ENR	CEPCI	Year	ENR	CEPCI	Year	ENR	CEPCI
1990	4,732	357.6	2001	6,334	394.3	2011	9,070	585.7
1991	4,835	361.3	2002	6,538	395.6	2012	9,308	584.6
1992	4,985	358.2	2003	6,695	402.0	2013	9,547	567.3
1993	5,210	359.2	2004	7,115	444.2	2014	9,807	576.1
1994	5,408	368.1	2005	7,446	468.2	2015	10,031	556.8
1995	5,471	381.1	2006	7,749	499.6	2016	10,443	541.7
1996	5,620	381.7	2007	7,971	525.4	2017	10,889	567.5
1997	5,826	386.5	2008	8,311	575.4	2018	11,186	619.2
1998	5,920	389.5	2009	8,570	521.9	2019		
1999	6,059	390.6	2010	8,801	550.8	2020		

$$\text{Cost for New York}\left(\frac{\text{Index for Toronto}}{\text{Index for New York}}\right) = \$900{,}000\left(\frac{109.9}{131.1}\right)$$

$$= \$754{,}462$$

6.3 ENGINEERING NEWS RECORD COST INDEX

The oldest and best-known index is the ENR Construction Cost Index (ENR-CCI), which is calculated using the cost of 2,500 lb of structural steel, 1.128 tons of cement, 1,088 board-feet (bf) of 2 × 4 lumber, and 200 h of common labor.

Table 6.2 gives the costs in 1913 when the ENR-CCI was first published. The index value was arbitrarily set at 100. It is a coincidence that the cost of the index commodities was almost exactly \$100. The initial index value would be set at 100, no matter what the actual cost is.

Table 6.2 also gives the cost of the same four commodities some years later, in year X. The cost for the same items in the same quantities had increased to \$1,200.88. The index value for year X is

TABLE 6.2 ENR Construction Cost Index

Item	Quantity	1913			Year X		
		Unit cost	Cost	% Total	Cost	$Cost_X$ / $Cost_{1913}$	% Total
Steel	2,500 lb	\$15/ton	\$37.50	38	\$155.00	4.13	13
Cement	1.128 tons	\$6.33/ton	\$7.14	7	\$24.18	3.90	2
Lumber	1,088 bf	\$0.01572/bf	\$17.10	17	\$153.04	8.95	13
Labor	200 h	\$0.19/h	\$38.00	38	\$868.60	22.86	72
Totals			\$99.74	100	\$1,200.82	12.04	100

$$I_X = I_{1913}(\text{Cost year } X\,/\,\text{Cost in 1913}) = 100(\$1{,}200.88\,/\,\$99.74) = 1{,}204$$

The items and quantities used to compute the index are fixed, but the rates of inflation will differ for each item. This causes the weight of each item in the index to change. In the example in Table 6.2, steel was 38% of the 1913 index but only 13% of the year X index. The weight of labor increased from 38% to 72%. The cost of the materials increased from \$62 to \$332, a factor of 5.38, while common labor increased by a factor of \$869/\$38 = 22.9.

The ENR Building Costs Index (ENR-BCI) replaces the 200 h of common labor with 66.38 h of skilled labor (bricklayers, carpenters, and structural ironworkers).

The ENR-CCI and ENR-BCI are not appropriate for pollution control process, but they will be suitable for parts of a treatment plant complex, such as administration buildings, garages and workshops, roads, and parking areas. The ENR-CCI is suitable for landfills, sewers, and other projects that are not heavy with process equipment.

6.4 CHEMICAL ENGINEERING PLANT COST INDEX

The Chemical Engineering Plant Cost Index (CEPCI), first published in 1964, was based on the components and weights given in Table 6.3. Water and wastewater treatment plants, like chemical

TABLE 6.3 Basis for the Chemical Engineering Plant Cost Index

Index Component	Weight of Component (%)
Equipment and machinery	
Fabricated equipment	22.6
Process machinery	8.5
Pipes, valves, and fittings	12.2
Process instruments and controls	4.3
Pumps and compressors	4.3
Electrical equipment and materials	3.1
Structural supports, insulation, and paint	6.1
Subtotal =	*61%*
Erection and installation labor	22
Buildings, material, and labor	7
Engineering and supervision	10
Total =	*100%*

plants, have pumps, piping, mixers, heaters, aerators, solid–liquid separations, and reactors. This index can be useful for water and wastewater treatment processes.

Table 6.3 shows the components of the CEPCI and the weight of each in the total index value. Equipment and machinery with necessary auxiliary and supporting materials make up 61% of the total index. The other 39% is for various labor components, including engineering and supervision.

6.5 CONCLUSION

The ENR-CCI is for the construction of projects with costs that are heavily weighted toward labor and materials (Portland cement, steel); highways are an example. ENR publishes a separate index for structures that has a different weighting of cement, steel, lumber, and skilled labor. Neither of these gives any weight to process equipment or machinery. The CEPCI includes pumps and other machinery that are part of wastewater treatment plants. The Marshal & Swift Index is for process equipment. The US EPA keeps an index for air pollution equipment, and at one time had indexes for wastewater treatment and sewer construction.

6.6 PROBLEMS

6.1 ENGINEERING NEWS RECORD INDEX

Calculate the ENR Index for the year X when

Structural steel cost \$0.30/lb
Portland cement cost \$90/ton
2 × 4 lumber cost \$0.255/board-foot
Common labor was \$18.70/h

6.2 UPDATING COSTS ESTIMATES

A project being planned is similar to a project that was completed several years ago that cost \$850,000. At that time the ENR Index was 4,012; today it is 5,375. What is the estimated cost of today's project?

6.3 ADJUST COST FOR LOCATION

A treatment plant in Colorado cost \$12,000,000. What is the cost of the same plant in New Jersey if the Colorado and New Jersey cost indexes are 980 and 1,140?

6.4 COST INDEX ADJUSTMENT ERRORS

Identify and explain three major sources of inaccuracy in cost index adjustments.

6.5 SLUDGE DIGESTION PROCESS INFLATION

Table P6.5 gives the index values that were used to adjust the costs of a new type of sludge digestion process from the date of construction (between 1977 and 1987) to 1990. Plot the cost index values. Calculate the percent increase and the index multiplier from year to year.

TABLE P6.5 ENR-CCI Adjustments

Year	**1977**	**1978**	**1979**	**1980**	**1981**	**1982**	**1983**
ENR Index	2,576	2,776	3,003	3,237	3,535	3,825	4,068
Year	**1984**	**1985**	**1986**	**1987**	**1988**	**1989**	**1990**
ENR Index	4,146	4,195	4,295	4,406	4,519	4,615	4,721

6.6 EQUIPMENT COST ADJUSTMENT

The cost of two pieces of equipment, one purchased in 2010 and one purchased in 2015, will be used to estimate the cost of a similar unit today. The cost index today is 560. Using the data in Table P6.6, derive an economy of scale equation ($C = KQ^M$) and estimate today's cost of an 80 m^2 unit.

TABLE P6.6 Equipment Cost-Capacity Data

	A	B
Area	70 m^2	130 m^2
Cost	$17,000	$24,000
Purchased	1990	1995
Cost Index	360	390

6.7 SEWAGE TREATMENT PLANT

The total cost of a sewage treatment plant with a capacity of 5 mgd completed in 2015 was $8 million for a town in Colorado. It was proposed to build a similar 9 mgd treatment plant in New Jersey for completion in 2022. For additional information given below, make a screening estimate of the cost of the proposed plant.

The total Colorado construction cost included $300,000 for site preparation that is not typical for similar plants. The variation of sizes for this type of treatment plants can be approximated by the exponential law with $M = 0.7$. The inflation rate was approximately 3% per year from 2010 to 2020. The location indexes of Colorado and New Jersey areas are 0.95 and 1.10, respectively, against the national average of 1.00. The installation of special equipment to satisfy the new environmental standard cost an extra $200,000 for the New Jersey plant. The site condition in New Jersey required special foundations which cost $500,000.

6.8 CHEMICAL ENGINEERING PROCESS COST INDEX

Table P6.8 lists the components of the Chemical Engineering Plant Cost Index. Calculate the new index value if cost inflation is 2% for fabricated machinery, 4% for process machinery, instruments, pumps and compressors, and 8% for labor. The cost of other cost categories does not change.

TABLE P6.8 CEPCI components

Index Component	**Weight of Component (%)**
Equipment and machinery	
Fabricated equipment	22.6
Process machinery	8.5
Pipes, valves, and fittings	12.2
Process instruments and controls	4.3
Pumps and compressors	4.3
Electrical equipment and materials	3.1
Structural supports, insulation, and paint	6.1
Subtotal =	61%
Erection and installation labor	22
Buildings, material, and labor	7
Engineering and supervision	10
Total	100

7 Economy of Scale

Doubling the size of a treatment system, process, or a machine does not double the cost. A large part of the cost to build a pipeline or sewer is excavation, sheeting, moving underground utilities, backfilling, repaving, and landscaping. These costs are almost the same, whether a 24-inch or an 18-inch pipe is installed, but the hydraulic capacity of the 24-inch pipe is almost double that of the 18-inch pipe. This is *economy of scale.*

7.1 COST AND CAPACITY

Many kinds of equipment, processes, and systems have a cost-capacity estimating function that reflects economy of scale:

$$C = KQ^M$$

where
- C = cost ($)
- Q = capacity (measured in units of volume, area, flow rate, power, etc.)
- M = economy of scale exponent
- K = cost for a capacity of one unit (measured in the same units as capacity)

The cost-capacity model for K = $10,000 and $M = 0.6$, as shown in Figure 7.1, is

$$C = \$10{,}000Q^{0.6}$$

M is the slope of the line on the log–log plot and Q is the cost of one unit of capacity. The exponent M has values near $6/10 = 0.6$ for many kinds of processes and equipment, and $C = KQ^M$ is often called the *Six-Tenths Rule*, even when M is not 6/10.

Lower values of M give more economy of scale. Table 7.1 shows how important this can be. For $M = 0.6$, one unit of capacity costs $10,000, and two units cost $15,157. The capacity is doubled for

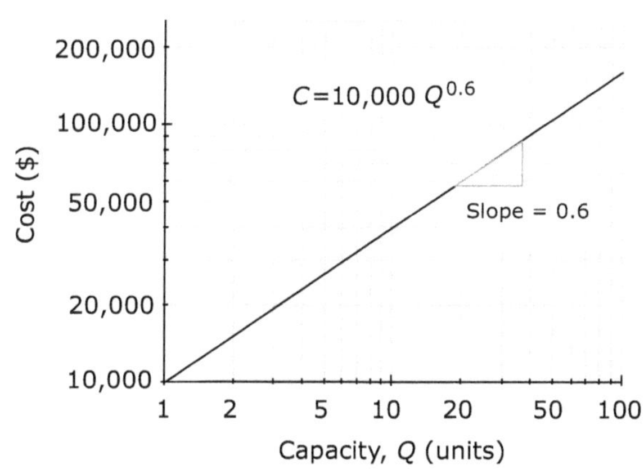

FIGURE 7.1 Cost-capacity curve for $M = 0.6$ ($C = \$10{,}000Q^{0.6}$).

TABLE 7.1 Economy of Scale for $M = 0.5$ to $M = 0.8$

Exponent	Units of Capacity			
M	1	2	3	4
0.5	$10,000	$14,142	$17,320	$20,000
0.6	$10,000	$15,157	$19,332	$22,974
0.7	$10,000	$16,245	$21,577	$26,390
0.8	$10,000	$17,411	$24,082	$30,314

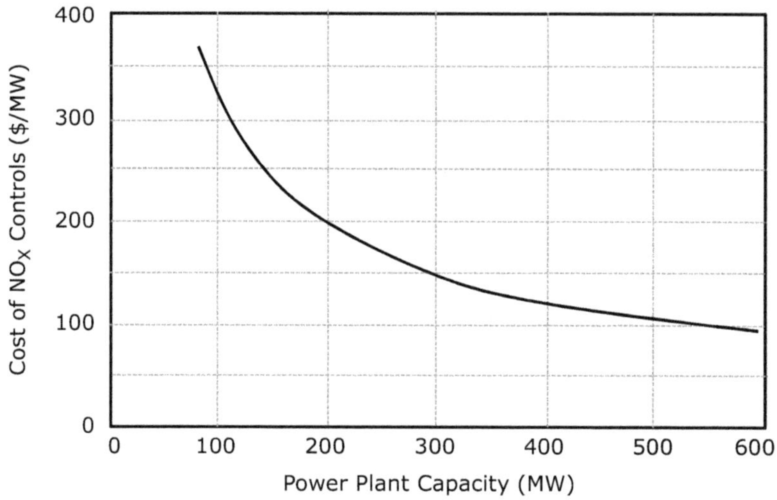

FIGURE 7.2 Cost-capacity curve for NO_X control (US$1999) (Friedel & Arroyo 2005).

a 50% increase in cost. At $M = 0.5$, the cost doubles when the capacity is raised from one to four units; at $M = 0.8$, four units cost three times more than one unit.

Figure 7.2 is a cost-capacity curve for NO_X control at a coal-fired boiler power plant. The economy of scale is more obvious by showing the unit price ($/MW) as a function of capacity.

7.2 VALUES FOR THE ECONOMY OF SCALE EXPONENT *M*

For cost estimating, the most important value is *K*, but it is the most difficult to find. Costs change with time and place. Cost indexes are used to adjust for this, but there is no method that makes them exactly comparable. The value of *M*, on the other hand, seems to be consistent over time and location.

Many chemical engineering books contain data and graphs in the form of the 6/10 rule. The US EPA published excellent cost-estimating manuals for wastewater treatment in the 1970s that were used for many years. Their purpose was to promote estimating consistency among applicants for the federal wastewater treatment plant construction grant program. More recently, the EPA has published manuals on air pollution control design and cost estimating (US EPA 2002)

Table 7.2 lists values of *M* for some wastewater treatment processes. There are similar values in Table 7.3 for process equipment. More data of this kind are given in Appendix B. The exponent may be valid for a limited range of capacity.

Some kinds of equipment and systems have little or no economy of scale. A value of *M* approaching 1.0 means that each unit of added capacity costs the same as all previous units. This can happen because of the way the capacity is defined. Small surface aerators, for example, have $M = 0.96$,

TABLE 7.2 Economy of Scale of Waste Treatment Processes (Various Sources)

Treatment Process	Units of Capacity	*M*
Wastewater treatment processes		
Activated sludge, aeration basin	Volume	0.50–0.70
Sedimentation: rectangular tanks	Area	0.56
Sedimentation: circular tanks	Area	0.66
Primary sedimentation	Area	0.60–0.76
Final clarifier	Area	0.57–0.76
Flotation	Volume	0.74
Agitated mixing tank reactor	Volume	0.5
Aqueous chrome reduction	Flow	0.38
Aqueous cyanide oxidation	Flow	0.38
Sludge handling processes		
Digestion (gas mixing)	Volume	0.34
Vacuum filter	Filter area	0.34
Centrifuge	Mass dry solids/h	0.81

TABLE 7.3 Economy of Scale of Equipment

Equipment	Units of Capacity	*M*
Wastewater process equipment		
Primary clarifier equipment	Area	0.32
Final clarifier equipment	Area	0.32
Surface aerators (small units)	Total hp	0.96
Surface aerators (large units)	Total hp	0.72
Air compressor (125 psig)	Airflow (cfm)	0.28
Centrifuge, solid bowl	Hp	0.73
Electric generators	kV-amp	0.71
Ion exchangers	Resin volume	0.85
Heat exchangers	Area	0.58
Pressure filter, plate, and frame	Area	0.58
Vacuum filter, rotary drum	Area	0.68
Motors, electric, 10–100 hp	Hp	0.80
Pumps, centrifugal (with motor)	(Flow)(pressure)	0.51
Air pollution control equipment		
Dust collector, cyclone	Airflow	0.8
Dust collector, cloth filter	Airflow	0.68
Dust collector, electrostatic precipitator	Airflow	0.75
Scrubbers	Airflow	0.85–0.95
Incinerators for combustible gases	Airflow	0.40
Miscellaneous equipment		
Cooling tower	Flow	0.69
Boilers	Steam production	0.50–0.6
Solid waste handling		
Hammermill	Mass flow of solids	0.67
Magnetic separator	Mass flow of solids	0.83–1.0

because capacity is defined as total horsepower. Total horsepower is increased by installing identical small units, each with the same cost. Another example is ion exchange, with capacity defined as total resin volume. A large system will have several parallel resin tanks with associated valves and piping, and this tends to reduce the economy of scale. Another reason could be that the cost of the resin itself is a large portion of the capital costs, and there is no economy of scale in the resin purchase price.

7.3 DESIGN INCENTIVES ARISING FROM ECONOMY OF SCALE

Economy of scale was shown in Chapter 3 to be a factor in planning for future growth. Here are some more examples of how it can affect design decisions.

Example 7.1 Joint Industrial-Municipal Treatment

Four industries need new treatment plants of capacities 4, 8, 10, and 20 units of flow, and the city needs a new treatment plant of capacity 50 units. One feasible option is to combine the wastewater from the four industries with city wastewater and build a single large facility with a capacity of 92 units. The cost of an individual facility or a combined facility is $C = KQ^{0.7}$. The value of K is unspecified, but that is not important if it is assumed to be the same for all options.

The cost of one large joint treatment plant is

$$C_{\text{Joint}} = (4+8+10+20+50)^{0.7} = 92^{0.7} = 23.7$$

The other option is for the city and the four industries to build separate treatment plants, which would cost

$$C_{5\text{ plants}} = 4^{0.7} + 8^{0.7} + 10^{0.7} + 20^{0.7} + 50^{0.7} = 2.6 + 4.3 + 5.0 + 8.1 + 15.5 = 35.5$$

Joint treatment saves 35.5 – 23.7 = 11.8 units of cost, which is one-third the cost for separate treatment plants.

There will be costs to connect the five locations, so other options may be attractive, but economy of scale creates a strong incentive to investigate sharing the cost of joint treatment.

The same concept works for smaller clusters, say two clusters:

$$50 + 4 = 54 \text{ units at a cost } 54^{0.7} = 16.3$$

$$\text{and } 8 + 10 + 20 = 38 \text{ units at a cost of } 38^{0.7} = 12.8$$

for a total of 25.9, and a cost reduction of 9.6 units compared to five individual plants.

As an exercise, suggest how the total cost should be allocated to the five partners in the joint treatment venture.

Example 7.2 Cost Savings from Pollution Prevention

The cost of the major equipment for a new treatment process will be $C = 100{,}000\ Q^{0.7}$, where Q is the design capacity (measured as flow, mass of solids, etc.). Table 7.4 shows that reducing the design capacity from $Q = 4$ to $Q = 3$ reduces the equipment cost from \$263,900 to \$215,000, a savings of 18%. Reducing from $Q = 4$ to $Q = 1$ reduces the equipment cost by \$163,900, or 65%.

A quick estimate of the total cost for a wastewater treatment plant (from Chapter 4) is

TABLE 7.4 Cost Savings from Flow Reduction

Design Capacity	Major Equipment Cost $C = \$100{,}000Q^{0.7}$	Savings on Equipment Cost Relative to $Q = 4$	
(Units)	($)	($)	(%)
$Q = 4$	$100{,}000(4)^{0.7} = 263{,}900$		---
$Q = 3$	$100{,}000(3)^{0.7} = 215{,}800$	48,100	18
$Q = 2$	$100{,}000(2)^{0.7} = 162{,}400$	101,500	38
$Q = 1$	$100{,}000(1)^{0.7} = 100{,}000$	163,900	65

$$\text{Total cost} = 3.45(\text{Cost of major equipment})$$

Reducing the flow from $Q = 4$ to $Q = 3$ is a savings of \$48,100 in equipment costs for a total savings in plant cost of 3.45(\$48,100) = \$165,945. Pollution prevention pays.

Example 7.2 has an interesting twist in that we expect lower M to give more savings when we are adding capacity. Each added unit of capacity costs less than the one before. When reducing the capacity, the economy of scale works in reverse and each unit of capacity removed saves less than the one before.

If $M = 0.6$ instead of 0.7, the financial incentive to reduce the design capacity is slightly reduced. $Q = 4$ costs \$229,739 and the savings of reducing capacity from $Q = 4$ to $Q = 1$ is \$130,000, or 56%. This is still a strong incentive to reduce the waste load, and the construction cost savings add to corresponding savings in operating costs.

Example 7.3 Case Study of Two Industries

Two competitors, A and B, were about the same size and both needed to make changes to meet wastewater discharge permits.

- Factory A built a wastewater treatment plant at a cost of \$1,000,000 for construction and \$30,000 per month for operation and maintenance. Over a ten-year life, factory A will spend \$1,000,000 + (120 months)(\$30,000/month) = \$3,600,000.
- Factory B did a waste minimization study and then a feasibility study for effluent treatment and disposal. The capital cost for factory B was \$800,000 (\$100,000 for the engineering study, \$200,000 to implement the waste minimization recommendations, and \$500,000 for waste treatment plant construction). The O&M cost is \$10,000 per month, but this is paid by savings of \$10,000 per month benefits from waste recovery. Factory B will spend \$800,000.

This comparison may not be entirely fair because the two factories may not have had the same opportunities for pollution prevention savings. It is necessary to search for ways to reduce pollution. If they can be found, cost savings will follow.

7.4 CONSTRUCTING THE COST-CAPACITY CURVE

The equipment cost-capacity curve can be developed by assembling costs from a manufacturer or from different projects. The curve for processes could be developed by designing and estimating several different sizes. Most often they are developed by fitting cost data from similar projects; *similar* because design details and construction conditions are never identical.

Example 7.4 Heat Exchanger Costs

The cost of a heat exchanger depends on the heat exchange surface area, the type of exchanger (shell and tube, crossflow, etc.), the material of construction (stainless steel, titanium, Monel, Hastelloy, etc.), and the operating pressure. Table 7.5 gives the cost data from the Matches' process equipment website (matche.com) that were used to construct the cost-capacity curves Figure 7.3).

The cost-capacity equations are

$$\text{Stainless steel:} \quad C = \$2{,}985A^{0.5}$$

$$\text{Carbon steel:} \quad C = \$2{,}213A^{0.4}$$

The reason for building heat exchangers with different kinds of steel relates to maintenance. Savings on capital cost can be quickly canceled if the useful life is shortened by excessive corrosion, or if the heat exchange capacity is diminished and more heat energy is required. Life cycle cost must be examined.

TABLE 7.5 Cost (US$ 2014) of Shell and Tube Heat Exchangers (150 psi), FOB Gulf

Area (ft²)	Carbon Steel	Stainless Steel
100	$18,400	$27,300
200	$24,200	$39,300
300	$28,400	$48,500
500	$34,700	$63,600
1,000	$45,700	$91,600
2,000	$60,100	$131,800
5,000	$86,400	$231,500

Source: Matches.

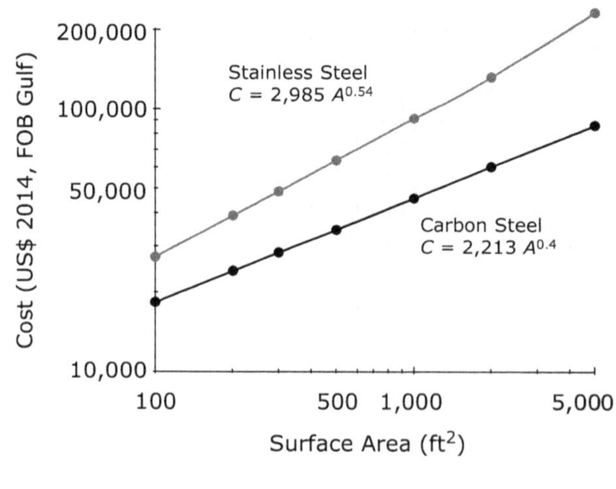

FIGURE 7.3 Cost-capacity curve for heat exchangers (US$ 2014, FOB Gulf).

Example 7.5 Cost-Capacity Curve Construction

The data in Table 7.6 are for 12 *similar* processes that were built between 2003 and 2015. A cost-capacity curve is to be constructed using costs that were adjusted to 2017 costs using the Chemical Engineering PCI with the 2017 index value of 585.7. Bringing the costs to a common year removes some of the variation and improves the accuracy of the cost-estimating model.

The adjusted cost for project 1 is

$$\text{Cost }2018\left(\frac{\text{CEPCI}_{2017}}{\text{CEPCI}_{2008}}\right)=\$289{,}000\left(\frac{585.7}{575.4}\right)=\$294{,}988$$

The original data are plotted in the left panel of Figure 7.4. The adjusted 2017 costs are shown in the right panel. The scatter is caused by the projects being *similar* but not identical, and the costs were adjusted for time but not location.

The adjusted 2017 cost-capacity model is

$$C=\$44{,}260Q^{0.31}$$

TABLE 7.6 Cost Data and Adjustments to 2017

Project	*Q*	*C* ($)	Year	CEPCI Index	Index Ratio	2017 Cost ($)
1	800	289,800	2008	575.4	1.0179	294,988
2	1,000	165,000	2002	395.6	1.4805	244,288
3	1,600	470,000	2011	585.7	1.0000	470,000
4	2,540	544,000	2004	576.1	1.0167	553,065
5	425	237,000	2010	550.8	1.0634	252,017
6	700	300,000	2003	402.0	1.4570	437,090
7	450	331,000	2004	444.2	1.3186	436,440
8	400	290,000	2009	521.9	1.1222	325,451
9	1,000	321,000	2006	499.6	1.1723	376,320
10	1,050	254,000	2002	395.6	1.4805	376,056
11	800	318,000	2009	521.9	1.1222	356,874
12	1,350	487,000	2015	556.0	1.0534	513,014
13	195	195,000	2006	499.6	1.1723	228,598

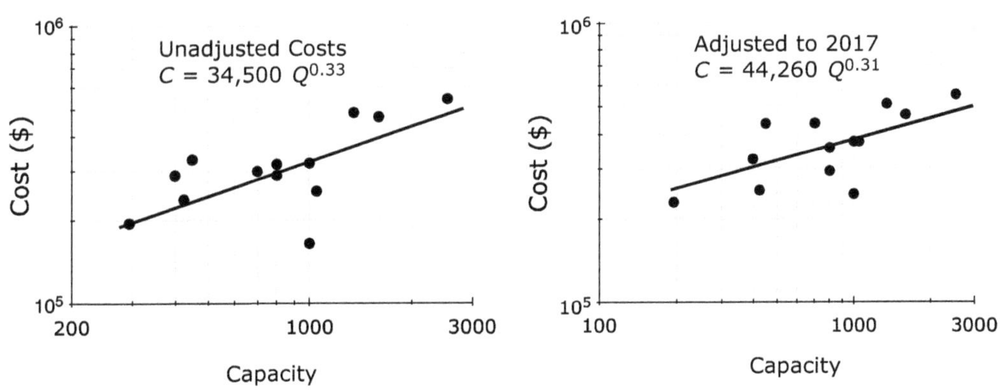

FIGURE 7.4 Cost-capacity curves for unadjusted and adjusted costs.

7.5 GRAPHICAL ESTIMATION OF *M* AND *K*

The economy of scale factor M is the slope of a plot of log(C) vs. log(Q), where log is the base 10 logarithm, as shown in Figures 7.1 and 7.4. One approach is to draw, by eye, a line that describes the data and to use two points, A and B, on that line to estimate

$$M = \frac{\log\left(C_A / C_B\right)}{\log\left(Q_A / Q_B\right)} = \frac{\log C_A - \log C_B}{\log Q_A - \log Q_B}$$

$$K = \frac{C_A}{Q_A^M} \text{ or } K = \frac{C_B}{Q_B^M}$$

K is the intercept at $Q = 1$. Many cost-capacity curves do not include $Q = 1$. Estimating K by extrapolating from higher values will magnify the estimating error, and that procedure is not recommended.

Example 7.6 Solid Bowl Centrifuge Sludge Dewatering

The capital costs of two solid bowl centrifuge installations are \$276,000 for a 75 L/min machine and \$550,000 for a 360 L/min machine. This includes a polymer addition system to feed 1.65 kg polymer per 1,000 kg of sludge solids (dry basis). Calculate the values of K and M for the solid bowl centrifuge.

Let C_A = \$276,000 and Q_A = 75 L/min
C_B = \$550,000 and Q_B = 360 L/min

$$\frac{\$276{,}000}{\$550{,}000} = \left(\frac{75}{360}\right)^M$$

$$M = \frac{\log(\$276{,}000 / \$550{,}000)}{\log(76/360)} = \frac{-0.29945}{-0.69897} = 0.43$$

$$K = \frac{C_A}{Q_A^M} = \frac{\$276{,}000}{76^{0.43}} = \$42{,}870\left(\text{rounded}\right)$$

$$C = \$42{,}870Q^{0.43}$$

The same result will be obtained using natural logarithms. Base 10 logs are used to be consistent with the log–log plots of the cost-capacity curve.

Changing the units of capacity, say from pounds to kilograms, or from gallons to cubic meters, will not change M but it will change K. The adjustment can be made without converting the data and refitting the model, as shown in Example 7.7.

Example 7.7 Converting *K* for Different Unit of Capacity

The cost of a process is $C = \$100{,}000Q^{0.7}$ with Q in million gallons per day (mgd). If flow is measured in cubic meters per day (m^3/d), the value of K must change to a new value, K', that gives the same cost for the equivalent amount of flow. *Note:* 1 mgd = 3,785 m^3/d.

$$C = \$100{,}000(1\ \text{mgd})^{0.7} = K'(3{,}785\ \text{m}^3/\text{d})$$

$$K' = \$100{,}000\left(\frac{1^{0.7}}{3{,}785^{0.7}}\right) = \frac{\$100{,}000}{319.65} = \$312.8$$

In mgd units C= \$100,000 $Q^{0.7}$
In m^3/d units C =\$313 $Q^{0.7}$ (rounded from \$312.8)
Cost for 1 mgd = \$100,000$(1)^{0.7}$ = \$100,000
Cost for 3,785 m^3/d = \$313$(3{,}785)^{0.7}$ = \$100,000

7.6 FITTING THE COST MODEL BY REGRESSION

An alternative approach is to fit the cost model by regression. (The method of least squares is explained in Chapter 16.) The regression calculation is done on the logarithms ($\log_{10}$). The log form of the equation is

$$\log(C) = \log(K) + M\log(Q)$$

The regression output will be of the form $y = a + bx$, where

$$a = \log(K) \quad \text{and} \quad b = M, \quad \text{giving } K = 10^a$$

Example 7.8 Estimating *K* and *M* by Linear Regression

The data in Table 7.7 were used to fit the log form of $C = KQ^M$ using linear regression.

$$\log(C) = \log(K) + M\log(Q)$$

TABLE 7.7 Example Equipment Cost Data

Q (Units)	*C* (\$1,000)	$\log_{10}(Q)$	$\log_{10}(C)$
400	56	2.6021	1.7482
425	67	2.6284	1.8261
450	62	2.6532	1.7924
700	80	2.8451	1.9031
800	80	2.9031	1.9912
800	98	2.9031	1.9031
1,000	105	3.0000	2.0212
1,000	94	3.0000	1.9731
1,050	103	3.0212	2.0128
1,350	115	3.1303	2.0607
1,600	142	3.2041	2.1523
2,540	155	3.4048	2.1903

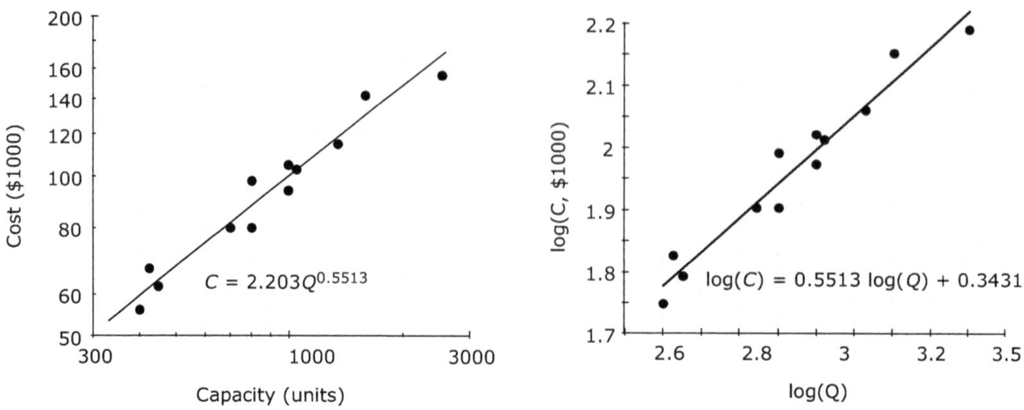

FIGURE 7.5 Estimating M and K linear regression on the log-transformed costs and capacities.

The left panel of Figure 7.5 is a log–log plot of the original data. On the right panel are the log transformed values. The plots are identical except for scaling.

The regression is done on the log-transformed values to get

$$\log(C) = 0.3431 + 0.5513\log(Q)$$

giving $M = 0.5513$ and $K = 10^{0.3431} = 2.203$

Because the costs are in thousands,

$$K = \$2,203$$

$$\text{and } C = \$2,203Q^{0.5513}$$

The values of $M = 0.5513$ and $K = \$2,203$ indicate more precision than is warranted by the data. The cost data are in even thousands ($71,000, $76,000, etc.). The predicted costs cannot be more accurate than the data. More honest estimates will be produced by rounding the predicted costs, or by rounding the model coefficients.

The calculated cost for 1,600 units is

$$C = \$2,203(1,600)^{0.5513} = \$128,660$$

which should be rounded to $129,000.

Rounding the coefficients of K and M to $2,200 and 0.55, respectively, gives a predicted cost for 1,600 units of

$$C = \$2,200(1,600)^{0.55} = \$127,260$$

which should be rounded to $127,000.

The costs differ by about 1.5%, which is small given the scatter in the data, so it is not important which method is used.

7.7 COST ESTIMATE SENSITIVITY TO THE VALUE OF *M*

The consequences of not knowing the true value of M are quite small. Suppose the true but unknown value is $M = 0.5$, and $M = 0.7$ is used to estimate the cost. For a capacity ratio of $Q_1/Q_2 = 2$, the resulting quantities are

$$2^{0.5} = 1.41 \quad \text{and} \quad 2^{0.7} = 1.62$$

The percentage error relative to the true value is

$$100(1.62 - 1.41)/1.41 = 15\%$$

Table 7.8 shows the errors for several values of M and capacity ratios when $M = 0.7$ is used instead of the true value of M (given in the left column of the table). If the true value is less than 0.7, the cost is overestimated, but only by a few percent unless the capacity ratio is large. Errors of 0.05–0.1 in the value of M are well within the margin of accuracy for this kind of cost estimate. If the value is unknown or uncertain, using $M = 0.6$ or 0.7 will produce some useful information.

7.8 COST OF STANDBY UNITS

Standby, backup, and emergency equipment are required, so service continues when one or more units are out of service for repair or maintenance. Reliability is improved by installing standby units or by keeping spare parts on hand so repairs can be made quickly. The choice depends on the cost or damage of being out of service, even temporarily.

This is a simplified model for standby capacity that assumes all units are of the same kind and capacity, and that all units are installed and capable of operation. Obviously, a designer has other options.

If N units are installed and one unit is on standby (out of service), the required capacity of the $N - 1$ operating units must be

$$q = \frac{Q}{N-1}$$

TABLE 7.8 Percentage Error if $M = 0.7$ Is Used Instead of the True M

True M	Capacity Ratio (Q_1/Q_2)					
	1.50	2.00	2.50	3.00	3.50	4.00
0.50	8%	15%	20%	25%	28%	32%
0.55	6%	11%	15%	18%	21%	23%
0.60	4%	7%	10%	12%	13%	15%
0.65	2%	4%	5%	6%	6%	7%
0.70	0%	0%	0%	0%	0%	0%
0.75	−2%	−3%	−4%	−5%	−6%	7%
0.80	−4%	−7%	−9%	−10%	−12%	13%
0.85	−6%	−10%	−13%	−15%	−17%	19%
0.90	−8%	−13%	−17%	−20%	−22%	24%

Assuming the usual cost-capacity model, the cost of each installed unit is

$$C = Kq^M = K\left(\frac{Q}{N-1}\right)^M$$

The total system cost is

$$C_N = NK\left(\frac{Q}{N-1}\right)^M = \frac{NKQ^M}{(N-1)^M}$$

Example 7.9 Standby Turbine Mixer

Suppose that the cost of a turbine mixer is $C = \$300H^{0.6}$, where H is horsepower (hp). A total of 200 hp is needed in operation at all times. The design requirement is for one standby unit – that is a minimum of two units installed with one operating. The number of installed units is not specified, so the designer is free to recommend installing more than two units. Present an analysis of how the total equipment cost changes as the number of installed turbine unit changes.

The cost of one 200 hp unit is

$$C = \$300(200)^{0.6} = \$7{,}207$$

The cost of two units of 200 hp, one being on standby, is

$$2(\$300)(200)^{0.6} = \$14{,}414$$

Installing three units of 100 hp, with 200 hp in operation and one on standby, costs

$$3(\$300)(100)^{0.6} = \$14{,}264$$

Three units, with two in operation and one on standby, is more reliable and flexible than two units. The standby unit can be rotated in and out of service. It also has a marginally lower cost.

For a more general solution, let

N = number of units installed
$N - 1$ = the number of units operating with 1 on standby
$h = H/(N - 1)$ = horsepower of each installed unit
H = 200 hp = operational requirement
Cost per unit = $\$300h^M$
Total cost of N units installed = $N(\$300h^M)$

Table 7.9 shows designs having two to eight installed units. The horsepower per unit has been rounded to commercially available values (there is no 66.67 hp unit for sale). When M is less than 0.6, indicating a large economy of scale, install a few larger units. For $M = 0.4$, install two units. At $M = 0.7$ and above, the minimum cost is for five installed units.

TABLE 7.9 Cost Per Installed Unit and Total Cost for a System with Operating Capacity

Number of Units	**2**	**3**	**4**	**5**	**6**	**7**	**8**
Operating Units	**1**	**2**	**3**	**4**	**5**	**6**	**7**
Capacity/Unit	**200**	**100**	**70**	**50**	**40**	**35**	**30**
M				**Cost per unit ($)**			
0.4	2,500	1,890	1,650	1,440	1,310	1,240	1,170
0.5	4,240	3,000	2,510	2,120	1,900	1,780	1,640
0.6	7,210	4,760	3,840	3,140	2,740	2,530	2,310
0.7	12,200	7,540	5,870	4,640	3,970	3,610	3,240
0.8	20,800	11,900	8,980	6,860	5,740	5,160	4,560
0.9	35,300	18,900	13,700	10,100	8,300	7,360	6,400
M				**Total cost of *N* units ($)**			
0.4	**5,000**	5,680	6,560	7,170	7,870	8,710	9,360
0.5	**8,480**	9,000	10,000	10,600	11,400	12,400	13,100
0.6	14,400	**14,300**	15,400	15,700	16,500	17,700	18,500
0.7	24,500	**22,600**	23,500	23,200	23,800	25,300	26,000
0.8	41,600	35,800	35,900	**34,300**	34,400	36,100	36,500
0.9	70,600	56,800	54,900	50,700	**49,800**	51,500	51,200

H = 200 hp and one unit in standby

It can be shown that the cost optimal number of installed units is $N^* = 1/(1 - M) = 1/(1 - 0.4) = 1/0.6 = 1.67$. This result will be rounded to an integer; $N^* = 1.67$ rounds to 2 units. For $M = 0.8$, $N^* = 5$ units, as shown in the table.

7.9 CONCLUSION

The economy of scale model known as the 6/10 rule is best known as a fast and simple way of estimating a project cost. The economy of scale *concept* offers a way to examine a variety of economic decisions, as the examples in this chapter are designed to show.

7.10 PROBLEMS

7.1 ECONOMY OF SCALE

A company has estimated the cost of two facilities as

5,000 units/h	$16,000,000
8,500 units/h	$23,500,000

Find the cost-capacity exponent and the cost of a 6,000 unit/h facility.

7.2 MATCHES EQUIPMENT COST ESTIMATES

Does the $C = KQ^M$ model fit the costs for the four types of equipment listed in Table P7.2?

Does M depend on the material of construction?

TABLE P7.2 Equipment Costs, US$ 2014, FOB Gulf Coast

Pumps and blowers				
Pump, centrifugal (horizontal)	**Cast iron**	**Bronze**	**Stainless 304**	**Hastelloy**
4-inch diameter	10,300	9,800	13,600	37,700
6-inch diameter	13,200	12,600	17,400	48,300
8-inch diameter	15,800	15,000	20,700	57,500
Blower, centrifugal	Carbon steel			
1,000 ft³/min	6,300			
5,000 ft³/min	10,600			
10,000 ft³/min	16,900			
50,000 ft³/min	88,300			
Air pollution control equipment				
Baghouse filter, large	(Carbon steel)			
2,000 ft³/min	16,000			
5,000 ft³/min	26,000			
10,000 ft³/min	41,700			
50,000 ft³/min	155,600			
Incinerator, direct flame				
Feed material	Low hazard	Corrosive	Hazardous	
20 million Btu/h	261,400	392,100	522,800	
50 million Btu/h	373,700	560,600	747,400	
100 million Btu/h	489,700	734,600	979,400	

Source: Matches Equipment Costs (http://www.matche.com/).

7.3 PUMP COSTS

Table P7.3 includes average cost of installed pumps with standard motors, including connections and anchors, based on normal industrial installations of 500 to 1,500 total horsepower.

TABLE P7.3 Pump Costs ($)

Power	Type of Pump		Electrical
(hp)	Drip Proof	Explosion Proof	480-Volt Wiring
10	870	1,225	960
15	1,085	1,650	1,250
20	1,300	2,075	1,575
25	1,550	2,550	1,900
30	1,800	3,000	2,225
40	2,250	3,925	2,800
50	2,675	4,800	4,200
75	4,175	8,900	5,000
100	5,675	12,400	6,525
125	7,175	15,500	8,025
150	8,725	17,650	9,750
200	11,900	21,375	12,700
250	15,200	24,200	15,800

Source: State of Michigan 2003.

The total electrical cost estimates for a 480-volt system are given. Costs include wiring, starters, and/or control panels. The cost for 230-volt power wiring will be up to 75% more than a 480-volt system in normal industrial installations.

Calculate the total costs (pump + electrical) for the two types of pumps. Plot the cost-capacity curves.

7.4 SLUDGE DEWATERING 1

Calculate K and M for the capital costs for two installations of a belt filter press and gravity thickener using the data in Table P7.4.

TABLE P7.4 Equipment Capital Costs

Belt Filter Press		Gravity Thickener	
Belt Width (W) (m)	Cost	Surface Area (A) (m^2)	Cost
1	$318,000	28	$166,000
2	$435,000	280	$394,000

7.5 SLUDGE DEWATERING 2

Estimate the cost of a gravity thickener of surface area A = 100 m^2 combined with a solid bowl centrifuge of capacity Q = 120 L/min. The cost-capacity equations are

$$\text{Gravity thickener } C = \$47{,}600\, A^{0.375}, \quad A = \text{surface area}\left(\text{m}^2\right)$$

$$\text{Centrifuge } C = 42{,}900 Q^{0.43}, \quad Q = \text{flow rate}\left(\text{L/min}\right)$$

7.6 ECONOMY OF SCALE: HEAT EXCHANGERS

A new heat exchanger with area = 100 m^2 costs $92,000. What is the cost for a similar unit with area = 50 m^2? The economy of scale exponent is M = 0.44.

7.7 MICROFILTRATION COST BENCHMARKING FOR LARGE FACILITIES

Use the data in Table P7.7 to construct graphs showing the total capital cost and the unit cost ($/1,000 gal/d) as a function of the design flow.

TABLE P7.7 Capital Cost for Microfiltration Equipment

Flow (mgd)	Cost ($)	Flow (mgd)	Cost ($)
0.03	22,500	2.8	308,000
0.13	41,600	3.6	576,000
0.7	210,000	9.5	456,000
1	110,000	10	390,000
1.8	176,400	19	627,000

7.8 TIME AND SIZE

The cost of two machines, one purchased in 2010 and one purchased in 2015, will be used to estimate the cost of a similar unit today. The cost index today is 560. Using the data in Table P7.8, derive an economy of scale equation and estimate today's cost of an 80 m² unit.

TABLE P7.8 Time and Capacity Data

	Machine A	Machine B
Area	70 m²	130 m²
Cost	$17,000	$24,000
Purchased	2010	2015
Cost Index	360	390

7.9 DESALTING WATER

Figure P7.9 shows the unit cost of water produced using reverse osmosis to desalt seawater (45,000 mg/L TDS) and brackish water (4,500 mg/L TDS). Convert the cost information in the graph that shows total cost as a function of average daily water production.

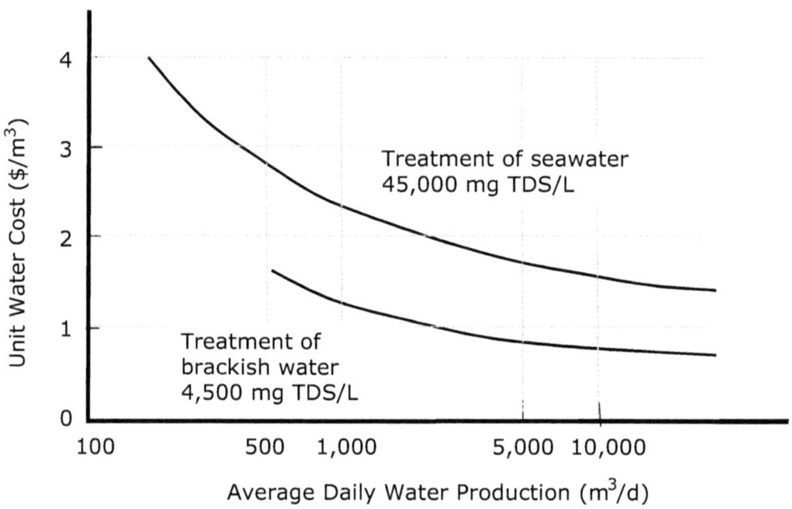

FIGURE P7.9 Desalinization costs.

7.10 POLLUTION PREVENTION SAVES

An industry plans to build a process of 20 capacity units, but before any equipment is purchased or any construction has started, the pollution control engineer discovers a way to reduce the needed capacity by 25%. The cost of installed equipment of capacity Q is

$$C = \$10{,}000Q^{0.7}$$

How much money will this pollution prevention intervention save the company?

7.11 AUTOTHERMAL AEROBIC DIGESTERS

Table P7.11 gives capital costs (US$ 1990) for autothermal aerobic digester systems constructed in Germany. Capacity is measured as kg/d of total solids processed. (a) Plot the data to show that the costs follow the form of the 6/10 rule. (b) Derive the log–log economy of scale models in the form of the 6/10 rule.

TABLE P7.11 Digester System Construction Cost (1990 US$); Cost Includes Heat Recovery Equipment, Process Retrofit to Existing Tanks, and Biofilter Odor Control

Facility	Capacity	Process Capital Cost (US$ 1990)			
	(kg/d)	Thickener	Storage	Digester	Total
1	800	71,000	33,000	191,000	295,000
2	1,000	76,000	36,000	132,000	244,000
3	1,600	105,000	47,000	318,000	470,000
4	2,540	161,000	62,000	330,000	553,000
5	425	71,000	27,000	154,000	252,000
6	700	71,000	32,000	333,000	436,000
7	450	71,000	32,000	217,000	436,000
8	400	65,000	27,000	233,000	325,000
9	1,000	76,000	36,000	264,000	376,000
10	1,050	76,000	36,000	225,000	376,000
11	800	71,000	33,000	253,000	357,000
12	1,350	95,000	42,000	376,000	513,000
13	291	N/A	29,000	200,000	229,000
14	899	N/A	39,000	321,000	360,000

7.12 ECONOMY OF SCALE: TURBINE MIXERS

The cost of individual turbine mixers is $C = 3{,}100H^{0.5}$, where H is horsepower (hp). There is a total online horsepower requirement of 1,000 hp. One spare unit (off-line) must be provided. All units, online and the spare, will be the same size in order to simplify maintenance. How much is the cost increased by providing the spare unit? Develop a design rule for how many mixers should be provided.

7.13 MULTIPLE UNITS

The cost for a complex of N machines having a total capacity Q, with each unit having capacity $q = Q/N$, is

$$C = \$500N\left(Q/N\right)^{M}$$

Show how the cost varies as a function of M for $N = 2$–4 for machines for systems that have total installed capacities of 1,500 and 9,000 units. Two machines are the minimum, so there is standby capacity when needed.

7.14 MINIMIZING THE COST OF MULTIPLE UNITS

The cost of a machine is $C = \$200Q^M$, where Q is the capacity. A total of 500 units of capacity is needed in operation at all times. Machines are available in capacities of 50, 75, 100, 200, and 250 units. The design specifies at least one standby unit; that is, a minimum of two units installed with one operating. The number of installed units is not specified, so the designer is free to recommend installing more than two units. The optimal number of installed machines depends on the exponent M. When M is large, there is little economy of scale and this encourages installing more small machines. The opposite is true when M is small. Develop a table of cost as a function of the number of installed machines and the value of M.

7.15 ECONOMY OF SCALE: TWO PROCESSES IN SERIES

A system will have two stages operating in series. The capacity of the system is Q (arbitrary units). The capacity of each stage is $Q = Q_1 = Q_2$. The equipment costs are

$$C_1 = 3Q_1^{0.8} \quad \text{and} \quad C_2 = 5Q_2^{0.4}$$

The largest capacity available is $Q = 20$ units. The estimating functions are valid within the range $1 \le Q \le 20$.

(a) Plot the system cost curve over the range Q = 1–20.
(b) Derive a system cost-estimating function in the form of the 6/10 rule.
(c) Specify the number and capacity of units to install for a total system capacity of 45 units. How much will the 45-unit system cost?

7.16 REGIONAL TREATMENT PLANT

A group of industries have started to discuss building a single large treatment plant that will be shared instead of building individual treatment facilities. How does economy of scale affect their decision? Does it matter how many industries are involved? Does the relative size of the cooperating industries matter?

7.17 JOINT TREATMENT 1

The cost for a wastewater treatment plant is $C = KQ^{0.7}$, where Q is the capacity in m^3/d. A group of industries have started to discuss building a single large treatment plant that will be shared instead of building individual treatment facilities.

(a) What is the cost function for n cooperating industries?
(b) How does economy of scale affect their decision?
(c) Does it matter how many industries are involved?
(d) Does the relative size of the cooperating industries matter?
(e) Devise a fair apportionment of the costs or the savings. Should this be according to contributed flow or load?
(f) If a government agency or industrial cooperative builds and runs the plant, how should it collect user fees to recover capital and operating costs?
(g) Plants have different costs due to different local soil conditions, waste strength, etc. How should this be handled in costing the joint treatment option?

7.18 JOINT TREATMENT 2

Figure P7.18 is a regional map that shows four locations that need wastewater treatment (marked by the octagons) and three possible treatment sites (marked by circles). One, two, or three treatment

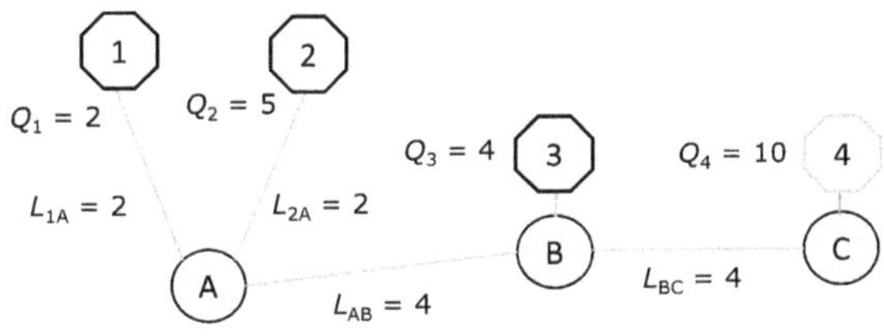

FIGURE P7.18 Regional map of wastewater sources and possible treatment plant sites.

plants can be built. For example, a plant at A could serve sources 1 and 2, or it could serve sources 1, 2, and 3, or it could serve all four sources. The cost of a treatment plant is $C_T = 10Q_T^{\,0.7}$. The cost of pipes to connect the treatment plants is $C_P = 2LQ_P^{0.4}$, where L is the length of the pipe and Q_P is the flow carried by the pipe. For example, a treatment plant at B serving all four sources would need pipes $L_{1A} = 2$ and $Q_1 = 2$, $L_{2A} = 2$ and $Q_2 = 5$, pipe $L_{AB} = 4$ and $Q = 7$, and pipe $L_{BC} = 4$ and $Q = 10$. Sites B and C are at locations 3 and 4, so the pipe lengths are negligible.

Evaluate the possible treatment schemes and see if the economy of scale savings from regional treatment will pay for the connecting pipes.

8 The Time Value of Money

Rational economic decisions are made after an economic analysis. Sometimes the analysis is a few calculations, and sometimes it is pages of spreadsheets. Guesswork and casual or emotional decisions are not acceptable.

8.1 CASH FLOW AND PROFIT

Cash flow is the difference between money coming in and money going out. It can be calculated weekly, monthly, or annually. All organizations keep an accounting of cash flow.

Financial decisions for all organizations involve:

- Capital investment = fixed assets, leased assets, working capital ($)
- Revenues and other benefits ($/y)
- Costs for raw materials, labor, supervision, etc. ($/y)
- Expenses for administration, technical services, etc. ($/y)
- Cash flow ($/y) = Revenues – Costs – Expenses

Profit-making organizations also have to deal with:

- Depreciation (an expense for tax purposes that does not affect cash flow) ($/y)
- Gross (before-tax) profit ($/y) = Revenues – Costs – Expenses – Depreciation
- Taxes owed
- Net (after-tax) profit = Gross profit – Taxes

Cash flow and net profit can be positive or negative. Public utilities do not earn a profit and they do not pay taxes, but they do have to track and manage cash flow. Revenues must cover their costs, or money must be borrowed.

8.2 PAYBACK TIME AND RETURN ON INVESTMENT

Two measures of profitability that are easy to calculate and easy to understand are *payback time* and *return on investment* (ROI). Their best use is for comparing a current method and a replacement method. The weakness of both is not taking into account cash flow and the time value of money.

Payback time is simple, and simplicity is attractive. A corporation might tell a plant manager, "Do not make any capital investment that has a payback time of more than 6 months." This is clear and simple. It makes perfect sense in a short time frame. It would not make sense to say, "Do not make a capital investment during the next 10 years unless the payback time is 6 months or less."

$$\text{Payback time (years)} = \frac{\text{Capital cost (\$)}}{\text{Annual net savings (\$/y)}}$$

Return on investment is the percentage of the capital investment that is recovered by savings or project revenue. This can be calculated per year or for some other time interval. *Return on investment* can be positive or negative.

$$\text{Return on investment (\%/y)} = \frac{\text{Annual savings (\$/y)}}{\text{Capital cost (\$)}} \times 100$$

TABLE 8.1 Payback Time and Return on Investment for Energy-Saving Projects at Appleton Paper

Recommendation	Project Cost ($)	Annual Savings ($/y)	Payback Time (months)	Return on Investment (%)
Recover heat from paper machine vents	1,500,000	1,015,000	18	68
Reduce silo temperatures	None	212,310	Immediate	100
Use direct-fired water heating for de-inking	135,000	162,164	9.5	120
Reduce compressed air costs	8,750	37,962	3	435
Use condenser water for steam makeup	10,000	12,995	9	130
Recirculate vacuum pump seal water	51,857	25,000	6	207

These are written as they would apply to a pollution prevention or control project that saves money by reducing the annual cost of operating costs, such as power and chemicals. In the normal business context, "savings" is annual net revenue.

Example 8.1 Energy Savings at Appleton Papers

Appleton Papers identified a series of energy-saving projects with an annual savings of 4.8 million kWh of electricity and nearly 150 trillion Btu of natural gas. Table 8.1 lists a few of the projects with their projected return on investment and payback times.

The payback time for the project to recover heat energy from paper machine vents is

$$\text{Payback time} = \frac{\text{Capital cost (\$)}}{\text{Annual net savings (\$/y)}} = \frac{\$1{,}500{,}000}{\$1{,}015{,}000/\text{y}} = 1.5\ \text{y}$$

The return on investment in the first year is

$$\text{Return on investment} = \frac{\text{Annual savings (\$/y)}}{\text{Capital cost (\$)}} \times 100$$

$$= \frac{\$1{,}015{,}000}{\$1{,}500{,}000} \times 100 = 68\%$$

Example 8.2 Payback Time for Spray Painting

An engineer wants to install new piping from a paint mix room to the spray booths to reduce the cleaning waste. An investment of $10,000 will eliminate 5,000 gal/y of semisolid waste. Each gallon of waste represents $3.00 of waste raw material and $0.45 waste disposal cost, including transportation and in-plant labor.

The annual savings are

Raw material (5,000 gal/y)($3.00/gal) = $15,000/y
Waste disposal (5,000 gal/y)($0.45/gal) = $2,250/y
Total annual savings $17,250/y

$$\text{Payback time} = \frac{\$10{,}000}{\$17{,}250/\text{y}} = 0.58\ \text{y} = 7\ \text{months}$$

Management should accept the proposal.

Example 8.3 Composting Payback Time

A proposed project will convert 40,000 T/y of organic waste (yard waste, food, etc.) to compost instead of putting it into a landfill. The cost of hauling to the landfill will be $40/T more than for hauling to the composting site. As an additional benefit, the compost product can be used to replace 20,000 T/y of topsoil that the city buys for $25/T. Compost not used by the city will be available at no cost to citizens.

Capital cost = $4,800,000
Operating and maintenance cost for composting = $1,250,000/y

Savings from not hauling and disposing of 40,000 T/y in the landfill

$$= (40{,}000 \text{ T/y})(\$40/\text{T}) = \$1{,}600{,}000/\text{y}$$

Savings from using compost instead of buying topsoil at $25/T

$$= (20{,}000 \text{ T/y})(\$25/\text{T}) = \$500{,}000/\text{y}$$

Total savings = $2,100,000/y
Net annual savings = Savings – O&M cost

$$= \$2{,}100{,}000 - \$1{,}250{,}000 = \$850{,}000$$

$$\text{Payback period} = \frac{\$4{,}800{,}000}{\$850{,}000/\text{y}} = 5.6 \text{ y}$$

8.3 THE IMPORTANCE OF CASH FLOW

Projects A and B each cost $1,200,000 and have a guaranteed three-year payback. This suggests that they are equally attractive, but the decision needs to be based on the annual cash flows, which are given in Table 8.2.

Project A returns nothing in years 1 and 2 and $1,200,000 at the end of three years. Project B returns $400,000 per year for three years.

Project B is better. Invest the $400,000 return from years 1 and 2 in some other venture, and you will have considerably more than $1,200,000 in your pocket at the end of three years.

You also want to know what happens after the payback is accomplished. Two possibilities are shown Table 8.3. In scenario 1, project A produces no further revenue, while project B continues to profit. In scenario 2, it is project A that prospers and surpasses B.

TABLE 8.2 Cash Flow for Technologies A and B

Year	Cash Flow	Project A	Project B
	Initial investment ($)	–1,200,000	–1,200,000
1	Net revenues ($)	0	400,000
2	Net revenues ($)	0	400,000
3	Net revenues ($)	1,200,000	400,000
	Total net revenues ($)	1,200,000	1,200,000
	Payback time	3 years	3 years

TABLE 8.3 Two Scenarios for Cash Flow ($) Beyond the Payback Time

Year	Cash flow	Scenario 1		Scenario 2	
		Project A	Project B	Project A	Project B
	Initial cost	−1,200,000	−1,200,000	−1,200,000	−1,200,000
1	Net revenue	0	400,000	0	400,000
2	Net revenue	0	400,000	0	400,000
3	Net revenue	1,200,000	400,000	1,200,000	400,000
4	Net revenue	0	400,000	500,000	0
5	Net revenue	0	400,000	500,000	0
	Total net revenue	1,200,000	2,000,000	2,200,000	1,200,000

Comparing projects A and B using only payback time hides more than it reveals. The cash flow pattern is important. A method is needed that looks at cash flow over time and brings the cash flows to equivalent values at time zero. This is done by calculating the present values, which is done in Section 8.5.

8.4 WHEN THERE IS NO PAYBACK

Many projects have no revenue or at least no revenue that is credited directly to the project. An example is a project where there is a choice between two pumps, with the purchase prices, annual O&M costs, and useful lifetimes, as shown in Table 8.4.

The pumps cannot be distinguished by the *simple* lifetime costs:

$$\text{Pump A} = \$40{,}000 + (10\text{y})(\$6{,}000/\text{y}) = \$100{,}000$$

$$\text{Pump B} = \$50{,}000 + (10\text{y})(\$5{,}000/\text{y}) = \$100{,}000$$

Some buyers might prefer pump A to pump B because of the lower initial cost. Some might prefer the lower O&M cost. Most will want an analysis that includes the cost of money over the lifetime of the pumps.

One way to do this is to imagine that money is borrowed to buy the pump and the loan is repaid in equal annual installments. A fair comparison then is the total of the annual loan repayment cost plus the annual O&M costs.

Another approach is to calculate the amount of money needed at time zero to make the annual O&M payments, assuming that this amount is held in a bank account and earns interest. This amount is added to the purchase price to give a lump sum for each pump. The lowest lump sum identifies the best buy.

The rest of this chapter is about these calculations.

TABLE 8.4 Pump Comparison

Cost Factor	Pump A	Pump B
Purchase price ($)	40,000	50,000
Annual O&M cost ($/y)	6,000	5,000
Pump lifetime (y)	10	10

8.5 FUTURE VALUE AND PRESENT VALUE

Suppose you can have $1,000 today or guaranteed $1,100 one year from today. There are three choices:

- Take $1,000 now because you must pay the electric bill tomorrow, or your business will be shut down. The $1,000 is there on the table. You can pick it up and pay your bills. The value of money in hand today is enormous. Interest obtained one year from now has no present value.
- Take $1,100 one year from now because you have no urgent need for $1,000 today. There is a guaranteed 10% gain on your $1,000 and you don't have to do anything or take any risk.
- Take $1,000 now and invest it with the expectation of earning 20% per year, to have $1,200 at the end of the year. *Expect 20%* implies risk. You might earn only 10% or you might lose part or all of your $1,000 investment.

These are all *rational financial decisions. Rational* means based on facts and reason, not on fear or emotion. Which you choose depends on how *you* judge the value (or worth) of money over time. *Value* means how you weigh the amount of money involved and your current financial situation.

When the money belongs to clients or customers of a utility, all decisions are made after an economic analysis. Sometimes the analysis is a few calculations and sometimes it is pages of spreadsheets. Guesswork and casual or emotional decisions are not acceptable.

The four basic calculations that consider the time value of money are as follows:

- Convert a *present value* to a *future value.*
- Convert a *future value* to *a present value.*
- Convert a *series of future costs* (equal or unequal) to an *equivalent present value.*
- Convert a *present value* to an *n-year series of equal future costs.*

8.6 STYLE IN ECONOMIC CALCULATIONS

Most books on engineering economics use diagrams of the cash flow as a guide to doing the calculations. For example, the diagrams in Figure 8.1 show the conversion of a present value *PV* to a future value *FV*, and the conversion of a series of uniform annual payments to a present value. Or, they can be for the reverse of these calculations. We shall not use these diagrams because, in our opinion, it is faster and just as clear to make a table of the values, keeping track of the signs, positive (+) for money coming in and negative (–) for money going out. This organizes the cash flow for spreadsheet

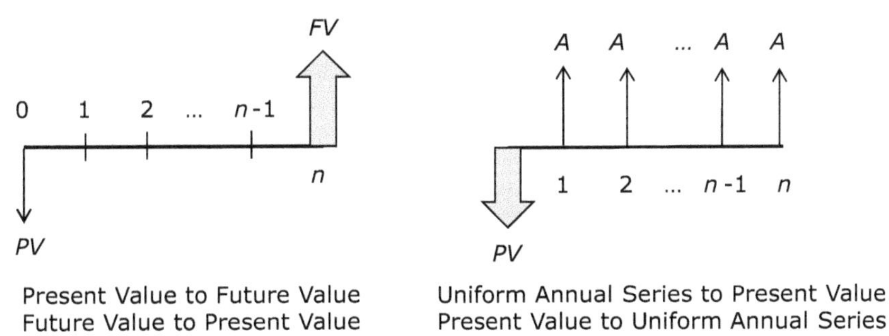

FIGURE 8.1 Diagrams for interpreting economic conversions of the time value of money.

calculations. For students and instructors who prefer the diagrams, that method can be found in many very good texts on engineering economics (e.g., Black & Tarquin 2018).

8.7 FUTURE VALUE

Money invested in a profitable project, or a bank savings account, will grow into a *larger future value FV*. The rate of growth of the account, or *interest rate*, is i percent per year, and this growth will continue for n years. By convention, interest is credited at the end of the year.

If the earned interest is reinvested at the same interest rate, the earning pattern is called *compounding*, or *compound* interest. This is the calculation for how much you need to invest to make a cash purchase at some time in the future, or the accumulated value of a savings bond.

The amount of money invested at time zero is the principal P, *which is also* the *present value* PV at time zero.

A bank account or an asset that has value PV at the start of year 1 and increases by i% per year will be worth: $FV_{i,1}$ at the end of year 1, $FV_{i,2}$ at the end of year 2, and $FV_{i,n}$ at the end n years, where

$$FV_{i,1} = PV + iPV = PV(1+i)$$

$$FV_{i,2} = PV(1+i)(1+i) = PV(1+i)^2$$

$$\vdots$$

$$FV_{i,n} = PV(1+i)^n$$

There are four variables: F, PV, i, and n. If any three are known, the fourth can be calculated.

Example 8.4 Simple Compound Interest

An amount P = \$15,000 will be invested for 10 years at an annual interest rate of 8%. The interest payment at the end of each year will be added to the previous year's balance to earn more interest as time goes on. This is *compound interest*. The pattern of growth is as follows:

End of Year 0	P = \$15,000
End of Year 1	$FV_{8\%,1}$ = \$15,000 + (0.08)(\$15,000) = \$15,000 + \$1,200 = \$16,200
End of Year 2	$FV_{8\%,2}$ = \$16,200 + (0.08)(\$16,200) = \$16,200 + \$1,296 = \$17,496
⋮	...
End of year 10	$FV_{8\%,10}$ = (\$15,000)(1.08)^{10} = \$15,000(2.1589) = \$32,384

Note that the percent interest rate is converted to decimal form in the calculations.

Example 8.5 Doubling Your Money

How long will it take to double your money if you can invest at 6% interest?

The investment at time zero is P, and the desired future value is $2P$.

$2P = P(1 + 0.06)^n$ or $2 = 1.06^n$

$\log(2) = n \log(1.06)$

$$n = \frac{\log(2)}{\log(1.06)} = 11.9 \text{ y}$$

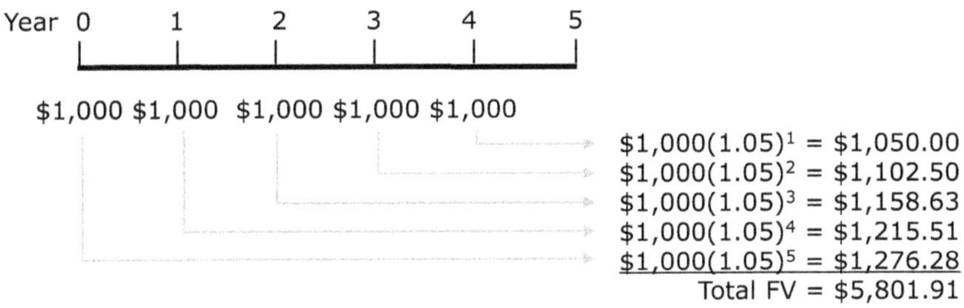

FIGURE 8.2 Future value after five years of equal $1,000 per year payments.

The Rule of 72 is a quick approximation for this calculation. Divide 72 by the interest rate (as a percent per year), and the result is approximately the number of years to double your money or double your liability for repayment.

An investment of $10,000 at 6% per year will grow to approximately $20,000 in 72/6 = 12 years. (The exact amount with annual compounding is $20,122.)

Example 8.6 Future Value of a Series of Investments

Figure 8.2 shows how the total future value accumulates for a series of five equal annual payments of $A = \$1{,}000$. An investment of $1,000 at the beginning of year 0 compounds in five years to

$$FV_{5\%,\,5y} = \$1{,}000(1.05)^5 = \$1{,}276.28$$

The five $1,000 investments will be worth $5,801.91 at the end of five years.

8.8 COMPOUNDING FOR *P* PERIODS PER YEAR

Interest can be compounded continuously, annually, or any interval in between. Compound interest for p periods per year at an annual interest rate i is calculated using a rate of i/p for np compounding periods. That is

$$FV_{i/p,np} = PV\left(i + \frac{i}{p}\right)^{np}$$

Example 8.7 Monthly Compounding

An amount $P = \$150{,}000$ will be invested at an effective annual interest rate of 12%, but the interest will be paid and compounded monthly. After 1 year, the amount is

$$FV_{12\%/12,\,1y} = \$150{,}000\left(1 + \frac{0.12}{12}\right)^{1(12)} = \$150{,}000(1.1268) = \$169{,}020$$

At simple interest, the amount would be $150,000(1 + 0.12) = $168,000.

Monthly compounding pays a nice bonus of $1,020. Remember that monthly compounding works against you when you must pay the future value, as when you are paying back a student loan or a mortgage.

Continuous compounding, the mathematical limit for earning interest, is

$$FV_{i,n} = PV\,e^{in}$$

where e is a mathematical constant (the base for natural logarithms). This form can be mathematically convenient but has little day-to-day use. Some banks use daily compounding. The one-year difference in earnings between daily and continuous compounding for a \$10,000 investment at 15% interest is less than \$1.00.

Continuous compounding	\$11,618.34
Daily compounding	\$11,617.98
Monthly compounding	\$11,607.55
Annual compounding	\$11,500.00

8.9 PRESENT VALUE

A *future value* (*FV*) is converted to *a present value* (*PV*) by a calculation called *discounting.* Discounting is the inverse of compounding interest. Given

$$FV_{i,n} = PV(1+i)^n$$

It follows that the present value of $FV_{i,\,n}$

$$PV_{i,n} = \frac{FV_{i,n}}{(1+i)^n}$$

Example 8.8 Convert Future Value to Present Value

Convert the ten-year future amount FV_{10} = \$32,384 to a present value P using i = 8% as the annual discount rate over ten years.

$$PV_{8\%,\,10} = \frac{\$32{,}384}{(1+0.08)^{10}} = \frac{\$32{,}384}{2.1589} = \$15{,}000$$

This is the inverse of Example 8.4.

Example 8.9 Discounting Multiple Future Values

An investment will repay \$1,000 at the end of years 1 through 5 at a discount rate of 5%. The present value of the payments at the end of each year and the total present value are shown in Figure 8.3.

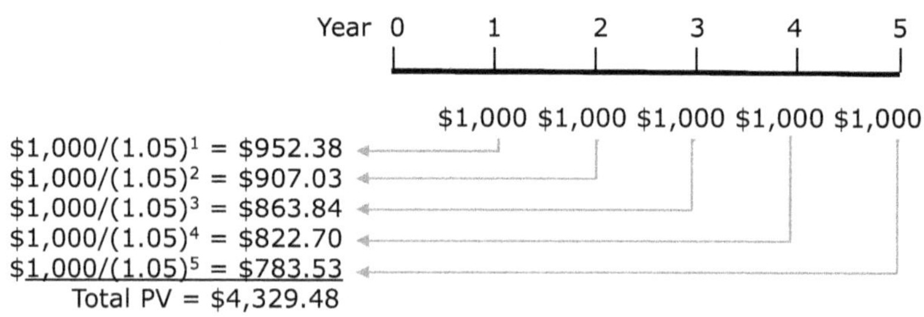

FIGURE 8.3 Present value of five future payments of \$1,000 per year.

Example 8.10 Storage Tank Construction and Maintenance

A storage tank can be constructed for $100,000. By convention, this is recorded as a capital cost at the start of year 1. The tank's useful life will be 30 years. There will be major maintenance costs of $20,000 after 10 and 20 years. Again, by convention, these are recorded as a lump sum payment at the *end of the year.*

The present values of the major maintenance costs at 10 and 20 years, using $i = 6\%$, are as follows:

$$PV_{6\%,\,10y} = \$20{,}000/(1+0.06)^{10} = \$20{,}000/1.7908 = \$11{,}178$$

$$PV_{6\%,\,20y} = \$20{,}000/(1+0.06)^{20} = \$20{,}000/3.2071 = \$6{,}236$$

The *total present value* for the project is

$$PV_{Total} = PV_{Capital} + PV_{6\%,\,10} + PV_{6\%,\,20}$$

$$= \$100{,}000 + \$11{,}178 + \$6{,}236 = \$117{,}413$$

Example 8.11 Storage Tank Present Value with Salvage

A storage tank can be constructed for $200,000. Maintenance is needed every five years. The maintenance costs are recorded as a lump sum payment at the *end of the year.* After 20 years of use, the tank can be sold for $20,000. The *salvage value* is cash coming into the project, so it is shown as a negative cost. The discount rate is 6% per year.

The calculations are given in Table 8.5. The total present value is

$$\$200{,}000 + \$7{,}743 + \$6{,}142 + \$5{,}007 - \$6{,}236 = \$212{,}386$$

As an exercise, convince yourself that putting this amount of money into an account that earns 6% per year will provide the funds to make all the payments and end with a zero balance.

8.10 NET PRESENT VALUE

Cash flow is the difference between revenue and costs. A capital investment is shown as a negative cash flow. Annual cash flows are recorded at the end of the year. For any given year, they can be positive or negative. Negative means that expenses exceeded revenues; the project lost money. Positive cash flow means the project earned money.

TABLE 8.5 Present Value for a Storage Tank with Maintenance and Salvage

Year	Cost Item	Cost	*PV* Factor $(1.06)^n$	*PV*
0	Capital cost ($)	200,000		200,000
5	Maintenance ($/y)	10,000	$1/(1.06)^5 = 0.74726$	7,472.58
10	Maintenance ($/y)	11,000	$1/(1.06)^{10} = 0.55839$	6,142.34
15	Maintenance ($/y)	12,000	$1/(1.06)^{15} = 0.41727$	5,007.18
20	Salvage ($)	–20,000	$1/(1.06)^{20} = 0.31180$	–6,236.00
		Total present value =		*$212,386*

TABLE 8.6 Present Value of Cash Flow

Year	Cash Flow ($)	*PV* Factor	Calculation	*PV* ($)	Net *PV* ($)
0	−80,000	1.00000		−$80,000	
1	−20,000	1/(1.1) = 0.90909	−20,000(0.90909) =	−18,182	−98,182
2	+20,000	$1/(1.1)^2$ = 0.82645	+20,000(0.82645) =	16,529	−81,653
3	+60,000	$1/(1.1)^3$ = 0.75131	+60,000(0.75131) =	45,079	−36,574
4	60,000	$1/(1.1)^4$ = 0.68301	+60,000(0.68301) =	40,981	4,407

The net present value (NPV) is the sum of the annual present values. *Net* is added as a descriptor because the cash flow can be positive or negative in any given year. This is sometimes called the total present value, with the *net* being taken for granted.

Keep track of the sign (+ or −) of the annual cash flows and make the present value calculation in the usual way.

Example 8.12 Present Value of Cash Flow

A waste reclamation process with a capital cost of $80,000 generates the annual cash flows shown in Table 8.6, along with their present values for a 10% discount rate.

The net present value is the sum of the annual present values:

$$\text{NPV} = -\$80,000 + \$18,182 + \$16,529 + \$45,079 + \$40,981 = \$4,047$$

At the end of three years, the net present value is negative. It is positive at the end of four years, so the payback time is about four years.

8.11 THE DISCOUNT RATE

Specifying the discount rate is the first step in making financial decisions. The calculations are then routine arithmetic. The *discount rate* is the value an individual or an organization puts on their money for the purpose of making economic decisions. Obviously, an organization's discount rate can change when their financial situation changes, just as banks change interest rates, and individuals change investments from savings accounts to certificates of deposit.

In public works projects, the discount rate is often the interest rate on bonds that are sold to pay for capital expenditures. If a municipality has a reserve fund (funds saved in a special account), the discount rate may be the current or anticipated interest rate used to calculate earnings on the reserve finds.

Industry and private business use discount rates that reflect the current options for investments. Every business has multiple investment opportunities to expand production, develop a new product, extend a marketing region, and so on. The discount rate is the *risk-adjusted rate* of return that the company could otherwise earn on any of its capital projects or investments. Risk includes the loss of purchasing power due to inflation.

Another name for the discount rate is the *opportunity cost of capital* because investing in project A decreases the opportunity to invest in project B.

The discount rate adjusts for inflation. Investors are encouraged to save rather than spend in times of inflation; that is, when the interest rate exceeds the inflation rate. The acceptable gap between earned interest and inflation rates is decided by the individual or corporate investor. Public works projects are sometimes delayed because of high interest rates on bonds.

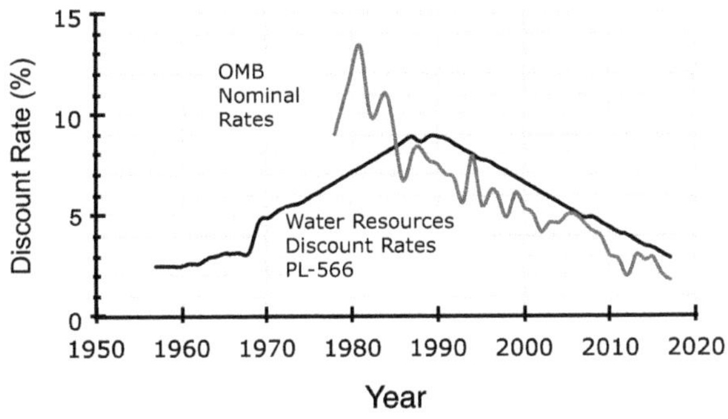

FIGURE 8.4 Discount rates used by the US Office of Management and Budget (OMB) and for Federal water resources management project.

The discount rate can also adjust for the risk of an investment. A deposit in a bank is the safest of all investments, and it has the lowest earnings. One expects to be offered higher earnings when there is some risk, because *risk* means the return on investment may be less than expected.

Public works investments are safe, and the interest rates are low. Loans are backed by the taxing power of the city or the power of utilities to charge customers for services.

Industrial investments carry a higher risk. Revenues expected from increasing production may not materialize because a competitor takes away sales. Industry uses a higher discount rate to account for risk.

Federal water resources projects are required to use the discount rate specified by the government funding agency. The government fixes the rates, so there will be consistency in how competing projects are evaluated.

The US Office of Management and Budget (OMB) has evaluated environmental projects using the discount rates shown in Figure 8.4. OMB rates change in increments of 0.1–0.2% (i.e., from 2.800% to 2.900%). Other government agencies have evaluated water resources project using slightly different rates. The water resources rates are adjusted on increments of 0.025–0.05% (i.e., from 3.500% to 3.375%).

8.12 CASE STUDY: COMPARING TWO PUMPS

Two pumps are compared by calculating their present values. The costs in Table 8.7 are on the same basis for both pumps. Purchase price includes shipping to the point of use and installation. Annual O&M cost includes electricity, labor, minor repairs, etc. Both pumps have a useful life of 10 years.

TABLE 8.7 Economic Summary for Comparison of Two Pumps

Cost Factor	Pump A	Pump B
Purchase price ($)	40,000	50,000
Annual O&M cost ($/y)	6,000	5,000
Useful life (y)	10	10
Discount rate (%)	8	8
Present value O&M costs ($)	40,260	33,550
Total present value ($)	80,260	83,550

The present values of the operating costs for the two pumps, evaluated for 10 years and $i = 8\%$, are as follows:

$$PV_{OC-A} = \frac{\$6,000}{1.08} + \frac{\$6,000}{1.08^2} + \frac{\$6,000}{1.08^3} + \cdots + \frac{\$6,000}{1.08^9} + \frac{\$6,000}{1.08^{10}} = \$40,260$$

$$PV_{OC-B} = \frac{\$5,000}{1.08} + \frac{\$5,000}{1.08^2} + \frac{\$5,000}{1.08^3} + \cdots + \frac{\$5,000}{1.08^9} + \frac{\$5,000}{1.08^{10}} = \$33,550$$

The *total present value* is the purchase price plus the present value of the operating costs, which gives

$$PV_A = \$40,000 + \$40,260 = \$80,260$$

$$PV_B = \$50,000 + \$33,550 = \$83,550$$

All other things being equal, pump A should be selected because it has the lowest present value.

This calculation can be done in a more direct way. The sum of the ten $1/(1.08)^n$ terms, for $n =$ 1–10, is 6.71008, so

$$PV_{OC\text{-}A} = 6.71008(\$6,000) = \$40,260$$

$$PV_{OC\text{-}B} = 6.71008(\$5,000) = \$35,550$$

A table of factors for making the conversion is given in Appendix B. They can only be used when the annual cash flow values are the same every year and they are only given for integer interest rates. Using tabulated factors is an artifact of the time when doing calculations was more cumbersome. The term-by-term calculation is a good method to practice. It is simple and what you are doing is clear to yourself and others. All you need to know is the interest rate, the number of years each annual cash flow is discounted, and whether the annual cash flow is positive or negative (coming in or going out).

Changing the discount rate shifts the relative importance of the initial investment and the time stream of annual cash flows. A higher discount rate reduces the present value of the annual costs, and also reduces the total present value, while the initial cost stays the same.

Add a new cost factor to the problem. Pump A must have major maintenance after 5 years, at a cost of \$5,000. This is in addition to the \$6,000 annual O&M cost. Pump B does not need this. Will this change the decision to buy pump B? (Table 8.8)

Nothing has changed for pump B from the prior comparison, so the total present value for pump B, for $i = 8\%$, is still

$$PV_B = \$50,000 + \$33,550 = \$83,550$$

Pump A has an additional cost of \$5,000 for major maintenance after 5 years. The present value for this cost is

TABLE 8.8 Comparing Two Pumps

Cost Factor	Pump A	Pump B
Purchase price (\$)	40,000	50,000
Annual O&M cost (\$/y)	6,000	5,000
Major maintenance cost at 5 years (\$)	5,000	0
Pump lifetime (y)	10	10

$$PV_{\text{MM-A}} = \frac{\$5,000}{(1.08)^5} = \$3,403$$

The total present value for pump A is

$$PV_{\text{A}} = \$40,000 + PV_{\text{MM-A}} + PV_{\text{OC-A}} = \$40,000 + \$3,403 + \$40,260 = \$83,663$$

The present values for pumps A and B are virtually equal, so non-monetary factors should be considered. For example, does one pump supplier have a better warrantee or more dependable service when replacement parts are needed?

8.13 INTERNAL RATE OF RETURN

The decision may change when the discount rate is changed. In the comparison of pumps A and B, the balance shifts to favor pump A at some discount rate between 8% and 10%, as shown in Table 8.9.

The discount rate where the present values of A and B are equal is called the *internal rate of return* (IRR). For this problem, the IRR = 8.25%. Pump A is favored if the discount rate is more than 8.25%. Below 8.25%, Pump B has the lower net present value

TABLE 8.9 Effect of Interest Rates on Pump Comparison

	i = 6%		*i* = 8%		*i* = 10%	
	Pump A	**Pump B**	**Pump A**	**Pump B**	**Pump A**	**Pump B**
Annual O&M ($/y)	6,000	4,000	6,000	4,000	6,000	4,000
PV O&M ($/y)	44,161	36,800	40,260	33,550	36,867	30,723
Special maintenance ($/y)	3,736	–	3,403	–	3,105	–
Capital cost ($)	40,000	50,000	40,000	50,000	40,000	50,000
Total *PV* ($/y)	87,897	86,800	83,663	83,550	79,972	80,723

8.14 THE PRESENT VALUE FOR PROJECTS WITH UNEQUAL LIFETIMES

A direct comparison of present values is valid when the projects have the same useful life. When they are not the same, there are two ways to solve the problem:

(1) Make the project lifetimes equal.
(2) Make the comparison using the annual costs by converting the purchase cost to a series of uniform annual costs. This will be explained in the next section.

Example 8.13 Comparison with Unequal Lifetimes

Pump 1 has an initial capital cost of $40,000 (installed) with a lifetime of 10 years. Pump 2 has a lower initial capital cost of $25,000 (installed), but with a shorter lifetime of 5 years. As shown in Table 8.10, there is a trade-off between initial cost and durability.

The project lifetimes are made equal at 10 years by replacing pump 2 after 5 years at a cost of $25,000. All O&M costs come at the end of the year

Table 8.11 shows the present value calculation year by year for i = 6%. Project 2 has the lowest total present value. The present value of the project 2 capital cost is only $3,681 more than that of project 1. This is because the $25,000 expense was deferred for five years.

The present values of the two projects would be equal if the annual O&M cost for project 2 were $5,400 per year.

TABLE 8.10 Project Costs and Lifetimes

Cost Factor	Project 1	Project 2
Purchase price ($)	40,000	25,000
Annual O&M cost ($/y)	6,000	4,000
Pump lifetime (y)	10	5

TABLE 8.11 Calculation of the Present Values

Year	Cost Factor	Costs ($)		*PV* Factor	Present Values ($)	
		Project 1	Project 2	i = 6%	Project 1	Project 2
0	Cost of pump	40,000	25,000		40,000	25,000
1	O&M	6,000	4,000	0.94340	5,660	3,774
2	O&M	6,000	4,000	0.89000	5,340	3,560
3	O&M	6,000	4,000	0.83962	5,038	3,358
4	O&M	6,000	4,000	0.79209	4,753	3,168
5	O&M	6,000	4,000	0.74726	4,484	2,989
5	Replace pump		25,000	0.74726		18,681
6	O&M	6,000	4,000	0.70496	4,230	2,820
7	O&M	6,000	4,000	0.66506	3,990	2,660
8	O&M	6,000	4,000	0.62741	3,764	2,510
9	O&M	6,000	4,000	0.59190	3,551	2,368
10	O&M	6,000	4,000	0.55839	3,350	2,234
				PV capital =	40,000	43,681
				PV O&M =	44,161	29,440
				Total *PV* =	84,161	73,122

Forcing equal project lifetimes can be an awkward way to solve the problem. Suppose they were 10 years for project 1 and 6 years for project 2. The present value evaluation would have to be over a 30-year period (30 being the lowest common denominator of 6 and 10).

The alternative and more convenient calculation is to convert the purchase cost to a series of *uniform annual costs.*

8.15 CONVERT A PRESENT VALUE TO SERIES OF UNIFORM ANNUAL COSTS

Repaying a loan or a mortgage over a fixed period with a series of uniform costs (monthly or annual) is called *amortization*. This is the familiar calculation of payments to buy a house or an automobile. The root *mort* means death, so a rough translation is "kill the mortgage" by paying it off. Part of each payment goes to reduce the unpaid amount of the loan (the principal) and part pays interest on the unpaid balance. As time goes on, the fraction allocated to principal increases and the fraction allocated to interest decreases.

The *total present value* of a sum of j uniform annual payments or costs, denoted by A, is the sum of the discounted annual future values. A convenient name for A is UNACOST, for *uniform annual cost.*

$$PV_{i,n} = \frac{A}{(1+i)^1} + \frac{A}{(1+i)^2} + \cdots + \frac{A}{(1+i)^{n-1}} + \frac{A}{(1+i)^n} = A\sum_{j=1}^{n} \frac{1}{(1+i)^j}$$

The sum of the discount factors is

$$\sum_{j=1}^{n} \frac{1}{(1+i)^j} = \left[\frac{(1+i)^n - 1}{i(1+i)^n}\right] = F_{AP,\,i,\,n}$$

$F_{AP,i,n}$ is called the *present value factor*, read as "factor to convert A to P." This is also called the PVF.

Converting a present value into a series of uniform annual payments is the inverse calculation.

$$A_{i,\,n} = \frac{PV}{\sum_{j=1}^{n} \frac{1}{(1+i)^j}} = PV\left[\frac{i(1+i)^n}{(1+i)^n - 1}\right] = PV \times F_{PA,\,i,\,n}$$

$F_{PA,i,n}$ is called the *capital recovery factor*, read as "factor to convert P to A." This is also called the CRF.

Table 8.12 gives some values for F_{PA} to convert a lump sum at time zero (PV) to uniform annual costs (A), and F_{AP} for the inverse of converting uniform annual costs (A) to a lump sum present value (PV). The factors are reciprocals of each other. More complete tables may be found in Appendix C.

Traditional books on engineering economics include tables for these factors and several others. These tables can be a convenience if the tables are at hand, but that is not always the case. The tables give factor values only for integer values of the discount rates. Integer rates are typical of textbook problems, but this is often not the case for real problems. This means that there is no great advantage to having tables of the factors because it is so easy to calculate values as needed. Excel has functions to do this, or a personal spreadsheet can be designed.

Many calculations can be done without using factors. Present values, for example, are easily done by making a table of cash flows, discounting each one, and summing the values to get the total present value. This works for any series of values, whether equal or unequal, and whether positive or negative. Factors can be applied only to a series of equal values.

TABLE 8.12 Factors for Converting between Present Values and Uniform Annual Costs

Year	Uniform Annual Cost to PV, F_{AP}				PV to Uniform Annual Costs, F_{PA}			
	Discount Rate				Discount Rate			
	4%	6%	8%	10%	4%	6%	8%	10%
5	4.45182	4.21236	3.99271	3.79079	0.22463	0.23740	0.25046	0.26380
6	5.24214	4.91732	4.62288	4.35526	0.19076	0.20336	0.21632	0.22961
7	6.00205	5.58238	5.20637	4.86842	0.16661	0.17914	0.19207	0.20541
8	6.73274	6.20979	5.74664	5.33493	0.14853	0.16104	0.17401	0.18744
9	7.43533	6.80169	6.24689	5.75902	0.13449	0.14702	0.16008	0.17364
10	8.11090	7.36009	6.71008	6.14457	0.12329	0.13587	0.14903	0.16275
11	8.76048	7.88687	7.13896	6.49506	0.11415	0.12679	0.14008	0.15396
12	9.38507	8.38384	7.53608	6.81369	0.10655	0.11928	0.13269	0.14676
13	9.98565	8.85268	7.90378	7.10336	0.10014	0.11296	0.12652	0.14078
14	10.56312	9.29498	8.24424	7.36669	0.09467	0.10758	0.12130	0.13575
15	11.11839	9.71225	8.55948	7.60608	0.08994	0.10296	0.11683	0.13147

Example 8.14 Blower Cost

A centrifugal blower to serve an activated sludge wastewater treatment process will cost \$2,000,000. The treatment plant can borrow money to buy the blower at 4% interest with a repayment period of 8 years. Payments will be made at the end of the year. The capital recovery factor, from Table 8.12, is

$$F_{PA,4\%,8y} = 0.14853$$

and

$$A = \$2{,}000{,}000(0.14853) = \$297{,}060 \text{ paid at the end of each year}$$

Example 8.15 Pump Comparison

This is the pump comparison problem of Example 8.13 with unequal lifetimes of 10 years for pump A and 12 years for pump B, using $i = 8\%$. The results are summarized in Table 8.13.

Annualized capital costs

Pump A Capital recovery factor = $F_{PA,8\%,10y} = 0.14903$

Annualized capital cost = \$40,000(0.14903) = \$5,961

Pump B Capital recovery factor = $F_{PA,8\%,12y} = 0.13269$

Annualized capital cost = \$50,000(0.13269) = \$6,635

Total annual costs = Annualized capital cost + Annual O&M cost

Pump A \$5,961 + \$6,000 = \$11,961

Pump B \$6,635 + \$4,000 = \$10,635

Pump B is preferred.

TABLE 8.13 Summary Cost Comparison of Two Pumps

Cost Factor	Pump A	Pump B
Capital cost ($)	40,000	50,000
Pump life (y)	10	12
Capital recovery factor (8%)	0.14903	0.13269
Annualized capital cost ($/y)	5,961	6,635
Annual O&M cost ($/y)	6,000	4,000
Total annualized cost ($/y)	11,961	10,635

Example 8.16 Amortization of Air Compressor Cost

An air compressor cost of \$110,000 will be paid with a 10-year loan at 8% interest. The annual payment is this cost amortized over 10 years at 8%. Show that the annual payment will repay the loan.

TABLE 8.14 Year-by-Year Repayment of a $110,000 Debt

Year	Balance Due Start of Year ($)	Interest Charge ($)	Balance Due Year End ($)	Payment (End of Year) ($)	Reduction in Principal ($)
1	110,000	8,800	118,800	16,393	7,593
2	102,407	8,193	110,599	16,393	8,201
3	94,206	7,536	101,743	16,393	8,857
4	85,349	6,828	92,177	16,393	9,565
5	75,784	6,063	81,847	16,393	10,331
6	65,453	5,236	70,690	16,393	11,157
7	54,297	4,344	58,640	16,393	12,050
8	42,247	3,380	45,627	16,393	13,013
9	29,234	2,339	31,572	16,393	14,055
10	15,179	1,214	16,393	16,393	15,179
Balance at the end of 10 years		0.00	0.00		

The capital recovery factor for 8% and 10 years is $F_{PA,8\%,10} = 0.14903$. The annual capital cost is

$$A = 0.14903(\$110{,}000) = \$163{,}393/\text{y}$$

Assume that the $110,000 is deposited in a bank where it earns 8% per year, with the required payments being made at the end of the year.

- Unpaid balance at the beginning of year 1 = $110,000
- Interest due at the end of year 1 = $8,800
- Unpaid balance at end of year 1 = $110,000+ $8,000 = $118,000
- Payment of end of year 1 = $16,393 (ignoring the pennies)
- Unpaid balance at beginning of year 2 = $118,000 – $16,393 = $102,407
- Of this payment, $8,800 was for interest and $16,393 – $8,800 = $7,593 was to reduce the principal.

This calculation is continued in Table 8.14, which shows that this uniform annual payment does retire the debt in 10 years.

Example 8.17 Case Study: Water Supply

A water supply case study in Chapter 4 described a city's three options for a new water supply. Options 1 and 2 were competitive; option 3 was not. The capital costs to meet the 2030 demands and the annual O&M costs are given in Table 8.15. These two kinds of costs can be combined to

TABLE 8.15 Options for Water Supply

Option	Capital Cost	Annual O&M Cost
1. Purchase from two wholesalers	$11,458,800	$2,675,500
2. Purchase all water from one wholesaler	$12,733,400	$2,699,300

get a life cycle cost or, alternatively, an annual cost that pays for the project. Use a 20-year project life and a 6% interest rate.

Life cycle cost is the present value of capital cost and the 20-year stream of O&M costs.
The PVF for 20-year at 6% = 11.4699.

$$\text{Option 1: O\&M } PV = \$30{,}687{,}700 \quad \text{Total } PV = \$42{,}146{,}500$$

$$\text{Option 2: O\&M } PV = \$30{,}960{,}700 \quad \text{Total } PV = \$43{,}694{,}100$$

Annual cost is the amortized capital cost plus the annual O&M cost.
The capital recovery factor is CRF = 0.08718.

$$\text{Option1: Annualized capital cost} = \$999{,}000 \quad \text{Annual cost} = \$3{,}574{,}500$$

$$\text{Option2: Annualized capital cost} = \$1{,}110{,}100 \quad \text{Annual cost} = \$3{,}809{,}400$$

8.16 BENEFIT–COST ANALYSIS

All projects aim to produce benefits that outweigh their costs. A formal benefit–cost analysis identifies, quantifies, and adds all the positive factors (the benefits) and the negatives (the costs). These studies can be controversial, first because of the difficulty in identifying and monetizing all of the benefits, and second because the benefits accrue over a long period of time and small changes in the discount rate can change the outcome. The benefit–cost notation suggests that projects can be compared using the simple ratio of benefits to costs, and the best project has the largest B/C ratio. The analysis is not that simple. The correct procedure uses the net present value (NPV) and not the *ratio* of benefits and costs, as shown in the next example.

These steps comprise a generic cost–benefit analysis:

- Define the goals and objectives of the project/activities.
- List alternative projects/programs.
- Select measurement(s) and measure all cost and benefit elements.
- Predict outcome of cost and benefits over relevant time period.
- Convert all costs and benefits into a common currency.
- Apply an appropriate discount rate and calculate net present value of project options.

Example 8.18 Benefit Cost Analysis of Two Projects

Table 8.16 shows two alternatives: project A and project B. The estimated net benefits for a 20-year project life have been converted to an equivalent annual cost. A 10% discount rate will be used.

Calculate the net present value for project A and project B. The first cost is treated as a negative benefit.

TABLE 8.16 Costs and Benefits for Two Projects

Cost or Benefit	Project A	Project B
Capital cost ($)	152,200	183,800
Estimated total benefits ($/y)	24,100	28,900
Maintenance ($/y)	2,000	3,000
Estimated net benefits ($/y)	22,100	25,900

The NPV of the projects over 20 years at 8% are calculated using $F_{PV} = 8.514$.

$$NPV_A = -\$152,200 + \$22,100(8.514) = -\$152,200 + \$188,159 = \$35,959$$

$$NPV_B = -\$183,800 + \$25,900(8.514) = -\$183,800 + \$220,513 = \$36,713$$

Project B is preferred because it has the larger NPV.

The incremental net present value of the differences is

$$NPV_B - NPV_A = \$36,713 - \$35,959 = +\$754$$

Another way to calculate this is

$$NPV_B - NPV_A = (\$183,800 - \$152,200) + (\$25,900 - \$22,100)(8.514) = +\$754$$

A NOTE ON THE BENEFIT/COST RATIO

Here is a mistake that is sometimes made with the benefit–cost ratio. It is wrong to make a simple comparison of the ratios of the two projects, which are

$$B/C \text{ of A} = \$188,159/\$152,200 = 1.236$$

$$B/C \text{ of B} = \$220,513/\$183,800 = 1.120$$

This seems to select project A, which was shown above to be wrong.

The *B/C* ratio should be calculated using the differences of the discounted benefits and the difference of the costs:

$$\Delta \text{ Costs} = \$183,800 - \$152,200 = \$31,600/y$$

$$\Delta \text{ Benefits} = \$25,900 - \$22,100 = \$3,800/y$$

$$\Delta \text{ Discounted benefits} = 8.514(\$3,800) = \$32,400/y$$

$$\frac{\Delta \text{ Discounted benefits}}{\Delta \text{ Cost}} = \frac{\$32,400/y}{\$31,600/y} = 1.024$$

This correctly selects project B.

The straightforward comparison with present values is easier and less prone to mistakes.

8.17 COST OF INFLATION AND INTEREST RATES

Inflation is the increase in the price of goods and services from year to year. Cost indexes (Chapter 6) are used to track inflation in the past, but using them to project future inflation is less reliable. The cost components of the CECPI and the ENR-CCI inflate at different rates, and so do chemicals, electricity, and labor.

An item that has an initial cost of A and inflates at the rate of r per year will cost $A_1 = A(1 + r)$ at the end of the year 1 (the start of year 2), and $A_2 = A(1 + r)^2$ at the end of year 2, and so on.

If the rate of inflation were known, the inflated annual costs could be estimated and then discounted to get the present value. The present value of a uniform annual series of costs that are inflating at the rate r and with the discount rate of i is

$$PV = \frac{A(1+r)^1}{(1+i)^1} + \frac{A(1+r)^2}{(1+i)^2} + \cdots + \frac{A(1+r)^{n-1}}{(1+i)^{n-1}} + \frac{A(1+r)^n}{(1+i)^n}$$

The difficulty is that long-term rates of inflation are not known, and they cannot be predicted with any reliability. Also, inflation is not constant rate from year to year. Therefore, the added detail may not improve the analysis.

A question is whether to inflate the costs and then discount them, as shown above, or to simply compare projects with a discount rate that will incorporate the risk of inflation. Either can be done, but adjusting the discount rate seems to be more practical and more common.

The situation changes when costs are projected for budgeting. Then the inflated costs are needed. Uncertainty in the interest rate is not a serious problem because budgets are plans for the near future.

An individual who lends money for repayment at a later point in time expects to be compensated for the time value of money; that is, for not having the use of that money while it is on loan. In addition, there should be compensation for the expected value of the lost purchasing power when the loan is repaid.

Example 8.19 Inflation and Loss of Purchasing Power

If $1,000 is borrowed for a year at 10% nominal interest, $1,100 must be paid back at the end of the year. This represents a 10% increase in purchasing power if average prices for goods and services are unchanged from the beginning of the year. However, if the prices of the goods and services purchased have increased by 25% over this period, there has been a real loss of about 12% purchasing power (1.0 – 1.1/1.25 = 0.12).

Interest rates reflect the rate of inflation. In times of inflation, when costs are rising, an investor needs an interest rate that exceeds the rate of inflation. If the interest rate is less than inflation over the term of the investment, the money returned to the investor has less purchasing power than money that was invested.

The *nominal interest rate i* is the amount charged by a lender to a borrower. It measures the sum of the compensations for inflation and risk, plus the time value of the money itself. The nominal rate of interest is also known as the *adjusted rate* or the *inflated rate*. Since the inflation rate over the course of a loan is not known initially, volatility in inflation represents a risk to both the lender and the borrower. In the United States, the nominal rate is based on the federal funds rate that is determined by the Federal Reserve Bank. The interest rates used in this chapter have been nominal interest rates.

The *real interest rate i** is the rate of interest an investor, saver, or lender receives (or expects to receive) after allowing for inflation. The *real interest rate* is approximately the *nominal interest rate* minus the *inflation rate r*. The real rate is also known as the *unadjusted rate* or *inflation-free rate*.

The *expected real interest rate* is the nominal interest rate minus the *expected* rate of inflation. This is not a single number, as different investors and different projects have different expectations and economic trajectories.

In the case of contracts stated in terms of the nominal interest rate, the real interest rate is known only at the end of the period of the loan, based on the realized inflation rate; this is called the ex-post real interest rate. Bonds can be inflation indexed.

If the amount borrowed is X_0, then X_1 is the amount to be repaid at the end of one time period and the real rate of interest is i^*:

$$i^* = \frac{X_1 - X_0}{X_0} \quad \text{and} \quad X_1 = X_0(1+i^*)$$

In the presence of inflation, the nominal interest rate i is the rate by which the borrower repays the investor. If X_0 is the amount borrowed, the amount owed using the nominal interest rate i is $X_1 = X_0(1+i)$. This equals the amount repaid using the real interest rate and rate of inflation r.

$$X_1 = X_0(1+i^*)(1+r)$$

giving a nominal interest rate of

$$i = (1+i^*)(1+r)-1 \; = \; i^* + r + ri^*$$

When i and r are small, ri^* can be ignored and the nominal rate of interest is approximately

$$i = i^* + r$$

where
i = nominal interest rate
i^* = real interest rate
r = expected inflation rate

Example 8.20 Real Interest Rate and Inflation

For example, if an investment has a locked in 5% nominal interest rate for the coming year and there is a 2% rise in prices, the real interest rate is 5% – 2% = 3%. On an economy-wide basis, the *real interest rate* is often considered to be the rate of return on a risk-free investment, such as US Treasury notes, minus an index of inflation, such as the rate of change of the Consumer Price Index (CPI) or Gross Domestic Product (GDP) deflator.

8.18 CONCLUSION

Payback time is easy to calculate and understand. Its weakness is the failure to take into account how the time sequence of costs and revenues should influence decisions. A more dependable analysis of lifetime costs uses the present value and future value of money, which depend on the interest rate that is paid on borrowing or could be earned on revenues.

A popular way of comparing investment options uses present values. Future values can be discounted to get the present value of that amount. A series of future values can be discounted to get the total present value of the series. If some future values are positive and some are negative, the sum of discounted values is called the net present value.

The present value of initial cost, for example, a capital equipment purchase, is the initial cost. The total present value is initial cost plus the total present value of the series of annual costs.

The second way to compare alternatives is to calculate the annual cost. The lump-sum initial cost is amortized to create a series of equivalent uniform annual costs, or UNACOST. The present value of the UNACOST series equals the initial cost. The annual cost is the UNACOST plus the annual O&M costs.

When competing projects have the same lifetime, it is convenient to make a direct comparison of their present values. For unequal lifetimes, it can be more convenient to convert the lump sum initial cost to a series of uniform annual payments and make the comparison based on annual costs.

8.19 PROBLEMS

8.1 RECYCLE WASTE SOLVENT

The initial investment for a still to recycle waste solvent is $7,700, and it will provide a net annual operations savings of $4,600. What is the payback period?

8.2 ACTIVATED SLUDGE AERATION SAVINGS

Table P8.2 shows the cost of electricity (euro/month) for an activated sludge process when it was operated under manual control for one year and with automated control of the blowers in the following year. The control system measures the dissolved oxygen in the aeration tanks and adjusts the aeration rate to minimize energy use by the blowers. The annual pattern of flows and waste loads follow the same ups and downs in both years.

(a) Calculate the annual savings derived from installing the control system.
(b) What investment is justified for an automated control system to have a payback period of five years?

TABLE P8.2 Aeration Cost (euro/month)

Month	Manual Control	Automated Control	Month	Manual Control	Automated Control
Jan	129,000	108,000	July	122,000	100,000
Feb	118,000	94,000	Aug	78,000	68,000
Mar	116,000	90,000	Sep	110,000	90,000
Apr	104,000	89,000	Oct	118,000	119,000
May	108,000	91,000	Nov	131,000	108,000
June	114,000	96,000	Dec	125,000	108,000

8.3 IN-VESSEL COMPOSTING COST

An in-vessel composting system that processes 4,000 T/y of organic waste (yard, food, etc.) to produce 3,500 T/y of finished compost has a capital cost of $2,800,000. The compost product can be used in place of 2,000 T/y of topsoil that would be purchased at $25/T. Operating costs are $125,000/y to cover the $70/T hauling cost to the landfill plus an additional $28/T landfill tipping fee to pay the landfill operating cost. The cost for hauling to the composting site is $30/T. (a) Calculate the annual savings for composting over disposal. (b) Calculate the payback period for investment in the composting system.

8.4 POLLUTION PREVENTION PAYS

The initial investment for a pollution prevention system is $10,000. The projected savings is $4,000 for the first year, $4,000 for the second year, $2,500 for the third year, $2,000 in the fourth year, and $2,000 for the fifth year. What is the payback period?

8.5 PAINT WASTE

An engineer wants to install two new dedicated pipe runs from a plant's paint mix room to the spray booths to reduce the line cleaning waste. This will cost $10,000 and eliminate 5,000 kg/y of waste. Also, each kilogram of waste represents $3.00 of wasted raw material. Waste disposal, including transportation and in-plant labor, costs $0.45/kg. Should management implement the proposed project?

8.6 TRANSMISSION FLUID MANAGEMENT

A transmission test department noticed red fluid entering the waste treatment facility. This was transmission oil that leaked from the fluid reclamation system piping. Degraded screens and clogged filters increased the fluid loss to the wastewater system.

The screens and filters were replaced and the leaking pipes were repaired. A preventive maintenance schedule was established. These adjustments are expected to save $4,000,000 per year in reclaimed transmission fluids. The cost of implementation was $288,000 and took four months. What is the payback time?

8.7 AYRES ROCK RESORT

The Ayres Rock Resort, an oasis in a semiarid Australian desert, has three hotels, five restaurants, three swimming pools, and an array of supporting shops and services. Energy costs at the resort average AU$ 1,200,000/y. Since 2000, the resort has implemented several energy reduction projects and estimates a reduction of 2 million kg of CO_2 emissions. Table P8.7 gives the cost and savings of several projects. What is the payback time for the listed projects?

TABLE P8.7 Ayers Rock Energy Reduction Projects

Energy Reduction Project	Cost (AU$)	Savings (AU$/y)
Fidelio computer system	20,000	(indirect)
Motion-sensor stairway lights	2,400	10,220
Movement detectors in showers and bathrooms	840	122,660
Replacement of incandescent light bulbs	4,800	151,200
Labor savings (50% less breakdown of equipment)	Included above	126,060
Totals =	28,040	410,140

8.8 TURBOCOMPRESSOR SAVES ENERGY

An Australian wastewater treatment plant serves a permanent population of 125,000, which doubles during the vacation season. It was expected that the treatment plant would reach the limits of its capacity during the 2015 holiday season. The short-term solution, one that required no new construction, was to improve the aeration system. This could be done by adding a second positive displacement (PD) blower or by adding a new turbocompressor (TC). The TC costs AU$ 140,000 more than the PD blower. Both machines can deliver 5,500 Nm^3/h of air at a pressure of 65 kPa and will operate 6,480 h/y. The price of electricity is AU$ 0.13/kWh. The power required is 115 kW for the TC and 170 kW for the PD blower. Calculate the annual savings in electricity cost and the payback time.

8.9 RECLAIM ANODIZING RINSE WATER

Anodized aluminum parts are cleaned in rinse systems that are supplied with city water. This water, along with acids used in the process, is dumped to the waste treatment system. Plant engineering recommends a reverse osmosis system to continuously clean the process water at 200 L/m. This will reduce city water use for rinsing by 98% and reduce the wastewater discharge by 90%. Also, 100% of waste acids can be recycled to the plating baths. The implementation cost was $278,500 and took three months. Savings are $539,400 per year. (a) What are the process factors that yield the cost savings? (b) What is the payback time?

8.10 AERATION BLOWER REPLACEMENT

An average of 7,200 cfm of air must be delivered to activated sludge aeration basins. Peak load oxygen demands are approximately 15,300 cfm of air. The system must have the capability for turndown to 3,000 cfm in order to maximize energy savings during low demand periods. Turbo blowers are recommended to replace the existing positive displacement blowers. The equipment cost is $1,470,000 and the total installed cost is $3,909,000. This is for five installed units, four normally in operation with one backup. The monthly electrical cost decreased from $128,000 to $77,800. What is the projected payback time? (*Source:* WEF Tech. Exhibition Conf. Proc., Orlando, FL)

8.11 GRIT REMOVAL MAINTENANCE

An existing grit removal basin must be cleaned 165 times per year at a cost of $430 per cleanout. Replacing the grit basin with a new solid separator at a cost of $80,000 will reduce this to 58 cleanings/y at a cost of $400 each.

Implementing pollution prevention measures at a capital cost of $131,000 will reduce the cleanouts for the existing system to 47/y, at the same $430 unit cost. Installing the new separator and implementing and adding the pollution prevention steps will reduce the number of cleanings to 5/y and produce additional savings of $40,000/y. Select the best option.

8.12 COMBINED HEAT AND ENERGY (CHP)

Combined heat and energy (CHP), also known as cogeneration, is a way to produce both heat and electricity. Table P8.12 compares a CHP option with an existing electrical generator. The cost of the CHP unit is $12,900,000. The cost of purchased electricity is $0.08/kWh. The cost of on-site thermal fuel and CHP fuel is $6/MM Btu. The added O&M cost for CHP is $750,000/y. Calculate (a) the annual savings in operating cost, (b) the simple payback period, and (c) the unit cost for electricity produced ($/kWh).

TABLE P8.12 Comparison of CHP with Existing Electrical Generator

Source of Energy	Base Case	CHP Case
Purchased electricity (kWh)	88,250,160	5,534,150
Generated electricity (kWh)	0	82,716,010
On-site thermal units (MM Btu)	426,000	18,872
CHP thermal units (MM Btu)	0	407,128
On-site thermal fuel (MM Btu)	532,000	25,500
CHP fuel (MMBtu)	0	969,845

8.13 HYDROGEN SULFIDE (H_2S) ODOR CONTROL

A community on the east coast of Central Florida withdraws an average of 7.06 mgd (2,576.9 million gal/y) of drinking water from deep wells in the Floridian Aquifer. The inland wells are freshwater and the coastal wells are saline. Both waters are high in chlorides and total dissolved solids (TDS), and both have unusually high levels of hydrogen sulfide (H_2S). The county operates two reverse osmosis (RO) plants that blend demineralized water with freshwater in a ratio between 3:1 and 4:1. The plant's finished water is stripped of hydrogen sulfide by forced-draft aerators. The off-gas flow of 13,000 cfm causes numerous complaints because of its rotten egg odor. Four odor control systems, listed in Table P8.13, are technically feasible. Calculate the present value of each system for 20 years at 8% interest.

TABLE P8.13 Cost of Water Treatment Systems

In-Line Oxidation with Chlorine		**Two-Stage Wet Scrubber**	
Capital cost ($)	200,000	Capital cost ($)	200,000
Operating costs ($/y)		Operating costs ($/y)	
Chlorine	202,878	Chlorine	86,126
Caustic	81,099	Caustic	68,797
Catalytic/Adsorptive carbon (95% removal)		**Ozone dissolution** (ozone from pure O_2)	
Capital cost ($)	185,000	Capital cost ($)	$185,000
Operating costs ($/y)		Operating costs ($/y)	
Rejuvenation water	54,259	Electricity	54,259
Electricity	8,500	Liquid oxygen (LOX)	27,130
Canister replacement	18,000	Deoxygenation air sparging	14,700

8.14 CLOSED-LOOP VAPOR DEGREASER

A company makes a variety of spring products from wire and sheet metal. The products must be cleaned before a final finish is applied. The company has used open-top vapor degreasers to clean the products with perchloroethylene (PERC). This system is causing violations of the Clean Air Act limits for volatile organics emissions. The company installed two batch closed-loop degreasers with in-line distillation to recover the PERC for reuse, coupled with a closed-loop cooling system that eliminated wastewater generation. PERC air emissions were reduced from 36,180 kg/y to 14 kg/y. The capital cost of the installed system was $300,000. Table P8.14 gives some process changes due to changing the degreasers. (a) Calculate the annual savings and the payback time. (b) Calculate the present value of the savings over 10 years at 5% interest. (c) Comment on other savings that can be expected from the new system. (d) The reduction in PERC purchased is more than the reduction in PERC air emissions. Suggest some reasons why this might occur.

TABLE P8.14 Comparison of Open-top and Closed-loop Degreasers

Cost Item	Old System	New System
PERC purchased	40,800 kg @ $0.95/kg	1,390 kg @ $1.14/kg
Hazardous waste generated	43 drums @ $190/drum	4 drums @ $190/drum
Non-cost factors		
PERC air emissions	36,180 kg	14 kg
Reduction in wastewater to evaporator	1,162 m^3	None

8.15 FUTURE VALUE

Calculate the future value at year 5 for the cash flows shown in Table P8.15 for $i = 6\%$. Cash flow occurs at the end of the year.

TABLE P8.15 Future Value of Cash Flow

Year	1	2	3	4	5
Cash flow ($)	1,000	2,000	3,000	4,000	5,000

8.16 PRESENT VALUE

Calculate the present value for the cash flows shown in Table P8.16 for $i = 6\%$. Cash flow occurs at the end of the year.

TABLE P8.16 Present Value of Cash Flow

Year	1	2	3	4	5
Cash flow ($)	5,000	4,000	3,000	2,000	1,000

8.17 AERATION SAVINGS

A treatment plant has a maximum daily dry weather flow of 16,300 m^3/d that is treated in two 390 m^3 oxidation ditches followed by secondary clarifiers. Each oxidation ditch operates with four 25-kW aerators 24 h/d, seven days a week. Waste-activated sludge from both oxidation ditches goes to an aerobic digester and then to two facultative lagoons. The two lagoons are 22,000 m^3 each with two 12 kW mechanical aerators to provide some mixing and keep the top layer aerobic to reduce odors.

The reliability of the lagoon aerators was poor, and they were to be replaced. The 24-h, 7-d operation of aerators in the oxidation ditches was unnecessary. The lagoon aerators were replaced with two solar-powered mixers. An on-line dissolved oxygen measurement system was installed in the oxidation ditches to maintain a constant DO level of 1.5–2.0 mg/L. Instead of four aerators running constantly, the system now operates a minimum of two aerators in each ditch at all times, with additional aeration cycling on and off as needed.

The project cost was $125,400. The electricity savings in the facultative lagoon aerator modifications were 31 kW and 276,698 kWh/y. The on-line DO control system produces savings of 48 kW and 427,488 kWh/y. The total cost savings for electricity are $80,700.

Calculate:

(a) The simple payback time, for a total project cost of $125,400.
(b) The treatment plant's payback period, taking into account a State of California Process Optimization Program that paid 60% of the project cost.
(c) The 10-year present value of the savings on electricity for a discount rate of 4%.

8.18 VAPOR DEGREASING SAVINGS

An open-top vapor degreaser was used to clean guide tubes for microwave transmission. The solvent was 1,1,1 trichloroethane. The company purchased an airless vacuum vapor degreasing system that uses perchloroethylene (PERC). Solvent purchases were reduced from 37,000 lb/y to 300 lb/y, at savings of $75,000/y. Hazardous waste was reduced by 75%, at savings of $3,500/y. Energy costs increased by $1,500/y. Regulatory costs went down by $9,000/y. Other cost savings were $9,000 labor, $2,000 waste oil elimination, and $2,500 miscellaneous. The capital cost of the project was $300,000. Calculate the (a) total cost savings, (b) payback period, and (c) net present value of the savings over 10 years at 10% interest.

8.19 AERATED STATIC PILE COMPOSTING

A compost operation processes 25,000 T/y of wood and yard waste to produce 15,000 T/y of finished compost. An aerated static pile compost operation has a capital cost of $3,000,000, including aerated compost pads, blowers, grinders, compost mixer, trommel screen, front end loaders, and offices. The annual O&M costs is $25/T. Offsetting this are savings from avoided solid waste disposal costs ($13/T), hauling to landfill ($5/T), and reduced landscaping costs avoided by using all the composted material as a substitute for topsoil which would cost $50/T. These costs are based on solid waste processed, not on compost produced. Hauling to the composting site is $50,000/y. What is the present value of the cost savings for $n = 20$ years and $i = 8\%$?

8.20 BLOWER SELECTION

Table P8.20 gives the capital cost and annual energy cost (US$ 2003) for four blower systems. One is to be selected to aerate an activated sludge treatment process. Calculate the net present value for each using a 15-year life and 8% interest.

TABLE P8.20 Blower System Costs

Blower Type	Capital Cost ($)	O&M Cost ($/y)
Multistage Centrifugal A	118,000	93,800
Multistage Centrifugal B	159,600	98,900
Multistage Centrifugal C	158,000	104,000
Positive Displacement	104,800	97,200

8.21 PRIMARY SETTLING AND AERATION COSTS

Most treatment plants have primary settling tanks to remove about 60% of the influent suspended solids and about 30% of the influent BOD. A city has considered not building these tanks. The increased BOD load would go to the activated sludge aeration tanks, and this would require using more air and more electric power for larger air blowers. The 2002 cost to construct primary settling tanks was $6,908,000. The present value of the increased aeration cost without primary settling was $4,595,000, calculated for 20 years at 8%. The costs in Table P8.21 were calculated in 2002 when the price of electricity was about $0.045/kWh. The 2019 cost was $0.14/kWh. The ENR Construction cost index values are 6,583 in 2002 and 11,311 in 2019. Would building primary sedimentation tanks in 2002 have been the correct economic decision? Would building primary sedimentation tanks in 2019 have been the correct economic decision? Assume the waste flows and loads are the same in 2002 and 2019.

TABLE P8.21 Cost of two options in 2002

Design Option	Present Value
(1) Construction of primary settling tanks	$6,908,000
(2) No primary settling – Additional power cost due to higher BOD load to aeration basin	$4,595,000

8.22 NO PRIMARY SEDIMENTATION

A city in semitropical south Florida had an $8,500,000 estimated construction cost for primary settling tanks. This is a higher cost than usual due to poor soils and the need for foundation piles to support the tanks. Furthermore, primary settling tanks are not popular for the following reasons: (a) BOD removal by settling is poor to fair, (b) odor problems associated with primary tanks may require added capital and O&M costs, and (c) more land is required.

If primary tanks are not installed, large solids and grit (sand) will be removed from the influent wastewater in an enclosed headworks facility. Odorous gaseous emissions in the headworks facility will be collected and treated in chemical scrubbers. The liquid stream will be discharged directly into the anoxic zone of the activated sludge process.

The BOD and suspended solids load that might have been removed by settling must be aerobically treated in the activated sludge process and this will increase the annual power cost for aeration. The estimated annual cost for electric power cost for aeration with primary settling is $1,500,000 and $2,000,000 without primary setting. Today's cost for electricity is $0.097kWh; this will escalate by 2% per year. Using 6% interest over 20 years, evaluate the proposal to eliminate the primary settling tank.

8.23 PUMPING COST TO LIFT WATER

Lifting 4,000 m^3/d of water 1 m costs $280 per year with a pump having 55% wire-to-water efficiency. What capital investment in a more efficient pump is equivalent to saving 1-m of head over 20 years at 6.5% interest?

8.24 PUMP SELECTION

For the data in Table P8.24 is pump A or pump B the better investment? The answer depends on the discount rate that is used to make the comparison. Evaluate using 4%, 6%, and 8%.

TABLE P8.24 Comparison of Two Pumps

Cost Component	Pump A	Pump B
Capital cost ($)	30,000	55,000
Service life (y)	5	10
Operation and maintenance ($/y)	1,200	1,000

8.25 PUMP COMPARISON AT DIFFERENT INTEREST RATES

Problem 8.24 showed that pump A was favored when the interest rate was 8%. Evaluate the pumps at interest rates above and below 8% to determine where pump B will be favored. Why does the pump selection change when the interest rate is changed?

8.26 OIL STORAGE TANK MAINTENANCE

An oil storage tank costs $66,000. Painting will cost $4,000 every 5 years and inside cleaning is $6,000 every 10 years. Calculate the present value, assuming a 30-year life and 8% interest.

8.27 TIMED PAYMENT FOR CONSTRUCTION

The contracted cost to build a storage tank is $1 million. Construction will take two years, and payment will be in three parts: $500,000 at the start of construction; $250,000 plus 8% interest accrued at the end of year 1, and a final payment of $250,000 plus accrued interest for two years. What is the amount of each payment? What is the total cost?

8.28 AIR COMPRESSOR MAINTENANCE

An air compressor costs $110,000. Oil is changed every six months at a cost of $4,000. An annual minor overhaul is $3,500, a major overhaul every 5 years is $30,000. The estimated compressor life is 10 years. Calculate the present value of the maintenance costs using 8% and six-month periods.

8.29 BID EVALUATION

Two companies have tendered bids that differ mainly in their terms for payment. Interest at 10% per year is charged on the unpaid balance starting from the date of purchase. Which company should be selected, assuming both provide equipment that meets the technical demands of the project? The terms are as follows:

Company A	Purchase price = \$150,000 Delivery in 11 months 20% down payment = \$30,000 Balance due 30 days after delivery = \$120,000
Company B	Purchase price = \$140,000 Delivery in 12 months 20% down payment = \$28,000 30% due six months after delivery = \$42,000 25% due one year after delivery = \$35,000 25% due three years after delivery = \$35,000

8.30 WATER STORAGE TANKS

An industry can provide water storage in an elevated steel tank adjacent to the plant or in a concrete standpipe located some distance away. The elevated tank will cost \$182,000; operating and maintenance costs will be 4% of the capital cost per year. The standpipe will cost \$204,000 plus 3% of the capital cost per year for O&M. Use a 10% rate of return and a 30-year project life and compare the two alternatives.

8.31 LOAN REPAYMENT SCHEDULE

Loans are usually repaid in equal annual payments, but other schedules can be negotiated. Three possible schemes to repay a loan of \$2,000,000 are offered. Schedule A is equal annual payments of \$234,920. Schedule 2 has an initial payment of \$300,000, and the payment decreases by \$8,000 per year. Schedule 3 has an initial payment of \$148,000 and the annual payment increases by \$8,000 each year. The total payments for the three schemes are \$4,698,400, \$4,480,000, and \$ 4,480,000. The interest rate is 10%. Which payment schedule do you recommend?

8.32 COMPARING FILTRATION SYSTEMS

Three filter installations are nearly equivalent for water purity. Using the data in Table P8.32 identify the lowest cost alternative based on the annual cost. Assume a discount rate of 8.1% and a 20-year project life.

TABLE P8.32 Filtration System Alternatives

	Rapid Sand Filters	Upflow Filters	Microstrainers
Capital cost (\$)	775,000	816,000	760,000
Electricity (\$/y)	17,410	18,140	15,460

8.33 CHEESE COMPANY FINED

A large cheese company paid California water regulators a \$4,000,000 fine for polluting water for years by illegally flushing wastewater from its plant. This was the largest fine ever paid to the Central Valley Regional Water Quality Control Board. The board had sought a \$4-million fine after regulators said the company had violated state water laws for nearly 16 years. If the cheese maker had invested X dollars in a pollution control facility 16 years ago, and if the annual cost of operation was 10% of the construction costs, what value of X would give a 16-year cost equivalent to $4 million. Use $i = 6\%$.

8.34 INFLOW/INFILTRATION REDUCTION

A city has determined that infiltration/inflow (I/I) can be reduced by 60% by rehabilitating 60 miles of sewers at an average cost of $125,000/mile. The current cost of wastewater treatment is $950/million gallons, and the future cost of treatment will escalate by 2%/y. The sewer repair will take three years to complete.

Table P8.34 gives the average monthly wastewater flows (mgd). These have been stable for a few years and are expected to be representative for the next 5 years. The rapid increase during the winter months is mainly due to increased infiltration/inflow. The wet months of December–April have excessive inflow/infiltration. The dry months of May–November are used to estimate the baseline dry weather flow (no excessive I/I). How many years will it take to pay back the cost of the sewer repair program?

TABLE P8.34 Average Monthly Wastewater Flows (mgd)

Jan	Feb	Mar	Apr	May	June	July	Aug	Sept	Oct	Nov	Dec
77.4	86.6	73.7	56.0	38.6	28.5	25.1	23.5	24.6	29.3	34.9	40.7

8.35 PUMPING STATION POWER SUPPLY

Table P8.35 gives data for two possible routes for a power transmission line to serve a remote pumping station. Compare the two options using an 8% discount rate.

TABLE P8.35 Power Transmission Line Data

Route	A	B
Length (mile)	15	5
Initial cost ($/mile)	5,000	25,000
Annual maintenance ($/mile-y)	200	400
Useful life (y)	15	15
Salvage value ($/mile)	3,000	5,000
Annual cost of power loss ($/mile)	500	500

8.36 CONVEYOR SYSTEM

Table P8.36 gives data for two types of conveyor systems that are being considered by a company that processes recycled solid waste. Assume a discount rate of 20% and compare the two systems.

TABLE P8.36 Conveyor System Design Data

	System A	System B
Initial investment ($)	10,000	24,000
Net salvage value ($)	2,000	0
Annual O&M cost ($/y)	1,800	1,200
Useful life (y)	10	15

8.37 BATTERY ACID TREATMENT

A battery manufacturer has been ordered to cease discharging mercury-laden acidic wastes to the city sewer. Three firms have provided quotes on the necessary pollution control equipment. The installed costs are quotations; the other costs are estimates, as given in Table P8.37. (A quotation is fixed by contract and will not change; an estimate will change as more information becomes available.) Evaluate using the annual cost for $i = 12\%$ and a useful life of 12 years.

TABLE P8.37 Three Systems for Battery Acid Cleanup

System	Installed Cost ($)	Annual O&M ($/y)	Annual Hg Sales ($/y)
A	35,000	1,400	2,000
B	40,000	1,200	2,200
C	80,000	1,600	3,600

8.38 SEWER SYSTEM FINANCING

The EPA is examining a proposal for a new wastewater treatment system. The system has an initial cost of $4,000,000, a life of 30 years, and zero expected salvage value. Expected annual operating costs are $200,000. Annual benefits to users are $780,000. The EPA has established a minimal acceptable rate of return of 8%. Determine the internal rate of return of this investment.

8.39 SEWER DESIGNS

Three sewer designs in Table P8.39 are being considered for a project that has a service life of 60 years. Option 1 installs reinforced concrete pipe (RCP) with a design capacity to serve for 60 years. Options 2 and 3 install smaller RCP and replace it with a larger pipe after 20 or 30 years. Replacement pipe costs include the removal of the older pipe. Assume that maintenance cost will be the same for all pipe sizes. The discount rate recommended by the Water Resources Council is 8%. This rate refers to current dollars. Which option is most economical?

TABLE P8.39 Sewer Design Options

Type	First Cost ($)	Replacement Year	Replacement Cost ($)
Option 1	800,000	60	–
Option 2	450,000	30	810,000
Option 3	400,000	20	600,000
		40	830,000

8.40 ION EXCHANGE WATER SOFTENING

The Apex softener company sold a water softener to a customer for a price $35,000. The customer wanted a deferred payment plan, and these two options were arranged:

(1) Full payment after 2 years, including 6% interest compounded quarterly
(2) 20% down payment with the balance paid in two equal payments at the end of years 1 and 2, payments to include interest at 8%, compounded quarterly

Which option is the most economical?

8.41 WATER SUPPLY RESERVOIR CAPACITY

A town will construct a water supply reservoir. A full capacity reservoir that costs $22,400,000 will serve for 30 years. An alternative plan is to build the reservoir in stages, $14,200,000 now and a second stage for $12,600,000 after 15 years. The cost estimates are in Table P8.41. Use a discount rate of 8% to compare the alternatives. Estimated costs are as follows:

TABLE P8.41 Reservoir Design Options

Alternative	Construction Cost ($)	Annual Maintenance ($/y)
Build full capacity now	22,400,000	100,000
Build in two stages		
Stage 1	14,200,000	75,000
Stage 2	15,600,000	100,000

8.42 OIL REFINERY WASTE DISPOSAL

An oil refinery must process its waste liquids before discharging to a stream. The estimated in-house capital cost is $450,000, with an annual O&M cost of $50,000/y. The equipment would have a useful life of 10 years. A contractor has offered to process the waste liquids for 10 years at a fixed price of $150,000/y, payable at the end of each year. The refinery manager considers 18% a suitable discount rate. Should he accept the proposal?

8.43 ALTERNATIVE SEWER DESIGN

The conventional sewer system design is to have gravity flow in sewers wherever possible and to provide pumping lift stations as needed at strategic locations. The alternative design is to use smaller pipes, with grinder pumps to prevent clogging, and to rely more on pumping and less on gravity flow. This allows a smaller diameter sewer pipe to be installed at shallower depths. The cost estimates are in Table P8.43. Compare the total annual costs for 25 years at 5% discount rate.

TABLE P8.43 Sewer Design Alternatives

	Conventional Design	Alternative Design
Capital cost ($)	6,830,000	5,300,000
Annual O&M cost ($/y)	100,000	178,000

8.44 WASTE PAPER DISPOSAL

A company pays a trucker $2,000 a month to haul waste paper and cardboard to the city dump. The material could be recycled by buying a $60,000 hydraulic press baler. The baler has an estimated useful life of 12 years. Annual operating costs will be $25/bale for the estimated 500 bales per year that would be produced. A waste paper company will pick up the bales at the plant and pay $2.40 per bale. If interest is 8%, would you recommend installing the baler?

8.45 GEAR BOX REPLACEMENT

The gearbox on an industrial mixer experienced severe gear tooth pitting in the first year of service and failures occurred in each subsequent year, costing thousands of dollars in production loss and

TABLE P8.45 Life Cycle Cost of the Original Equipment (Geitner & Galster 2000)

Year	Cost Category	Cost ($)
0	Capital	160,000
1	Maintenance	26,887
2	Maintenance	64,867
3	Maintenance	53,930

pollution control costs. The original gearbox cost $160,000. The maintenance costs for the first three years are given in Table P8.45. Use the average O&M cost for years 1, 2, and 3, which is $48,561/y, to project O&M costs for years 4–10.

A new gearbox was specified with wider gears and a bigger gearbox. As a result, the motor had to be moved several inches. The cost of the new gearbox and motor relocation was $294,333. The new gearbox ran five years with minor maintenance costs ($500/y) before needing major maintenance costing $25,000 in year 5. Maintenance costs in years 6–10 were $1,000/y. Compare the projected cost performance of the original and new gearboxes for a 10-year period.

8.46 TOXIC SOLVENT HANDLING

A company is currently incinerating a toxic solvent at a total annual cost of €207,400. Two other options are being considered:

(1) Send the solvent to an external waste management firm for recycling, with an annual cost of €149,800.
(2) Purchase an improved incineration system for €76,000, plus €50,600 in auxiliary equipment, and €43,000 for the installation. Start-up would be immediately after purchase. The system must be renewed at the end of year 5, at a cost escalation of 20%. The annual operating cost will be €104,000 per year. Compare annual costs of the options using $i = 10\%$.

8.47 SOLVENT RECOVERY ECONOMICS

An industry currently spends $20,000/y for solvent and $15,000/y for solvent disposal. They expect that these costs will increase by 5% per year.

As an option, they are considering a solvent recovery/recycle system that will cost $80,000 to purchase and install. This will reduce the cost of purchased solvent to $1,800/y, and the cost of solvent disposal to $1,200/y. The annual O&M cost is $12,000/y. The firm will borrow the capital cost and make equal payments for three years at 12%/y interest. Assume a ten-year life of the equipment. There is no salvage value at the end of ten years. The company's internal discount rate is 15%. Make the financial calculation to determine whether the company should buy the solvent recovery system.

8.48 NET PRESENT VALUE

An initial investment of $10,000 will bring annual cash flows of $4,000 at the end of years 1 and 2, $2,500 in year 3, and $2,000 in years 4 and 5. What is the net present value of the project, calculated at a 15% discount rate?

8.49 NEUTRALIZATION OF ACID WASTES

The Shanahan Valley Chemicals neutralization plant can treat 22.4 m^3/h of waste sulfuric acid using either lime (CaO) and caustic (NaOH) at the costs given in Table P8.49. Assume the annual

TABLE P8.49 Neutralization Options

Cost Factor	NaOH	CaO
Construction ($)	110,000	150,000
Annual Chemical Costs ($/y)	98,000	69,600
Other O&M costs ($/y)	4,000	6,000
Sludge Disposal Cost ($/y)	0	20,000

costs are paid at the end of the year. Evaluate the two treatment processes for a ten-year life and 10% interest. Compare the annual costs.

8.50 ACTIVATED CARBON ADSORPTION COST

A petrochemical wastewater flow of 96 m^3/h contains 1 mg/L of BTEX compounds, with a benzene concentration of about 40 ppb. This is treated by adsorption in an activated carbon (AC) column that contains 6,800 kg of activated carbon. The activated carbon (AC) usage rate is 0.04 lb AC/m^3 treated. The cost estimates are in Table P8.50. The system will operate 6,000 h/y. (a) Calculate the annual carbon usage rate (lb AC/y). (b) Complete the annual cost estimate using a 10-year life and 8% interest).

TABLE P8.50 Activated Carbon Cost Estimates

Capital Cost Item	Cost ($)
Carbon adsorption equipment	165,000
Installation	15,000
Site work and foundation	20,000
Pump upgrades and miscellaneous work	25,000
Total capital cost =	225,000
Annual cost items ($/y)	
Maintenance (8% of capital cost)	
Electrical energy (62,500 kWh @ $0.08/kWh)	
Carbon exchange ($2/kg)	
Freight ($2,500 per exchange; 3 exchanges/y)	

8.51 BOILER COST EVALUATIONS

The three quotations in Table P8.51 are for a 3,000 kW steam boiler that will burn No. 2 oil. Operation will be 4,000 h/y. One gallon of No. 2 oil has a fuel value of 10.7 kWh/L and costs $0.80/L. Evaluate the proposals on the present value (i = 6% for n = 20 years).

TABLE P8.51 Quotations for Price and Guaranteed Efficiency

Vendor	Quoted Price (Includes Freight)	Guaranteed Efficiency (at 100% Load)
Vendor A	$1,000,000	86%
Vendor B	$910,000	84%
Vendor C	$860,000	82%

8.52 GRAVITY BELT THICKENER

The annual cost of maintenance for gravity belt thickeners (GBT) is \$23,500/y for labor plus \$56,000 every two years to replace the belt. Calculate the present value for operating the gravity belt filter at 5% interest over 15 years.

8.53 FLUE GAS DESULFURIZATION

A 600-MWe (megawatt electric) high-sulfur bituminous coal-fired boiler emits 50,000 T/y of sulfur dioxide (SO_2). A flue gas scrubber (FGS) will be installed to reduce the sulfur emissions to less than 1,000 T/y (98% removal efficiency). Either a wet limestone scrubber (WLS) or a wet buffered lime scrubber (WBLS) will achieve this removal rate. Gypsum ($CaSO_4$) is a salable by-product of flue gas desulfurization. Table P8.53 gives the capital and annual costs associated with each of the alternative devices. Compare the two systems for a lifetime of 20 years using an interest rate of 7%.

TABLE P8.53 Flue Gas Desulfurization Costs for Dry and Wet Scrubbers (million dollars)

Cost Component	WLS	WBLS
Capital cost (\$)	180	200
Annual O&M costs		
Fixed O&M costs[a] (\$/y)	1.80	2,0
Reagent (\$/y)	3.75	1.2
Auxiliary power (\$/y)	1.15	1.3
Parasitic power[b] (\$/y)	0.375	0.95
Annual gypsum sales (\$/y)	0.600	1.2

(a) Estimated at 1% of capital cost.

(b) In many systems, a pollution control device causes a loss in productive capacity by creating obstructions in the flue, temperature losses, or other physical changes that affect performance. These losses are collectively termed "parasitic power" losses.

8.54 COMPOST SYSTEM

A turned-windrow compost system with a runoff collection pond will be constructed. The runoff collection pond will require 17,000 yd^3 of excavation at a cost of \$2/yd^3. The pad area includes space for raw materials storage, active composting, curing, and storage of cured compost. The purpose of the composting pad is to provide a surface that can withstand the abuse of the front-end loader. The cost of \$0.07/ft^2 of pad area includes removing topsoil, purchasing the material, and building the pad. The pad dimensions are given in Table P8.54. Calculate (a) the construction cost of the project, and (b) the annualized cost for 20 years at an interest rate of 10%.

TABLE P8.54 Composting Pad Dimensions

Process Component	Width (ft)	Length (ft)	Area (ft^2)
Raw materials storage	1,000	80	80,000
Curing storage area	365	130	47,450
Compost pad area	800	130	104,000
Total pad area			231,450

8.55 INDUSTRIAL PARK WASTEWATER TREATMENT

The largest industrial park in Nevada has an area of 8,000 ha, with an expected buildout area of 1,100 ha. The average sanitary sewage flow is estimated as 2,300 m^3/d, with a peak day flow of 4,600 m^3/d. Influent wastewater contaminant levels will be similar to municipal wastewater. Treated effluent contaminant levels must not exceed 30 mg/L of BOD_5, 30 mg/L of TSS, and 10 mg/L of TKN.

Tables P8.55a and P8.55b give the construction and O&M costs for a sequencing batch reactor (SBR) plant, which was selected as the most cost-effective of several biological wastewater treatment options. The estimated construction cost is $10,840,000, with an estimated annual operating cost of $255,000. This estimate is based on the traditional design-bid-construct delivery method. Assuming a 20-year lifetime and a 6.2% discount rate,

(a) What is the annualized cost ($/y)?
(b) What is the present value of the project?

TABLE P8.55a Cost Estimates for the SBR Treatment Plant

Item	Description		Estimated Cost ($)
1	Pump station with submersible pumps		344,000
2	Headworks building with screen and grit removal		772,000
3	Odor control		37,500
4	SBR (includes post-aeration tank)		1,548,500
5	Two aerobic sludge digesters with fine bubble aeration		523,000
6	Three blowers, 1,000 cfm with building		236,000
7	UV disinfection		273,000
8	Mechanical sludge dewatering and building a storage area		664,000
9	Chemical storage and feed equipment		30,000
10	Sludge transfer pump		14,000
11	Non-potable water system		85,000
12	New maintenance/administration building		485,000
13	Influent/effluent automatic samplers		25,000
14	Standby power generator to WWTP		150,000
15	Electrical		750,000
16	Piping		545,000
17	Rapid infiltration basin		672,000
18	Site work and fencing		400,000
19	SCADA system		300,000
		Subtotal	7,854,000
	Contingency (20%)		1,570,800
		Subtotal construction cost	9,424,800
	Engineering services (10%)		942,480
	Construction management (5%)		471,240
		Subtotal engineering & CM	1,413,720
		Total	10,838,520

TABLE P8.55b Summary Annual Operational Costs for the SBR Treatment Plant

Annual Operating Cost Item	Cost ($)
Common equipment power cost (920k kWh/y @ $0.10/kWh)	92,000
Power for aeration and pumping (270k kWh/y @ $0.10/kWh)	27,000
Sludge disposal (600 wet tons/y @ $60/wet ton)	36,000
Labor (one full-time operator)	100,000
Estimated annual operating costs =	255,000

Common equipment power cost includes headworks equipment (influent pumps, grit removal pumps, miscellaneous trash/debris handling rakes/augers), blowers for aerobic digesters and effluent post-aeration basin, waste-activated sludge pumps, UV disinfection panels, and miscellaneous, administration building, operational costs including all instrumentation, and control process–related equipment.

Return activated sludge pump power costs are included in the SBR power.

8.56 MICROTURBINE ELECTRIC GENERATION FROM METHANE I

Calculate the present value of electricity produced in 10 years from a microturbine generator that will burn biogas from anaerobic sludge digesters that treat 18,000 lb/d of volatile solids (VS). The digestion process will convert 50% of the volatile solids to gas at a conversion of 15 ft^3 of gas per lb of VS destroyed. Each microturbine will generate 30 kW at a digester gas feed rate of 13.13 ft^3/min. Assume a reasonable price for electricity and interest rate based on your location.

8.57 PURCHASE NITROGEN CREDITS

A watershed nutrient control plan requires an existing treatment plant to reduce the effluent nitrogen concentration from 30 to 6 mg/L. The existing plant has been paid off. The annual maintenance cost is $333,000/y. Adding nitrogen removal capability for a design flow of 4,000 m^3/d will cost $6,500,000 to build and an additional $175,000/y to operate. The upgrade can be postponed by buying nitrogen credits; that is, to pay someone to remove the nitrogen elsewhere. The concept is that money can be saved by continuing to use an otherwise good facility while postponing the capital expense of the new facility by purchasing credits. Credits can be purchased directly from neighboring treatment facilities or from local farmers who can reduce nitrogen inputs to streams. Use a discount rate of 4.125% and assume a 20-year lifetime for the upgraded facility to determine the break-even price for buying nitrogen credits.

8.58 PIPE REPAIR PROGRAM

A city pumps 21,200 m^3/d (5.47 mgd) from the water treatment plant, but bills its customers for only 15,100 m^3/d. The difference of 6,100 m^3/d, 29% of the water pumped, is lost due to leakage in the water distribution system. Approximately 80% of US cities have a leakage rate lower than this. More than half of the cities surveyed have a leakage rate less than 15%; 38% of cities surveyed lost less than 10%. Clearly, the system can be improved. The variable cost of producing and delivering water is $0.32/$m^3$, so the utility is paying $1,760/d for water that is not delivered and not billed. Reducing the leakage by 50% will save $880/d.

Table P8.58 gives the total length and the estimated volume lost per day per kilometer of pipe (m^3/d-km). The average loss rate for the system is 21.86 m^3/d-km. A survey is underway to produce

TABLE P8.58 Leakage in a Water Distrbution System

Average Loss Rate (m^3/d-km)	Length (km)	Leakage (m^3/d)
8	12	96
10	15	150
12	23	276
14	50	700
16	22	352
18	34	612
20	14	280
22	21	462
24	28	672
26	16	416
28	15	420
30	22	660
35	15	525
40	12	480
Total	299	6,101

more detailed information, to include pipe age, kind of pipe, and pipe diameter, but this will not be finished for one year. In the meantime, a preliminary budget is needed. This requires establishing a policy for pipe repair and replacement.

The leak detection survey will identify sections of high leakage rates and that the goals can be met by repairing some fraction of the installed pipe. For planning purposes, the estimated cost for repairs is $28,000/km of installed pipe, averaged over all pipe sizes and working conditions. Because not all pipes will be repaired, it will be assumed that the leakage rate after repairs will be 10 m^3/d-km.

Four criteria have been proposed to establish a repair and replacement program:

(1) Repair all pipes that are leaking at more than 12 m^3/d-km
(2) Repair all pipes that have a payback time of less than 10 years.
(3) Reduce the leakage by 50%.
(4) Repair pipes for which the 10-year present value at 10% interest exceeds the repair cost.

The water utility commission wants a target based on costs and savings. Recommend a policy to the utility commission, including a repair schedule assuming a maximum budget of $300,000 per year.

8.59 PUMPING SYSTEM WITH A PROBLEM CONTROL VALVE

A pump transports a process fluid that contains some solids from a storage tank to a pressurized tank. A heat exchanger heats the fluid, and a fluid control valve (FCV) regulates the rate of flow into the pressurized tank to 80 m^3/h. The control valve currently operates between 15% and 20% open and with considerable cavitation noise from the valve. It appears the valve was not sized properly for the application. It was discovered that the pump was oversized for 110 m^3/h instead of 80 m^3/h. This causes a larger than intended pressure drop across the FVC, and it fails every 10–12 months due to erosion caused by cavitation.

The cost of each repair is €4,000. One proposal is to replace the existing valve with one that can resist cavitation. Make a life cycle cost analysis on these four alternative solutions:

(a) Install a new control valve that can accommodate the high-pressure differential. The cost is €5,000.
(b) Trim the pump impeller to reduce the pressure drop across the current valve. The cost is €2,250.
(c) Install a 30 kV variable speed drive (VSD) to vary the pump speed and achieve the desired process flow. The flow control valve can be removed. The cost is €21,500.
(d) Leave the system as it is and repair the FCV every year.

The cost estimates and operating data for each alternative are in Table P8.59. The process operates at 80 m^3/h for 6,000 h/y. The current energy cost of €0.16/kWh will inflate at 4% per year. Ignore inflation for other annual costs. The project life is 8 years. The discount rate is 8%.

TABLE P8.59 Data for Alternative Solutions to Faulty Fluid Control Valve Problem

Cost Item	Change FCV	Trim Impeller	Install VSD	Repair FCV Annually
Capital cost (€)	50,000	22,500	215,000	None
Routine maintenance (€/y)	1,000	1,000	2,000	1,000
Repair every 2 years (€)	4,000	4,000	4,000	4,000
Other annual costs	0	0	0	5,000
Operating conditions				
Pump head (m)	71.7	42.0	34.5	71.7
Pump/motor efficiency (%)	75.1	72.1	77.0	75.1

8.60 EXPANSION PLAN

A landfill is planned for an ultimate capacity at the end of a 50-year lifetime. The population of the service area has been growing by 3,500/y. Solid waste production has been growing at 3,000 T/y. A new program is planned to divert 30% of the solid waste from the landfill. The growth has been linear and this trend will be used to plan the construction of the project. There is a modest large economy of scale in the early part of the project, which is acquiring the land and providing the basic infrastructure (roads, leachate collection, gas collection, etc.). The cost is $C = KQ^{0.75}$, where Q can be measured as mass or volume of solid waste delivered to the landfill (K will be different for mass and volume). There is no economy of scale for the cost of operating the landfill, which includes opening cells, placing the waste, closing the cells, installing gas collection wells, etc. (a) Should the project be developed in stages? If construction is done in two stages, inflate the cost of the second stage by 1% per year. Consider plans based on 3%, 4%, and 5% interest rates.

9 Depreciation and Asset Valuation

Depreciation is an allowance for the decrease in value of a property over a period of time due to wear and tear, deterioration, and normal obsolescence. The intent is to distribute the capital cost of an asset over the period of time that begins when the asset is put into operation and ends when it is retired from service or when the cost of the asset is fully recovered, whichever comes first.

The depreciable life and the actual life of an asset may be different, and the actual value may be more or less than the depreciated value. The actual life is determined by the asset's capability to function as and when needed. Age is a factor, but the operating environment (e.g., soil condition for sewers and underground piping), routine maintenance, pro-active maintenance, and reactive maintenance will affect both condition and failure probability.

Asset valuation has two sides, one is financial and one relies on engineering expertise and experience. Most of this chapter is about the financial aspects, mainly depreciation to establish a book value for assets and tax depreciation. The final section is an introduction to the evaluation of asset conditions, failure risks, and projecting replacement and repair costs.

9.1 DEPRECIATION

Non-profit organizations, which include water, wastewater, and municipal solid waste collection and disposal utilities, depreciate tangible property because they are required to maintain a *valuation* of all assets (i.e., buildings, sewers, pumps, trucks, etc.). One valuation is the *book value*, which is the original cost minus accrued depreciation. Another is *replacement cost*, which is the cost to reproduce or replace an asset today to match the functionality of the asset when it was new.

Book depreciation is used by an organization for internal financial accounting to track the value of an asset or a property over its life. Book depreciation can be calculated using any method. Book depreciation is used to report net income and make pricing decisions.

Public utilities can be purchased by non-public companies and book value will be one consideration in setting a fair purchase price. *Fair market value* is the price of an asset that would change hands between a willing buyer and a willing seller, both fully aware of the facts, and with no compulsion to buy or sell by a specific date.

Depreciation and taxes are irrevocably linked for business because depreciation is an accounting tool to reduce taxable income. Explaining taxes is beyond our scope, so straight-line depreciation will be used to explain the basic concepts. The Modified Accelerated Cost Recovery System (MACRS) depreciation will be discussed briefly. Accelerated depreciation gives greater deductions in the early life of an asset.

Tax depreciation is used by a corporation or business to determine taxes due based on current tax laws of the government entity (country, state, province, etc.). Even though depreciation itself is not a cash flow, it can result in actual cash flow changes because of the amount of tax depreciation. For US tax purposes, depreciation must be calculated using MACRS.

9.2 PUBLIC WORKS ASSET DEPRECIATION

Public utilities are required to value their assets and to have a plan for asset replacement and repair. The wear-out or replacement rates of equipment and infrastructure may differ due to variations in construction, environment, or the physical and chemical characteristics of the wastewater. Faster depreciation rates should not be used if they are the result of imprudence, poor maintenance, or negligence by the wastewater company. Tables 9.1 and 9.2 give the useful life of assets that are used for depreciation.

TABLE 9.1 Average Useful Life Used for Depreciation of Wastewater Treatment Facilities

Depreciable Asset	Average Life (y)	Depreciable Asset	Average Life (y)
Structures and improvements	30	Treatment and disposal equipment	20
Power generation equipment	20	Plant sewers	20
Collection – force mains	50	Outfall sewer lines	30
Collection – gravity sewers	50	Other plant and miscellaneous equipment	15
Special collecting structures	50	Office furniture and equipment	15
Services to customers	50	Computers and software	5
Flow-measuring devices	10	Transportation equipment	5
Flow-measuring installations	10	Stores equipment	25
Reuse services	50	Tools, shop, and garage equipment	20
Receiving wells	30	Laboratory equipment	10
Pumping equipment	20	Power-operated equipment	20
Reuse distribution reservoirs	40	Communication equipment	10
Reuse transmission & distribution systems	40	Miscellaneous equipment	10

TABLE 9.2 Typical Useful Life for Selected Infrastructure Assets

Facility	Life (y)	Facility	Life (y)
Roads		**Wastewater**	
Pavement substructure	50–100	Gravity sewer lines	80–100
Wearing surfaces	10–20	Maintenance holes	20–50
Curb and gutter	50–80	Pumping station – structures	50
Footpaths	15–50	Pumping station – electrical	15
Bridges	30–80	Risers	25
Culverts	50–80	Treatment plant – structures	50
Bus shelters	20	Treatment plant – electrical	15–25
Bike paths	50	**Water Supply**	
Street lighting	20	Storage tanks	50–80
Traffic signals	10	Treatment plant – structures	60–70
Drainage		Treatment plant – electrical	15–25
Drains (underground)	50–80	Water lines	65–95
Culverts	50–80	Pumping station – structures	60–70
Maintenance holes	20–50	Pumping station – electrical	25
Detention basins	50–100	**Solid Waste Facilities**	
Pumping station structures	50	Landfills (depends on fill rate)	–
Pumping station electrical	25	Transfer stations	20
		Garbage collection vehicles	6

9.3 STRAIGHT-LINE DEPRECIATION

Straight-line depreciation distributes the cost of an asset in equal annual charges over its expected useful life of n years. The annual depreciation is

$$D = \frac{I - S}{n}$$

where
D = annual depreciation charge ($/y)
I = investment ($)
S = salvage value ($)
n = number of years

If the asset is put into service mid-year, the depreciation for that year is half the annual depreciation charge. That half-year of depreciation is made up at the end of the asset life. For example, a five-year asset with no salvage value is depreciated by 20% per year. Using the *half-year convention*, the depreciation would be 10%, 20%, 20%, 20%, 20%, and 10%, giving a total of 100% of the asset value. Under the half-year convention, a property placed in service is considered to be in use one-half year regardless of when during the year the property was placed in service.

Example 9.1 Straight-Line Depreciation

A $240,000 piece of equipment will be depreciated over seven years. The salvage value is $30,000. The total amount that can be depreciated is $240,000 – $30,000 = $210,000. The annual depreciation is

$$D = \frac{\$240{,}000 - \$30{,}000}{7\ \text{y}} = \$30{,}000/\text{y}$$

9.4 BOOK VALUE OF ASSETS

Book value is the original investment minus the accumulated depreciation. The market value of a piece of equipment is likely to be lower than the book value because new technology is better and often cheaper. The depreciated cost (book value) is not expected to equal the replacement cost, but it is a useful accounting tool.

Example 9.2 Depreciation and Book Values of Assets

The book values of a sampling of assets shown in Table 9.3 are calculated by subtracting the accrued depreciation (accumulated depreciation) from the cost when new.

TABLE 9.3 Depreciated Values for Property and Equipment

Asset	Value When New ($)	Accrued Depreciation ($)	Book Value ($)
Equipment	2,000,000	1,300,000	700,000
Vehicles	200,000	160,000	40,000
Office equipment	175,000	75,000	100,000
Land	500,000	0	500,000
Total	2,875,000	$535,000	1,340,000

9.5 ASSET EVALUATION

The regulatory book value of an asset has little to do with its economic value. *Valuation* is an analysis to determine the monetary value of an asset (or liability). *Evaluation* looks at the functionality and operational characteristics of an item but does not provide a monetary value. A utility needs to know what assets it owns, where they are located (GPS mapping is used for this), their condition (remaining useful life), initial costs, current value, and replacement costs. This is essential for planning major rehabilitation and replacement expenses. The information also supports applications for loans and grants (Chapter 10). Figure 9.1 shows the steps in the evaluation process.

Figure 9.2 indicates a general relation between asset condition and the probability of failure. Asset condition evaluation is designed to detect deterioration in physical condition or performance and prevent failures or expensive repairs.

Replacement cost is the amount of money required to replace an existing asset with an equally valued or similar asset at the current market price.

Table 9.4 is one way of classifying assets according to their current condition and by their remaining useful life. Some systems use numerical categories.

This concept can be expanded to assign a grade according to the expected occurrence and severity of the failure.

$$\text{Criticality} = (\text{Occurrence})(\text{Severity})(\text{Detectability})$$

- Occurrence (frequency) is an estimate of how frequently a failure may occur. This can be scored on a scale from 1 to 5, with 1 being low risk of failure and 5 being recurrent or certainty of failure.
- Severity (or consequence) is an assessment of damage or inconvenience with respect to equipment, process, or consumer.
- Detectability scores how easy or difficult it is to detect the symptoms of impending failure. This can range from a high detection probability which is good to a low detection probability, or failure without a warning.

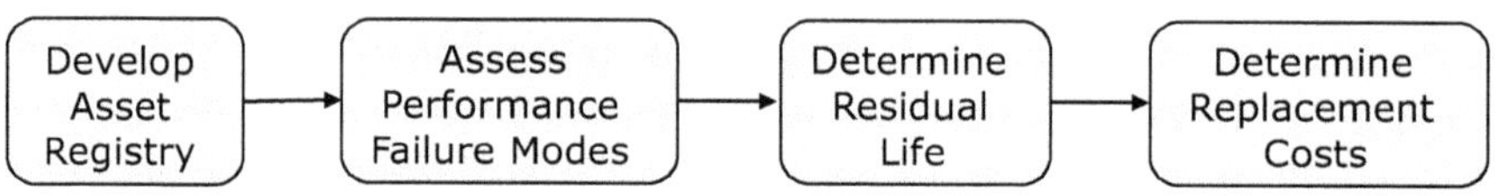

FIGURE 9.1 Steps in asset evaluation.

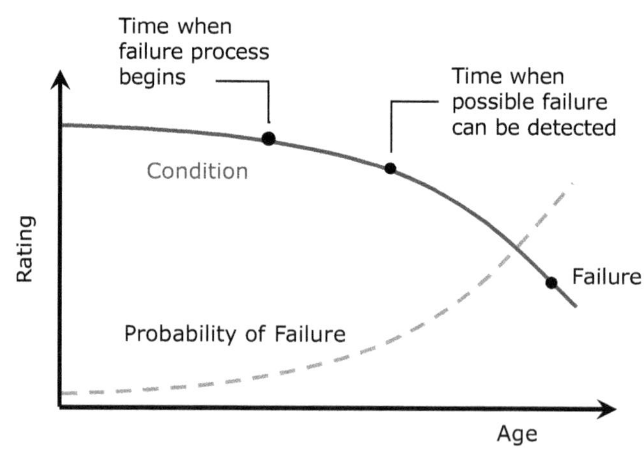

FIGURE 9.2 Probability of failure is related to asset condition.

TABLE 9.4 Evaluation of Assets According to Remaining Lifetime

Condition Grading	By Condition	Remaining Useful Life
Very Good	New or nearly new; operable and well maintained. Only planned maintenance required	>90%
Good	Superficial wear and tear. Minor maintenance required plus planned maintenance r	70–90%
Fair	Asset showing sign of deterioration. Significant maintenance required	30–50%
Poor	May fail in the next 1–2 years. Significant renewal/rehabilitation required	10–30 %
Very poor	Physically unsound. Beyond rehabilitation	<10%

These categories are subjective so experience and expert knowledge of the system are required to execute this program.

Example 9.3 Sewer Asset Condition Assessment

The sewer system of a city covers 20 square miles and provides service to 12,000 sewer connections serving a population of approximately 46,000. The sewer system comprises more than 17,500 assets. Sewers are the largest category in terms of cost, but lift stations and other appurtenances have a much larger number of assets. This is a relatively young sewer system (90% of the sewers have more than 50% of their useful life left), so most ratings are good to very good, as shown in Table 9.5. The replacement cost for the poor and very poor sewers is $5,498,000. The percentages for asset condition are calculated from the replacement costs.

Example 9.4 Criticality of Risk

Criticality depends on the probability of failure and the severity of the consequences of failure. A chlorine pump delivers a solution of hypochlorite for effluent disinfection.

Failure is likely. The pump has failed many times in the past ten years due to several factors. The system has spare parts and a spare pump. A failure may have major consequences because there is no continuous monitoring of the chlorine level in the effluent. This would mean failure to meet the effluent limit for coliform bacteria. The risk is mitigated by having spare parts and a spare pump, but the chlorine level could be low for several days before the problem is discovered.

TABLE 9.5 Assessment of Sewer System Condition and Replacement Cost (2015 $1,000)

		Asset Condition (%)				
Asset	2015 Replacement Value ($1,000)	Very Good	Good	Fair	Poor	Very Poor
Sewers	274,887	65	32	1	1	1
Lift stations	10,616	64	36	0	0	0
Other	5,934	0	40	60	0	0
Total system	291,437	63.7	32.3	2.2	0.9	0.9
Asset		Replacement cost ($1,000)				
Sewers	274,887	178,677	87,964	2,749	2,749	2,749
Lift stations	10,616	6,794	3,822	–	–	–
Other	5,934		2,374	3,560		
Total system	291,437	185,471	94,159	6,309	2,749	2,749

TABLE 9.6 Ten-Year Renewal Plan with Project Costs

Asset	Remaining Life (y)	Planned Renewal Year	Renewal Cost ($)
Pump station 1	0	2013	50,000
Pump station 2	1	2014	23,400
Treatment process	1	2014	500,000
Pump station 3	2	2015	29,400
Manholes	3	2016	16,200
Treatment plant	4	2017	16,000,000
Treatment plant	5	2018	340,000
Manholes	5	2018	314,900
Manholes	6	2019	67,800
Electrical building	7	2020	21,500
Manholes	7	2020	151,000
Manholes	8	2021	10,900
Manholes	9	2022	65,300
Pump instruments	9	2022	13,800
Total			17,604,200

Example 9.5 Ten-Year Replacement Plan

Table 9.6 shows a ten-year plan that evaluates the remaining useful life of wastewater system assets. The projected one-year total cost is $17,604,200. This will be the basis for budgets and financing plans.

9.6 ACCELERATED DEPRECIATION

The US tax code requires companies to use the *Modified Accelerated Cost Recovery System* (MACRS). The code specifies the economic life for all types of assets (Table 9.7) and gives tables of factors to calculate the annual depreciation.

Table 9.7 gives the factors for the half-year convention for MACRS. Note that with the half-year convention, MACRS changes to the straight-line method when that method provides an equal or a greater deduction. For a five-year asset, that happens in the fourth year, as shown in Table 9.8.

TABLE 9.7 Description of Assets, Class Life, and MACRS Recovery Period

Class of Asset	Class Life (y)	MACRS Recovery period (y)
Information systems	6	5
Automobiles, taxis	8	5
Light general-purpose trucks	4	3
Industrial steam and electric generation systems	22	15
Construction assets	16	10
Manufacture of pulp and paper	13	7
Manufacture of chemicals and allied products	9.5	5
Manufacture of cement	20	15
Natural gas production plant	14	7
Liquefied natural gas plant	22	15
Municipal wastewater treatment plant	24	15

TABLE 9.8 MACRS Depreciation Factors (%) Switching to Straight-Line Depreciation (Half-Year Convention)

Year	Asset Lifetime (y)			
	3	5	7	10
1	33.33	20.0	14.29	10.0
2	44.45	32.0	24.49	18.0
3	14.81	19.2	17.49	14.4
4	7.41	11.52	12.49	11.52
5		11.52	8.93	9.22
6		5.76	8.92	7.37
7			8.83	6.55
8			4.46	6.56
9				6.55
10				6.55
11				3.28

Publication 946 (2018), Internal Revenue Service, US Department of Treasury, Washington, DC.

There are other tables (e.g., first-quarter and last-quarter start-up) and professional advice is needed so that the correct tables are used. Note that with the half-year convention, a five-year asset is allowed some depreciation the sixth year to catch up. The final entry brings the total to 100%, so 100% of the allowable depreciation is used.

MACRS is based on the declining balance of the asset value.

$$V_e(n) = V_b(n)(1-f)$$

$$V_e(n+1) = V_e(n)(1-f)$$

$$V_e(n+2) = V_e(n+1)(1-f)$$

where

V_e = value of the asset at the end of year 1
V_b = value of the asset at the beginning of year 1
f = depreciation factor

Example 9.6 MACRS Depreciation of a Three-Year Asset

An asset purchased for $100,000 that falls into the three-year MACRS category would be depreciated as shown in Table 9.9.

TABLE 9.9 MACR's Depreciation Rates

Year	MACRS Rate (%)	Depreciation Expense
1	33.33	(0.3333)($100,000) = $33,330
2	44.45	(0.4445)($100,000) = $44,450
3	14.81	(0.1481)($100,000) = $14,810
4	7.41	(0.0741)($100,000) = $7,410
Totals	100	$100,000

TABLE 9.10 Depreciation Calculations

Year	MACRS Depreciation – Tabled Factors		
	Rate (%)	**Calculation**	**Depreciation**
2021	20.00	\$100,000(0.20)	\$20,000
2022	32.00	\$100,000(0.32)	\$32,000
2023	19.20	\$100,000(0.192)	\$19,200
2024	11.52	\$100,000(0.1152)	\$11,520
2025	11.52	\$100,000(0.1152)	\$11,520
2026		\$100,000(0.0576)	\$5,760

TABLE 9.11 Calculation Using the Declining Balance Formula

Year	Balance	Calculation	Depreciation
2021	\$100,000	(\$100,000/5)(2.0)(0.5)	\$20,000
2022	\$100,000 – \$20,000 = \$80,000	(\$80,0000/5)(2.0)	\$32,000
2023	\$80,000 – \$32,000 = \$48,000	(\$48,000/5)(2.0)	\$19,200
2024	\$48,000 – \$19,200 = \$36,480	(\$28,800/5)(2.0)	\$11,520
2025	\$36,480 – \$11,520 = \$6,910	(Note A)	\$11,520
2026		\$11,520(0.5) (Note B)	\$5,760

Note A: MACRS declining balance changes when the straight-line method provides an equal or a greater deduction. Deduction under 200% declining balance MACRS for 2025 would be \$6,910. This is lower than \$11,520 under the straight-line method over the remaining recovery period.

Note B: Due to half-year convention, depreciation is charged for half-year in 2026.

Example 9.7 MACRS Depreciation of a Five-Year Asset

Table 9.10 shows the depreciation calculations for a five-year asset using the half-year convention. The original value of the asset is \$100,000. There is no salvage value.

Straight-line depreciation would be (\$100,000/5 y) = \$20,000/y. An option under MACRS is the 200% declining balance method. This doubles the first year's deduction to \$40,000. The half-year convention allows half of this, which is \$20,000. The balance to start year 2 is \$100,000 – \$20,000 = \$80,000. Apply the same formula but for a full year.

MACRS changes when the straight-line method provides an equal or a greater deduction.

This happens in year 2025. The MACRS declining balance deduction would be \$6,910. This is lower than depreciation under the straight-line method over the remaining recovery period, which is \$11,520 as shown in Table 9.11.

9.7 DEPRECIATION AND TAXES

This discussion is presented in the context of US Federal tax regulations, but the important ideas are generic.

Corporations pay federal income tax based upon net taxable earnings (gross income less tax deductions). Starting in 2018, the US corporate tax rate is flat 21%.

Example 9.8 Tax Savings Due to Depreciation

A company buys a \$21,000 piece of equipment that earns them \$10,000 per year in gross revenues. The tax law allows the purchase price of the equipment to be depreciated over a seven-year

TABLE 9.12 Equipment Depreciation

Income or Depreciation	With Depreciation	Without Depreciation
Before-tax income ($)	10,000	10,000
Subtract depreciation ($)	–3,000	0
Taxable income ($)	7,000	10,000
Subtract income tax ($) (21% rate)	–1,470	–2,100
After-tax income ($)	8,570	7,900

lifetime. Straight-line depreciation gives an annual deduction of $21,000/7 = $3,000 per year. This reduces taxable income by $3,000 per year, from $10,000 to $7,000. The income tax on $7,000, for a 21% tax rate, is $1,470. The after-tax income is $10,000 – $1,470 = $8,550.

Without depreciation, 21% tax would be paid on $10,000, and the after-tax income would be $10,000 – $2,100 = $7,900.

The depreciation reduces taxes and increases the company's profit by $650, as shown in Table 9.12.

The tax laws allow accelerated depreciation (the depreciation would be more than calculated with the straight-line method), and this would create an even larger reduction in the tax bill and a larger after-tax profit in the early years.

States may not levy corporate taxes, but they may impose other taxes that are similar. State and local taxes are deductible before the calculation of the federal tax. If T_S is the incremental state tax rate and T_F is the incremental federal tax, both expressed as decimals, then the combined incremental rate T_C is

$$T_C = T_S - (1 - T_S)T_F$$

Example 9.9 Combined Tax Rate

If the federal rate is T_F = 21% and the state rate is T_S = 7%, then the combined tax rate is 26.5%.

$$T_c = 0.07 + (1 - 0.07)(0.21) = 0.265$$

9.8 CASE STUDY: COPPER RECOVERY FROM PLATING WASTEWATER

An industry will invest $750,000 in a process to recover up to 200 T/y of copper from waste electrolysis solutions. The price of copper changes over time. The price of copper in 2017 was $5.90/kg (it was 50% higher in 2012). There is no way to predict future prices, so we use a constant $5.90/kg. The amount of recovered copper and the income are given in Table 9.13.

The cash flow for year zero is negative $750,000 to reflect payment of the capital cost. Income, fixed and variable operating costs, and interest paid are given in Table 9.14. The calculations treat these as end-of-year payments. Fixed operating costs are $100,000/y. Variable operating costs are $3.80/kg of recovered copper. Annual interest payments are $60,000. These values are used to calculate the net cash flow (before taxes) for the project that is shown in Table 9.14.

The net cash flow for years 1–5 is income minus operating costs and interest. For year 1, this is

$$\text{Net cash flow (year1)} = \$708,000 - \$100,000 - \$456,000 - \$60,000 = \$92,000$$

Table 9.15 shows how depreciation comes into the calculation. Straight-line depreciation gives an annual depreciation of $750,000/5 y = $150,000/y. This is not a cash flow. No money comes into

TABLE 9.13 Revenues from the Proposed Copper Recovery Schemes (@ $5.90/kg copper)

Year	Copper Recovered (kg/y)	Revenue ($/y)
1	120,000	708,000
2	160,000	944,000
3	200,000	1,180,000
4	200,000	1,180,000
5	200,000	1,180,000

TABLE 9.14 Net Cash Flow (in $) before Taxes

Year	Capital Cost	Income	Operating Costs		Interest	Net Cash Flow before Taxes
			Fixed	Variable		
0	750,000	–	–	–	–	–750,000
1		708,000	100,000	456,000	60,000	92,000
2	–	944,000	100,000	608,000	60,000	176,000
3	–	1,180,000	100,000	760,000	60,000	260,000
4	–	1,180,000	100,000	760,000	60,000	260,000
5	–	1,180,000	100,000	760,000	60,000	260,000

TABLE 9.15 Net Cash Flow (in $) after Taxes

Year	Net Cash Flow before Taxes	Depreciation	Profit	Taxes (21%)	Net Cash Flow after taxes	Net Present Value
0	–750,000	0	0		–750,000	–750,000
1	92,000	–150,000	–58,000	0	92,000	85,185
2	176,000	–150,000	26,000	5,460	170,540	146,211
3	260,000	–150,000	110,000	23,100	236,900	188,059
4	260,000	–150,000	110,000	23,100	236,900	174,129
5	260,000	–150,000	110,000	23,100	236,900	161,230
			Total net present value =			*4,813*

or leaves the company. This is an accounting tool that reduces taxable profits by claiming that a tangible asset has declined in value.

Assume taxes are 21% of taxable profit.

$$\text{Net Profit} = \text{Netcash flow before taxes} - \text{Depreciation allowance}$$

$$\text{Taxes} = 21\% \text{ of positive net profit}$$

$$\text{Net cash flow after taxes} = \text{Net cash flow before taxes} - \text{Taxes}$$

The net present value after taxes, for an 8% discount rate, is $4,813.

9.9 CONCLUSION

Depreciation is an accounting tool that is used to decrease the book value of assets over time as they age and wear out. In municipalities and utilities, this is part of budgeting and planning for asset replacement and major maintenance.

In industry and business, depreciation is an accounting tool for reducing taxes. The depreciation of physical equipment can be deducted from pre-tax profits, so taxes are paid on a smaller amount. Depreciation is not cash flow. It does not move money into or out of the firm.

9.10 PROBLEMS

9.1 STRAIGHT-LINE DEPRECIATION

A $420,000 piece of equipment will be depreciated over eight years. The salvage value is $40,000. Calculate the annual depreciation.

9.2 DEPRECIATION COST OF A MACHINE

The purchase price of a new machine is $75,000. Other costs shown in Table P9.2 bring the total to $81,900. Depreciate the asset on a straight-line basis for eight years, assuming a salvage value of $9,900.

TABLE P9.2 Machine Depreciation Costs

Purchase price	$75,000
Freight	$900
Installation labor	$2,500
Site preparation	$3,500
Cost of machine (cost basis)	$81,900

9.3 TAX SAVINGS DUE TO DEPRECIATION 1

A company buys a $48,000 piece of equipment that earns $10,000 per year in gross revenues. The tax law allows this to be depreciated over a six-year lifetime. Use straight-line depreciation. The tax rate is 21% of taxable income. Calculate, with and without depreciation, the after-tax income.

9.4 STRAIGHT-LINE BOOK DEPRECIATION

An asset that was depreciated over a five-year period by the straight-line method has a book value of $62,000 with a depreciation charge of $26,000/y. Determine the first cost and assumed salvage value.

9.5 DEPRECIATION BY PUBLIC UTILITIES

Public utilities are required to value their assets and have a plan for asset replacement and repair. What are the lifetimes used to depreciate these assets? Flow-measuring devices, laboratory equipment, vehicles, pumping equipment, treatment plant structures.

9.6 SOLAR ENERGY PROPERTY

An industry will install solar energy to heat a treatment process. Solar energy equipment is eligible for a cost recovery period of five years. Show the depreciation schedule for an initial investment of $400,000.

9.7 INDUSTRIAL WASTEWATER TREATMENT

An industry-owned pollution control plant can be depreciated to reduce taxable income. The initial investments for some plant components are in Table P9.7. Develop a seven-year depreciation schedule using straight-line depreciation.

TABLE P9.7 Pollution Control Equipment Costs When New

Structures	$100,000
Power generation equipment	$60,000
Pumping equipment	$80,000
Treatment and disposal equipment	$900,000
Transportation equipment	$50,000
Computers and software	$18,000

9.8 TAX SAVINGS DUE TO DEPRECIATION 2

A company buys a $48,000 piece of equipment that has the pre-tax income shown in Table P9.8. The useful lifetime is six years. Use straight-line depreciation. The tax rate is 21% of taxable income. Calculate, with and without depreciation, the annual after-tax income. Calculate the net present value for the project, using 10% as the discount rate.

TABLE P9.8 Pre-tax Income from Equipment

Year	1	2	3	4	5	6
Pre-tax income ($)	10,000	12,000	15,000	20,000	20,000	20,000

9.9 DEPRECIATION AND AFTER-TAX INCOME

A company purchased equipment for $75,000 and estimates that the equipment will have a useful life of five years. At the end of its useful life, the company expects to sell the equipment for $5,000. Use straight-line depreciation with a five-year life.

(a) Calculate the annual taxable income.
(b) Calculate the net (after-tax) income for a tax rate of 21%.
(c) Calculate the total present value of five years of after-tax income for $i = 15\%$.

9.10 ACCELERATED (MACRS) DEPRECIATION

Redo Problem 9.9 using MACRS depreciation, assuming the project qualifies for a cost recovery period of five years.

9.11 CHROME RECOVERY

An industry will invest $800,000 in a process to recover up to 150 T/y of chromium from waste electrolysis solutions. The current (2020) price of chromium is about $10/kg. Prices can be volatile and unpredictable, so $10/kg is used for the lifetime of the project. The amount of recovered chromium and the income over a five-year project period is given in Table P9.11. Fixed operating costs

TABLE P9.11 Revenues and Costs for the Proposed Chromium Recovery Schemes ($10/kg chromium)

Year	Capital Cost ($)	Chromium Recovered (kg/y)	Income ($/y)	Operating Costs ($/y) Fixed	Operating Costs ($/y) Variable	Interest ($/y)	Net Cash Flow Before Taxes ($/y)
0	800,000						−800,000
1		90,000	900,000	120,000	585,000	75,000	120,000
2		120,000	1,200,000	120,000	780,000	75,000	225,000
3		150,000	1,500,000	120,000	975,000	75,000	330,000
4		150,000	1,500,000	120,000	975,000	75,000	330,000
5		150,000	1,500,000	120,000	975,000	75,000	330,000

are $120,000/y. Variable operating costs are $6.50/kg of recovered chromium. Annual interest payments are $75,000. These values are used to calculate the net cash flow (before taxes) in Table P9.11. Calculate the annual depreciation (straight line), pre-tax profit, taxes paid, after-tax profit, and the net present value for the project for an 11% discount rate.

9.12 CHROME RECOVERY

Redo Problem 9.11 using MACRS depreciation, assuming the chrome recovery process is classified as a five-year economic lifetime.

9.13 NEW PROJECT: INTERNAL RATE OF RETURN

The revenue and variable O&M costs for a new waste recovery project are given in Table P9.13. The capital investment is $70,000. Depreciation is over five years (straight line). Federal, state, and local taxes are 40%. The cost of capital is 12%. The company's minimum hurdle rate (required rate of return) is 20%.

(a) Calculate the pre-tax revenue, the taxable revenue, and the after-tax revenue for each year.
(b) Calculate the after-tax present value for $i = 12\%$, for the hurdle rate of 20%.
(c) Calculate the internal rate of return (IRR) that makes the net present value zero.

TABLE P9.13 Costs (in $) for New Waste Recovery Project

Year	0	1	2	3	4	5
Capital cost	70.0					
Revenues		50.0	71.5	90.0	127.5	160.0
Variable cost		27.5	39.2	49.5	58.7	73.6
Depreciation		14.0	14.0	14.0	14.0	14.0

9.14 EVALUATION OF SEWER ASSETS

Table P9.14a gives the condition of sewer mains in a town that is creating a repair and replacement program for all their assets. The assets classified as very poor will be replaced next year. The poor assets will be replaced in years 2 and 3. All replacement will be PVC pipe. Use the construction costs in Table P9.14b to make a schedule of expenditures. Escalate the costs by 2% per year.

TABLE P9.14a Evaluation of Sewer Assets

Material	Size (inch)	Length (ft)	Useful Life (y)	Length of Sewer (in ft) by Condition				
				Very good	Good	Fair	Poor	Very poor
PVC	≥6	1,684	75–100	779	920			
	12–15	662,909	75–100	496,354	165,782	920		
	≥15	43,143	75–100	35,322	7,786			
Total		707,863		532,455	174,488	920		
DI	≤6	66	60	14	0			
	8–12	59,303	60	33,899	21,810	3,593		
	≥15	12,208	60	3,214	8,296	701		
Total		71,577		37,127	30,155	4,295		
HDPE	12–5	1,969	70	1,969				
Concrete	≥15	17,503	75		17,419	84		
AC	12–15	10,315	50				6,569	3,746
Totals		809,402		571,551	222,063	5,299	6,569	3,746

PVC = polyvinyl chloride; DI = ductile iron, HDPE = high-density polyethylene, AC = asbestos concrete

TABLE P9.14b Construction Cost for PVC Sewer ($/lineal foot) (Sammamish Water & Sewer District 2016)

Diameter (inch)	Sewer Cost	Construction Cost	Diameter (inch)	Sewer Cost	Construction Cost
8	120.00	285.74	20	216.67	448.07
10	137.50	314.78	24	253.33	509.14
12	150.00	336.38	30	291.67	576.04
15	179.17	384.33	36	330.00	643.29
16	182.22	390.49	48	413.33	788.76
18	197.50	416.35	60	483.33	915.59

10 Financing Capital Costs

Every organization needs to invest in fixed assets such as land, buildings, process structures, and major equipment. These assets are often expensive and have a long useful life, and it makes sense to finance their purchase over a period of time. The most common way to raise funds is by selling general obligation bonds, revenue bonds, or special assessment bonds. Other ways to raise capital are government grants and loans, connection fees, special assessments, short-term borrowing, and service charge revenues.

A *bondholder* owns a bond issued by a *borrower.* The borrower *issues* a bond to obtain money for construction and promises to repay the *bondholder.* The borrower may be a municipality, a public utility, or a privately owned business or industry.

10.1 TYPES OF BONDS

General obligation (GO) bonds are the traditional way for public agencies to raise funds. The power behind GO bonds is the municipality's guaranteed ability to collect taxes to repay the bondholders. The full faith and credit of the municipality are pledged toward repayment. Because municipal taxes back the bond, there is little risk to the bondholder of default and, as a result, interest rates are lower than for other types of bonds. It is generally easier and less costly to issue a general obligation bond than a revenue bond or special assessment bond. However, a municipal GO bond adds to the debt ratio of the issuing municipality and often requires a positive vote by the citizens in the municipality to authorize the spending.

Revenue bonds are backed by the revenue from a specific commodity or project, such as water bills and user service charges. Revenue bonds are not a drain on the general taxing powers of the municipality and do not count against direct debt.

Revenue bonds require a more complex financial analysis and more complex legal covenants to assure potential bond purchasers that revenues will be sufficient to repay the bond. To provide this assurance, a revenue bond will normally require that revenues exceed repayment needs by 25–50%. This is termed *coverage.*

Special assessment bonds are used less often and are normally associated with a project that is limited and well-defined. The repayment is made through special assessments against the properties that benefit from the project. If the special assessment is not paid, the payment obligation becomes a lien against the benefiting property. Assessments are determined by various measures of benefit, such as front footage or area.

10.2 DEBT STRUCTURING

Debt repayment must be planned to maintain stable rates and budgeting flexibility. New debt repayments can be structured to *wraparound* old debts in order to level off future payments. Other reasons for special repayment schedules are to accommodate expected changes in revenue growth, to ensure stable user rates or to lower front-end costs.

One common refinement in a bond issue is a *call provision,* which allows the borrower to repay the debt after a defined number of years but prior to when the bond was scheduled to mature. A borrower would exercise a call provision if interest rates dropped and the issuer could save money by selling a new bond to pay off the old bond. The call provision is a risk because a bond is called when interest rates have decreased, which makes it difficult for the repaid bond holder to invest at the same rate of return. The interest rate on the bond issue usually will be greater if it contains a call provision that returns the capital with a premium added in exchange for the debt retirement.

A bond issue might also allow bondholders to redeem or add to their bonds in the form of additional debt at a stated interest rate. A *callable bond* can be repaid or redeemed early. There is a *call protection period*, and there may be a *call premium* paid to compensate the bond holder for loss of future income.

Example 10.1 Callable Bond

A callable bond with a face value of $10,000,000 and a yield of 6.5% per year was issued on January 1, 2017. The bond holder (buyer) of course knew about the call option and agreed that 6.5% yield rate was high enough to accept the early repayment provision.

The call protection period is five years, so after five years (usually with a 30-day notice) the bond issuer can exercise the call option and pay back $10,000,000 with no interest paid after redemption.

Short-term cash requirements can be satisfied through traditional bank borrowing or through the use of notes. A *note* is a legal document that serves as an IOU from a borrower to a creditor, usually for terms much shorter than are usual for bonds. Notes are used to bridge cash needs between bond issues and are often termed bond anticipation, grant anticipation, or tax anticipation notes. Notes are marketed similar to bonds and will often include an underwriting expense. Borrowing directly from a major bank often provides funding flexibility, allows tailoring the terms to specific needs, and saves money on fees.

The lender has some costs for arranging and issuing the loan or bonds. These costs often are added to the amount required for use, and the *total loan amount* is greater than the amount needed to pay the capital cost of the project. These costs include the following:

- *Debt service reserve* (DSR) is an amount put aside to be available in the event the borrower is unable to make a monthly payment on time. The DSR earns interest which is credited toward the monthly payments, and the principal of the reserve fund is used to help offset the final year's debt service payments. The rate used to calculate the return on the DSR is the same as the bond rate.
- The cost of financing is approximately 2% of the loaned amount. This can be paid at closing or financed.
- Municipal bonds are typically issued in denominations of $5,000 or greater. The amount borrowed will be rounded up to the nearest $5,000.
- The *total loan amount* is used as the basis in determining the payments. This gross payment may be reduced by the interest earned by the debt service reserve.

Example 10.2 Cost of Issuing A Bond

The funds needed for a project are $1,000,000. The term of the loan is 20 years, and the interest rate is 4%. These are typical costs.

- Debt service reserve = $115,000 (to be financed)
- Amount to be financed = $1,000,000 + $115,000 = $1,150,000
- Cost of issuance = 2% of $1,115,000 = $22,300 (to be financed)
- Rounding amount = $2,700 to give $25,000
- Total bonds issued = $1,000,000 + $115,000 + $25,000 = $1,140,000
- Issuance cost (portion not financeable) = $2,700
- Gross monthly payment = $6,908
- Less reserve interest = $380
- Net monthly payment = $6,528

There are other fees such as bonding counsel fees and underwriter expenses.

10.3 BOND YIELD

The *par value* of a bond, also known as the *face value*, is the amount of money that the bond issuer agrees to repay the bondholders (bond owner) at the maturity date of the bond. A bond is a written promise that the amount loaned to the issuer will be repaid.

The *coupon rate* is the agreed interest rate paid by its issuer for the term of the bond. For example, a bond with a face value of $1,000 that pays $25 semiannually ($50 annually) has a coupon rate of 5%. The coupon rate is commonly referred to as the *interest rate.*

Bond yield is the amount of return an investor will realize on the bond. *Nominal yield* is calculated by dividing the amount of interest paid by the *par value* (face value) of the bond. *Current yield* is the interest paid divided by the current market price of the bond. Current yield is inverse to the bond price: as bond prices increase, bond yields fall. The original investor receives the same amount of income, but the yield is calculated using a different basis.

Example 10.3 Bond Yield Computation

A bond with a face value of $10,000 and a maturity (life) of six years with a coupon rate of 8% pays $800 annually (at the end of the year). At the end of six years, the bond holder is returned the $10,000 investment. The total interest paid over six years is

$$6(\$800) = \$4,800.$$

The present value of the six interest payments is $3,698.

Example 10.4 Nominal Bond Yield

An investor purchased a bond with a par value of $10,000 that pays 10% per year. The nominal yield on this bond is the annual interest earned divided by the face value:

$$\text{Nominal yield} = \frac{\$1,000}{\$10,000} = 0.1 \text{ or } 10\%$$

If the bond price were to fall to $9,000, the investor will still receive $1,000 per year of interest, but the *current yield* becomes

$$\text{Current yield} = \frac{\$1,000}{\$9,000} = 0.111 \quad \text{or } 11.1\%$$

The current yield is irrelevant if the original bond owner holds the bond to maturity. It is important when bonds are bought and sold before maturity.

10.4 YIELD OF MUNICIPAL AND CORPORATE BONDS

Public purpose bonds used for government projects are always tax exempt. That is, tax is not paid on the interest earned. Municipal bonds (*munis*) are of this type. Their yields are lower than private purpose bonds because the tax benefits are priced into the bond.

Corporate bonds fund projects that benefit a private entity. They are not tax exempt, and they carry a higher interest rate to compensate for the taxes that must be paid on the interest.

Example 10.5 After-Tax Bond Yields

An investor wants to purchase a $10,000 bond. The options are a tax-free municipal bond that yields 7% and a taxable corporate bond that yields 9.75%, interest to be paid annually. Table 10.1 shows the net income for the interest-free municipal bond and the taxable corporate bond.

TABLE 10.1 Net Earnings for Non-taxable Municipal Bond and the Taxable Bond

Marginal Tax Rate	Municipal Bond Yield ($/y)	Taxable Bond Yield ($/y)	
		Pre-Tax	Post-Tax
10	700	975	878
12	700	975	858
22	700	975	760
24	700	975	741
32	700	975	663
35	700	975	634
37	700	975	614

The municipal bond has a net income of (0.07)($10,000) = $700, regardless of tax bracket of the bond owner. The corporate bond has a pre-tax income of $975, but taxes reduce that by 10–37%, depending on the tax bracket. The post-tax corporate earnings exceed the municipal bond until the investor is taxed at the 32% bracket. The municipal bond has a higher net yield if the bondholder pays taxes at a marginal rate of 32% or higher. At lower taxation rates the corporate bond is the better choice.

The *tax-equivalent yield* is the rate of interest earned after taxes are paid. An investor's income tax rate determines the tax-equivalent yield. A tax-exempt bond and a taxable bond that offer the same interest rate will have different after-tax yields. The relationship is

$$R_M = R_C(1-t)$$

where

R_M = interest rate of a municipal bond (%)
R_C = interest rate of a comparable corporate bond (%)
t = tax rate of the bond owner (decimal fraction)

The *marginal tax rate t* at which an investor is indifferent between holding a corporate bond yielding R_C and a municipal bond yielding R_M is

$$t = (R_C - R_M) / R_C$$

In the United States, as of 2018, there are seven marginal tax-rate brackets for personal income: 10%, 12%, 22%, 24%, 32%, 35%, and 37%. Investors who own bonds as an investment are usually in the top brackets.

Example 10.6 Equivalent Yield of Municipal and Corporate Bonds

A corporate bond pays 10% per year to a bondholder who pays taxes at the rate of t = 25%. The municipal bond interest rate that generates equal after-tax income is

$$R_M = R_C(1-t) = 10\%(1-0.25) = 7.5\%$$

TABLE 10.2 Equivalent Corporate Yield Rate

Marginal Tax Rate (%)	Equivalent Corporate Rate (%)	Interest ($)	Taxes ($)	Net Yield ($)
10	7.78	778	78	700
12	7.95	795	95	700
22	8.97	897	197	700
24	9.21	921	221	700
32	10.29	1,029	329	700
35	10.77	1,077	377	700
37	11.11	1,111	411	700

Table 10.2 shows the corporate bond interest rates that give a net yield of $700 per year, which is equal to a $10,000 tax-free bond that pays 7% interest per year. The equivalent rate depends on the marginal rate of taxation. For example, the equivalent rate is 9.215% for a bondholder who is taxed in the 24% tax bracket. The bond earns $921 in interest income, which is taxed at 24%, for a tax of $221, and a net income of $921 – $221 = $700. This is the same yield as the municipal bond with a tax-free rate of 7%.

10.5 SINKING FUNDS

A *sinking fund* contains monies set aside over a period of time to fund a future capital expense or repay a long-term debt. The sinking fund account can be funded through service charges, special assessments, or taxes.

Sinking funds are also used to collect funds for specific maintenance that occur at intervals such as painting, roof repairs, pump rehabilitation, or diffuser replacements. Major mechanical components wear out and need replacement. A reserve/replacement fund is a special account that can pay for major repairs without resorting to or requesting emergency funding or leaving a failed component out of service until the next budget year. These funds can be accumulated and maintained as a sinking fund.

Good accounting practice, and often bond covenants, requires that monies are deposited into a separate account well in advance of the principal and interest payment dates to ensure that payment will be made on time. If the monies are not in place by a specified time, the bond issuer is usually required to use an alternative source of revenue, such as a property tax or special assessment, to accumulate the necessary funds. Separate auditing of the sinking fund account helps ensure that funds are accumulated and expended appropriately.

Sinking fund implies uniform end-of-year payment of amount R which will accumulate to the required total at the time of the last payment. The payment per period needed, R, to accumulate a future value FV in a sinking fund at the end of year n is

$$R = FV\left(\frac{i}{(1+i)^n - 1}\right)$$

Payments made at the beginning of the year are held longer and earn more interest. The future value of an annuity R equal annual beginning-of-the-year deposits is

$$FV = R\left[\frac{(1+i)^n - 1}{i}\right](1+i)$$

Rearranging to get the formula for a beginning-of-the-year sinking fund gives

$$R = FV\frac{1}{(1+i)}\left[\frac{i}{(1+i)^n - 1}\right]$$

Example 10.7 Sinking Fund: End-of-Year Payments

The annual payment to accumulate $900,000 in a sinking fund in five years at an interest rate of 4.8% is

$$R = FV\left(\frac{i}{(1+i)^n - 1}\right) = \$900,000\left(\frac{0.048}{(1+0.048)^5 - 1}\right) = \$163,529$$

The accumulation of funds for a $163,529 end-of-year deposit earning 4.8% interest is shown in Table 10.3 and Figure 10.1. The values are rounded to the nearest dollar.

10.6 STATE REVOLVING LOAN FUNDS

The US Environmental Protection Agency supports a Clean Water State Revolving Fund (CWSRF) for pollution control projects and Drinking Water State Revolving Fund for water utilities. The two funds, which have similar goals, guidelines, and procedures, provide grants to all 50 states and Puerto Rico to capitalize state loan programs. The funds are self-perpetuating. Loan repayments finance the next set of projects.

TABLE 10.3 Accumulation of $900,000 in a Sinking Fund that Pays 4.8% per Year

Year	End of Year Deposit ($)	Annual Interest ($)	End of Year Balance ($)
1	163,529	0	163,529
2	163,529	7,849	334,908
3	163,529	16,076	514,513
4	163,529	24,697	702,739
5	163,529	33,731	900,000

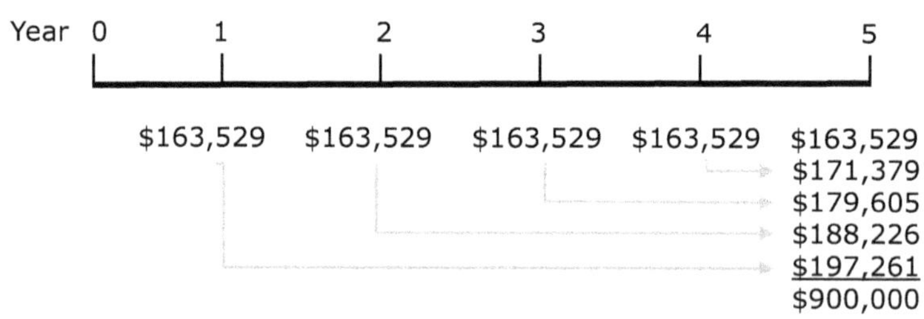

FIGURE 10.1 Accumulation of $900,000 over five years in a sinking fund that pays 4.8% interest per year.

The Drinking Water State Revolving Fund, established in 1996, has a federal investment of over $19.1 billion, to provide more than $35.4 billion to water systems through 2017. The state provides a 20% match to:

- Improve drinking water treatment
- Fix leaky or old water distribution systems
- Improve source of water supply
- Replace or construct finished water storage tanks
- Other infrastructure projects needed to protect public health

Over the past 20 years, the cumulative federal outlay for the Clean Water Revolving State Fund has been $23.6 billion, and the return has been $54.6 billion in disbursements, the difference being state contributions and interest earned on loans. This is a return of $2.31 for each federal dollar invested. Federal money is augmented by state contributions at an approximate 90% federal to 10% state ratio.

States can target specific communities and environmental needs by awarding grants with no repayment, loans with interest rates from 0% to market rates, or some combination of loans and grants. For example, the State of Wisconsin Clean Water Fund and the Safe Drinking Water Loan programs had an official interest rate (present worth discount rate) of 3.525% in 2019. The 3.525% was considered the *market rate*. Projects for municipalities of less than 100 population or a median household income (MHI) less than 65% of the Wisconsin median value could borrow money at 0% interest. Municipalities of less than 10,000 population with MHI less than 80% of the Wisconsin median could borrow at 33% of the market rate, or 1.196%. Projects not meeting the special needs criteria could borrow at 55% of the market rate or 1.994%

Loan terms can be customized to meet the needs of small and disadvantaged communities or to provide incentives for certain types of projects through additional subsidization, such as grants, principal forgiveness, and negative interest rate loans.

Eleven types of projects are eligible to receive Clean Water SRF assistance:

- Construction of publicly owned treatment works
- Implementation of a non-point source pollution management program
- Construction, repair, or replacement of decentralized wastewater treatment systems that treat municipal wastewater or domestic sewage
- Management, reduction, treatment, or recapture of stormwater or subsurface drainage water
- Reducing the demand for publicly owned treatment works capacity through water conservation, efficiency, or reuse
- Development and implementation of watershed projects
- Reducing energy consumption at publicly owned treatment works
- Projects for reusing or recycling wastewater, stormwater, or subsurface drainage water
- Security measures at publicly owned treatment works
- Technical assistance to plan, develop, and obtain financing for CWSRF eligible projects and to assist each treatment works in achieving compliance with the CWA

Example 10.8 Madison Metropolitan Sewerage District Clean Water Fund Loans

The Madison Metropolitan Sewerage District (MMSD) has received 23 loans for a total of $218.1 million under the Clean Water Fund program. The MMSD is a financially self-sufficient organization and does not qualify for grants or subsidized interest rates.

The six loans it held in 2015 are listed in Table 10.4. Funds are withdrawn from the approved loans as needed to complete the projects.

TABLE 10.4 Madison Metropolitan Sewerage District Clean Water Fund Loans in 2015

Project	Amount ($)	Rate (%)	Date of Issue	Final Payment
Nine Springs 11th Addition	50,362,380	2.518	Feb. 22, 2012	May 1, 2031
Pump station 18	14,773,549	2.643	Sep. 25, 2013	May 1, 2033
Process control upgrade	4,746,580	2.465	Nov. 27, 2013	May 1, 2033
Pump station 18 force main	12,362,184	2.72	Feb. 26, 2014	May 1, 2033
Pump stations 11 and 12 Rehabilitation	10,663,025	2.26	Feb. 25, 2015	May 1, 2034
Maintenance facility	12,094,707	2.25	May 27, 2015	May 1, 2035

TABLE 10.5 Projects Costs for Ashland, Wisconsin

Wastewater treatment plant	$7,300,000
Conveyance facilities	$3,100,000
Third St. utility and street	$1,200,000
Bayfront sanitary sewer improvements	$100,000
Total construction cost	$11,700,000

Example 10.9 Ashland, Wisconsin, Wastewater Facility Upgrade

The Ashland, Wisconsin, wastewater facilities, shown in Table 10.5, were improved using a $4,100,000 grant from the Farmers Home Administration and a Clean Water Fund loan of $7,600,000 for 20 years at 1% interest.

Example 10.10 Utah State Revolving Fund (SRF) for Water Supply

The Utah SRF evaluates a community's need for funds and *ability to pay* based on the median adjusted gross household income (MAGI). The Utah state average MAGI was $45,000 in 2017. The fund's goal was for a community's average monthly water bill to not exceed 1.73% of its MAGI.

Table 10.6 describes two communities that each needs $3,232,000 to replace the aging water storage tanks, water mains, chlorination systems, and other appurtenances. Revenue from water bills will be used to repay the bonds or loans over a 30-year period.

Community B has a MAGI of $70,569, and the average annual water bill is 0.9% of MAGI. This community qualifies for revolving fund loans but does not need special grants to afford the system upgrade.

TABLE 10.6 Financial Characteristics of Communities A and B

Community	A	B
Customers	86	86
MAGI ($)	37,000	70,569
% of state MAGI	82	156
Current monthly water bill ($)	53.40	53.40
Annual bill as % of MAGI	1.73	0.9

TABLE 10.7 State Grants and Loans for Community A

Financial Factor	Amounts
Project cost ($)	3,232,000
Grant (interest free)($)	969,000
Loan amount ($)	2,263,000
Payback period of loan (y)	30
Effective interest rate (%)	0
Monthly loan payment ($) (360 payments)*	6,286
Current average water bill ($/month-customer)	53.40
Average water bill increase ($/month-customer)**	73.09
New average monthly water bill ($/month-customer)	126.49
Annual average water bill ($/y-customer)	1,517.808
% of MAGI	4.1

* $6,286 = $2,263,000/360
** $73.09 = ($6,286/mo.)/(86 customers)

Community A has roughly half the MAGI of community B, only $37,000, which means that it has only half the ability to repay loans. The average community A water bill of $53.40 is already at 1.73% MAGI. Because of this, it qualifies for an interest-free grant and a low-interest loan, but with the maximum allowable state assistance, the goal of 1.73% MAGI cannot be met.

The maximum grant limit from the state is $969,000, leaving community A in need of a $2,263,000 loan. If the loan is interest free (0% interest), and the loan repayment period is 30 years (360 months), the community's monthly loan repayment will be $6,286. This divided among 86 customers is $73.09 per month, which would be added to the current average water bill of $53.40. The new average water bill would be $126.49, or 4.1% of MAGI. This is not sustainable for community A. The calculations are summarized in Table 10.7.

A grant of $2.263 million would be required to reduce the average monthly water bill to 1.73% of MAGI. The State Revolving Fund cannot provide enough financial assistance for community A to reach this goal. Perhaps the project scope can be scaled back, or the community can find other sources of grant funds.

10.7 WATER INFRASTRUCTURE FINANCE AND INNOVATION ACT (WIFIA)

The Water Infrastructure Finance and Innovation Act (WIFIA) of 2014 offers water and wastewater projects low, fixed interest rates and flexible financial terms. Eligible borrowers are government entities, partnerships and joint ventures, corporations and trusts, and Clean Water and Drinking Water State Revolving Fund (SRF) programs. Projects must be creditworthy and have a dedicated source of revenue, which includes taxes, user fee payments, transfers pledged from state or local governments, dedicated taxes, a municipal general obligation pledge, and revenues that are pledged for the purpose of retiring debt on the project.

WIFIA loans can be combined with private equity, revenue bonds, corporate debt, grants, and State Revolving Fund (SRF) loans. A single fixed interest rate based on the US Treasury rate on the date of loan closing is established at the closing date. The interest rate will be equal to or greater than the US Treasury rate of a similar maturity at the date of closing.

A borrower may receive multiple disbursements over several years at the same fixed interest rate. Borrowers can customize repayments to match the anticipated revenues and expenses for the life of the loan. This flexibility provides borrowers time to phase in rate increases to generate revenue to repay the loan.

The maximum final maturity date from substantial completion is 35 years. Repayment may be deferred for up to five years after substantial completion of the project. Projects are expected to be completed within seven years of closing, but there is no specific maximum project length. Disbursement schedules are negotiated prior to the closing of a loan.

The minimum project size is $20 million for large communities and $5 million for small communities (population of 25,000 or less). The maximum portion of eligible project costs that WIFIA can fund is 49% and total federal assistance may not exceed 80% of a project's eligible costs.

Thirty-nine projects were selected in fiscal year 2018 with a total loan amount of $5 billion to public and private entities. The total water infrastructure supported was $10 billion. The projects were in 16 states and 22 million people were assisted.

Example 10.11 City of Waukesha, Wisconsin, Water Utility

In 2018 the City of Waukesha, Wisconsin, obtained a 35-year $116 million WIFIA loan to transition Waukesha's water supply from radium-contaminated groundwater to Lake Michigan water. The total cost of the project is $286,218,000. This project will bring Waukesha into compliance with state drinking water standards for radionuclides by September 1, 2023.

The project will obtain Lake Michigan water from the Milwaukee Water Works and return treated sewage effluent to Lake Michigan. The project includes:

- Water supply connection
- A 15.2-mgd water supply pumping station
- A 13.5-mile, 30-inch water supply pipeline
- Two 10-million-gallon reservoirs
- A 15.75-mgd booster pumping station
- Chemical feed system
- Distribution system connection
- Improvements to the existing distribution system
- A 12-mgd return flow pumping station to Lake Michigan
- A 23.5-mile, 30-inch return flow pipeline to Lake Michigan
- Return flow facilities at Root River

Example 10.12 San Diego Water Purification Plant

The City of San Diego, CA, will invest $1.4 billion to construct a state-of-the-art water purification facility that will produce 30 mgd of drinking water. A WIFIA loan will finance $614 million of that amount. Because the WIFIA program offers loans with low-interest rates, the City expects to save up to an estimated $184 million compared to a typical bond issue. Project construction and operations are expected to create 480 jobs, with construction beginning in 2019 and targeted for completion in 2023.

10.8 CONCLUSION

The most common way for municipalities and utility districts to finance capital investments is to issue bonds. The typical repayment period is 20–30 years. General obligation bonds, backed by the taxing power of municipalities, are popular because of their low interest rates. Utilities and utility district bonds are backed by and repaid with service fees that assure the necessary funds will be available when needed.

The State Revolving Loan and the Federal WIFIA programs can provide loans or grants for a variety of water supply and water pollution control projects. Terms can be customized to fit special needs of the specific projects.

10.9 PROBLEMS

10.1 BOND YIELD COMPUTATION

The owner of bond with a face value $20,000 that is due in eight years has a 6% yield. The bond owner needs money and will sell the bond for a fair price. Bonds offered today pay 4% interest, so a bond with 6% yield is attractive to buyers.

(a) The seller defines a *fair price* as the present value of the bond, plus a premium of 2% of the face value. Calculate the fair price.
(b) Calculate price for the purchaser to earn a 7% rate of return.

10.2 BOND YIELD

An investor purchased a bond with a par value of $20,000 that yields $1,600/y. (a) Calculate the nominal yield. (b) Calculate the current yield if the bond price falls to $18,000. (c) Calculate the actual yield if the bond value increases to $23,000. (d) Is the current yield important or interesting if the bond is held to maturity?

10.3 CALLABLE BOND

A small firm pays $800 per year on a five-year $10,000 bond that is due in three years. The bond is callable with 30 days' notice and a call premium of 2% of the face value. What is the cost of the bond if it is held to maturity? What is the cost if it is called now?

10.4 BOND MANAGEMENT

Pollution control facilities have been financed with the three long-term bonds described in Table P10.4. Bond issues B and C are callable now. If the bond is called, the face value is paid to the lender plus the call premium. Interest rates are now 3.2%, which means that the 2.8% bond has become more valuable and should be held. There may an advantage to call bond issues B and C. Calculate the cost and make a recommendation.

TABLE P10.4 Long Term Bond Data

Bond Issue	Par Value	Interest Rate	Date of Issue	Date of Maturity	Type of Bond	Call Premium
A	$10,000,000	2.8%	2005	2025	Municipal	None
B	$10,000,000	4.2%	2008	2028	Municipal	One-year interest
C	$20,000,000	6.6%	2015	2035	Municipal	5% of par value

10.5 TAXABLE BOND YIELDS

Calculate the taxable bond yield rate that provides net income equal to a tax-free bond if income is taxed at 22%, 32%, or 37%.

10.6 AFTER TAX YIELD

An investor who pays income tax at the 24% level owns a tax-free municipal bond that pays 4.5% interest and a taxable corporate bond that pays 7.0%. The face values of the bonds are the same. Which bond has the highest net yield?

10.7 SINKING FUND 1

A sinking fund will be established to pay \$400,000 for a major repair project at the end of four years. Deposits will be made at the end of the year and will earn 3% interest. Calculate the annual deposit.

10.8 SINKING FUND 2

Payments will be made at the beginning of the year for four years to accumulate \$800,000 to pay for a project that will start at the end of year 4. The interest rate is $i = 6\%$. Calculate the annual deposit.

10.9 STATE REVOLVING FUND 1

A community with 600 customers needs \$1,750,000 to build stormwater control infrastructure. The current average annual bill for wastewater is \$500/y. The goal is to arrange the project financing so that the average bill does not exceed \$725/y. Grants and interest-free loans are available when needed; otherwise, loans are made at 3.5% interest for 30 years. The maximum grant is \$750,000. Propose a financial plan for the community.

10.10 STATE REVOLVING FUND 2

A community with 500 customers needs \$2,750,000 to replace aging infrastructure. The goal is for the average annual bill to not exceed \$778 per household. The current average annual bill is \$550. Grants and interest-free loans are available when needed; otherwise, loans are made at 3.5% interest for 30 years. The maximum grant is \$750,000. Propose a financial plan for the community.

10.11 CLEAN WATER FUND LOANS

A metropolitan sewerage district holds the six loans in 2015 listed in Table P10.11. All the loans were for 20 years at different interest rates as shown. The loans are repaid with uniform annual payments. Calculate the schedule of annual payments, starting from 2021.

TABLE P10.11 Clean Water Fund Loans

Project	Date of Issue	Amount ($)	Rate (%)
1. Plant addition	2008	50,362,380	2.518
2. Pump station	2010	14,773,549	2.643
3. Process control upgrade	2012	4,746,580	2.465
4. Force main	2013	12,362,184	2.720
5. Pump stations rehabilitation	2013	10,663,025	2.260

10.12 WASTEWATER AND RECLAIMED WATER PROJECT FUNDING 1

A city is financing a wastewater treatment plant expansion and a reclaimed water facility as described in Table P10.12. An EPA grant of \$3,791,200 has been awarded for the wastewater treatment part of the project. A replacement reserve must be established for 5% of the loan. This will be paid over six years (1/6 paid per year). A debt service reserve equal to one year of loan payments is required, also payable over six years. Assume all the needed funds will be obtained by borrowing at 3% interest. Each project has a 20-year life. (a) Calculate the annual cost for each project and for the total. (b) What is the revenue required per year to pay for the project?

TABLE P10.12 Project Costs ($)

Cost Category	Wastewater Expansion	Reclaimed Water	Total
Facility	2,324,600	326,800	2,651,400
Infrastructure	1,679,400	588,900	2,268,300
Engineering	600,600	137,650	738,250

10.13 WASTEWATER AND RECLAIMED WATER PROJECT FUNDING 2

The project described in Problem 10.12 is supported by an additional grant of $1,000,000 from the state revolving loan fund (SRL), divided equally between the wastewater treatment plant and the reclaimed water plant. The balance will be financed at 3% over 20 years. The replacement reserve and the debt service reserve are still required. (a) Calculate the annual cost for each project and for the total. (b) What is the revenue required per year to pay for the project?

11 Financial Management for Operations

Financial management is budget management. A budget is a short-term forecast of spending and revenues. It establishes a financial roadmap for an upcoming project or a financial year and provides goals or constraints on spending. It is a decision-making and cost-tracking tool and should not be more complicated than necessary to provide those two functions.

Budgeting is not an exact science because not all tasks or costs can be accurately predicted. It is not required or expected that a budget will be met exactly. The temptation to inflate a budget to always show a positive closing balance undermines a good budgeting process. Likewise, end-of-year spending to fully consume a budget category is equally poor management. A budget is a plan, not a code of financial restrictions. Some budget categories will be exceeded and others under spent.

The degree of detail included in a budget should match the accuracy needed by the user to confidently track costs and to make good spending decisions. The amount of detail should relate to the total budget. If the budget is $15,000,000, it may not be necessary to define individual expenditures under $500 as long as the smaller costs are captured somewhere in the budget. If detailing all expenditures helps make good purchasing or operating decisions, then such detail is justified.

11.1 TYPES OF BUDGETS

Zero-based budgeting starts from a zero base, and each expenditure must be justified for each new period. This makes little sense when predicting expendable costs because it ignores the valuable and instructive historic cost patterns.

Line item budgets are built from past experience and projections of how expenditures will change from one year to the next. They will include growth for services, inflation of prices, salary increases, and known projects to repair and replace facilities, or to construct new facilities. These are used by utilities.

11.2 BUDGETS AND BUDGET CATEGORIES

Budget categories are created to suit the application. Water and wastewater utilities will have similar categories. Solid waste management programs will be different.

Table 11.1 is the annual budget for operations at a wastewater treatment plant. New capital improvements are not included, but it does include debt service to pay for past capital projects. There are three general categories:

- Administration = $667,675,
- Collection system = $977,605
- Treatment/disposal systems = $4,368,750

The total operating budget is $6,014,030. Add to this the annual debt service of $2,002,255 to get the total required annual revenue of $8,016,285. This is to be paid with user service charges. Debt service is the cost of loan repayments.

For most projects and organizations, salary costs are 30–85% of the total budget, and care is taken to accurately predict and track all direct and indirect salary-related costs. *Direct salary*

TABLE 11.1 Annual Budget for Wastewater Treatment Plant Operations

Budget Category	Budget ($)
Administration	667,675
Collection system	
User charge/pretreatment	136,400
General services	4,800
Pumping stations	752,655
Interceptors	83,750
Collection subtotal	977,605
Treatment and disposal systems	
General salaries	1,260,580
General supplies	16,400
Power	1,079,000
Replacement parts	250,000
Chemicals	8,200
Vehicle operation	44,650
Radio expense	2,800
Training expense	30,200
Computer expenses	23,050
Laboratory expenses	158,060
Contract services	24,600
Effluent diversion	147,460
Sludge reuse	1,240,600
Utilities and miscellaneous	83,150
Treatment subtotal	4,368,750
Total required operating revenue	6,014,030
Total required debt service revenue	2,002,255
Total required revenue	8,016,285

expenses include social security tax, worker's compensation coverage, pensions, health insurance, and life insurance costs. *Indirect salary expenses* include vacation time, sick leave usage, training, research, and other non-billable hours.

Most businesses will track *indirect expenses* such as administrative expenses, building heat and light, automotive expenses, personnel relations, safety and fire protection, property and liability insurance, postage, telephone, taxes, etc. These costs are often assigned against salaries for budgeting or billing purposes. Anyone developing or reviewing a budget must know where these costs are allocated and how these costs will be recovered.

The expenditure side of the budget is usually more complicated since it must predict where the money will be going over the selected time frame. It can be set up to track costs by any number of allocations such as project tasks, unit processes, or functional areas. Functional areas include administration, engineering, construction, operation, and maintenance. Within these areas, costs will be allocated to salaries, benefits, supplies, chemicals, parts, services, inventory, etc. Each organization or project will have a distinct set of cost categories that aid decision-making and cost-tracking.

Expendables, a major category in most operation budgets, include fuel, chemicals, power, and supplies. Expendables are easily budgeted because there is a history related to the volume or amount of material that is treated. Anticipated production and unit cost increases combine to give a reasonably close budget estimate.

Non-recurring costs are *task specific*. An example is equipment replacement and major maintenance expenses. Different equipment may be involved in each project or budget year. Since the equipment is normally specific to a certain process, the costs will also be specific. Zero-based budgeting is suitable for these task-specific costs because it forces identification of the actual costs and efforts involved in completing the task and this defines the project work plan.

The revenue side of a budget is easier to construct, but harder to control. There are well-defined revenues such as interest on investments and rent, and less-defined sources that depend on services provided or products sold. Since the object of every budget is to match the revenues to the expenditures, the rates for services or product pricing must be set to collect sufficient revenue to offset the expenditures. The receipt of revenues will not match the timing of expenditures, and a separate cash flow budget is needed to make sure bills can be paid when due.

11.3 CAPITAL AND DEBT SERVICE COSTS

Debt service is the money paid to the lending authority to cover the payment of principal and interest on loans and bonds. Debt service needs to be controlled. Overborrowing may force service rates to be raised beyond users' willingness to pay. Also, excessive debt service can affect the utilities' future access to low interest rates as bond ratings can be affected by the amount of existing debt.

Debt service coverage is the net revenues divided by the annual debt service payment. This measures the portion of net revenues that is allocated to debt. Debt service payments cannot be neglected or postponed, so the ratio must be greater than 1 or the utility will lose money or have make up the difference out of reserve funds. A practical minimum ratio is 1.2 for many lending institutions to obtain a reasonable interest rate loan. A more typical value for wastewater utilities with A-rated bonds is 1.9, while for AAA-rated bonds, the median value is 2.7 (Fitch ratings, 2013).

Example 11.1 Debt Service Coverage

A utility has annual gross revenues of $180,000,000 against O&M expenses of $80,000,000. The annual debt service is $77 million.

Net revenue = $180,000,000 – $80,000,000 = $100,000,000
Debt service coverage = $100,000,000/$77,000,000 = 1.3

The coverage factor of 1.3 would be acceptable to most lending institutions, even though it is below the more typical 1.9 for A-rated utilities.

11.4 CONTINGENCIES

If there is significant uncertainty regarding the estimated costs in the budget total or in a specific cost category, a contingency can be added to cover expected but unknown costs. This will be more important if there are no cash reserves to rely upon should the budget be exceeded, or if raising additional funds to cover extra expenses midyear is very difficult or expensive, or if exceeding the budget results in a significant adverse reaction from some constituency.

Use of contingencies is not a sign of a carelessly prepared budget. There are usually some unidentified needs with unknown costs. At the same time, contingencies must not be used as a substitute for thoughtful and detailed budget preparation. The more the certainty and thought that go into the preparation of a budget, the less need for contingencies.

11.5 ACCOUNTING

While budgeting is an inexact planning tool based on projected revenues and expenditures, accounting is a precise tracking of actual revenues and expenditures. The accounts must balance, and this demands care and diligence in tracking ongoing financial activities.

Accounting reveals at year's end how well revenues and expenditures were projected for the year. The result will be a report that reflects the financial condition of the organization. An audit should be conducted by an independent auditor to confirm that the financial statements are free of misstatements.

Detailed daily cost accounting is a necessary part of every project and business. The accounting is segmented into areas like accounts receivable, accounts payable, payroll processing, asset management, inventory control, and maintenance management.

11.6 ACCRUAL ACCOUNTING

There are a number of variations of accounting methodologies. The two most used are *accrual accounting* and *cash accounting.*

Accrual accounting reports revenues on the income statement when they are earned. Expenses are reported when the expense occurs, not when the cash is paid. The result of accrual accounting is an income statement that better measures the profitability of a company during a specific time period.

In accrual basis accounting, when a business purchases a long-lived asset, such as a vehicle, a building, or a piece of equipment, it does not immediately write off the full cost as an expense. The cost is spread over the expected life of the asset using an accounting procedure known as depreciation. This adheres to accrual accounting's *matching principle*, in which expenses are matched to the revenue generated by those expenses.

Cash accounting is the opposite. Revenues and expenditures are reported on the income statement when the cash is received or paid out.

Example 11.2 Utility Bills

A utility bought electricity and chemicals and paid employees in December to provide services that customers received in December. However, the December costs are billed to the customers in January. To have the proper amounts on the utility's financial statements, there needs to be an adjusting entry to increase revenues that were earned in December and the receivables that the utility has a right to as of December 31.

Example 11.3 Contract Maintenance

A utility pays $6,000 every six months for contract maintenance (one bill in January and one in July). Under cash accounting, the utility would record a $6,000 expense in January and a $6,000 expense in July. Using the accrual method, it would instead record a $1,000 expense each month for the whole year. That is, the expense is matched to the period in which it was incurred: $1,000 for January, $1,000 for February, and so on.

Private businesses and utilities have traditionally used *accrual accounting.* More recently, the standards for financial reporting for state and local governments have been revised to require full-accrual accounting on their capital investments and holdings.

Accrual accounting virtually eliminates discrepancies or errors because each financial transaction is recorded as it occurs. Since everything is recorded all the time, an audit is available at any time. The accuracy level is higher because, unlike cash accounting, accrual accounting follows a double entry system. As a result, it is always clear how one account is reduced and another account is increased.

Accrual accounting requires that all assets be assigned a value and an expected useful life. From that information, a depreciation schedule can be developed and used to predict when future expenditures need to be made to replace worn-out assets. Most often straight-line depreciation schedules are used for simplicity, but other approaches can be used if the value of the asset does not decrease linearly. The depreciation schedule also provides the business with a systematic way to collect money based on the schedule so that monies are available when the asset needs replacement.

Proper assignment of the asset value and useful life can make a significant difference in when money is received, and the total amount of money accumulated for replacement of that asset. It can be difficult and speculative to assign a useful life to a particular asset that will be close to the actual life. An engineer's judgment, coupled with operating experience, is necessary to make good estimates of asset values and useful lifetimes.

11.7 CASH ACCOUNTING

Cash accounting is simpler than accrual accounting, but less often used in business. A cash accounting system is similar to a personal budgeting system. There is less need to accurately value all assets, estimate asset lives and track depreciation of all assets, tasks that take considerable time and effort.

Operating revenue for a selected length of time is determined by examining the needs over that time period. Money for replacement is collected based on known needs instead of according to a depreciation schedule. Because the needs are immediate and the replacement is done within the budget term, the costs for that replacement are more accurate than long-term replacement needs. A process or a piece of equipment is a candidate for replacement when the cost of replacement is less than the present value of projected future maintenance costs.

Because a reserve of cash is not being generated based on the depreciation of assets, the organization must stay current on maintenance and replacement so as to avoid significant unexpected expenditures. A cash reserve, often as much as 30–50% of the annual budget, is necessary to prevent significant increases in the service fees or annual budget if a large, unexpected expenditure becomes necessary.

11.8 FINANCIAL STATEMENTS

Financial statements are written records that convey the business activities and the financial performance of a company or public utility. Financial statements include the balance sheet, cash flow statement, and income statement. The financial statement will indicate whether a particular company is financially positioned to perform on a particular project. They should be audited by an independent agency to ensure accuracy.

An example financial statement is shown in Table 11.2. The assets and liabilities must balance. Assets include current assets and long-term assets. *Current assets* are either cash or assets that are expected to naturally convert to cash within a period of 12 months. The single largest current asset for most firms is *accounts receivable.* This is money owed to the firm or utility (outstanding invoices), but not yet paid by clients.

Long-term assets are items such as real estate, vehicles, computers, equipment, and office furniture. The current value of a long-term asset is the depreciated value, which is the value when new minus accrued deprecation. The table shows the value of property when new as $2,000,000 and the accrued depreciation of $1,300,000, giving a current value (book value) of $700,000.

TABLE 11.2 Example Balance Sheet

Assets	Budget ($)
Current assets	
Cash in bank	200,000
Accounts receivable	3,200,000
Accounts receivable retainage held	200,000
Notes receivable	20,000
Inventory	150,000
Prepaid expenses	20,000
Total current assets	3,790,000
Long-term assets	
Equipment – value when new	2,000,000
Accrued depreciation – equipment	(1,300,000)
Vehicles – value when new	200,000
Vehicles – accrued depreciation	(160,000)
Office equipment – value when new	175,000
Office equipment – accrued depreciation	(75,000)
Land	500,000
Total long-term assets	1,340,000
Total assets	5,130,000
Liabilities	
Current liabilities	
Accounts payable	900,000
Retainage payable	200,000
Payroll deductions	10,000
Federal withholding	120,000
State withholding	3,000
Accrued salaries	6,000
Accrued interest	6,000
Current portion – long term debt	60,000
Installment payment	10,000
Note – current payment	60,000
Capital lease	800,000
Total current liabilities	2,175,000
Long-term liabilities	
Bank note	300,000
Long-term debt	500,000
Total long-term liabilities	800,000
Total liabilities	2,975,000
Owner's equity	
Common stock	60,000
Retained earnings	2,595,000
Shareholders distribution	(500,000)
Total stockholder equity	2,155,000
Total equity and liabilities	5,130,000

Liabilities are also characterized and grouped as either current or long-term, depending on whether the debt or obligation is due and payable within 12 months or beyond 12 months.

Owner's equity is sometimes referred to as the net worth or book value of the firm:

$$\text{Owner's Equity} = \text{Total Assets} - \text{Short-term Liabilities} - \text{Long-term Liabilities}$$

11.9 BALANCE SHEET RATIOS AND INDICATORS

The *debt-to-worth* ratio is total liabilities divided by owners' equity. This ratio is a primary measure of organization's financial health that banks consider when extending credit. A debt-to-worth ratio of 2.0 or less is generally considered reasonable and safe. A debt-to-worth ratio of 4.0 or more signals creditors that a firm may be a credit risk. A firm in this condition is said to be highly *leveraged* (the amount of debt in the firm relative to the owners' equity stake).

Another balance sheet measure is *current ratio*, defined as current assets divided by current liabilities. A current ratio of 2.0 or higher is desirable. This would say that for every dollar of debt obligation due over the coming 12 months, the firm has \$2 of cash (or assets that will convert to cash over that period) to meet its obligation. Creditors consider the current ratio as a key indicator of how likely a firm is to make its future payments in a timely fashion. The higher the current ratio, the more *liquidity* the firm is said to have.

Example 11.4 Balance Sheet Analysis

The data in Table 11.2 can be used to calculate these indicators of the organization's financial condition.

$$\text{Owner's equity} = \text{Total assets} - \text{Short-term liabilities} - \text{Long-term liabilities}$$

$$= \$5{,}130{,}000 - \$2{,}175{,}000 - \$800{,}000 = \$2{,}155{,}000$$

$$\text{Debt-to-worth ratio} = \frac{\text{Total liabilities}}{\text{Owner's equity}} = \frac{\$2{,}175{,}000 + \$800{,}000}{\$2{,}155{,}000} = 1.38$$

$$\text{Current ratio} = \frac{\text{Current assets}}{\text{Current liabilities}} = \frac{\$3{,}790{,}000}{\$2{,}175{,}000} = 1.74$$

11.10 CONCLUSION

Financial management is about understanding budgets, cash flow, and accruals of revenue and expenses. This chapter should assist engineers who need to work with accountants and other financial management specialists.

11.11 PROBLEMS

11.1 A PERSONAL BUDGET 1

Put yourself in your senior year of high school and make a zero-based budget for the first year of college. The budget is zero-based because you have no direct historical experience on which to build a budget. The budget should be assets (savings, income, loans, gifts) and liabilities (tuition, travel, room and board, books, etc.). Assets and liabilities must balance.

11.2 A PERSONAL BUDGET 2

Today you have experience managing a budget for education. Make a budget for next year based on this experience. Show last year's assets and liabilities and the projections for next year.

11.3 DEBT COVERAGE 1

What is the debt service converge for a utility with revenues of \$18 million, O&M expenses of \$8 million, and debt service of \$7 million?

11.4 DEBT COVERAGE 2

A utility that has net revenue of \$22,000,000 and debt of \$9,000,000 wants to increase its debt by \$8,000,000. What is the current debt service coverage? What will be the debt service coverage if the utility takes on the additional debt? How will its ability to borrow and potential bond rating be affected?

11.5 ACCOUNTING

A utility pays \$69,000 every six months for chemical reagents (one bill in January and one in July). (a) Under cash accounting, when will the utility record the expense \$69,000 and in what amount? (b) Using accrual accounting, how will the expenses be recorded?

11.6 BALANCE SHEET

Table P11.6 list assets and liabilities typical of those in balance sheets. Organize the entries into current assets and long-term assets and give the amount for the total assets. Do the same for the liabilities. Total assets should equal total liabilities. Check that this is correct.

TABLE P11.6 Balance Sheet

Assets	Amount ($)	Liabilities	Amount ($)
Cash in bank	200,000	Accounts payable	900,000
Accounts receivable	3,200,000	Retainage payable	200,000
Equipment – value when new	2,000,000	Payroll deductions and withholdings	130,000
Equipment – accrued depreciation	1,300,000	Accrued salaries	80,000
Vehicles – value when new	200,000	Current loan payment	210,000
Vehicles – accrued depreciation	160,000	Current bank note payment	160,000
Office equipment – value when new	175,000	Capital lease	800,000
Office equipment – accrued depreciation	75,000	Bank note	560,000
Land	500,000	Long-term debt	1,700,000

11.7 BALANCE SHEET ANALYSIS

Use these financial data to calculate the owner's equity, debt-to-worth ratio, and current ratio for a consulting firm.

Current assets = \$2,100,000
Long-term assets = \$2,800,000
Current liabilities = \$2,375,000
Long-term liabilities = \$800,000

11.8 EQUITY AND WORKING CAPITAL

Use the data in Table P11.8 to calculate for 2012 and 2011, the company's (a) equity (net worth) for 2012 and 2011, (b) working capital, and (c) current ratio. Table values are in US$.

TABLE P11.8 Company's Balance Sheet

	2012	2011
Assets		
Current assets	2,305,078	1,877,676
Retention money	6,124,992	5,837,658
Accounts receivable	1,743,663	1,659,415
Material inventory	942,765	761,763
Costs and estimated earnings in excess of billings on work in progress	581,221	486,472
Prepaid expenses and others	1,000,026	1,062,968
Total current assets	12,697,745	11,685,952
Fixed assets		
Property and equipment	2,655,000	2,580,000
Construction plant	806,200	800,000
Vehicles/trucks	414,560	310,000
Furniture and fixtures	69,500	60,100
Total depreciable assets	3,945,260	3,750,100
Less accumulated depreciation	2,051,900	1,663,400
Net fixed assets	1,893,360	2,086,700
Total assets	14,591,105	13,772,652
Liabilities		
Current liabilities		
Accounts payable	3,930,309	3,481,330
Accrued expenses	1,441,215	1,076,450
Notes payable	588,149	358,817
Retention money payable	835,495	551,763
Billings in excess of costs and estimated earnings on work in progress	560,847	495,167
Other current liabilities	323,232	213,478
Total current liabilities	7,679,247	6,177,005
Long-term liabilities	1,480,513	1,901,445
Total liabilities	9,159,760	8,078,450
Equity (net worth)		
Capital stock	2,000,000	2,000,000
Additional paid-in capital	800,000	800,000
Retained earnings	2,631,345	2,894,202
Total equity	5,431,345	5,694,202
Equity + total liabilities	14,591,105	13,772,652

12 Utility Service Revenues and Rate Making

Most service-based organizations pay their operating costs and debt service with fees and user charges collected from their customers. Rate making is the process of establishing the pricing structure that will provide the required revenue. Rates may be subject to consumer and legal challenge and, therefore, must be equitable, logical, and defensible.

The fundamental requirements for rate making are as follows:

- Rates are set to raise the total revenues needed to operate the utility.
- Customers are expected to pay in proportion to their use and benefit.

Rates are usually adjusted annually as part of the annual budget cycle. Multi-year rate periods are more likely to be used when an overseeing regulatory agency, like a public service commission, approves rate changes and substantial documentation is required to support a rate change.

Most of this chapter is about wastewater treatment. Water utilities charge customers according to the volume of water that they use. Water use is metered so the volumes are known and it is easy to understand how charges are calculated.

Wastewater charges are more complicated because pollutant loadings and volume affect the cost of treatment. Wastewater collection and pumping are a function of volume. Wastewater volume is not measured for individual discharges, but it is some fraction of water usage, and wastewater charges are often tied to metered water sales for residential customers. Industries will pay additional charges based on pollutant strength and flow if they discharge wastewater to a city for treatment.

The examples in this chapter are intended to explain the fundamental concepts, and they are not general rules for setting rates.

12.1 REVENUES AND EXPENSES

Revenues are all monies received by the utility from fees, user charges, interest, rents, and other sources. User service charges for the collection and treatment of wastewater account for almost all of a system's total revenue. Other revenue sources are:

- Connection fees – charges for new customers to connect to the existing system
- Taxes and assessments – special front footage or other annual charges
- Interest earnings – interest earned from investments like checking accounts, savings accounts, and treasury bills
- Other revenues, including sales of treatment by-products such as soil amendments, digester gas, and penalties charged for not paying the sewer bill on time.

Expenses are costs of operating, maintaining, paying debt service, and replacing equipment. Replacement costs are for obtaining and installing equipment, accessories, or appurtenances during the useful life of the treatment works necessary to maintain the design capacity and performance.

12.2 WATER AND WASTEWATER RATES

Water and wastewater services, like a two-way toll bridge, are paid for coming and going. Turning lake, river, or groundwater into safe drinking water has a cost. Cleaning the resulting wastewater for its return to the environment adds another cost.

Customers are sometimes surprised to find their sewer bill is as much or more than their water bill. Some reasons for cities having such different drinking water rates relate to the quality of the drinking water source, the extent of treatment needed, and the cost of electricity. Groundwater requires minimum treatment, sometimes only disinfection. River water usually requires coagulation, settling, filtration, and disinfection. Hard water and brackish water are treated by removing dissolved minerals.

Wastewater treatment costs are determined mainly by the effluent levels required by the discharge permit that specifies which pollutants must be removed and the allowable effluent concentrations. Sludge treatment and disposal are expensive, and some cities cannot use the lower cost options such as application to farmland.

The geography and topography of the service area influence the costs of water distribution and wastewater collection. Drinking water is delivered through pressurized pipelines. It can move uphill as well as downhill. This means that water lines can be built in already-cleared road rights-of-way and they can be constructed at minimum depth below the ground surface, often above any rock layers.

Most sewer systems have some pumping stations and force mains (pressurized lines), but the preferred design is for sewage to flow downhill by gravity as much as possible. In rolling terrain, sewer lines must sometimes be built deep beneath the ground surface, sometimes into hard rock. Trench excavation is the largest part of the cost of building sewers and pipelines. The deeper the pipe, the higher the cost of construction. Gravity sewers are built in low-lying areas, so lateral sewers from buildings will be at a higher elevation. Rights-of-way must be acquired and cleared, adding to overall costs.

12.3 EXAMPLES OF UTILITY BILLING METHODS

Cities usually bundle the charges for drinking water, wastewater collection and treatment, and may also bundle solid waste collection and disposal fees. Bills are calculated from unit costs: $/m^3 or $/1,000 gal of water; $/kg or $/lb of biochemical oxygen demand (BOD) or total suspended solids (TSS), or $/T of solid waste. Deriving these unit costs is the essence of rate making and setting user charges.

The next example is for residential customers. Commercial establishments and industries are billed using different rates and formulas.

Example 12.1 Water, Sewer, Stormwater, and Urban Forestry Rates

Residential customers in the City of Madison, Wisconsin, get a monthly bill for water, sewer service, landfill, stormwater, and urban forestry. The City of Madison has its own water utility. It is a customer of the Madison Metropolitan Sewerage District (MMSD) which bills the city for wastewater treatment and certain costs for wastewater collection and pumping. The city adds its own wastewater-related costs for local sewers, pumping stations, administration, etc., to the amount it must pay MMSD and bills its individual customers. Thus, there are two coordinated rate-setting systems: MMSD's for billing the city, and the city's for billing its residents and businesses.

The 2019 rates for single family homes and duplexes include a fixed monthly base charge per connection (meter charge) and a volume-based charge, as given in Table 12.1.

The monthly water and sewer charges for a household using 5,000 gallons of water per month were as follows:

Water: Fixed base charges (per 5/8-inch meter) = $11.74

$$\text{Volume-based charge} = (\$3.41/1{,}000\ \text{gal})(3{,}000\ \text{gal}) + (\$4.55/1{,}000\ \text{gal})(2{,}000\ \text{gal})$$

$$= \$19.33$$

TABLE 12.1 Residential Water Rates for Madison, Wisconsin

Volume	\$/1,000 gal
First 3,000 gal	\$3.41
Next 3,000 gal	\$4.55
Next 3,000 gal	\$5.46
Next 5,000 gal	\$7.85
Over 14,000 gal	\$9.40

$$\text{Total monthly water bill} = \$11.74 + \$19.33 = \$31.07$$

Sewer: Fixed base charges (per 5/8-inch meter) = \$14.19

$$\text{Volume-based charge} = (\$3.03/1{,}000\text{ gal})(5{,}000\text{ gal}) = \$15.15$$

$$\text{Total monthly sewer service charge} = \$14.19 + \$15.15 = \$29.34$$

In addition, there are monthly charges for landfill, stormwater, and urban forestry

Many industries do not want to own and operate a wastewater treatment plant, nor do they want to deal with state regulations for discharge permits, monitoring, and monthly reporting. Instead, they pay to discharge wastewater to municipal sewers. If the strength is not more than typical municipal sewage, the fee is the same as for residential users. If the concentrations of BOD, TSS, nitrogen, or phosphorus exceed typical municipal sewage concentrations, there is a surcharge to cover the additional cost of treatment.

Example 12.2 Wastewater Charges for an Industry

Calculate the wastewater charges for an industry that discharges to the municipal sewer system a flow of 5,000 m^3/d at a strength of 500 mg BOD/L, 500 mg TSS/L, and 1 mg TP/L. There will be charges for BOD and TSS because the concentrations exceed the municipality's typical wastewater values of 200 mg/L BOD and 210 mg/L TSS. There will be no charge for nitrogen or phosphorus because they are below the average municipal concentration.

The BOD charge will apply to 500 – 200 = 300 mg BOD/L and the TSS charges will apply to 500 – 210 = 290 mg TSS/L. For convenience, convert the concentrations to 0.3 kg BOD/m^3 and 0.29 kg TSS/m^3.

The rates that apply are a volume charge of \$0.032/m^3 plus \$0.40/kg BOD and \$0.20/kg TSS.

$$\text{Charge for volume} = (\$0.032/\text{m}^3)(5{,}000\text{ m}^3/\text{d}) = \$160/\text{d}$$

BOD charge is based on an average daily load of

$$= (5{,}000\text{m}^3/\text{d})(0.3\text{ kg BOD/m}^3) = 1{,}500\text{ kg BOD/d}$$

$$\text{BOD charge} = (1{,}500\text{ kg BOD/d})(\$0.40/\text{kg BOD}) = \$600/\text{d}$$

TSS charge is based on an average daily load of

$$= (5{,}000\text{m}^3/\text{d})(0.29\text{ kg TSS/m}^3) = 1{,}450\text{ kg TSS/d}$$

$$\text{TSS charge} = (1{,}450\text{ kg TSS/d})(\$0.20/\text{kg TSS}) = \$290/\text{d}$$

Total charge = \$160/d + \$600/d + \$290/d = \$1,050/d

Assuming a 30-day billing period, the industrial wastewater treatment charge is

(\$1,050/d)(30 d/month) = \$31,500/month

Charges for solid waste collection and hauling to the landfill are usually managed by the municipalities or by private contractors. Revenues can be collected by taxes, at least for residential users, or by user charges. Reclamation and recycling charges will be included. Leaf collection may be added as a separate charge.

Landfill operation may be handled by the municipality or by the county, which often operates a regional landfill. Major costs are to provide daily cover and the final closure of cells. Collection and treatment of leachate is another cost. Gas collection, processing, and utilization may be a cost or revenue stream, depending on how the resource is managed.

Rates must be set for single-family homes, multi-family residences, and non-residential properties that generate municipal-like material. Wastes that are not from normal household activities are regulated by state law. Hazardous wastes typically do not go into municipal or county landfills. Demolition debris has the potential to contain asbestos, lead, mercury, and other substances that should not go into a municipal landfill. Clean demolition debris might be allowed in a sanitary landfill, subject to local regulations and tipping fees.

Example 12.3 Solid Waste Service Charges

The city of Cincinnati, Ohio, charges fees for solid waste collection and disposal, including leaf collection, via real estate property taxes. The annual charges for a single-family home in 2017 were:

Collection and hauling (tipping fee) = \$56/ton

Base charge =(0.88429 tons/household)(\$56/ton) = \$49.52/household

Landfill disposal charge = \$29.71/household

Leaf vacuuming charge = \$97.99/household

Direct costs = \$49.52 + \$29.71 + \$97.99 = \$177.22/household

Indirect costs = \$195.88/household

Total charges = \$373.10/household

Indirect costs cover county-based systems disposal costs.

Multi-family residential and non-residential properties are charged for larger quantities of solid waste, with proportionally larger indirect costs.

12.4 ALLOCATION OF COSTS TO CUSTOMER CLASSES

The rest of this chapter is about wastewater collection and treatment systems. The *sewerage system* of a wastewater treatment utility is defined as the wastewater collection system, the treatment works, and the final disposal of its treated effluent and waste solids.

The collection system includes local sewers, trunk sewers, interceptor sewers, pumping stations, and appurtenances. A small town or city may own and operate all of these facilities. If there is a regional sewerage district, the district will own and operate the treatment plant, the interceptors, and some of the pump stations, while the cities and towns that belong to the district own the rest of the infrastructure (local sewers, pumping stations, etc.). The owner/operator will levy charges on users to pay for operation and maintenance, and also for new construction or replacements.

The fundamental principle of rate making is to recover the costs of providing services from those who benefit from those services. The degree of service provided is different for residential, commercial, and industrial customers, and rate structures must be devised for the different classes of service.

Commercial users include hotels, motels, restaurants, and retail stores. The commercial class may be further segregated into large or small users, based on average wastewater volume. If there are administrative advantages or fairness issues associated with cost allocations, the commercial class could be further divided into categories such as bakeries, gasoline stations, restaurants, etc. Each user is unique, yet the administrative burden of characterizing the waste from every user to ensure an "exactly proportional" bill would not be sensible. Each rate maker must decide what balance makes sense between the cost of administering the user charge system and fairness to the customers.

An *industry*, for purposes of user charges, is any space or area occupied for the purpose of manufacturing that results in the discharge of industrial wastes into the municipal treatment works. The industrial classification normally includes customers who discharge very large volumes or high pollutant loads. Such users include food processors, packing plants, industrial production facilities, and similar users.

Industries are charged a share of the municipal treatment cost on the basis of the volume and strength of the wastewater discharge. Individual monitoring of industrial customers is often justified based on the size of the bill and the potential for the discharge to have a direct and noticeable effect on the conveyance or treatment system. This assures fairness to the industry and the municipality. It also encourages industries to reduce wastewater volume and strength by practicing pollution prevention and control.

Specialty wastes include septage, holding tank wastes, grease pit discharges, landfill leachate, and other hauled wastes. A discharger located outside of the municipal jurisdiction will be charged more than one within the municipal or district boundaries.

12.5 COST-CAUSATIVE FACTORS

Total costs are allocated to *cost-causative factors* that can be linked to a capital or operating cost of a process or system. Costs can be apportioned to collection, processing, and discharge or delivery, as suggested by Table 12.2.

A water utility has identifiable costs for raw water supply, water purification, and treated water delivery, the latter to include construction and maintenance of water mains, pumping stations, and

TABLE 12.2 Major Cost Centers for Drinking Water, Wastewater, and Solid Waste Systems

Drinking Water	Wastewater	Solid Waste
Intakes, wells	Interceptors and sewers	Collection and hauling
Raw water pumping	Pumping stations	Recycling center
Water treatment plant	Treatment plant	Landfill
Pumping and distribution	Sludge disposal	Incinerator
	Effluent outfall	Leachate collection and treatment
		Landfill gas collection, processing, and distribution

storage tanks. Water users are billed according to the volume of water used. The cost per unit of water generally decreases as the volume used increases. Special rates may be provided for large users because, for example, the cost of delivery is less than for residential users.

Refuse volume and mass are factors for a municipal refuse collection and disposal. Charges for collection and hauling can be considered separate from landfill operation. Refuse composition is a factor for sorting and recycling, and for segregating and handling specialty wastes (e.g., toxic and hazardous materials, solvents), which can include leachate collection and treatment and collection, purification, and beneficial use of landfill gas.

Wastewater collection and treatment processes are determined by volume and the type and strength of pollutants. Some parts of the system, an interceptor sewer, for example, are billed according to flow rate or volume. The costs of wastewater treatment and sludge disposal are divided between volume and the mass load of BOD, TSS, and other pollutants.

12.6 WASTEWATER TREATMENT COST ALLOCATION

The most used cost-causative factors for wastewater rate making are as follows:

- *Average volume of wastewater.* Operating and capital costs for pumping are assigned to the volume. The electrical cost for pumping is related to volume. The cost of chemicals is related to the average daily flow at the treatment plant and would be appropriately billed to a volume parameter.
- *Pollutants.* The design of many wastewater treatment unit processes is based on the concentration and mass of pollutants such as suspended solids, BOD, phosphorus, and nitrogen. For example, primary settling tanks are designed based on flow and they function to remove suspended solids. It is reasonable to assign capital and operating costs to flow and suspended solids. The activated sludge process design and operation are related to flow, BOD removal, phosphorus removal, and nitrogen conversion, and costs are allocated to these parameters. Electrical costs for activated sludge aeration or mixing are distributed between flow, BOD, and phosphorus and nitrogen removal.
- *Phosphorus.* Phosphorus can be removed by chemical precipitation or by a modification of the activated sludge process. It leaves the system as part of the sludge solids, so the cost will be apportioned between the removal process and solids handling.
- *Solids handling and sludge treatment.* Solids are removed from wastewater as grit, screenings, and sludge from settling tanks. Grit and screenings are handled as solids. Sludge is 95–99% water and the solids have a high organic fraction. Solids removed in the settling tanks must receive additional treatment. The activated sludge process converts BOD to cell mass (biosolids) that becomes part of the solids handling stream. Costs for solids thickening, digestion, and ultimate reuse are assigned to suspended solids and BOD, with part of the cost being assigned to the source of the solids within the treatment system.
- *Infiltration/inflow* (I/I). Sewer systems are of two kinds – separate and combined. These terms refer to how stormwater is handled. Separate means that stormwater is collected in storm sewers that are not connected to sanitary sewers. Combined sewers collect stormwater and domestic wastewater. Separate is preferred and is used in all new construction. Many large older cities have combined systems and to rebuild them can be prohibitively expensive. *Infiltration* enters the sewer system from groundwater. *Inflow* enters from surface water sources, usually associated with rainfall and snow melt. If the source of *I/I* can be identified, it should be eliminated or the costs for handling it charged to the areas from where it originates. If the source is not identifiable, the costs are assigned to a customer parameter.
- *Number and types of customers.* Administrative costs and laboratory costs are more dependent on the number of customers than the strength or volume of wastewater. Customer charges are similar to the meter charge from a water or electrical utility.

- *General.* There may be special circumstances and charges associated that cannot be neatly defined by cost-causative factors. It is the responsibility of the utility to determine the monetary value of "need" and "benefit" as they relate to a specific customer or customer class.

Example 12.4 Cost Allocation for Interceptor Force Main and Pumping Station

Interceptor sewers have a design life of 50 years or more, and they are designed for peak flow. The peak flow may not materialize for many years, depending on the rate of development of the service area. This means that there is excess capacity for much of the project life. The portion of the cost associated with current users is charged to current users. The excess capacity cost is distributed among all customers.

The design peak flow is 5 mgd and the current peak flow is 2 mgd, so there is 3 mgd of excess capacity.

The annual interceptor debt service is $1,000,000 per year, and the annual O&M cost is $500,000. Debt service is charged 60% to customers (for excess capacity), 35% to flow, and 5% to suspended solids (SS). The O&M cost allocation is 80% to flow and 20% to suspended solids. BOD is irrelevant to the interceptor design and function, but it becomes important at the treatment plant. Table 12.3 shows the allocations by percentage and by dollars.

An allocation of debt service costs and operating costs can be made in a similar manner for every facility or process in the collection system or at the plant for which debt has been incurred.

12.7 CASE STUDY: HYPOTHETICAL SEWERAGE DISTRICT

This case study, described as a hypothetical sewerage district, is based on a project for a large sewerage district. Many details have been simplified or omitted, but the load estimating procedure and the quantities are realistic. The district is required to remove BOD, TSS, nitrogen, and phosphorus. To make the example more concise, the only pollutants considered are BOD and TSS.

The treatment plant uses the activated sludge process. The activated sludge effluent is disinfected by ultraviolet radiation. The waste solids are applied to agricultural lands as a soil supplement. The total service area has 489 miles of sewers and interceptors. Table 12.4 presents the design loadings

TABLE 12.3 Allocation of Debt Service and O&M Costs to Flow, Solids, and Customers

Cost	Total Cost ($)	Percent Allocation			Cost Allocation ($)		
		Flow	SS	Customers	Flow	SS	Customers
Debt service	1,000,000	35	5	60	350,000	50,000	600,000
O&M costs	500,000	80	20	0	400,000	100,000	
				Totals	750,000	150,000	600,000

TABLE 12.4 Wastewater Operating Characteristics

Operating Parameter	Flow	BOD		TSS	
	(mgd)	(mg/L)	(lb/d)	(mg/L)	(lb/d)
Design capacity	64	243	130,000	210	112,000
Effluent limits	64	30	16,000	30	16,000
2019 influent load	42	323	113,000	304	106,500
2019 discharge permit	42	25	8,800	29	10,200

TABLE 12.5 Waste Load Contributions by User Class

Source	Flow (mgd)	BOD (lb/d)	TSS (lb/d)	Connections (each)
Estimated				
Residential[a]	31.2	38,300	60,600	45,000
Commercial – employees	2.5	4,200	5,300	2,700
Industrial – employees	0.5	850	1,060	
Industrial – process waste	8.6	74,300	38,600	180
Total estimated waste load	42.8	117,650	105,560	
Observed	42.0	113,000	106,500	47,880
Estimated as percent of observed	102%	104%	99%	

[a] Includes dry weather infiltration.

and the current (2019) loadings of the treatment plant. The design is for 64 mgd; the current loading is 42 mgd. The plant operates below the designed capacity, and the effluent quality is well below the discharge permit limits for BOD and TSS.

The estimated and observed wastewater contributions by residential, commercial, and industrial users are given in Table 12.5. The estimates will be explained.

Residential users include single-family houses and apartments up to four units. The estimated total connected residential population of the service area is 202,000. Using typical per capita contributions of flow, BOD and TSS give an average total residential load of:

$$\text{Flow} = (100 \text{ gal/cap-d})(202{,}000 \text{ persons}) = 20.2 \text{ mgd}$$

$$\text{BOD load} = (0.167 \text{ lb BOD/cap-d})(202{,}000 \text{ persons} = 33{,}700 \text{ lb BOD/d}$$

$$\text{TSS load} = (0.209 \text{ lb TSS/cap-d})(202{,}000 \text{ persons}) = 42{,}300 \text{ lb TSS/d}$$

Infiltration and inflow (*I/I*) are included in the residential flow. Ninety-two percent of infiltration and inflow are associated with sewers in residential areas; this flow has been assigned to the residential class. *I/I* is estimated to be 11 mgd. Infiltration is clean water. Inflow is not, and its strength is estimated at 50 mg/L for BOD and 200 mg/L for TSS. This gives the total *I/I* load on the sewer system of flow = 11 mgd, 4,600 lb BOD/d, and 18,300 lb/d.

An inflow of 11 mgd in a total flow of 42 mgd is a burden on the treatment facilities, not so much because of the volume but because it is delivered as peak flow during storm events.

Commercial users are businesses, industries, and institutions that discharge primarily domestic wastewater. The 2,700 commercial connections are charged at the residential rates. The daily commercial domestic flow was estimated at 18 gal per employee per day. There are an estimated 141,000 commercial employees.

$$\text{Flow} = (18 \text{ gal/emp-d})(141{,}000 \text{ employees}) = 2.54 \text{ mgd}$$

$$\text{BOD} = (8.34)(2.54 \text{ mgd})(200 \text{ mg BOD/L}) = 4{,}200 \text{ lb BOD/d}$$

$$\text{TSS} = (8.34)(2.54 \text{ mgd})(313 \text{ mg TSS/L}) = 5{,}300 \text{ lb TSS/d}$$

Industrial Users. The service area has 180 industrial contributors. Twelve were classified as major industries with large flows and/or high strength wastewater, generally discharging more than 300,000 gpd of process waste or more than 500 lb/d of BOD and TSS. They accounted for 80%

of the total process flow and 95% of the industrial process wastewater BOD and TSS load. These industries will be charged according to their measured flows and waste loads.

Smaller industries can be certified, so process waste strengths and charges are based on typical values for their industrial class. The advantage is that monitoring is not required.

The total industrial process load was estimated to be

$$\text{Process flow} = 8.6 \text{ mgd}$$

$$\text{BOD} = 74{,}340 \text{ lb/d}$$

$$\text{TSS} = 38{,}000 \text{ lb/d}$$

Industrial users have 28,100 employees that generate 18 gal/cap-d of flow that is assumed to have the strength of residential domestic sewage. The domestic waste load for industrial employees is:

$$\text{Flow} = (18 \text{ gal/emp-d})(28{,}100 \text{ employees}) = 0.51 \text{ mgd}$$

$$\text{BOD} = (8.34)(0.51 \text{ mgd})(200 \text{ mg BOD/L}) = 850 \text{ lb BOD/d}$$

$$\text{TSS} = (8.34)(0.51 \text{ mgd})(250 \text{ mg TSS/L}) = 1{,}060 \text{ lb TSS/d}$$

The *required revenue from user fees* for operation is $27,000,000. The residential, commercial, and industrial users must be charged in proportion to their contributions of flow, BOD, and TSS. Table 12.6 shows the allocation of operating costs to treatment processes and pollution parameters.

TABLE 12.6 Allocation of Operating Costs to Treatment Processes

Cost Basis	Total Cost ($1,000)	Cost Allocation ($1,000)			
		Flow	BOD	TSS	Equivalent Meters
Collection system	1,100	0	0	0	1,100
Primary treatment	2,300	700	0	1,600	
Grit and screenings	1,600	500	0	0	1,100
Activated sludge	4,500	1,400	3,100	0	0
Disinfection	400	400	0	0	0
Organic solids handling	7,600	0	2,300	5,300	0
Reuse	2,700	0	800	1,900	0
General plant	6,800	0	0	0	6,800
Totals =	27,000	3,000	6,200	8,800	9,000
Average daily flow (10^6 gal)		42			
Waste loads (lb/d)			113,000	106,500	
Equivalent meters (each)					47,880
Total annual flow (10^6 gal)		15,330			
Total annual waste load (lb/y)			41,245,000	38,872,500	

Unit Costs based on:

Volume = $195.69/mgd = $0.196/1,000 gal

Pollutant load = $0.150/lb BOD and $0.226/lb TSS

Equivalent meters = $188/y = $15.66/month.

TABLE 12.7 Hypothetical Monthly Bills for an Average Residential, Commercial, and Industrial Customers

Factor	Units	Residential	Commercial	Industrial
Flow	(gal/d)	400	8,000	700,000
BOD	(mg/L)	200	200	2,100
TSS	(mg/L)	250	250	980
Flow	(gal/month)	12,000	240,000	21,000,000
BOD	(lb/month)	20	400	367,794
TSS	(lb/month)	25	500	171,637
Equivalent meters	(Each)	1	6	50
Flow	($/month)	2.35	46.97	4109.59
BOD	($/month)	3.01	60.18	55,287.25
TSS	($/month)	5.66	113.28	38,855.42
Equivalent meters	($/month)	15.66	93.98	783.21
Total	($/month)	26.69	314.41	99,035.47

Costs that are not linked to flow, BOD, or TSS are charged to all users according to equivalent meters. *General plant*, which includes administration and laboratory, workshops, storage, and other activities that are not linked directly to flow, BOD and TSS are charged to meters. A standard residential water meter is 5/8 inch (20 gal/min). Multi-family housing, commercial clients, and industries have larger meters and may be assigned multiple *standard meters*, so they are charged correspondingly more.

The unit costs given in the bottom line of the table are calculated from the total required revenue for each of the four cost measures divided by the annual waste load.

Table 12.7 uses the unit costs in Table 12.6 to calculate monthly bills for hypothetical residential, commercial, and industrial customers.

These user fees do not include debt service to repay loans that purchased capital equipment. Debt service is not an operating cost as defined in Table 12.6, but it is a cost that must be recovered as part the user fee system. The next case study includes debt service.

12.8 CASE STUDY: REGIONAL SEWERAGE DISTRICT

This case study shows more budget categories and introduces revenue billing to pay debt service. The data (from 1995) are for a regional sewerage district that serves four cities, six villages, eight districts, and two government institutions. These 20 customers receive a quarterly bill from the sewerage district. The cities, villages, and districts pass this cost along to their customer classes using whatever rate structure they feel is appropriate.

Table 12.8 gives flow and waste loads. These are based on pumping records and quarterly monitoring of wastewater composition at key locations in the collection system. (*Note:* These costs and flows are real, but they are not current.)

Annual wastewater volume = 12,953 million gal
Annual interceptor infiltration = 1,962 million gal
Annual BOD load = 25,789,075 lb
Annual TSS load = 17,979,533 lb
Annual total N load = 2,836,415 lb
Total equivalent meters = 76,108
Total customers = 61,047

TABLE 12.8 Average Flows and Waste Loadings to the District Wastewater Treatment Plant

Customer	Flow (gal/d)	BOD (lb/d)	TSS (lb/d)	Total N (lb/d)	Equivalent Meters	Customers
City A	1,450,000	1,400	2,500	255	3,320	2,060
City B	26,700,000	42,000	39,500	6,150	56,760	46,550
City C	1,750,000	21,000	2,100	390	4,660	3,550
City D	1,200,000	1,750	1,350	275	3,360	2,700
Villages	2,300,000	4,325	3,800	697	7,710	6,185
Districts	1,988,369				10	
Government institutions	98,250	180	9	4	288	2
Daily loads	35,486,619	70,655	49,259	7,771	76,108	61,047

Table 12.9 is the budget for expenses and revenues that are not derived from user charges. The total cost for each budget category has been allocated to the cost-causative factors. The cost allocations of most utilities are not this elaborate, but it is instructive to look at a budget in more detail. A very good cost accounting systems is needed that can be used by an engineer with deep knowledge of the system.

The total revenue required for treatment plant operation is $6,229,088. Revenue from rents, interest, and other non-user fee sources is $328,972. The total revenue required from user charges is $5,900,116.

Annual debt service is $1,999,572 to repay loans that were used to build interceptor sewers, pumping stations, and expansions of the treatment plant. The facilities that were constructed with each loan are known, as are their costs, so debt service allocation is straightforward. The required user charge revenue for operations is $5,900,116 (from Table 12.9) plus the $1,999,572 debt service gives a total of $7,899,688 total user charge revenue for the budget year.

Table 12.10 converts the revenue requirements to unit charges for the six cost categories. For example,

$$\text{Unit charge for volume} = \frac{\$2{,}279{,}072}{12{,}953 \text{ million gal}} = \$175.95/\text{million gal}$$

$$\text{Unit charge for BOD} = \frac{\$1{,}486{,}772}{25{,}789{,}075 \text{ lb BOD}} = \$0.05765/\text{lb BOD}$$

The annual billing rate for city B is

Volume charge	($175.95/million gal)(9,746 million gal/y) = $1,714,802/y
BOD charge	($0.05765/lb BOD)(15,330,000 lb BOD/y) = $883,794/y
SS charge	($0.08542/lb TSS)(14,417,500 lb TSS/y) = $1,231,532/y
Nitrogen charge	($0.4318/lb N)(2,244,750 lb N/y) = $969,284/y
Equivalent meters	($11.8166/meter)(56,760 meter) = $670,712/y
Customers	($7.7637/customer)(46,550 customers) = $361,399/y
Total annual bill	$5,831,523/y

TABLE 12.9 Allocation of Budget Categories to Cost-Causative Factors

Budget Category	Amount	Allocations ($) to Cost-Causative Factors					
		Volume	BOD	TSS	Total N	Meters	Customers
Expenses							
Administration	$667,675	–	–	–	–	667,675	–
Collection system							
User charge/ pretreatment	$136,400	–	–	–	–	136,400	–
General services	$4,800	4,320	–	480	–	–	–
Pumping stations	$752,655	677,390	–	75,266	–	–	–
Interceptors	$83,750	75,375	–	8,375	–	–	–
Collection system Subtotal =	*$977,605*	*757,085*	–	*84,121*	–	*136,400*	–
Treatment and disposal							
General salaries	$1,260,580	288,673	343,382	281,614	323,087	23,825	–
General supplies	$16,400	3,608	4,625	3,477	4,248	443	–
Power	$1,079,000	385,203	363,623	56,108	249,249	24,817	–
Replacement parts	$250,000	55,000	70,500	53,000	64,750	6,750	–
Chemicals	$8,200	5,199	869	631	1,501	–	–
Vehicle operation	$44,650	10,359	12,949	9,734	9,198	2,411	–
Computer expenses	$53,250	13,313	13,313	13,313	13,313	–	–
Laboratory expenses	$158,060	–	52,687	52,687	52,687	–	–
Contract services	$24,600	49	3,370	9,668	3,346	8,167	–
Effluent diversion	$147,460	147,460	–	–	–	–	–
Sludge reuse	$1,240,600	–	297,744	744,360	198,496	–	–
Utilities and miscellaneous	$83,150	17,628	22,617	23,781	17,794	1,330	–
Subtotal =	*$4,365,950*	*926,491*	*1,185,678*	*1,248,371*	*937,667*	*67,743*	–
Outlay expenses	$123,750	22,028	24,750	44,550	22,399	10,024	–
Community pump stations	$94,108	84,697	–	9,411	–	–	–
Total revenue required	*$6,229,088*	*1,790,317*	*1,210,428*	*1,386,453*	*960,065*	*881,842*	–
Revenue sources							
Servicing pumping	$94,108	84,697	–	9,411	–	–	–
Pretreatment	$3,400	–	–	–	–	–	3,400
Septage treatment	$15,000	–	3,150	7,800	1,500	2,400	–
Sludge reuse program	$20,000	–	4,800	12,000	3,200	–	–
Rent	$19,764	4,348	5,573	4,190	5,119	534	–
Interest	$175,000	50,225	33,950	39,025	26,950	24,850	–
Miscellaneous	$1,700	488	330	379	262	241	–
Total revenue =	*$328,972*	*139,758*	*47,803*	*72,805*	*37,181*	*28,025*	*3,400*
Revenue required from user charges for operating costs	$5,900,116	1,650,524	1,162,624	1,313,648	922,885	853,817	(3,400)

TABLE 12.10 User Charges to Pay for Operation and Debt Service

Cost Item	Total	Volume	BOD	TSS	Total N	Equivalent Meters	Customers
Revenue requirements							
Operations ($)	5,900,116	1,650,542	1,162,624	1,313,648	922,885	853,817	−3,400
Debt service ($)	1,999,572	628,530	324,148	222,150	301,881	45,523	477,341
Total ($)	*7,899,688*	*2,279,072*	*1,486,772*	*1,535,798*	*1,224,766*	*899,340*	*473,941*
Annual loadings (million gal and lb)		12,953	25,789,075	17,979,535	2,836,415	76,108	61,046
User charge unit costs ($/million gal and $/lb)							
Operating cost		$127.43	$0.04508	$0.07306	$0.32537	$11.2185	($0.0557)
Debt service		$48.52	$0.01257	$0.01236	$0.10643	$0.5981	$7.8194
Total user charges		*$175.95*	*$0.05765*	*$0.08542*	*$0.43180*	*$11.8166*	*$7.7637*

12.9 JOINT TREATMENT

Joint treatment of municipal and industrial wastewater is common. Cooperation depends on finding a fair allocation of costs. These are some reasons why this may require negotiation:

Design period. The design period for municipal wastewater treatment facilities is usually 20 years and the design period for interceptor sewers is usually longer. Industry typically plans five to ten years ahead. If industry constructs its own treatment plant, it would finance only the facilities needed for its present load plus short-term growth. It would not build excess capacity. The municipal facility, in contrast, will have excess capacity in treatment and collection works.

Peak capacity. Municipal interceptor sewers and treatment works may be designed for a peak-to-average flow variation of 3 to 1 or 4 to 1. An industry that operates 24 h/d, 7 d/week, may have a peak-to-average flow ratio of 1.25.

Implementation timing. The municipal planning-design-construction period may be three to five years. During this time, costs escalate. Industry can do a private project in 1–1.5 years.

Design features.

- Municipal facilities are designed for longer life spans and lower maintenance. The cost of paying interest on a 20-year bond debt and annually retiring a portion of that debt costs more than the industry would pay for a project that is built faster and planned to serve for a shorter lifetime.
- Aesthetics: Industry does not have to consider the aesthetics of the treatment plant. The average cost of architectural treatment, site development, and landscaping ranges from 5% to 10% of the total cost of a municipal project. Industry will spend much less than this on a private project.
- Utility services: Industrial complexes can usually furnish normal utility services to an industrial treatment facility at a minimum cost. Electric distribution systems, water supply, and compressed air are often available. Electricity is purchased at wholesale rates. A municipal plant may be located in a remote area, requiring utility services to be brought some distance, and at a substantial cost. In contrast, industrial plants would likely only require some additional utility capacity such as an expanded substation or larger booster pump to support the upgrade.

- Degree of redundancy: Municipalities are subject to regulations in the area of redundancy and fail-safe systems. Industry has different options. One is to reduce or cease production when there is a loss of treatment capability. And, industry often has a unique maintenance situation since the entire operation may be shut down one or two weeks a year in order for major maintenance to be performed.
- Staffing: Municipal treatment works are usually semiautonomous with a staffing structure that is independent of other municipal operations. An industry adding a treatment plant may already have the necessary engineering and maintenance services, as well as the purchasing and accounting services.
- Sludge disposal: A large portion of the costs of municipal systems is associated with sludge treatment and disposal. Industry may have different options, especially if the sludge does not contain putrescible matter or pathogens.
- Disinfection: The entire municipal wastewater effluent flow must be disinfected. Some industrial wastewaters do not need disinfection, particularly those that are pretreated for discharge into a municipal sanitary sewer.

12.10 CONCLUSION

An equitable rate structure for public utilities is designed to accommodate the local situation. The only givens are that revenues must at least meet expenses, debt service must be kept current, and the rates must be fair and equitable. Users pay in proportion to their use of and benefits from the utility that provides service.

Residential and commercial customers have more uniform requirements than industry, which often have large volume or strong wastewater. Industries may prefer to pay for the right to discharge to a municipal sewer rather than deal with discharge permits and effluent monitoring and reporting. The surcharges are based on discharges that exceed the normal residential pollutant concentrations. Industries can reduce surcharge payments by reducing the volume and strength of the wastewater that is discharged. This can be done by pretreatment or pollution prevention and control.

12.11 PROBLEMS

12.1 WATER AND SEWER CHARGE

The monthly bill for water and sewer services includes

- A fixed water charge of \$2.80 per customer account
- A fixed sewer charge of \$3.10 per customer account
- Usage charge – residential users pay a usage charge according to Table P12.1.

TABLE P12.1 Residential Water and Sewer User Charges

Units of Usage	Usage per Month		
kgal (1,000 gal)	<9	9–16.5	>16.5
Billing rate (\$/kgal)	\$1.71	\$2.81	\$5.53
100 ft^3 (Ccf)	<12	12–22	>22
Billing rate (\$/Ccf)	\$1.28	\$2.10	\$4.14
Cubic meters (m^3)	<34.8	34.8–63.8	>63.8
Billing rate (\$/m^3)	\$0.44	\$0.72	\$0.42

A customer is a residence, multi-unit apartment, or commercial establishment that is served by one meter. The average US residential customer uses about 9,000 gal/month (11 Ccf or 34.8 m^3).

Three measures are shown for usage. Some US utilities have water meters that measure in units of one hundred cubic feet (Ccf = 100 ft^3). Others measure in gallons or 1,000 gal. Most of the world will use a metric measure, either liters or cubic meters.

Calculate the monthly bill for customers that use (a) 11 Ccf, (b) 10,000 gal/month, (c) 24,000 gal/month, and (d) 14 m^3/month.

12.2 STORMWATER MANAGEMENT BILLING RATES

A common rate basis for stormwater management is impervious area associated with a typical residential property, or an equivalent residential unit (ERU). This measure can be developed for single-family homes, multi-family homes, apartments, manufactured or mobile homes, and condominiums or townhouses. The average impervious area for a single-family home is 2,000–3,000 ft^2. For this problem, the stormwater management fee will be added to the monthly utility bill based on ERUs served by the utility. The number of ERUs in year 1 is 61,000; this increases by 1,000/y.

Table P12.2 gives the estimated budget items for the first three years of the project. Salaries account for almost half the cost. Mapping and new equipment are large costs in year 1. Construction commences in year 2 and increases in year 3. Add $10,000/y for contingencies. Available revenue collected from outside the stormwater billing program is $500,000/y. Calculate the total budget and the monthly billing rate per ERU.

TABLE P12.2 Annual Stormwater Management Billings ($)

Budget Item	Year 1	Year 2	Year 3
Salaries	1,332,900	1,458,400	1,458,400
Expenses	195,650	244,650	244,650
Equipment	747,750	179,900	100,000
Construction	–	500,000	1,000,000
Master Plan	75,000	200,000	100,000
Mapping	400,000	250,000	50,000
Billings	40,000	42,000	44,000

12.3 WATER, SEWER, AND STORMWATER RATES

A city charges customers for water, sewer service, landfill, stormwater, and urban forestry. Calculate the monthly utility bill for a residence that uses 12,000 gal/month for the 2018 rates, which are given in Table P12.3. This bill will include these fixed fees:

TABLE P12.3 Residential Water Rates

Volume-Based User Charge	$/1,000 gal
First 3,000 gal	2.84
Next 3,000 gal	3.26
Next 3,000 gal	3.60
Next 5,000 gal	4.50
Over 14,000 gal	5.07

Fixed base charges = \$6.00 water meter charge

+ \$2.00 fire protection charge

+ \$5.50 stormwater fee + \$14.00 sewerage base charge

+ 0.50 landfill charge = \$28.00/month

12.4 COST OF TREATMENT VERSUS BILLING RATE

A city has an annual water pollution control budget of \$500,000 to treat an average flow of 1,000,000 gal/d. All the users discharging to the sanitary sewer system are on a public water supply. The water company records, which are considered accurate, indicate that a total flow of 800,000 gal/d is being sold to users who are connected to the sewer system. Wastewater treatment costs are recovered via the water bills. For simplicity, assume that user charges will be based on volume. Calculate (a) the actual cost of treating the wastewater in \$/1,000 gal. (b) The billable rate based on the volume of water being sold to users of the sewer system.

12.5 SIMPLE COST ALLOCATION

A process has an annual (amortized) capital cost of \$800,000 and an annual operating cost of \$800,000. The capital cost allocation will be based 80% on flow and 20% on suspended solids. Some BOD removal occurs, but this is coincidental with the removal of solids. The annual O&M cost allocation is based 50% on flow, 40% on suspended solids, and 10% on BOD. Construct a table to show the allocated costs.

12.6 INDUSTRIAL CHARGES

An industry will be charged for discharge to a city sewer if the BOD exceeds 150 mg/L, if the TSS exceeds 250 mg/L, and if TP exceeds 6 mg/L. The waste load averages 200,000 gal/d at a strength of 500 mg BOD/L, 500 mg TSS/L, and 1 mg TP/L. The discharge fees are as follows:

Volume = \$1.337/kgal BOD = \$0.1764/lb

TSS = \$0.0866/lb TP = \$3.1347/lb

Calculate the total daily and monthly user charges.

12.7 REGIONAL WASTEWATER TREATMENT

A regional wastewater treatment plant serves a city and five suburbs. The total revenue required to cover wastewater treatment costs is \$2,838,033. The total annual domestic wastewater flow is 2,214 million gal, or 2,214,000 thousand gallons. The total annual flow is 3,333 million gal. The average concentrations for the domestic wastewater are 160 mg BOD/L, 250 mg TSS/L, and 8 mg TP/L. The BOD, TSS, and TP loads that are billable to user groups are given in Table P12.7a. The allocation of costs for wastewater treatment are in Table P12.7b. Debt service has been allocated to wastewater volume. This is a simplification, but more detail does not help explain the basic ideas of rate making. Calculate the charges for volume (\$/1,000 gal), BOD (\$/lb), TSS (\$/lb), and TP (\$/lb).

TABLE P12.7a Annual Billable Waste Loadings

Wastewater Source	Volume (10^6 gal)	BOD (lb)	TSS (lb)	Total P (lb)
Domestic wastewater	2,214	2,954,362	4,616,190	144,718
Industrial	1,108	1,401,176	1,223,918	12,295
Septage and hauled wastes	11	380,981	361,802	4,383
Total	3,333	4,736,519	6,201,910	164,396

TABLE P12.7b Allocation of Annual Wastewater Treatment Costs

Cost Factor	Total Cost ($)	Allocated Costs ($/y)			
		Volume	BOD	TSS	TP
O&M	1,907,325	275,329	773,845	487,212	370,939
Replacement	300,000	120,000	62,000	65,000	53,000
Debt	630,708	630,708			
Total	2,838,033	1,026,037	835,845	552,212	423,939

12.8 WASTEWATER COLLECTION SYSTEM COST ALLOCATION

The costs of operating and maintaining the wastewater collection system in Problem 12.7 are given in Table P12.8. There are two categories for cost allocation – volume and users. The fixed cost would usually be assigned to customers or connections. For simplicity, the total collection cost shall be allocated to volume. Customers get a combined bill for water and wastewater. Water is billed per 1,000 gallons used and wastewater charges need to be put on the same basis. Calculate the charges for volume (including collection cost), BOD, TSS, and TP on a volumetric basis, i.e. $/1,000 gal.

TABLE P12.8 Allocation of Wastewater Collection Costs

Cost Factor	Total Cost ($/y)
Sewer maintenance	1,040,634
Sewer replacement	178,000
Pumping O&M	265,880
Sewer billing	315,000
Total O&M	1,799,514
Inflow/infiltration	280,000
Interceptor debt service	325,000
Total	2,404,514

12.9 SURCHARGES FOR HIGH-STRENGTH WASTE

A city has an annual water pollution control budget of $500.000, broken down by process as given in Table P12.9. These annual costs include all labor, indirect, energy, and chemical costs. The chief

TABLE P12.9 Annual Operating Costs and Allocations

Unit Process	Annual Cost ($/y)	Allocation by Factor (%)		
		Volume	BOD	TSS
Collection system	75,000	100	0	0
Grit and screening	8,000	0	0	100
Primary clarifier	20,000	80	0	20
Aeration and secondary clarifier	100,000	100	0	
Disinfection	50,000	100	0	0
Sludge processing	150,000	0	50	50
Administrative	62,000			
Reserve fund	35,000			
Total	500,000			

engineer will apportion the operating costs to the various processes based on volume, BOD, and TSS. The average BOD and TSS concentrations are 250 and 280 mg/L, respectively. The average flow is 3,875 m³/d. Administrative costs will be allocated in proportion to the operating cost for each category. Reserve fund allocations are $10,000 to volume, $15,000 to BOD, and $10,000 to TSS based on expected maintenance costs.

(a) Allocate the unit process costs to volume, BOD, and TSS.
(b) Calculate the cost to treat 1 kg of BOD and 1 kg of TSS.
(c) Use the calculated costs to bill an industry that discharges 58 m³/d of wastewater with TSS = 600 mg/L and BOD less than 250 mg/L.

12.10 USER CHARGE ALLOCATION

A regional wastewater treatment plant serves a city and five suburbs. These six user groups are charged to recover the anticipated annual operating cost of $2,847,033. These customers may bill their users on the basis of some other formula, which is usually related to drinking water purchases. The loads that are billable to user groups are given in Table P12.10a. Table P12.10b gives the allocation of costs to O&M, facilities replacement, and debt service.

(a) Calculate the unit costs for volume ($/m³) and mass of pollutant ($/kg).
(b) Combine the four unit charges into a single rate per cubic meter, based on average concentrations for the domestic wastewater of 257 mg BOD/L, 352 mg TSS/L, and 8 mg TP/L.

TABLE P12.10a Annual Billable Waste Loadings

Wastewater Source	Volume (m³)	BOD (kg)	TSS (kg)	Total P (kg)
Domestic wastewater	8,402,851	1,319,253	2,083,403	63,950
Industrial	4,204,329	636,898	556,326	5,589
Septage and hauled wastes	42,884	173,173	164,455	1,992
Total	12,650,063	2,129,324	2,804,185	71,530

TABLE P12.10b Allocation of Annual Wastewater Treatment Costs

Cost Factor	Total Cost ($)	Allocated Costs ($)			
		Volume	BOD	TSS	TP
O&M	1,916,325	275,329	773,845	487,212	370,939
Replacement	300,000	120,000	62,000	65,000	53,000
Debt	630,708	630,708			
Total	2,847,033	1,026,037	835,845	552,212	423,939

12.11 WASTEWATER SURCHARGES

An industry is paying $627,500 to discharge untreated wastewater to a municipal wastewater treatment plant. The user charge can be reduced by pretreatment to remove BOD before discharging to the city sewer. The BOD charge is zero if the BOD is 200 mg/L or less. Pretreatment will not change user charges for wastewater volume. The annual cost of building and operating a pretreatment plant is

$$C = \$40,000(T)^{0.5}$$

where T = hydraulic detention time of the treatment process (minutes).
The percentage of BOD remaining in the effluent is

$$R = \exp(-0.025T)$$

What level of pretreatment is recommended?
The current waste loads and fees are calculated below:

Average flow = 10,000 m^3/d for (5 d/wk)(50 wk/y) = 250 d/y
Annual discharge to municipality = (10,000 m^3/d)(250 d/y) = 2,500,000 m^3/y
Current volume charge = $0.075/$m^3$
Annual volume charge = ($0.075/$m^3$)(2,500,000 m^3/y) = $187,500
Current BOD charge = $0.08/kg BOD on BOD concentration in excess of 200 mg/L
Average BOD = 2,200 mg/L = 2.2 kg/m^3
User charge based on BOD = 2,200 mg/L − 200 mg/L = 2,000 mg/ L = 2.0 kg/m^3
Average daily BOD load = (2 kg/m^3)(10,000 m^3/d) = 20,000 kg/d
Annual BOD load = (2,000 kg/d)(250 d/y) = 5,000,000 kg/y
Annual BOD charge = ($0.08/kg)(5,000,000 kg/y) = $400,000/y
Total annual charges = $587,500

12.12 RATE MAKING: NITROGEN AND PHOSPHORUS REMOVAL

A treatment plant that serves 3,770 equivalent residential units (ERU) does not remove nitrogen (N) or phosphorus (P). The average monthly bill is $18.65/ERU. This will change soon when the plant will be required to remove nutrients (N and P). The effluent limits have not been established, but one of the three cases listed in Table P12.12b will be imposed. The influent nutrient concentrations are TKN = 40 mg/L and TP = 5 mg/L. These will not change over the life of the project. Current and project future (20-year) wastewater loadings are given in Table P12.12a.

The capital costs in Table P12.12b are for the overall plant upgrade needed, including nutrient removal. There will be no allocation of the nutrient removal costs to a flow category. Likewise, the annual O&M cost is the increment that is added to the base O&M cost of the existing plant; the base is the current cost without nutrient removal. These added O&M costs are for biosolids hauling and disposal, ferric chloride, polymer, and increased electricity use. There is no added cost for the increased flow included in this analysis because that is part of the base cost; it is the same with or without nutrient removal.

(a) Convert the capital cost to a uniform annual cost for a 20-year life and 5% interest.
(b) Allocate the total annual costs to the mass influent loads of TKN and TP. The cost allocations are
  Case 1 100% to TP
  Case 2 20% to TP and 80% to TKN
  Case 3 100% to TP
(c) For the first year of the project, estimate the annual and monthly cost increase per ERU, the project monthly bill, and the percent increase.
(d) The median gross annual income (MAGI) is \$28,700/ERU. The state affordability threshold is 1.4% pf MAGI. Do the projected bills meet this threshold?

TABLE P12.12a Costs for the Three Cases

Loading Factor	Current Load	Projected Load
Flow (m^3/d)	3,480	4,844
BOD (kg/d)	906	1,500
TSS (kg/d)	854	1,187
TKN (kg/d)	140	194
TP (kg/d)	17.5	24.2

Note: TKN is total Kjeldahl nitrogen (ammonia plus organic N). TP is total phosphorus.

TABLE P12.12b Class 4 Estimates of Capital Costs

Category	Effluent Limits		Removal		Capital Cost (\$)	Annual O&M Cost[a] (\$/y)
	TP (mg/)	TKN (mg/L)	TP (%)	TKN (%)		
Case 1	1.0	No limit	80	0	629,000	95,800
Case 2	1.0	20	80	50	10,907,000	20,500
Case 3	0.1	No limit	98	0	5,998,000	106,600

Note: TKN is total Kjeldahl nitrogen (ammonia plus organic N). TP is total phosphorus.
[a]Incremental O&M costs are the annual differential costs relative to the base line O&M cost of the existing plant.

12.13 INDUSTRIAL WASTEWATER CONTROL PROGRAM

A city monitors and regulates industries that discharge to the city sewer system. Table P12.13 shows the revenues collected from the industries against expenses for the last three years. Unfortunately, the billings recovered only a fraction of the actual costs. (a) Calculate the unrecovered costs, as $ and % of cost recovery. (b) Expenses and revenues are far out of balance. Why is this system unfair to the city's customers? (c) Suggest reasons why the "cost recovery" system is not working. Suggest possible adjustments to the system.

TABLE P12.13 Industrial Wastewater Control Program Revenues and Estimated Unrecovered Billable Costs (in $) for Three Fiscal Years

Revenue or Expense	Year 1	Year 2	Year 3	Total
Permitting fees	211,067	211,067	204,992	614,730
Monitoring fees	280,046	280,046	260,357	808,043
Violation fees	23,483	23,483	20,153	64,886
Miscellaneous revenues	None	None	3,003	3,003
Expenses	3,190,876	3,190,876	3,465,149	9,793,999

12.14 SIMPLE WASTEWATER FEE SCHEDULE

A city has adopted a simple fee schedule that has a basic monthly fee (fixed fee) and a flow-based monthly fee. The city provides local wastewater collection services and a wastewater management district operates the regional collection and treatment facilities. Customers pay a combined fee. The basic fee is $13.55/month. The variable fees, as $/1,000 gal of water used, are given in Table P12.14. The values in parentheses, i.e., 400, are mg/L of combined BOD and TSS. Residential users would be in the low-strength category. (a) Calculate the monthly bill for a residence that uses 5,000 gal/month. (b) The simplicity is attractive. Are there disadvantages or elements that might be considered unfair?

TABLE P12.14 Flow-Based Monthly Fees ($/1,000 gal)

		Wastewater Strength (BOD + TSS)				
Service	All Residential Customers	Low (400 mg/L)	Medium (800 mg/L)	High (1,200 mg/L)	Very High (1,600 mg/L)	Super-High (>1,600 mg/L)
District	2.694	3.622	5.276	7.487	9.703	11.916
City	2.445	2.445	2.445	2.445	2.445	2.445
Total	5.139	6.067	7.721	9.932	12.148	14.361

12.15 SEWER USE RATE CHARGE STUDY

A city has a fixed wastewater charge of $60.26 per quarter per residence. A variable charge per 1,000 gal of water used by a residence is added to the fixed charge. Water use is more than the wastewater discharged by 10–20%. The first step toward setting the variable charge is to determine how much the wastewater utility spends. The first step will be expressed as $/1,000 gal of wastewater received at the treatment plant. Table P12.15a gives the treatment plant flow and loads. Table P12.15b gives the annual expenses, including a contribution to the equipment replacement fund and debt repayment, and how the cost of each category is allocated to flow and pollutant. (a) Calculate the cost for each budget category that is allocated to flow and to each pollutant. (b) Calculate the cost per 1,000 gal of flow, and the cost per pound of each pollutant. (c) Calculate a total cost per 1,000 gal that can serve as a basis for setting the variable charge.

TABLE P12.15a Wastewater Flows and Loads

Wastewater Characteristic	Concentration (mg/L)	Average Flow and Loadings	
Influent flow rate		413,000 gpd	150,745,000 gal/y
BOD	211	721 lb/d	263,165 lb/y
TSS	271	926 lb/d	337,990 lb/y
TKN	25	86 lb/d	31,340 lb/y
TP	5	17.2 lb/d	6,286 lb/y

TABLE P12.15b Wastewater Utility Annual Expenses

Expense Category	Operating Expense ($/y)	Allocation (%)				
		Flow	BOD	TSS	TKN	TP
Sewer general + maintenance	184,668	25	25	25	20	5
Meter reading + maintenance	10,290	100	0	0	0	0
Sewer equipment repair	7,572	100	0	0	0	0
Administrative	30,276	25	25	25	20	5
Accounting	27,017	25	25	25	20	5
Operating expenses	195,970	20	20	20	20	20
Interest expenses	7,405	20	20	20	20	20
Depreciation	188,706	20	20	20	20	20
Bond issuance	1,318	100	0	0	0	0
Equipment replacement fund	60,000	30	25	25	10	10
Debt principal payments	15,000	30	25	25	10	10
Total	728,222					

13 Financial Management for Engineering Projects

"A failure to plan is planning to fail" is an adage that applies to the financial success of a consulting firm and its individual engineering projects. A consulting firm's key plan is the annual budget. For projects, the key is the project work plan that includes the project contract, scope of work, budget, schedule, assigned staff, and strategies that will be used to manage the technical quality and cost. The project budget translates general plans into a detailed script that coordinates the individual parts. Adhering to the budgetary guidelines raises the expectation that the goals will be achieved. When things don't go well, the budget is the tool that identifies and focuses departures from the script. It is the benchmark for measuring success or failure in meeting goals and objectives and facilitates corrective action.

13.1 THE PROJECT SCHEDULE

A project schedule organizes the sequence of activities according to the duration of each activity. It is developed after the project scope of work and activities have been defined.

Sequence is how the activities follow one another. If activities A and C are to be completed in sequence, then A is the predecessor of C, and C cannot start until A is complete. Table 13.1 gives the predecessor of each activity, and also the start time, duration, and finish time for a hypothetical project with seven activities. The project is complete when activity G is complete.

A Gantt chart, Figure 13.1, is one way to show the start and finish times for the project activities. The numbers on the bars are the finish times for the activities.

13.2 BUDGET FOR A DESIGN PROJECT

Every engineering study or design project will have a budget that defines the financial goals and constraints of that project. *Measurable* major work activities and milestones are identified, so they can be tracked in the business cost accounting system. Without this oversight, it is easy to underestimate the effort required for project management, quality assurance, and expenses.

Table 13.2 is a budget plan for an engineering design project. This will control the in-house work. The major elements of design work are organized according to the design drawings and documents that must be produced. Experience from past projects provides a record of labor hours per sheet of drawings that can be used for budgeting. The project manager prepares a list of drawings based on the project details and the firm's standards for drawing production. Estimating an average of 75 h/sheet and a salary of $60/h for labor gives a budget estimate of $4,500/sheet.

13.3 PROJECT COST CONTROL

The essential task of project management is tracking the periodic estimates of the percent completion of each work activity and the subsequent estimates of the earned (billable) income. This alerts the project manager to potential cost overruns and shortages.

The project manager submits the estimates of percent complete at the end of each accounting period. These measures must be made from the outset of the project. It may be too late to implement any effective cost control measures after more than 25% of the budget is spent.

TABLE 13.1 Hypothetical project activity durations, start times, and end times

Activity	Predecessor	Duration	Start	Finish
Start		0	0	
A	Start	4	0	4
B	Start	6	0	6
C	A	5	4	9
D	A	7	4	11
E	C	6	9	15
F	D	4	11	15
G	E	5	15	20

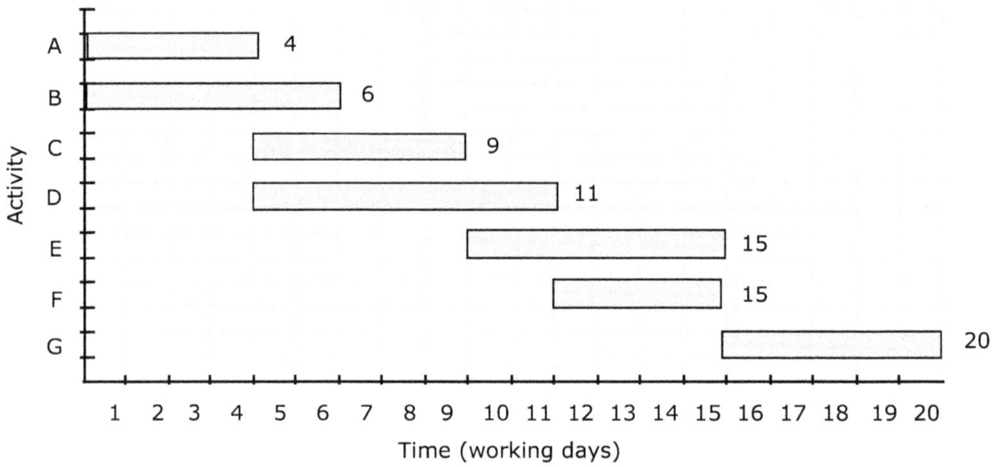

FIGURE 13.1 Gantt Chart for a hypothetical project.

Some estimates of percent complete are easy to track. Assessing the completion of project management costs and expenses is relatively easy because they usually are expended at a constant rate across the duration of the project. That is, their percent complete after three months of a nine-month project will be 33% (see Table 13.3).

Other tasks need a different strategy to estimate completion. For example, if the work calls for the development of four alternative designs and one of them has been completed and *drafted on paper*, that task would be 25% complete. The important qualifier is *drafted on paper*. The project manager must not use the percent of budget spent to estimate percent of work complete. Percent complete means what is finished and on paper.

A good approach is to meet with the task leader and mutually decide how much of the task is done on paper. These are *not* legitimate answers:

> *"The budget for the task is 20 hours and I have charged 10 hours, so I must be 50% complete."*
>
> *"I've been scheduled to work on this for four weeks and I have been working on it for two weeks, so I must be 50% complete.*

Woe to the project manager who has spent 80% of the budget and has only 40% of the work complete.

TABLE 13.2 Budget Plan for a Design Project[a]

Design Activity	No. of Sheets	Budget ($)
Project management[b]		375,000
Predesign study		375,000
Subtotal		750,000
Final Design (Items A–G)		
A. General information	40	180,000
B. Primary treatment systems	35	157,500
C. Secondary treatment systems	70	315,000
D. Solids handling systems	55	247,500
E. Support facilities	65	292,500
F. Electrical	75	337,500
Total drawings	340	1,530,000
G. Specifications		225,000
Subtotal =		1,755,000
Quality assurance		
A. Senior review		225,000
B. Revisions		150,000
Expenses[c]		375,000
Subtotal =		750,000
Total =		**3,255,000**

[a] O&M Manual, services during bidding and construction are under separate contracts.
[b] Includes work plan, technical control, and cost control.
[c] Includes laboratory testing, travel, copying, telephone, etc.

Table 13.3 is a budget for a hypothetical engineering study that is to be completed in nine months. Over-budget tasks are **bold**-flagged to indicate a negative variance from the budget. This can be a task leader problem, or it can be client-induced, such as when additional alternatives are requested. When additional work is requested, the cost control measure is to negotiate a revised project scope with increased design fees to accommodate the additional work.

The "Develop Alternatives" task has a cost overrun of $14,610. A project manager might think, "It's a small variance, I can make it up by working harder and faster." This will be a challenge. The tasks that have not started have a budgeted cost of only $130,000, so there is not much opportunity to catch up by doing these tasks faster and under budget. Whatever the cause, the tasks that are underway and behind budget need to be brought into control.

The total over-budget cost for the first three months of work is $114,602 – $97,670 = $16,932. The overrun has resulted in an overall percent complete of only 19.5% through one-third of the project schedule.

$$\text{Percent complete} = 100 \times \frac{\text{Earned income}}{\text{Budget cost}}$$

$$= 100 \times \frac{\$96{,}670}{\$500{,}000} = 19.5\%$$

If the remaining tasks are performed at the same rate, the estimated project cost at completion will be

$$\text{Projected project overrun} = \frac{\text{Cost overrun to date}}{\text{Fraction complete}}$$

$$= \frac{\$16{,}192}{0.195} = \$86{,}830$$

TABLE 13.3 Project Manager's Assessment of a Nine-Month Engineering Study at the End of the third Month

Task	Budget Cost ($)	Cost to Date ($)[a]	Percent Complete (%)[b]	Earned Income ($)[c]	Variance ($)[d]
Project management[e]	60,000	22,450	33	20,000	**−2,450**
Scope of work					
Project definition/goal	15,000	12,230	100	15,000	2,770
Develop alternatives	75,000	52,110	50	37,500	**−14,610**
Environmental review	75,000	2,102	5	3,750	1,648
Community outreach	95,000	5,150	5	4,750	**−400**
Analysis, selecting best	35,000	0	0	0	0
Draft report	35,000	0	0	0	0
Subtotal	*330,000*	71,592	18.5[f]	61,000	**−10,592**
Quality assurance (QA)					
Senior review	30,000	0	0	0	0
Revisions	30,000	0	0	0	0
Subtotal QA	*60,000*				
Expenses[g]	50,000	20,560	33	16,670	**−3,890**
TOTAL	500,000	114,602	19.5[f]	97,667	**−16,392**

[a] Actual costs to date from cost accounting report.
[b] Project team's assessment of work completed.
[c] Earned income = (% Complete)(Budgeted cost).
[d] Variance = Earned income − Cost to date.
[e] Includes work plan, project instructions to staff, and cost control.
[f] Percent complete = Earned income/Budget cost.
[g] Includes travel, copying, telephone, reproduction, etc.

There is a clear need to identify the problem and develop a strategy to bring the projected costs under control.

13.4 BUDGET FOR A CONSULTING FIRM

Table 13.4 shows an annual budget summary for a hypothetical consulting office that had $28,860,000 in gross revenue against $25,907,000 in salaries and expenses, giving a gross profit of $2,953,000. This is a return of 11.4%.

Direct labor expenses include the salaries of professionals who charge work directly to their projects. There will be other expenses that are charged directly to the project, such as travel.

Indirect expenses, including fringe benefits associated with the direct labor plus the expenses associated with the operation and maintenance of the organization (general overhead), are recovered by applying an overhead multiplier to the billable direct salary costs.

The information in Table 13.4 can be used to develop the overhead rate. Here it will be assumed that all costs in the table are allowable overhead.

The total indirect costs (fringe benefits plus general overhead) are 147.9% (say 148%) of the direct labor cost. The overhead rate is the direct labor cost multiplied by 1.48. To breakeven, the firm must bill and collect 248% of direct labor billings. This does not include profit.

TABLE 13.4 Annual Budget for a Hypothetical Engineering Firm

Description	Cost ($)	% of Direct Labor
Direct labor – salaries	*10,450,000*	*100*
Indirect costs		
Fringe benefits		
Paid leave	975,000	9.3
FICA	1,488,000	14.2
Unemployment insurance	147,000	1.4
Employee bonus	294,000	2.8
Group insurance	450,000	4.3
401(k) employer match	317,000	3.0
Workers' compensation	39,000	0.4
Subtotal fringe benefits	3,710,000	35.5
General overhead		
Indirect labor (salaries + benefits)	6,000,000	57.4
Rent	675,000	6.5
Maintenance and repairs	260,000	2.5
Travel	237,000	2.3
Insurance	465,000	4.4
Telephone	328,000	3.1
Printing	280,000	2.7
Utilities	350,000	3.3
Taxes	238,000	2.3
Depreciation	324,000	3.1
Subscriptions/memberships	132,000	1.3
Advertising/public relations	45,000	0.4
Employee training/recruitment	220,000	2.1
Professional fees	255,000	2.4
Interest	416,000	4.0
Bad debts	327,000	3.1
Computer	475,000	4.5
Vehicles	375,000	3.6
Supplies and miscellaneous	345,000	3.3
Subtotal general overhead (GO)	11,747,000	112.4
Subtotal indirect costs (fringe + GO)	15,457,000	147.9
Total expenses	25,907,000	247.9

13.5 BILLING RATE DEVELOPMENT

The *billing rate* is

$$\textit{Billing Rate} = \textit{Raw Labor Cost} + \textit{Overhead} + \textit{Profit}$$

Raw labor cost is the direct pay to employees assigned to the project, where *raw* means that overhead and profit are not included. Overhead, the ratio of indirect costs to direct labor costs was 1.48 for the budget in Table 13.4. Profit is added to get an hourly billing rate. The profit percentage is a negotiable item that will increase with the complexity of the project.

TABLE 13.5 Hourly Billing Rate Development ($/h)

Cost Item	Billing Factor
Overhead rate	1.48
Raw labor cost	1.00 + Overhead rate = 1.00 + 1.48 = 2.48
Profit	10% of Raw labor cost = (0.10)(2.48) = 0.248
Subtotal with profit	2.48 + 0.248 = 2.73
Hourly billing rate	2.73 × Hourly raw labor cost

TABLE 13.6 Typical Values for Hourly Billing Rates and Percent of Work Time Billed by Job Classification

Job Classification	Billing Rate ($/h)	% of Time Billed
Principals	185	55
Project managers	155	80
Staff and project engineers	135	85
CAD specialists	110	90
Administrative staff	77	15–20

The billing rate developed from the Table 13.4 budget is summarized in Table 13.5. The amount of contract-based billable time will be the hourly billing rate multiplied by the direct labor hours.

$$\text{Billable time (\$/month)} = [\text{Hourly billing rate (\$/h)}] \times [(\text{Billable direct labor (h/month)})]$$

To this will be added agreed expenses beyond those included in the overhead rate. The agreed expenses might include direct expense associated with field testing (drilling, etc.) or a project office. A multi-year contract might include a provision for annual salary increases.

The number of billable hours per employee varies with the kind of work they do on a project. Project managers, designers, and CAD specialists work 80–90% of their time directly on the project. Secretaries, accountants, and others who directly support individual projects can bill against the project. Time that is not directly billable is recovered through the overhead charges. Table 13.6 gives some typical percentages for work time billed by job classification. These will vary from company to company and from project to project.

Example 13.1 Billing Rate

The monthly payroll for an office with 110 employees who are paid on average $70/h, and work 160 /month, is

$$(110 \text{ employees})(\$70/\text{h-employee})(160 \text{ h/month}) = \$1{,}232{,}000/\text{month}$$

If the average direct labor billable hours are 68% of the total 160 h/month, the direct labor billings will be

$$0.68(\$1{,}232{,}000/\text{month}) = \$837{,}760/\text{month}$$

Applying a 2.6 overhead multiplier gives a billable total income for the month of

$$2.6(\$837{,}760/\text{month}) = \$2{,}178{,}176/\text{month}$$

The *effective multiplier* is net fee income divided by direct labor:

$$\text{Effective multiplier} = \frac{\text{Net fee income}}{\text{Raw direct labor}}$$

Net fee income is total income less all direct project expenses other than payroll. Stated another way, net fee income is the portion of gross fees billed that the firm keeps after subtracting costs of consultants and other project-specific expenses incurred to complete projects. In essence, the effective multiplier measures the firm's efficiency at converting direct labor spent completing projects into revenue dollars.

The current (2019) average effective multiplier achieved by US consulting firms is about 2.90. For every dollar of direct labor spent on projects, the firm generates about $2.90 of net fee income.

Example 13.2 Effective Multiplier

A company has direct raw labor cost of $870,000/month. Using an effective multiplier of 2.9, the net income will be

$$\text{Net income} = 2.9(\$879{,}000/\text{month}) = \$2{,}549{,}100/\text{month}$$

13.6 METRICS FOR MANAGING FINANCIAL STATUS

The financial status of a firm determines the opportunities for growth by means of opening new offices, introducing new lines of service, and hiring specialized technical staff. Financial management is the key to maintaining cash flow and profitability. Cash flow is monitored to keep borrowing in check and accounts payable current, and to maintain a good credit line. The business must also have responsive and accurate billing, active collection efforts, and good screening and selection of clients.

Overhead burden must be kept at acceptable levels. Non-billable management jobs are kept to a minimum. Senior management and firm principal's salaries are kept in line with market forces.

Carrying unprofitable ventures and situations hurts the employee atmosphere and it hurts profits. Unprofitable business lines are either fixed or discontinued.

The process of quickly and efficiently recording costs, billing the customer, and receiving payment have a dramatic impact on profits. Errors in coding vouchers or timesheets, or delays in generating invoices, can be devastating to a firm's cash flow.

The *average collection period* (ACP), or *day sales outstanding*, is a measure of the time it takes to collect on an invoice and convert a receivable into cash. ACP is the total *accounts receivable* divided by *average daily billings* at the end of the month, including subcontractors and reimbursables. *Accounts receivable* is uncollected charges owed the firm by their clients.

Example 13.3 Average Collection Period

An engineering firm has average monthly receipts of $802,000, and the accounts receivable are $1,818,000.

The average collection period (ACP) is

$$\text{ACP} = \frac{\$1{,}818{,}000}{\$802{,}000/\text{month}} = 2.27 \text{ months} = 68 \text{ d}$$

Working in the other direction, if monthly receipts are \$802,000 and the ACP is 2.27, the average uncollected billings are

$$(\$802{,}200/\text{month})(2.27\ \text{months}) = \$1{,}818{,}000$$

Reducing the average collection period increases profits. For example, reducing the ACP by 15%, from 68 d to 59 d, would reduce the outstanding accounts receivable to

$$(\$802{,}000/\text{month})\left(\frac{59\ \text{d}}{30\ \text{d/month}}\right) = \$1{,}558{,}000$$

This would increase annual revenue by \$1,818,000 – \$1,558,000 = \$260,000.

If the company has been paying interest of 10% to cover expenses, this increase in revenue represents a \$26,000/y saving in interest. If the annual interest paid were \$120,000 with the 68 d collection period, it will be \$94,000 with a 59 d collection period. This is a reduction in interest payments of about 21.7%.

13.7 CONCLUSION

Very few projects or annual budgets are completed as originally envisioned. There are always surprises. However, project surprises can be minimized/mitigated with a good work plan that has cost controls that will alert the project manager to problems, so adjustments can be made in a timely manner. Likewise, a firm's year-end performance can be maximized by monitoring key performance metrics and making the adjustments indicated.

13.8 PROBLEMS

13.1 GANTT CHART 1

Table P13.1 lists seven tasks. Some tasks can be done concurrently. Others cannot be started until their predecessor task is complete. For example, task 3 cannot begin until task 1 is complete. The project start will be on Monday. No work will be done on Saturday or Sunday. Draw the Gantt chart for the project.

TABLE P13.1 Work Tasks to be Scheduled

Task	Predecessor	Duration (d)
Start		0
1	Start	4
2	Start	5
3	1	4
4	2	5
5	4	5
6	4	5
7	6	5
Finish		0

13.2 INDUSTRIAL PARK WASTEWATER TREATMENT SCHEDULING

An industrial park will construct wastewater treatment facilities in two phases, using a 4,600 m^3/d (1.2 mgd), average day flow, and a 9,200 m^3/d (2.4 mgd peak day flow for each phase). The traditional design-bid-build method of project delivery schedule in Table P13.2 shows an estimated 33 months from the date of a notice to proceed with the plant's design to having the first phase fully operational. The owner wants a more aggressive approach to complete the project's design and construction within 18 months.

(a) Explain how the project can be compressed from 33 months to 18 months.
(b) If the cost estimate was based on the traditional delivery sequence of design-bid-construct, will the costs need to be adjusted to accommodate the new delivery plan?

TABLE P13.2 Design-Bid-Build-Startup Schedule

Scope of Work Item/Task	Duration (months)
Preliminary site surveys and soils investigations	1
Initial NDEP approval of design criteria	2
Preparation of construction drawings and specifications	8
Public solicitation of competitive general contractor bids	2
Bid opening and award of construction contract	2
Project construction	12
Final construction inspection/issue notice of project completion	2
Plant start-up and operator training	4
Total project duration period	33

13.3 CONSTRUCTION COST ESCALATION –1

Figure P13.3 shows the scheduled time for the purchase and installation of four mechanical items. The price quotations were obtained at the beginning of the year 2020. The price to be paid is the quoted price plus escalation at 4%/y. Revise the equipment prices to correspond with the scheduled work.

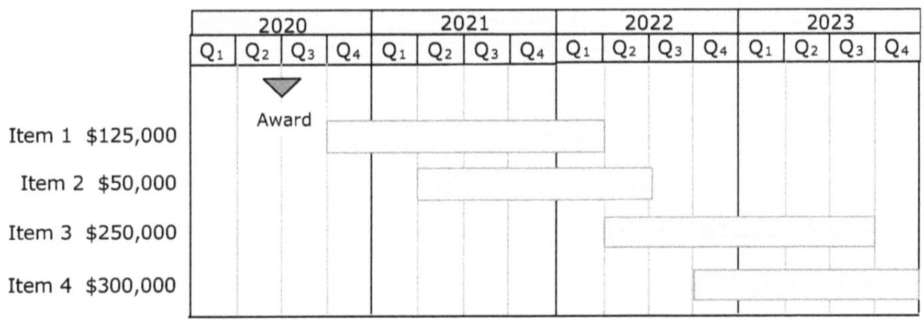

FIGURE P13.3 Purchase and Installation Schedule for Four Pieces of Equipment.

13.4 CONSTRUCTION COST ESCALATION –2

Figure P13.4 shows the scheduled start-time and end-time for work referenced to a base time when the total project cost was estimated to be \$25,000,000. The horizontal bars, from top-down, are for engineering, materials, construction labor, and subcontracts. The costs are

Engineering = 15% = \$3,750,000
Equipment and materials = 50% = \$12,500,000
Construction labor = 25% = \$6,250,000
Subcontract = 10% = \$2,500,000

The costs will increase before the activities are started. All times for escalation are taken at the cost centroid of the activity – the time when half the cost has been spent. This is not necessarily half-way through the activity timeline. The cost centroids are:

Engineering = 3.5 y
Equipment and materials = 3.75 y
Construction labor = 4.75 y
Subcontract = 4.75 y

The rates of escalation are 3%/y for engineering, 5%/y for materials, 4%/y for construction labor, and 3%/y for subcontracts.

Calculate the escalated cost estimates for the four cost categories.

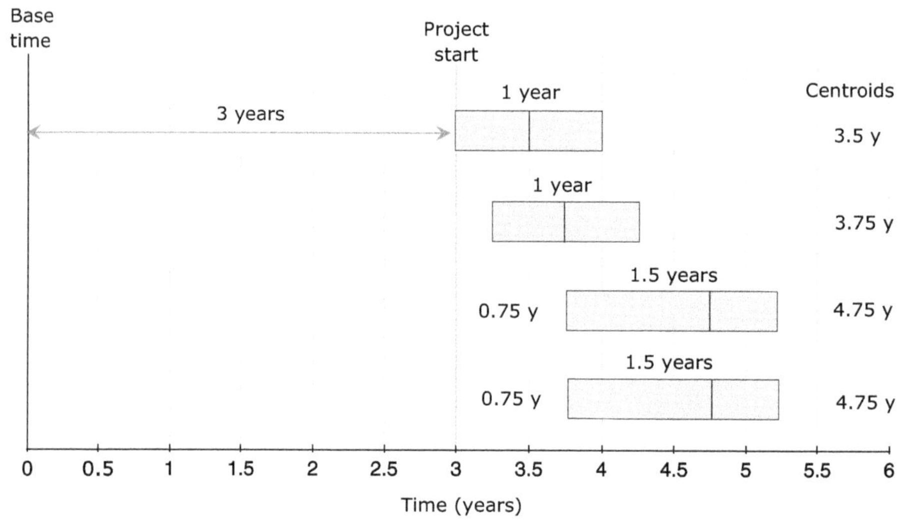

FIGURE P13.4 Time Schedule for Project Work Categories.

13.5 ESCALATION OF SCHEDULED PROJECT PAYMENTS

Table P13.5 lists the estimated costs (million \$) of eight activities with start-time, end-time, and duration. Figure P13.5 displays this information. Estimated costs are in current dollars. Since the

TABLE P13.5 Work Schedule

Activity	Start	End	Duration (months)	Cost (million $)
A1	0	6	6	0.5
A2	2	10	8	2.0
A3	6	26	20	6.8
A4	6	22	18	8.2
A5	8	18	10	12
A6	10	30	20	21
A7	12	20	8	8
A8	18	40	22	28

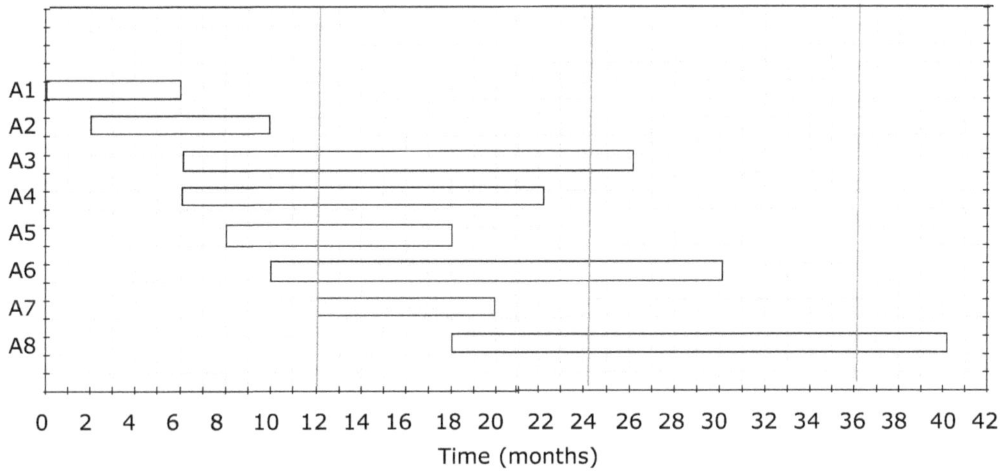

FIGURE P13.5 Gantt Chart for Project Work Schedule.

duration of this project is more than three years, the costs need to be escalated to the time when the activity will be accomplished. Escalation is the provision for increases in the cost of equipment, material, labor, etc., due to continuing price changes over time. Use a 3.6%/y escalation rate and compound the cost monthly from time zero to the midpoint of the activity. For activity 1, the midpoint is three months.

13.6 FINANCING WORKING CAPITAL FOR A PROJECT

Table P13.6 lists the cumulative monthly costs incurred by an engineering firm and the corresponding monthly revenue. Calculate the cost to the firm to borrow the working capital to finance the project for an annual rate of interest of 6%.

TABLE P13.6 Cumulative Project Expenditures and Revenue

End of Month	Cumulative Expenditures ($)	Cumulative Revenue ($)
0	0	0
1	12,000	0
2	20,000	0
3	54,000	0
4	90,000	14,000
5	130,000	40,000
6	180,000	100,000
8	220,000	130,000
9	270,000	210,000
10	290,000	300,000
11	290,000	320,000
12	290,000	340,000

13.7 ENGINEERING STUDY COST CONTROL

Table P13.7 shows the project manager's assessment of ten-month engineering study at the end of the third month. Use these data to calculate earned income, the variance of earned income and cost to date, and the estimated percent completion for the project. If project management does not improve, there will be a cost overrun. Assuming no improvement in work efficiency, calculate the projected cost overrun and the projected final project cost.

TABLE P13.7 Project Cost Control for an Engineering Study

Task	Budgeted Cost ($)	Cost to Date ($)	Estimated % Compete
Project management	72,000	26,491	30
Scope of work			
Project definition/goal	18,000	14,431	100
Develop alternatives	90,000	61,490	40
Environmental review	90,000	2,480	5
Community outreach	114,000	6,077	5
Analysis, selecting best alternative	42,000	0	0
Draft report	42,000	0	
Subtotal	396,000	84,479	
Quality assurance (QA)			
Senior review	36,000	0	0
Revisions	36,000	0	0
Subtotal QA	72,000	–	0
Expenses	60,000	24,261	30
Totals	600,000	135,230	

13.8 ENGINEERING PROJECT COST CONTROL

Table P13.8 shows the project manager's assessment of an engineering study at the end of the third month of a ten-week project. Use this data to calculate earned income, the variance of earned income and cost to date, and the estimated percent completion for the project. Is the project well managed? What is the expected profit or cost overrun?

TABLE P13.8 Cost Control Assessment for an Engineering Study

Project Task	Budgeted Cost ($)	Cost to Date ($)	Estimated % Complete
Project management	240,000	125,000	50
Scope of work			
Project definition	60,000	42,805	100
Develop alternatives	300,000	274,000	100
Process design	1,200,000	1,115,000	90
Drawings and specs	800,000	364,000	50
Report preparation	320,000	132,000	40
Subtotal	2,680,000	1,927,805	73.4
Quality assurance (QA)			
Senior review	120,000	30,000	20
Revisions	120,000	8,000	5
Subtotal QA	240,000	38,000	12.5
Expenses	200,000	129,000	70
Totals	3,360,000	2,219,805	67.2

13.9 CONSULTING OFFICE BUDGET ANALYSIS

A consulting office has total fees = $10,300,000, total gross revenue = $13,100,000, and expenses as shown in Table P13.9. Calculate the net revenue for the office.

TABLE P13.9 Annual Budget Summary for a Consulting Office

Total salaries ($)	**5,300,000**	**Non-billable expenses ($)**	
		401(k) contributions	80,000
Billable expenses ($)		Taxes	500,000
Auto	275,000	Equipment lease and maintenance	145,000
Travel	450,000	Professional fees, conferences, etc.	257,000
Telephone	230,000	Insurance: general and group	750,000
Job costs	2,000,000	Rent	900,000
Other	335,000	Office operation	173,000
Total billable expenses	3,290,000	Miscellaneous	500,000
		Total non-billable expenses	3,305,000

13.10 BALANCE SHEET ANALYSIS

Use these financial data to calculate the owner's equity, debt-to-worth ratio, and current ratio for a consulting firm.

Current assets = $2,100,000
Long-term assets = $2,800,000
Current liabilities = $2,375,000
Long-term liabilities = $800,000

13.11 EQUITY AND WORKING CAPITAL

Use the data in Table P13.11 to calculate for 2012 and 2011, the company's (a) equity (net worth) for 2012 and 2011, (b) working capital, and (c) current ratio.

TABLE P13.11 Company's Financial Data

	2012	2011
Assets ($)		
Current assets	2,305,078	1,877,676
Retention money	6,124,992	5,837,658
Accounts receivable	1,743,663	1,659,415
Material inventory	942,765	761,763
Costs and estimated earnings in excess of billings on work in progress	581,221	486,472
Prepaid expenses and others	1,000,026	1,062,968
Total current assets	12,697,745	11,685,952
Fixed assets ($)		
Property and equipment	2,655,000	2,580,000
Construction plant	806,200	800,000
Vehicles/trucks	414,560	310,000
Furniture and fixtures	69,500	60,100
Total depreciable assets	3,945,260	3,750,100
Less accumulated depreciation	2,051,900	1,663,400
Net fixed assets	1,893,360	2,086,700
Total Assets	14,591,105	13,772,652
Liabilities ($)		
Current liabilities		
Accounts payable	3,930,309	3,481,330
Accrued expenses	1,441,215	1,076,450
Notes payable	588,149	358,817
Retention money payable	835,495	551,763
Billings in excess of costs and estimated earnings on work in progress	560,847	495,167
Other current liabilities	323,232	213,478
Total current liabilities	7,679,247	6,177,005
Long-term liabilities	1,480,513	1,901,445
Total liabilities	9,159,760	8,078,450
Equity (net worth $)		
Capital stock	2,000,000	2,000,000
Additional paid-in capital	800,000	800,000
Retained earnings	2,631,345	2,894,202
Total equity	5,431,345	5,694,202
Equity + total liabilities	14,591,105	13,772,652

13.12 EVALUATING PERCENT COMPLETION

Use the accounting data in Table P13.12 to calculate the percent completed for projects A and B by three methods:

$$\text{(a) \% Completed} = \frac{\text{Cost incurred}}{\text{Cost incurred + Forecasted cost}}$$

$$\text{(b) \% Completed} = \frac{\text{Cost incurred}}{\text{Original estimated cost}}$$

$$\text{(c) \% Completed} = \frac{\text{Payments received}}{\text{Cost incurred}}$$

TABLE P13.12 Project Cost Data

Financial Data	Project A Costs ($)	Project B Costs ($)
Contract amount	15,000,000	15,000,000
Original estimated cost	14,400,000	14,800,000
Amount billed to date	10,700,000	10,700,000
Payments received to date	10,000,000	10,630,000
Cost incurred to date	11,450,000	10,550,000
Forecasted cost to complete	3,000,000	4,100,000
Costs paid to date	9,400,000	9,600,000

13.13 OVERHEAD RATE

Table P13.13 gives the direct labor and indirect costs for a hypothetical engineering firm. Determine the overhead rate based on direct labor salaries.

TABLE P13.13 Budget Summary for a Hypothetical Engineering Firm

Description	Cost ($)
Direct labor – salaries	8,882,500
Indirect costs	
Fringe Benefits	2,968,000
General overhead	10,337,360
Subtotal indirect costs	13,305,360
Total expenses	22,187,860

13.14 BILLING RATES

A consulting firm has an overhead rate of 1.95 and wants a profit of 10% of raw labor. What is the hourly billing rate?

13.15 MONTHLY BILLING RATE

Calculate the monthly payroll for an office with 200 employees who are paid an average of $90/h, and work 160 h/month. The average billable hours are 70% of the total. The overhead rate is 2.85. Calculate the monthly payroll, direct labor billings, and the billable total income.

13.16 AVERAGE COLLECTION PERIOD

The average monthly receipts are $750,000, and the accounts receivable is $1,200,000. (a) What is the average collection period (ACP)? (b) How much will the accounts receivable be reduced if the average collection period is reduced to 30 days?

13.17 PROJECT PROFIT

Table P13.17 shows the budget for a project. The overheard factor is 1.53. Calculate the gross profit, and projected net profit, and profit as a percentage of the project fee. Direct labor is $22,900.

TABLE P13.17 Project Budget ($)

Project fee	100,000
Reimbursables	2,000
Sub-consultants	40,000
Net fee	68,000
Direct expenses	3,000

14 Optimization of Linear Models

Optimization is about comparisons. It is about trade-offs. It may be trading bigger trucks against fewer drivers, one location against another, alternative transportation routes, or different process designs. In process design, the major trade-off is between capital investment and operating costs, with energy and chemical costs having a dominant role.

A key to formulating an optimization problem is recognizing what can be traded. If a system component is required and fixed, there is no give and take and that factor will not be part of the optimization. If a cost item is small, relatively speaking, it can be ignored. The same is true if a cost is independent of the cost of other major elements. For example, electrical service to a pumping station is essential. Its cost cannot be ignored in the project budget, but it can be ignored in a pumping station optimization problem. The reason is that the cost of the electrical equipment is virtually independent of the type and size of piping and pumps that will be installed. This means that optimization problems deal only with the costs that change when some element of the design is changed.

14.1 MATHEMATICAL OPTIMIZATION

Describing a problem with mathematical functions will summarize the relevant information and bring a new level of understanding of the problem. The problem could be summarized, for example, by using a schematic diagram, which may be best for presentations to clients. But the language of mathematics is more precise for engineering analysis. An additional advantage is that the mathematical model can be manipulated to evaluate design changes that may make the system more effective.

Standard algebraic methods for solving simultaneous equations, calculus, and algorithmic techniques can be used to manipulate a mathematical model. In general, we do not know what solution method will be used when we start to formulate a model. The method used for a particular problem depends largely on the structure of the mathematical model.

The mathematical statement of an optimization problem consists of two parts. The first part states an objective, such as maximize profits, maximize production, minimize wastewater volume, or minimize the size of work crews. This statement is called the *objective function*. It is the weighted sum of decision variables (design variables).

The second part is a set of constraints that define a feasible solution. Most of these will be design equations, such as

$$Q = vA, \quad y = \frac{x}{1+kt}, \quad t = V/Q$$

that come from material and energy balances, process chemistry, and reaction kinetics.

Other constraints arise from considerations of safety or technical limitations:

- The thickness of a wall must be 15–25 mm thick.
- The pH in a chemical reactor must be between 8 and 9.
- The flow cannot exceed 1 m^3/s.
- The mass loading must be less than 1,200 kg/h.

And, there will be constraints on system performance:

- The effluent concentration of pollutant X cannot exceed 0.5 mg/L.
- The solids concentration of the digester feed must be 3–4% total solids.
- The temperature shall be 35°C, at least 90% of the time.

Linear optimization problems, also known as *linear programming* (LP) problems, have an objective function and constraints that are all linear equations or inequalities. All decision variables must be positive; there is no negative wall thickness or flow rate.

For example, the objective is to maximize $Z = X_1 + 2X_2$, where X_1 is constrained to the range between 0 and 2, and X_2 is constrained to the range between 0 and 4.

The linear programming statement of this problem is

Objective function: Maximize $Z = X_1 + 2X_2$

Subject to: $X_1 \leq 2$ and $X_2 \leq 4$

$X_1 \geq 0$ and $X_2 \geq 0$

The solution is to make X_1 and X_2 as large as possible without violating the constraints. That is, X_1 can be 2 and no larger and X_2 can be 4 and no larger, giving

$$\text{Max } Z = X_1 + 2X_2 = 2 + 2(4) = 10$$

14.2 LINEAR OPTIMIZATION PROBLEMS

A linear optimization problem has the following general formulation, which looks like torture by subscripts. This complexity disappears when the variables and equation have a technical context. This will be shown by examples.

Objective function: $\text{Maximize (or minimize) } Z = c_1X_1 + c_2X_2 + \cdots + c_nX_n$

Constraints

$$a_{11}X_1 + a_{12}X_2 + \cdots + a_{1n}X_n = (\text{or} \leq, \text{ or} \geq)\ b_1$$

$$a_{21}X_1 + a_{22}X_2 + \cdots + a_{2n}X_n = (\text{or} \leq, \text{ or} \geq)\ b_2$$

$$\cdots$$

$$a_{m1}X_1 + a_{m2}X_2 + \cdots + a_{mn}X_n = (\text{or} \leq, \text{ or} \geq)\ b_m$$

$$X_i \geq 0, \quad i = 1, 2, \ldots, n$$

The X_1, X_2, ..., X_n are non-negative decision (design) variables. Most constraints will include only some of the decision variables. The c_j are cost coefficients. The a_{ij} are technical coefficients specific to the problem. The b_i are the limiting values for the quantity defined by the constraint equation.

14.3 THE GEOMETRY OF LINEAR PROGRAMMING

From here on, the notation will be simplified by writing X1 instead of the subscripted X_1 and variables will not be in italics.

The basic concept of linear optimization can be explained geometrically using an example with two decision variables, A and B.

Maximize $Z = A + 2B$

Subject to: $B \leq 6$

$A + B \leq 10$

$A \geq 0$ and $B \geq 0$

The constraints define a *feasible region*. The left panel of Figure 14.1 shows the feasible region for this problem. In a linear programming problem, each boundary of the feasible region is a straight line. The constraints are satisfied inside the shaded region or on its border, and the constraints are violated in the white space outside the shaded region.

The right panel of Figure 14.1 shows the objective function, which is also a straight line. At A = 10 and B = 0, Z = 10, at A = 0 and B = 5, Z= 10, and every combination of A and B on that line is a feasible solution with Z = 10. This line does not define the optimum solution because Z can be increased by shifting the line toward the northeast corner until it reaches the *vertex* of A = 4, B = 6, which gives Z = 16. A *vertex* is the intersection of two or more constraint equations.

Figure 14.2 shows how the feasible region might be modified by adding more constraint equations. The left panel shows the feasible region when the constraint $2A - B \leq 8$ has been added.

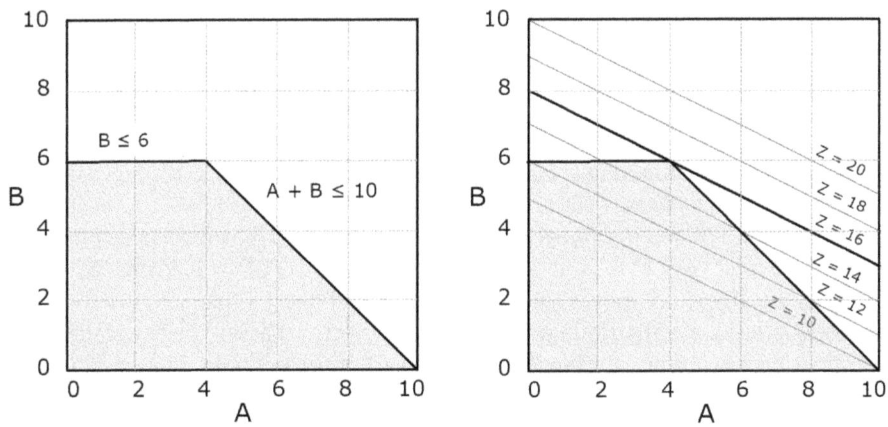

FIGURE 14.1 Geometry of a linear programming problem.

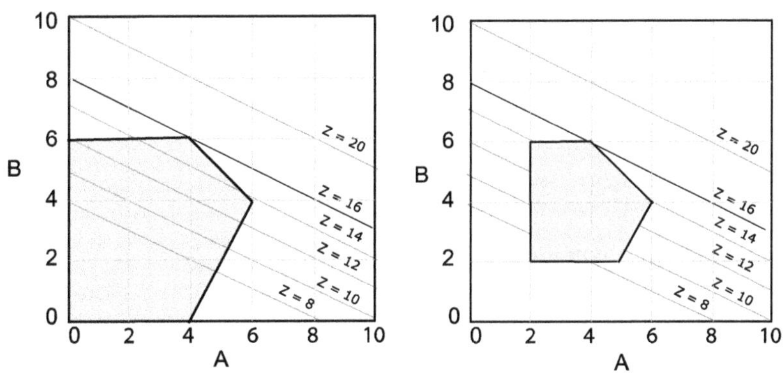

FIGURE 14.2 Modified feasible regions with the same solution of $Z = 16$.

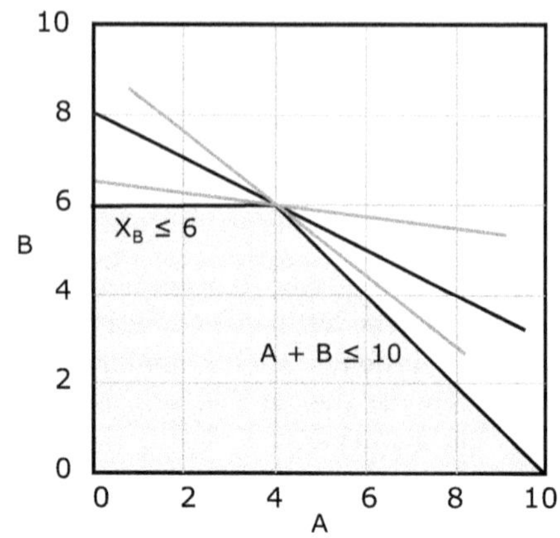

FIGURE 14.3 Changing the objective function may not change the solution.

Additional constraints of A ≥ 2 and B≥ 2 have been added in the right-hand panel. For this example, the size and shape of the feasible region are changed (made smaller), but the solution remains the same. Changing one of the original constraints, say making B ≤ 8, would change the optimal value of the objective

Figure 14.3 shows that the same solution can be obtained for objective functions that have a different cost coefficients (i.e., a different slope). The optimal value of Z changes, but the solution (A = 6, B = 4) does not.

The interesting feature of linear programming problems is that the optimal solution, whether maximum or minimum, and regardless of the number of variables, is always at a vertex of the feasible region. If the objective function is parallel to a constraint, it will pass through two vertices and all solutions on or between the vertices are optimal solutions.

The Simplex method of solving linear programming problems is wonderfully efficient. It starts with a feasible solution. The obvious one for Figure 14.1 is A = 0, B = 0. From the initial feasible solution (vertex), the algorithm moves to another vertex and checks that solution for optimality. If that vertex is not optimal, the algorithm moves to another vertex that will be better than the previous one. It never goes backward, and it never has to check all of the possible solutions.

The mathematical details are elegant and interesting, but it is not necessary to learn them because software is available that will translate algebraic equations into a solution. An excellent software package is LINGO (LINDO Systems, Chicago, IL), which can handle thousands of variables and constraint equations.

14.4 FORMULATING THE LINEAR PROGRAMMING PROBLEM

Formulating the problem means defining the variables and writing equations. Several examples are given. The first step is to define the variables. In some problems, like Example 14.1, this is very natural, and most people will tend to make the formulation as shown. This is not the case for Example 14.2.

Example 14.1 Waste Management

A company has studied its manufacturing operations with the goal of reducing liquid and solid waste production. The plant produces only three products: A, B, and C. Net profits are \$1 per unit of A, \$0.80 per unit of B, and \$1.30 per unit of C. There are minimum sales commitments of 5,000 units/d of A and 3,000 units/d of B.

TABLE 14.1 Waste Management Data

Product	Profit ($/unit)	Liquid Waste (m^3/unit)	Solid Waste (kg/unit)
A	1.00	2.0	1.0
B	0.80	1.5	1.2
C	1.30	2.6	0.9

These three products produce different amounts of waste, as shown in Table 14.1. Total waste production is to be limited to no more than 50,000 m^3/d of liquid waste and 9,000 kg/d of solid waste.

Product C has the largest unit sales price, but it produces the largest volume of liquid waste, so there is a tension between sales and waste-handling costs. As soon as the production of A reaches 5,000 units/d and B reaches 3,000 units/d, the profits will be maximized by producing as much C as possible without violating the waste-handling constraint.

Let A = units of A produced
B = units of B produced
C = units of C produced

Maximize Profit $Z = 1\,A + 0.8\,B + 1.3\,C$ Dollars per day ($/d)

Subject to:

$A \geq 5{,}000$	Production requirement (units/d)
$B \geq 3{,}000$	Production requirement (units/d)
$2A + 1.5\,B + 2.6\,C \leq 50{,}000$	Liquid waste limit (m^3/d)
$1\,A + 1.2\,B + 0.9\,C \leq 9{,}000$	Solid waste limit (kg/d)
$A \geq 0 \quad B \geq 0 \quad C \geq 0$	Non-negativity constraints

The solution is

Max Z = $7,977.78/d
A = 5,000 units/d
B = 3,000 units/d
C = 444.44 units/d

This produces 15,654 m^3/d of liquid waste and 9,000 kg/d of solid waste. Production and profit are limited by the constraint on solid waste disposal.

The optimization algorithm reports the mathematical solution values with several digits after the decimal and this level of precision is not justified. The "engineering" solution should be reported with the rounded values. In this case, round to three significant figures and report 15,700 m^3/d of liquid waste instead of 15,654 m^3/d.

A sensitivity analysis can be done on any of the costs or coefficients. Suppose the solid waste disposal system can be expanded to handle 9,000, 10,000, or 20,000 kg/d. Table 14.2 shows how many units of A, B, and C can be produced and the corresponding profit. These are calculated by increasing the right-hand side of the solid waste constraint. Notice that the profit increases by $1.44 for every extra kg/d of solid waste that can be handled.

Suppose now that the mass of solid waste associated with producing product C can be reduced from 0.9 kg/unit to 0.2 kg/unit. The solution is then

Max Z = $10,000/d
A = 5,000 units/d B = 3,000 units/d C = 2,000 units/d

The mass of solid waste produced is 9,000 kg/d.

TABLE 14.2 Sensitivity of Profit with Respect to the Limit on Solid Waste Production

Limit on Solid Waste (kg/d)	Profit ($/d)	Production (units/d)		
		A	B	C
9,000	7,977	5,000	3,000	444
10,000	9,422	5,000	3,000	1,555
20,000	23,867	5,000	3,000	12,666

14.5 SOLVING PROBLEMS WITH LINGO

For most problems in this book, it is assumed that students have favorite applications and they are welcome to use whatever is convenient. An excellent choice for optimization is LINGO, a modeling language and optimizer software, produced by LINDO Systems.

LINGO is easy to use because mathematical expressions are written in familiar language and syntax. Linear optimization requires that all variables are non-negative (greater than or equal to zero). LINGO assume this to be true, so it is not necessary to write the non-negativity constraints in the program.

The basic rules for writing a LINGO program are as follow:

- Each line ends with a semicolon (;)
- The objective function is MAX = X + 2*Y; or MIN = X + 2*Y;
- The last line of the program is END (no semicolon)
- Comment lines start with an exclamation point (!) and end with a semicolon (;)
- Addition, subtraction, multiplication, division, and exponentiation are +, –, *, /, and ^.
- LINGO does not use ≤ or ≥. Instead of ≤, write <=; instead of ≥, write >=.
- LINGO accepts < as being ≤.

Additional commands will be given later.

This is the LINGO program for Example 14.1.

```
! Objective Function;
      MAX = 1*A + 0.8*B + 1.3*C;
! Constraints;
! Production Requirements;
      A >= 5000;
      B >= 3000;
! Liquid waste limit;
      2*A + 1.5*B + 2.6*C <= 50000;
! Solid waste limit;
      1*A + 1.2*B +0.9*C <= 9000;
  END
```

The solution report gives

Maximum profit = $7,977.78/d

which is found for

A = 5,000 units/d   B = 3,000 units/d   C = 444.44 units/d

If the units of production are continuous, say gallons or kilograms, then C = 444.44 is an acceptable solution. If the units are discrete, like C = the number of pumps manufactured, then C must be an integer because a fraction of a pump makes no sense. The same is true if C were the number of people assigned to a job.

If C were pumps, 444.44 could be rounded down to 444, and there is a small and unimportant change in the profit, and A and B do not change because they are at their maximum limits.

In a larger problem, where several or all variables must be integers, it will not be possible to find the solution by rounding. Integer and mixed integer programming are used in such cases. This is explained in section 14.9.

The LINGO solution report includes more than the values of the objective and the variables. The additional information is not essential for learning how to formulate and solve linear programming problems. It is useful for users who deal with larger problems. Appendix D provides an explanation.

14.6 ADDITIONAL EXAMPLES OF PROBLEM FORMULATION

Here are two problems that are linear, but do not seem to be at first glance.

Example 14.2 Solid Waste Reclamation

A reclamation center produces three salable products, A, B, and C, by mixing four kinds of solid waste materials. The required product outputs are:

Product A ≥ 2,000 units
Product B ≥ 2,000 units
Product C ≤ 5,000 units

The raw materials need some pretreatment before being blended into the products. The amount of each material that is available and the pretreatment costs are

Material 1 = 3,000 units	\$3/unit
Material 2 = 2,000 units	\$6/unit
Material 3 = 4,000 units	\$4/unit
Material 4 = 1,000 units	\$5/unit

There is some flexibility in the mix for each grade, but quality standards specify the maximum or minimum mass fraction of a particular material in the mix. For products A and B, the mass fraction is fixed for one of the materials. The product specifications and the net revenue (\$/unit) are given in Table 14.3.

TABLE 14.3 Reclaimed Solid Waste Product Specifications

Product	Material	Mass Fraction	Selling Price (\$/unit)	Manufacturing Cost (\$/unit)	Net Revenue (\$/unit)
A	1	≤30%	8.50	3.0	5.5
	2	≥40%			
	3	≤50%			
	4	=20%			
B	1	≤50%	7.00	2.5	4.5
	2	≥10%			
	3	=10%			
C	1	≤70%	6.50	2.5	4.0

Define M1A = mass of material 1 in Product A
M2A = mass of material 2 in Product A
And so on.

The problem is to

Maximize Net Revenue = REVENUE – PRETREAT
REVENUE = 5.5*A + 4.5*B + 4.0*C
PRETREAT = 3*M1 + 6*M2 + 4*M3 + 5*M4
Constraints on raw materials:

Mass of Material 1 used	$M1 = M1A + M1B + M1C \leq 3{,}000$
Mass of Material 2 used	$M2 = M2A + M2B + M2C \leq 2{,}000$
Mass of Material 3 used	$M3 = M3A + M3B + M3C \leq 4{,}000$
Mass of Material 4 used	$M4 = M4A + M4B + M4C \leq 1{,}000$

Constraints on production:

Units of Product A produced	$A = M1A + M2A + M3A + M4A \geq 2{,}000$
Units of Product B produced	$B = M1B + M2B + M3B + M4B \geq 2{,}000$
Units of Product C produced	$C = M1C + M2C + M3C + M4C \leq 5{,}000$

The fraction of material 1 in Product A shall not exceed 30%. The constraint is

$$\frac{MA1}{MA1 + MA2 + MA3 + MA4} \leq 0.3$$

or $MA1 \leq 0.3(MA1 + MA2 + MA3 + MA4)$
which can be rearranged to give a linear constraint.

$$0.7\ MA1 - 0.3\ MA2 - 0.3\ MA3 - 0.3\ MA4 \leq 0$$

Constraints on the composition of Product A

$$0.7\ MA1 - 0.3\ MA2 - 0.3\ MA3 - 0.3\ MA4 \leq 0 \text{(grade A, material 1)}$$
$$-0.4\ MA1 + 0.6\ MA2 - 0.4\ MA3 - 0.4\ MA4 \geq 0 \text{ (grade A, material 2)}$$
$$-0.5\ MA1 - 0.5\ MA2 + 0.5\ MA3 - 0.5\ MA4 \leq 0 \text{(grade A, material 3)}$$
$$-0.2\ MA1 - 0.2\ MA2 - 0.2\ MA3 + 0.8\ MA4 = 0 \text{ (grade A, material 4)}$$

Constraints on the composition of Product B

$$0.5\ MB1 - 0.5\ MB2 - 0.5\ MB3 - 0.5\ MB4 \leq 0 \quad \text{(grade B, material 1)}$$
$$-0.1\ MB1 + 0.9\ MB2 - 0.1\ MB3 - 0.1\ MB4 \geq 0 \quad \text{(grade B, material 2)}$$
$$-0.1\ MB1 - 0.1\ MB2 + 0.9\ MB3 - 0.1\ MB4 = 0 \quad \text{(grade B, material 3)}$$

Constraint on the composition of Product C

$$0.3\ MC1 - 0.7\ MC2 - 0.7\ MC3 - 0.7\ MC4 \leq 0 \quad \text{(grade C, material 1)}$$

Listed below is the LINGO program for this problem.

```
! Objective Function;
! Net Revenue = REVENUE - PRETREAT;

MAX = REVENUE - PRETREAT;

REVENUE = 5.5*A + 4.5*B + 4.0*C;
```

```
PRETREAT = 3*M1 + 6*M2 + 4*M3 + 5*M4;

! Define variables;
! Products;
A = M1A + M2A + M3A + M4A;
B = M1B + M2B+ M3B + M4B;
C = M1C + M2C + M3C + M4C;

! Raw Materials;
M1  = M1A + M1B + M1C;
M2  = M2A + M2B + M2C;
M3  = M3A + M3B + M3C;
M4  = M4A + M4B + M4C;

! Constraints on raw materials;
! Not all the material must be used;
M1  <= 3000;
M2  <= 2000;
M3  <= 4000;
M4  <= 1000;

! Subject to constraints on production;
A >= 2000;
B >= 2000;
C <= 5000;

! Constraints on product mixture A;
 0.7*M1A - 0.3*M2A - 0.3*M3A - 0.3*M4A <= 0;
-0.4*M1A + 0.6*M2A - 0.4*M3A - 0.4*M4A >= 0;
-0.5*M1A - 0.5*M2A + 0.5*M3A - 0.5*M4A <= 0;
-0.2*M1A - 0.2*M2A - 0.2*M3A + 0.8*M4A = 0;

! Constraints on product mixture B;
 0.5*M1B - 0.5*M2B - 0.5*M3B - 0.5*M4B <= 0;
-0.1*M1B + 0.9*M2B - 0.1*M3B - 0.1*M4B >= 0;
-0.1*M1B - 0.1*M2B + 0.9*M3B - 0.1*M4B = 0;

! Constraints on product mixture C;
0.3*M1C - 0.7*M2C - 0.7*M3C - 0.7*M4C <= 0;

END
```

The LINGO solution is

Net Revenue = \$4,500
REVENUE= \$46,500
PRETREAT = \$42,000

The solution is

A = 3,667	B = 2,000	C = 4,333	
M1 =3,000	M2 = 2,000	M3 = 4,000	M4 = 1,000
M1A = 0	M2A = 1,467	M3A = 1,467	M4A = 733
M1B = 1,000	M2B = 533	M3B = 200	M4B = 267
M1C = 2,000	M2C = 0	M3C = 2,333	M4C = 0

Thus, the product compositions that maximize net revenue are:

A = 0 units of M1 + 1,467 units of M2 + 1,467 units of M3 + 733 units of M4
= 3,667 units
B = 1,000 units of M1 + 533 units of M2 + 200 units of M3 + 267 units of M4

= 2,000 units
C = 2,000 units of M1 + 0 units of M2 + 2,333 units of M3 + 0 units of M4
= 4,333 units

Example 14.3 WASTEWATER TREATMENT AND STREAM POLLUTION

High levels of a toxic substance have been observed in a river. Environmental authorities have set a standard of 2 g/m^3 (2 mg/L) as the maximum allowable concentration at all locations in the river and 0.05 g/m^3 at the estuary. The data for the three major dischargers and the three river segments are in Table 14.4. The cost of treatment depends on the mass of pollutant in the wastewater feed to the treatment plant. Figure 14.4 and Table 14.5 summarize the mass balance equations used in the model. Table 14.6 summarizes the solution.

TABLE 14.4 Data for Dischargers and River Segments

Discharger	Flow Q (m^3/d)	Concentration C (g/m^3)	Mass M (g/d)	Treatment Cost ($/g)
Above Plant 1	500	0.1	50	
Plant 1	100	40	4,000	0.9
Plant 2	75	60	4,500	1.2
Plant 3	200	50	10,000	2.4
Segment	**Distance D (km)**	**Decay coefficient k (per km)**	**Exponent kD (No units)**	**Decay Factor e^{-kD} (No units)**
2–3	10	0.2	2	0.135
4–5	15	0.2	3	0.0498
6–7	20	0.05	1	0.368

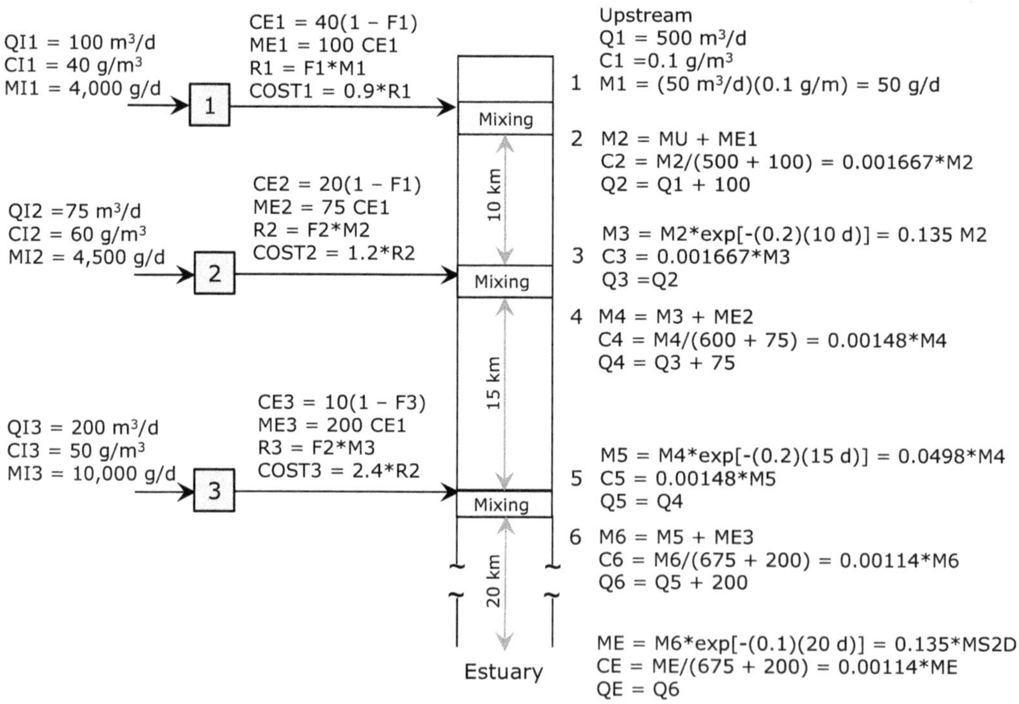

FIGURE 14.4 Definition of the river pollution problem (C = concentration, M = mass).

TABLE 14.5 Equations and Constraints for the River Model

Constraint Equation	Explanation
COST1 = 0.9*R1	Cost at plant 1 ($/d)
COST2 = 1.2*R2	Cost at plant 2 ($/d)
COST3 = 2.4*R3	Cost at plant 3 ($/d)
M1 = 50	Mass of toxic above discharge 1 (g/d)
MI1 = 4,000	Mass of toxic in plant 1 influent (g/d)
MI2 = 4,500	Mass of toxic in plant 2 influent (g/d)
MI3 = 10,000	Mass of toxic in plant 3 influent (g/d)
CE1 = 40 − 40*F1	Concentration of plant 1 effluent (g/m³)
CE2= 60 − 60*F2	Concentration of plant 2 effluent (g/m³)
CE3 = 50 − 50*F3	Concentration of plant 3 effluent (g/m³)
ME1 = 100*CE1	Mass of toxic in plant 1 effluent (g/m³)
ME2 = 75*CE2	Mass of toxic in plant 2 effluent (g/d)
ME3 = 200*CE3	Mass of toxic in plant 3 effluent (g/d)
R1 = MI1 − ME1	Mass of toxic removed at plant 1 (g/d)
R2 = MI2 − ME2	Mass of toxic removed at plant 2 (g/d)
R3 = MI3 − ME3	Mass of toxic removed at plant 3 (g/d)
M1 = 50	Mass of Toxic above outfall 1
M2 = M1 + ME1	Mass of toxic at 2 (g/d)
M3 = 0.135*M2	Mass of toxic at 3 (g/d)
M4 = M3 + ME2	Mass of toxic at 4 (g/d)
M5 = 0.0498 M4	Mass of toxic at 5 (g/d)
M6 = M5 + ME3	Mass of toxic at 6 (g/d)
ME = 0.368*M6	Mass at estuary (g/d)
C2 = 0.001667*M2	Concentration at 2 (g/m³)
C3 = 0.001667*M3	Concentration at 3 (g/m³)
C4 = 0.00148*M4	Concentration at 4 (g/m³)
C5 = 0.00148*M5	Concentration at 5 (g/m³)
C6 = 0.00114*M6	Concentration at 6 (g/m³)
CE = 0.00114*C6	Concentration at estuary (g/m³)
C1, C2, C3, C4, C5 ≤ 2	Water quality limits in river (g/m³)
CE ≤ 0.05	Water quality limit at the estuary (g/m³)
F1, F2, F3 ≥ 0.5	Technical limit – Minimum removal = 50%
F1,F2, F3 ≤ 0.98	Technical limit – Maximum removal = 98%
All variables ≥ 0	Non-negativity constraints

TABLE 14.6 Solution for the Wastewater Treatment and River Pollution Problem

Treatment Plant Data			
Removal efficiency (%)	F1 = 0.71	F2 = 0.74	F3 = 0.83
Cost ($/d)	COST 1 = 2,565	COST2 = 3,972	COST3 = 19,951
Influent concentration (g/m³)	CI1 = 40	CI2 = 60	C3I = 60
Effluent concentration (g/m³)	CE1 = 11.5	CE2 = 15.8	CE3 = 8.4
Mass in influent (g/d)	MI1 = 4,000	MI2 = 4,500	MI3 = 10,000
Mass in effluent (g/d)	ME1 = 1,150	ME2 = 1189	ME3 = 1,687
Mass removed (g/d)	R1 = 2,850	R2 = 3,311	R3 = 8,313
River Data			
Masses in stream (g/m³)	M1 = 50	M2 = 1,200	M3 = 162.4
	M4 = 1,351	M5 = 67.3	M6 = 1,754
Mass in estuary (g/m³)	ME = 646		
Concentrations in stream (g/m³)	C2 = 2.0	C3 = 0.27	C4 = 2.0
	C5 = 0.1	C6 = 2.0	CE < 0.002

Define F1 = fraction of toxic to be removed in treatment plant 1
F2 = fraction of toxic to be removed in treatment plant 2
F3 = fraction of toxic to be removed in treatment plant 3
MI1 = mass of toxic flowing into treatment plant 1 (g/d)
MI2 = mass of toxic flowing into treatment plant 2 (g/d)
MI3 = mass of toxic flowing into treatment plant 3 (g/d)

The treatment process cannot remove less than 50% of the influent pollutant nor more than 98%. Within this range the cost of removing 1 g of pollutant is constant at the value given in Table 14.4. The masses of pollutant removed, R (g/d), at each plant and the costs of removal are as follows:

R1 = F1*M1 (g/d) $0.5 \le R1 \le 0.98$
R2 = F2*M2 (g/d) $0.5 \le R2 \le 0.98$
R3 = F3*M3 (g/d) $0.5 \le R3 \le 0.98$
COST1 = 0.9 R1 ($/d)
COST2 = 1.2 R2 ($/d)
COST3 = 2.4 R3 ($/d)

The total cost is

TOTAL = COST1 + COST2 + COST3

How much treatment should each discharger provide so that the total cost is minimized while all water quality constraints are satisfied?

The mass of pollutant decays exponentially in the river. Exponentials, which are not allowed in linear programming, can be replaced with numerical constants to give linear equations, as shown below.

$$M3 = M2\ e^{-(0.2)(10)} = M2\ e^{-2} = 0.135\ M2$$

$$M5 = M4\ e^{-(0.2)(15)} = M4\ e^{-3} = 0.0498\ M4$$

$$ME = M6\ e^{-(0.05)(10)} = M6\ e^{-1} = 0.368\ M6$$

This LINGO Program introduces a new feature. Multiple commands can be put on the same line if they are separated by semicolon (;).

```
! OBJECTIVE;
MIN = COST1 + COST2 + COST3;
      COST1 = 0.9*R1; COST2 = 1.2*R2; COST3 = 2.4*R2;
! Effluent Concentrations;
      CE1 = 40 - 40*F1; CE2= 60 - 60*F2; CE3 = 50 - 50*F3;
! Influent Mass Load;
      MI1 = 4000; MI2 = 4500; MI3 = 10000;
! Effluent Mass Discharge;
      ME1 = 100*CE1; ME2 = 75*CE2; ME3 = 200*CE3;
! Mass Removed;
      R1 = MI1 - ME1; R2 = MI2 - ME2; R3 = MI3 - ME3;
! Mass in Stream;
      M1 = 50; M2 = M1 + ME1; M3 = 0.135*M2;
      M4 = M3 + ME2; M5 = 0.0498*M4;
      M6 = M5 + ME3; ME = 0.135*M6;
! Concentrations of Toxic in Stream;
      C2 = 0.001667*M2; C3 = 0.001667*M3;
      C4 = 0.00148*M4; C5 = 0.00148*M5;
      C6 = 0.00114*M6; CE = 0.00114*C6;
! Toxic concentration must be less than 2;
      C2 <= 2; C3 <= 2; C4 <= 2;
      C5 <= 2; C6 <= 2; CE <= 2;
! Removal efficiency between 50% and 98%;
```

```
F1 >= 0.5; F2 >= 0.5; F3 >= 0.5;
F1 <= 0.98; F2 <= 0.98; F3 <= 0.98;
    END
```

$$\text{Minimum cost } (\$/\text{d}) = \$2{,}565 + \$3{,}972 + \$19{,}951 = \$26{,}489$$

14.7 TRANSPORTATION PROBLEMS

The transportation problem is a special form of linear programming problem. Materials, such as wastewater, dewatered sludge, or solid waste, can be shipped from i sources to j possible destinations. Figure 14.5 is a diagram of shipment from three sources to four destinations. Shipments go only from source to destination.

The general statement of the transportation model uses the following definitions of variables:

i = source index, $i = 1, 2, \ldots, n$
j = destination index, $j = 1, 2, \ldots, m$
X_{ij} = number of units shipped from source i to destination j
C_{ij} = cost of shipping one unit from source i to destination j
a_i = number of units available at source i
b_j = number of units needed at destination j
$a_1 + a_2 + \ldots + a_n = \Sigma a_i$ = total number of units shipped from all sources
$b_1 + b_2 + \ldots + b_m = \Sigma b_j$ = total number of units received at all destinations

For example,

X_{11} = amount shipped from source 1 to destination 1
X_{12} = amount shipped from source 1 to destination 2
C_{11} = cost of shipping one unit from source 1 to destination 1
C_{12} = cost of shipping one unit from source 1 to destination 2
And so on.

The problem is *balanced* if the number of units of available resources from all sources equals the number of units required by all of the destinations. This is the balanced material balance equation:

$$\sum a_i = \sum b_j$$

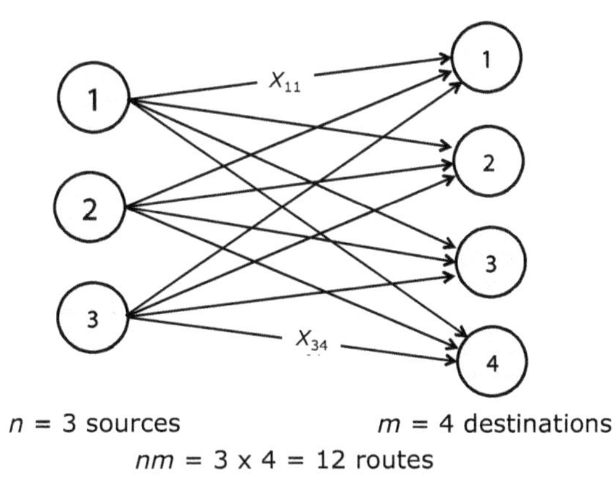

FIGURE 14.5 Transportation from three sources to four destinations.

Unbalanced means that all available resources are not shipped. This can happen because there is limited demand or because the destinations have insufficient capacity. This is the unbalanced material balance equation.

$$\sum a_i \leq \sum b_j$$

Here is the model to minimize the total cost of shipping the required units from $n = 3$ sources to $m = 4$ destinations (users). The total number of units available for shipping is $a_1 + a_2 + a_3$. The total number of units that must be delivered is $b_1 + b_2 + b_3 + b_4$. The delivery orders must be satisfied, but not all the available units must be shipped. That makes this an unbalanced problem.

The objective function is

Minimize $$Z = C_{11}X_{11} + C_{12}X_{12} + C_{13}X_{13} + C_{14}X_{14}$$
$$+C_{21}X_{21} + C_{22}X_{22} + C_{23}X_{23} + C_{24}X_{24}$$
$$+C_{31}X_{31} + C_{32}X_{32} + C_{33}X_{13} + C_{34}X_{14}$$

Subject to:

$X_{11} + X_{12} + X_{13} + X_{14} \leq a_1$		Total units shipped from source 1
$X_{21} + X_{22} + X_{23} + X_{24} \leq a_2$		Total units shipped from source 2
$X_{31} + X_{32} + X_{33} + X_{34} \leq a_3$		Total units shipped from source 3
$X_{11} + X_{21} + X_{31} = b_1$		Total units delivered to destination 1
$X_{12} + X_{22} + X_{32} = b_2$		Total units delivered to destination 2
$X_{13} + X_{23} + X_{33} = b_3$		Total units delivered to destination 3
$X_{14} + X_{24} + X_{34} = b_4$		Total units delivered to destination 4
$0 \leq X_{ij}$	for all i and j	Non-negativity constraints

14.8 CASE STUDY: SLUDGE DISPOSAL

Three treatment plants serve two cities that are located on the flood plain of a river; mountains overlook the cities. This geography makes sludge disposal difficult and expensive. The treatment plants may ship sludge to any of four disposal sites, which include two farmland disposal sites (LD1 and LD2), one landfill (LF), and one incinerator (IN).

The District Manager and an engineer are working on a plan to dispose of the sludge at the lowest cost.

E I have organized the information on sludge production and the disposal capacity at the four disposal sites in Table 14.7. The total available capacity, 168 T/d, equals the sludge production that creates the demand for disposal. The combined costs for sludge treatment and transportation are given in Table 14.8. The 60 T/d of sludge produced at TP1 can be

TABLE 14.7 Sludge Production and Treatment Capacities

Facility	Production (T/d)	Facility	Disposal Capacity (T/d)
TP1	60	LD1	38
TP2	40	LD2	50
TP3	68	LF	20
		IN	60

TABLE 14.8 Transportation and Treatment Costs ($/T)

	Disposal Sites			
Sources	**LD1**	**LD2**	**LF**	**IN**
TP1	40	52	58	60
TP2	30	56	68	72
TP3	70	34	38	80

shipped to any of the four disposal sites. The costs are $40/T to LD1, $52/T to LD2, $58/T to LF, or $60/T to the incinerator.

M Good work! These tables condense a lot of complex information. Show me how to use it to work out the best transportation schedule for sludge disposal.

E I would say, "First come, first served." Let TP1 choose what it likes, then let TP2 choose, and so on.

M Will that give us the lowest overall cost? Why should we give TP1 first choice?

E Because it won't matter who gets first choice. It will work out to the same cost.

M Really? You will have to show me that is correct.

E It's like this. There is 60 T/d of sludge produced at TP1. This can be shipped to any of the four sinks: to LD1 for $40/T, to LD2 for $52/T, to LF for $58/T, or to the incinerator for $60/T. The least cost option is 38 T/d to LD1 and 22 T/d to LD2.

$$\text{Total cost for TP1} = (38\ \text{T/d})(\$40/\text{T}) + (22\ \text{T/d})(\$52/\text{T}) = \$2{,}664/\text{d}$$

That's the best TP1 can do. It's the same for the others.

M Well, someone has to use the incinerator, because demand for disposal equals our capacity for disposal. Given the first choice, TP1 will avoid incineration, as will everyone else who is given the first choice. Here's the weakness in your idea. TP1 can use the incinerator for $60/T, because its location gives it a cost advantage for that disposal site. TP2 and TP3 have to pay $72/T or $80/T. Should TP1 have the choice to push incineration onto one of the other plants? This is a District operation and we have to balance this among the three treatment plants.

E Let me check my idea. TP1 goes first and chooses LD1 and LD2. TP2 goes next. It would prefer LD1, but that was taken by TP1, so it goes to LD2 and then LF. TP3 would prefer LD2 but that has been taken, so it is forced to use LF and IN (incinerator). The total cost of this scheme is $10,152/d.

M It is a feasible solution (Table 14.9), but I can reduce it by ($20/T)(60 T/d) = $1,200/d by switching treatment plants 1 and 3 at the incinerator. Maybe there are other profitable switches.

TABLE 14.9 A Feasible, But Not Optimal, Solution

	LD1	**LD2**	**LF**	**IN**	***Supply***
TP1	$40/T	$52/T	$58/T	$60/T	
	38 T/d	22 T/d			**60 T/d**
TP2	$30/T	$56/T	$68/T	$72/T	
		28 T/d	12 T/d		**40 T/d**
TP3	$70/T	$34/T	$38/T	$80/T	
			8 T/d	60 T/d	**68 T/d**
Capacity	**38 T/d**	**50 T/d**	**20 T/d**	**60 T/d**	**168 T/d**

E There must be. Switching the incinerator input of 60 T/d from TP3 to TP1 requires a few other transports to be shifted. The whole matrix has to be rebalanced. Maybe it will be better – maybe not.

M There is a nugget of a good technique here. The demands from our three treatment plants change from week to week. It would be great to have a scheme for quickly rescheduling. Two years from now, we will have a greater capacity at LD2 and LF and two new land disposal options. So, the problem will become larger and more complicated. Keep working.

E I'll see you soon with an improved method.

(*Note:* The Engineer did not have a workable method to get an optimal solution, but was close. There is a clever procedure, called the Northwest Corner Rule, that can be used to find the minimum by shifting allocations in the kind of table the engineer was using. Using LINGO is easier.)

A day later

E I have a new approach – linear optimization using LINGO. This is easy to use and easy to change the conditions of the problem. There is no trial-and-error shifting of allocations in the tables. We can reroute every day as sludge production changes, or if a disposal site has excess capacity, or if haul costs change because of road conditions, detours, construction, or weather.

M I look forward to learning about this. Explain.

E There are 3 sources and 4 disposal sites, so there are 12 ways sludge can move from source to disposal. All routes do not have to be used. Define the variables:

X1 = tons transported from TP1 to LD1
X2 = tons transported from TP1 to LD2
X3 = tons transported from TP1 to LF
X4 = tons transported from TP1 to IN
Y1 = tons transported from TP2 to LD1
Y2 = tons transported from TP2 to LD2
Y3 = tons transported from TP2 to LF
Y4 = tons transported from TP2 to IN
Z1 = tons transported from TP3 to LD1
Z2 = tons transported from TP3 to LD2
Z3 = tons transported from TP3 to LF
Z4 = tons transported from TP3 to IN

The objective is to minimize the total cost. The objective function is the sum of costs for the 12 possible routes. All routes are included but not all will be used. There are constraint equations to specify the amount of sludge coming from each source and going to each disposal site. The total at the source must equal the total received at the disposal site. This constraint can be modified if disposal capacity exceeds sludge production. But now production and disposal are balanced. The LINGO program is given below:

```
! Objective = Minimize cost;
   MIN = 40*X1 + 52*X2 + 58*X3 + 60*X4
       + 30*Y1 + 56*Y2 + 68*Y3 + 72*Y4
       + 70*Z1 + 34*Z2 + 38*Z3 + 80*Z4;
! Constraints for delivery from the three sources;
      X1 + X2 + X3 + X4 = 60;
      Y1 + Y2 + Y3 + Y4 = 40;
      Z1 + Z2 + Z3 + Z4 = 68;
!Constraints on delivery to the four disposal sites;
      X1 + Y1 + Z1 = 38;
```

```
    X2 + Y2 + Z2 = 50;
    X3 + Y3 + Z3 = 20;
    X4 + Y4 + Z4 = 60;
END
```

The minimum cost is $7,244/d.
The optimal shipments are in Table 14.10.

M There is a 2 T/d haul from TP2 to LD2. That is not ideal. Can we explore changes to get rid of that?

E The LINGO format makes it very easy to change the problem. Let's investigate forcing Y2 to be at least 5 tons. Just add the constraint Y2 >= 5. The results are in Table 14.11. The minimum cost increases by $144/d, to a total of $7,400. A small amount of sludge (3 T/d) from TP1 is redirected from IN to LD1. This is balanced by 3 T/d of sludge from TP3 going to IN instead of to LD2.

M That is not an improvement. Now there are two small loads, one of them to the incinerator.

E A better solution might be to increase the capacity at LD2 by 5 T/d to 55 T/d, without changing the sludge production at any of the three treatment plants. The disposal capacity of 173 T/d will then exceed the 158 T/d demand for disposal. The disposal capacity will need to be increased to accommodate growth and having excess disposal capacity will be advantageous.

The excess capacity problem is modeled by changing the constraints on delivery to the four disposal sites from $<$ to $\leq$ because production and disposal capabilities are no longer equal. The new LINGO program looks like this.

TABLE 14.10 Optimal Shipments (T/d) from the Three Treatment Plants to the Four Disposal Sites – First Iteration

Source	Disposal Sites				
	LD1	LD2	LF	IN	Totals
TP1	X1 = 0	X2 = 0	X3 = 0	X4 = 60	60
TP2	Y1 = 38	Y2 = 2	Y3 = 0	Y4 = 0	40
TP3	Z1 = 0	Z2 = 48	Z3 = 20	Z4 = 0	68
Totals	38	50	20	60	

TABLE 14.11 Optimal Shipments (T/d) from the Three Treatment Plants to the Four Disposal Sites – Second Iteration

	Disposal Sites				
Source	LD1	LD2	LF	IN	Totals
TP1	X1 = 3	X2 = 0	X3 = 0	X4 = 57	60
TP2	Y1 = 35	Y2 = 5	Y3 = 0	Y4 = 0	40
TP3	Z1 = 0	Z2 = 45	Z3 = 20	Z4 = 3	68
Totals	38	50	20	60	

TABLE 14.12 Optimal Transports (T/d) from the Three Treatment Plants to the Four Disposal Sites with Increased Capacity at LD2

	Disposal Sites				
Source	**LD1**	**LD2**	**LF**	**IN**	**Totals**
TP1	X1 = 0	X2 = 5	X3 = 0	X4 = 55	**60**
TP2	Y1 = 38	Y2 = 2	Y3 = 0	Y4 = 0	**40**
TP3	Z1 = 0	Z2 = 48	Z3 = 20	Z4 = 0	**68**
Totals	**38**	**55**	**20**	**55**	

```
! Objective = Minimize cost;
MIN = 40*X1 + 52*X2 + 58*X3 + 60*X4
      + 30*Y1 + 56*Y2 + 68*Y3 + 72*Y4
      + 70*Z1 + 34*Z2 + 38*Z3 + 80*Z4;
! Constraints for delivery from the three sources;
      X1 + X2 + X3 + X4 = 60;
      Y1 + Y2 + Y3 + Y4 = 40;
      Z1 + Z2 + Z3 + Z4 = 68;
! Constraints on delivery to the four disposal sites;
      X1 + Y1 + Z1 <= 38;
      X2 + Y2 + Z2 <= 55;
      X3 + Y3 + Z3 <= 20;
      X4 + Y4 + Z4 <= 60;
END
```

The new results are in Table 14.12. Again, the solution changes in small ways. The minimum cost is $7,204/d, which is $40/d less than the original solution. The 2 T/d from TP2 to LD2 is still part of the solution and this needs to be resolved. On the positive aside, each TP has to transport sludge to just two of the four disposal sites.

M I see the power of this tool for day-to-day operations and for planning. We can change shipping plans daily if that seems more practical. It is also a great planning tool because we can refine our plans for expanding the disposal sites.

14.9 INTEGER PROGRAMMING AND MIXED INTEGER PROGRAMMING

Any model that requires one or more integer values in the solution is an *integer programming* (IP) model. (Some authors make a distinction between integer programming, where all variables are required to be whole numbers [integers], and *mixed integer programming,* where only specified variables are integers.)

There are two types of integer variables – *general* and *binary.* A general integer variable is required to be a whole number. A binary integer variable is further required to be either zero or one.

General integer variables are used where rounding of fractional solutions is problematic. If a model indicates 4.5 operators on a shift, it does matter whether 4 or 5 operators are assigned. Rounding is not a problem if the model indictates producing 500,200.5 units; it does not matter whether you produce 500,201 or 500,200.

Binary variables are used to model Yes/No types of questions, such as

Produce A/Don't produce A
Incur a fixed cost/Don't incur a fixed cost.

Integer programming does not work by rounding because that typically will lead to an infeasible or suboptimal solution. For example, the model

$$\text{Max} = X$$
$$X + Y = 25.5$$
$$X \leq Y$$

gives the optimal solution that X = Y = 12.75. If X is to be integer, rounding up to X = 13 increases the objective function, but this is an infeasible solution because it gives Y = 12 and X is constrained to be less than Y. Rounding to X = 12 gives Y = 12.5 and this does work.

The computing algorithm, called *branch-and-bound*, enumerates and evaluates all combinations of the integer values. This seriously complicates the problem. If there are many variables, it is best, wherever possible, to avoid integer variables.

LINGO assumes all variables are continuous, unless told otherwise. To define a *general integer variable*, use the function @GIN. The syntax is @GIN(variable name). This can be used anywhere in a model where you would normally enter a constraint.

A *binary integer variable*, also called a 0/1 variable, is required to be either zero or one. The function that does this is @BIN and the syntax @BIN(variable name).

In summary,

@GIN(X) makes X an integer
@BIN(X) makes X either one or zero.

Example 14.4 A Simple Integer Programming Problem

Redo Example 14.1 with C defined as an integer. The new program (with reorganized comments) is as follows:

```
! Objective Function;
   MAX = 1*A + 0.8*B + 1.3*C;
! Constraints;
   A >= 5000;                        ! Production Requirement for A;
   B >= 3000;                        ! Production Requirement for B;
   2*A + 1.5*B + 2.6*C <= 50000;     ! Liquid waste limit;
   1*A + 1.2*B+0.9*C <=9000;         ! Solid waste limit;
!Define C as an integer;
   @GIN(C);
END
```

The solution is Profit = $7,977.60/d
at A = 5,000 units/d   B = 3,000 units/d   C = 444 units/d

The next two examples use binary integers as off/on or in/out switches. A binary integer can have only two values, 1 or 0.

Example 14.5 LOCATING WASTE PROCESSING STATIONS

A county wants to build the minimum number of waste processing stations to serve six cities such that no city is more than 15 miles from a processing station. There is a potential site for a waste disposal station in each city, but it is not required that each city shall have a station. The distances between cities are given in Table 14.13. How many stations should be built and where they should be located?

The key to solving the problem is defining the variables and writing the constraints.

Define the binary integer variables X1, X2, X3, X4, X5, and X6 as

$X_i = 1$ if a station is to be built in location *i*
$X_i = 0$ otherwise

The objective is to minimize the total number of stations that are built:

$$\text{MIN } Z = X1 + X2 + X3 + X4 + X5 + X6$$

Table 14.14 lists the sites that are within 15 miles of each city. There are only two potential sites within 15 miles of city 1, and they are site 1 in city 1 and site 2 in city 2. Therefore, either site 1 or site 2 must be included in the solution. The constraint that requires this is

$$X1 + X2 \geq 1$$

This ensures that either X1 = 1 or X2 = 1, and that X1 = X2 = 0 is impossible. Thus, at least one station will be built within 15 miles of City 1.

Similarly, to ensure that at least one station will be located within 15 miles of city 2 requires a station at site 1, site 2, or site 6. The requirement is

TABLE 14.13 Distance (miles) between Cities in Service Area

	To					
From	Site 1	Site 2	Site 3	Site 4	Site 5	Site 6
City 1	0	10	30	30	30	20
City 2	10	0	25	35	20	10
City 3	20	25	0	15	30	25
City 4	30	35	15	0	15	25
City 5	30	20	30	15	0	14
City 6	20	10	20	25	14	0

TABLE 14.14 Potential Processing Sites within 15 miles of Each City

Source	Processing Site		
City 1	1	2	
City 2	1	2	6
City 3	3	4	
City 4	3	4	5
City 5	4	5	6
City 6	2	5	6

$$X1 + X2 + X6 \geq 1$$

Similar constraints are needed for cities 3–6.

The complete model is

$$\text{MIN} \quad X1 + X2 + X3 + X4 + X5 + X6$$

Constraints	$X1 + X2 \geq 1$	City 1 constraint
	$X1 + X2 + X6 \geq 1$	City 2 constraint
	$X3 + X4 \geq 1$	City 3 constraint
	$X3 + X4 + X5 \geq 1$	City 4 constraint
	$X4 + X5 + X6 \geq 1$	City 5 constraint
	$X2 + X5 + X6 \geq 1$	City 6 constraint
	X1, X2, X3, X4, X5, X6	Binary (1,0) integers

You may use LINGO to confirm that stations should be built at sites 2 and 4. Site 2 will serve cities 1, 2, and 6, while site 4 will serve cities 3, 4, and 5.

Example 14.6 Regional Hazardous Waste Processing

Hazardous waste will be collected at seven cities, identified as A, B, C, …, G, and sent to regional processing plants, of which there can be two or three. It is not required to use all three sites. It is not feasible to bring all the waste to one large processing plant. At the preliminary stage of design, the candidate processing sites are A, C, and G. Waste originating at A, C, and G can be processed at zero shipping cost, but the processing cost is the same as any other waste that is delivered there. The quantities of wastes originating at each city are given in Table 14.15.

Figure 14.6 shows the cost function that applies at all processing sites. There is a fixed fee of K_i if 1 ton is treated at site i. From 1 ton to the limiting capacity of the site, the cost is b_i. The amounts processed (ton/y) at each site are TA, TC, and TG, resulting in costs of TCA, TCC, and TCG.

TABLE 14.15 Sources of Hazardous Waste

City	A	B	C	D	E	F	G
Quantity (ton/y)	40	56	5	1.7	4.8	3.7	8.8

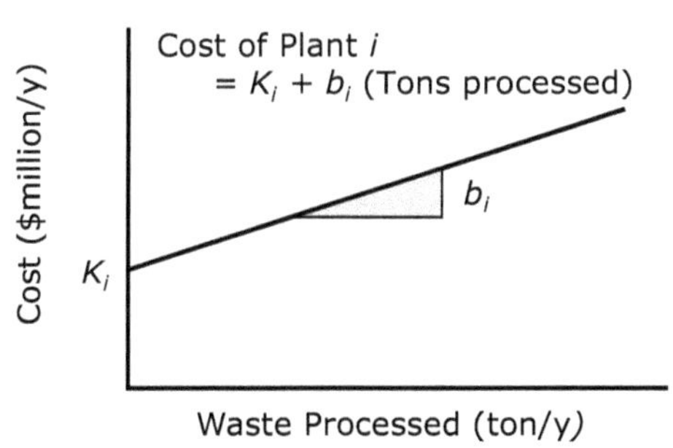

FIGURE 14.6 Cost function for hazardous waste processing

Processing costs, in million dollars per year, are

Plant A	TCA = 25 + 0.1 TA
Plant C	TCC = 22 + 0.1 TC
Plant G	TCG = 15 + 0.08 TG

Define the amounts (X) of waste shipped to each site as follows:
XAA = mass of waste shipped from city A to site A (ton/y)
XAC = mass of waste shipped from city A to site C (ton/y)
XAG = mass of waste shipped from city A to site G (ton/y)
XBA = mass of waste shipped from city B to site A (ton/y)
XBC = mass of waste shipped from city B to site C (ton/y)
XBG = mass of waste shipped from city B to site G (ton/y)

and so forth.

The minimum travel times between locations, based on Figure 14.7, are in Table 14.16. The transportation cost is \$0.02/T-min. For example, the cost for shipping one T of XAC from A to C is

$$(\$0.02/\text{T-min})(95 \text{ min}) = \$1.9/\text{T}$$

Constraints on amount shipped from source to processing sites A, C, and G:

From A	XAA + XAC + XAG = 40 T/y
From B	XBA + XBC+ XBG = 56 T/y
From C	XCA + XCC + XCG = 5 T/y
and so on	

FIGURE 14.7 Map of the hazardous wastes service area. Values are travel times in minutes.

TABLE 14.16 Shortest Travel Time (min)

	From						
To	**A**	**B**	**C**	**D**	**E**	**F**	**G**
A	0	40	95	80	100	130	145
C	95	55	0	105	100	105	55
G	145	105	55	110	45	60	0

Constraints for shipping to the processing sites:

XAA + XBA + XCA + XDA + XEA + XFA + XGA = TA
XAC + XBC + XCC + XDC + XEC + XFC + XGC = TC
XAG + XBG + XCG + XDG + XEG + XFG + XGG = TG

TA, TC, and TG are quantities of waste (ton/y) processed at sites A, C, and G.
The maximum processing capacities are

$TA \leq 100$ ton/y $TC \leq 60$ ton/y $TG \leq 60$ ton/y

A binary integer variable is used to include or exclude a processing site. Define

dA = 1	if processing (any amount) is done at A
dA = 0	otherwise
dC = 1	if processing (any amount) is done at A
dC = 0	otherwise
dG = 1	if processing (any amount) is done at A
dG = 0	otherwise

This means the cost models are

At A	$TCA = K_A + b_A\, TA$	if dA = 1
	$TCA = 0$	if dA = 0
At C	$TCC = K_C + b_C\, TC$	if dC = 1
	$TCC = 0$	if dC = 0
At G	$TCG = K_G + b_G\, TG$	if dG = 1
	$TCG = 0$	if dG = 0

The mixed integer LINGO formulation has 33 variables (3 are integers), and 21 constraints:

```
! Objective;
MIN = TransC + ProcC;
! Transportation Cost;
! Calculate the mass of waste transported (TMass) in min-ton/y;
   TMass = 0*XAA + 95*XAC + 145*XAG
      + 40*XBA + 55*XBC + 105*XBG
      + 95*XCA + 0*XCC + 55*XCG
      + 80*XDA + 105*XDC + 110*XDG
      + 100*XEA + 100*XEC + 45*XEG
      + 130*XFA + 105*XFC + 60*XFG
      + 145*XGA + 55*XGC + 0*XGG;
! Convert to transportation cost ($/y) using $0.02/ton-min;
      TransC = 0.02*TMass;
! Plant Processing Cost;
      TCA = 25*dA + 0.1*TA;      !Plant A;
      TCC = 22*dC + 0.1*TC;      !Plant C;
      TCG= 15*dG + 0.08*TG;      !Plant G;
      ProcC = TCA + TCC + TCG;
!Source Constraints;
      XAA + XAC + XAG = 40;
      XBA + XBC + XBG = 56;
      XCA + XCC + XCG = 5;
      XDA + XDC + XDG = 1.7;
      XEA + XEC + XEG = 4.8;
      XFA + XFC + XFG = 3.7;
      XGA + XGC + XGG = 8.8;
! Processing Requirements;
```

```
        XAA + XBA + XCA + XDA + XEA + XFA + XGA = TA;
        XAC + XBC + XCC + XDC + XEC + XFC + XGC = TC;
        XAG + XBG + XCG + XDG + XEG + XFG + XGG = TG;
  ! Limits on Processing Capacity;
        TA <= dA*80;
        TC <= dC*60;
        TG <= dG*60;
  ! Allow for two or three processing sites;
        dA + dC + dG <=3;
  ! Define Binary Variables;
        @BIN( dA);
        @BIN( dC);
        @BIN( dG);
END
```

The optimal LINGO solution is

Total annual cost = Transportation cost + Plant processing cost
$134.8 = $83.6 + $51.2 (millions per year)
Plant A capacity = 80
Plant C capacity = 0 (Plant C is not needed)
Plant G capacity = 40
Waste going to A is from A and B = 40 + 40 = 80 T/y
Waste going to G is from B, C, D, E, F, and G = 16 + 5 + 1.7 + 4.8 + 3.7 + 8.8 = 40 T/y

14.10 CONCLUSION

The variety of problems that can be described using only linear equations is surprising, especially to engineers who are accustomed to using algebraic and differential equations. It is also fortunate because the solutions are so easy to calculate once the problem is properly formulated, which is often a matter of defining the variables.

This chapter has emphasized formulation, along with an introduction to the LINGO software for solving the problems.

A sensitivity analysis was illustrated for Example 14.1. It is easy, in a small problem, to change cost-coefficients and other values in the problem statement to see how the objective and the solution changes. *Solution* means the final values of the reported variables. Often the minimum cost or maximum profit will change, but the solution stays the same.

In large problems, it is not feasible to manually manipulate all the coefficients and constraints, so the linear programming output provides some additional information in the form of *reduced cost*, *slack and surplus*, and *dual cost*. These are explained in Appendix D.

14.11 PROBLEMS

14.1 CEMENT MANUFACTURING

A company requires a minimum of 36,500 m^3 of sand and gravel to make a concrete mix. The mix design requires a minimum of 14,900 m^3 of sand and no more than 19,000 m^3 of gravel in making the concrete mix. Sand and gravel can be extracted at two sites. The raw material from site A produces 35% sand and 65% gravel. Site B produces 48% sand and 52% gravel. The unit delivery costs (including the cost of raw materials and the hauling costs) are $745 and $820/$m^3$ from A and B, respectively. How much material from each site should be used to minimize the cost of the concrete mix?

14.2 ALLOCATION OF FUNDS

A city can spend up to $25 million, $15 million from Federal funds and $10 million from State funds, to control water pollution by street cleaning and sewer flushing. Street cleaning generates 40 jobs per million dollars of spending; sewer flushing generates 30 jobs per million in spending. The Federal grant will finance 50% of the cost of street cleaning and 75% of the cost of sewer flushing. The remaining costs will be paid by the state. The city would like to create as many jobs as possible. Determine how the city should spend the money, assuming the environmental benefits are acceptable, whichever spending plan is adopted.

14.3 AIR POLLUTION CONTROL

A company must reduce air pollutant emissions by at least the following amounts:

Particulates	60,000,000 kg/y
Sulfur oxides	150,000,000 kg/y
Hydrocarbons	125,000,000 kg/y

Two sources of pollution are blast furnaces and open-hearth furnaces. The reduction in emissions (1,000,000 kg/y) from the maximum feasible use of an abatement method is given in Table P14.3a. The total annual costs per unit of reduction (1 unit = 1,000,000 kg/y) use of each abatement method are in Table P14.3b. Applying one unit of technology A to pollution control on blast furnaces will remove 12 units of particulates, 35 units of SO_2, and 37 units of hydrocarbons. Determine the minimum cost pollution control strategy.

TABLE P14.3a Maximum Units of Pollution Removed (million kg/y)

Pollution	Technology A		Technology B		Better Fuels	
	Blast Furnace	Open-Hearth Furnace	Blast Furnace	Open-Hearth Furnace	Blast Furnace	Open-Hearth Furnace
Particulates	12	9	25	20	17	13
Sulfur oxides	35	42	18	31	56	49
Hydrocarbons	37	53	28	24	29	20

TABLE P14.3b Cost ($) per Unit

Technology	Blast Furnace	Open-Hearth Furnace
Technology A	8	10
Technology B	7	6
Better fuels	11	9

14.4 WASTE MINIMIZATION IN REBAR CUTTING 1

A construction company has a supply of reinforcing bars delivered to the project site. The bars are 5.5 m in length and need to be cut into smaller lengths in these quantities: 5,000 pieces of 2-m length, 2,250 pieces of 3 m, and 9,600 pieces of 1.6 m. Cutting is to be done on-site. Instructions must be given to the machine operator to minimize the amount of scrap that is produced. Scrap is the short lengths of the 5.5 m bars that are unusable at this project site. For example, cutting a 5.5 m bar into one 2-m bar and one 3-m bar will produce 0.5 m of scrap. What is the cutting plan that will minimize the quantity of scrap produced. Assume there is no penalty from producing excess numbers of bars of usable lengths (2.0, 3.0, or 1.6 m bars).

14.5 SLUDGE DISPOSAL

Five treatment plants (TP) may ship sludge to any of five disposal sites, which includes two landfills (A and B) and two land disposal sites (C and D). The mass of sludge, sludge disposal capacity, and cost of disposal change with the season. Tourists increase the sludge production at some plants, weather limits disposal capacity, and transportation times change. The sludge production and disposal capacities given in Table P14.5a and the transportation and treatment costs given in Table P14.5b are typical for one season. Disposal capacity exceeds sludge production. The optimal disposal schedule for this season is important, but more important is a systematic way to reoptimize the schedule as conditions change. What is the optimal disposal schedule?

TABLE P14.5a Sludge Production and Treatment Capacities

Source Facility	Sludge Produced (T/d)	Disposal Facility	Sludge Disposal (T/d)
1	40	A	67
2	60	B	73
3	40	C	40
4	68	D	50
Total	208		230

TABLE P14.5b Transportation and Disposal Costs ($/T)

Sources	Treatment Site			
	A	B	C	D
1	40	80	70	30
2	40	76	68	32
3	70	34	38	80
4	36	66	54	34

14.6 SHIPPING WITH STORAGE

Figure P14.6 shows a transportation network with two waste sources (S1 and S2), a distribution center (DC), and two waste processing centers (W1 and W2). Unit costs are shown on the links. All waste material produced must be shipped to a processing center. What is the best shipping plan?

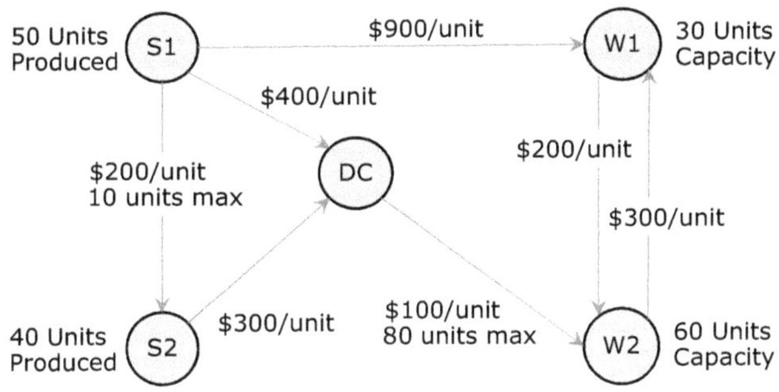

FIGURE P14.6 Transportation network for waste material.

14.7 DREDGING ALLOCATION

A company has contracted to remove 1,900,000 m³ of silt and clay from a harbor (Dredge sections 3 and 4) and an additional 1,100,000 m³ of sand from its approach channel (Dredge sections 1 and 2). The quantities of sand to be removed are 400,000 m³ from section 1 and 700,000 m³ from section 2. The quantities of silt/clay to be removed are 800,000 m³ from section 3 and 1,100,000 m³ from section 4. There are three possible dredge spoil disposal areas. For environmental reasons, only sand may be deposited in area 3 (which is along the beach). Disposal areas 2 and 3 have limited capacities of 800,000 and 650,000 m³, respectively. Disposal area 1 is large enough to accommodate all the dredged material, if this is desirable. The dredged sand occupies 10% more space than the volume removed from the channel. The clay–silt mixture in sections 3 and 4 will occupy 30% more space than the material in place in the harbor. It is estimated that 18,500 m³ of sand can be removed each day from section 1 if it is pumped to area 1; 10,000 m³/d if it is pumped to area 2; and so on. These estimates are shown in Table P14.7. Minimize the total dredging time.

TABLE P14.7 Estimates of Daily Production (1,000 m³/d)

Dredge Section	Spoil Disposal Area		
	1	2	3
1	18.5	10.0	20.0
2	12.0	17.0	15.5
3	25.0	28.0	*
4	22.0	25.0	*

14.8 CHANNEL DREDGING BIDDING STRATEGY

A company plans to tender a bid for a dredging project. The unit prices multiplied by the quantities are the bid cost for each item. The sum of these is the project bid price. Table P14.8 shows a bid derived in this fashion. The $2,520,000 is considered a reasonable bid. This is the price the company wants if its bid is successful.

The company would like to increase the unit prices of work that is done early in the project and balance this with decreased prices on work that is done later. The unit prices can be altered so long

TABLE P14.8 Costs for Channel Dredging

	Item	Duration (months)	Completion (months)	Unit Price ($)	Estimated Quantity	Estimated Amount ($)
1	Mobilization	2	2	240,000	Lump Sum	240,000
2	Debris removal	4	6	240,000	Lump Sum	240,000
3	Levee construction	8	14	600,000	Lump Sum	600,000
4	Soil excavation	12	26	1.80/m^3	500,000 m^3	900,000
5	Rock excavation	8	34	12.00/m^3	25,000 m^3	300,000
6	Demobilization	2	36	240,000	Lump Sum	240,000
					Total =	2,520,000

as the total is not changed. For example, the unit price for mobilization could be increased and the demobilization price could be decreased. This will reduce the firm's dependence on external financing during the project. Also, overpricing items with quantities that were underestimated on the bid form, while underpricing overestimated items, could yield a greater profit. In short, the company would like to prepare a bid using optimal unit prices that would improve its cash flow and maintain the $2,520,000 bid total. The acceptable rate of return for the company is 1% per month throughout the period under consideration. A reasonable objective is to maximize the present value of anticipated project revenue.

The work proceeds sequentially. Item 2 begins when Item 1 is complete, and so on. Payment accrues after each item is completed and at a rate consistent with the unit price, except for Item 1 which allows for 60% payment for item 1 upon mobilization and the remaining 40% upon completion at two months.

The acceptable rate of return for the company is 1% per month throughout the 41-month project period. A reasonable objective is to maximize the present value of anticipated project revenue.

The company knows that the paid quantity for dredged soil will be less than the estimated amount because 100% of the bed overdepth is rarely achieved. Assuming 90% of the bid quantity is reasonable, the paid quantity for Item 4 = 0.9(500,000 m^3) = 450,000 m^3.

14.9 LAKE POLLUTION CONTROL

A new water mandate requires removing at least 58,000 kg of the baseline pollution going into a lake. Moreover, airport and city managers want to participate in the pollution removal program by removing at least 60% of their baseline pollution allocations per year. The pollution processing plants at all three rivers need to remove between 20% and 95% of their pollution load. Removal costs are given in Table P14.9. What is the optimal percent removal allocation?

TABLE P14.9 Removal Costs and Pollution Values for Water Pollution Control Problem

Source	Removal Cost ($/kg)	Pollution to Lake (kg)
River A	146	27,500
River B	145	21,000
River C	149	24,500
City	215	13,200
Airport	203	18,900

14.10 INDUSTRIAL PRETREATMENT AND POLLUTON CONTROL

A factory has a maximum capacity of 100,000 kg of product per week. The product sales price is $2/kg. The production cost is $1.50/kg produced. The factory produces waste at the rate of 3 kg/kg of product, contained in wastewater at a concentration of 2 kg/m^3. The factory's waste treatment plant operates at a constant efficiency of 85% removal of the pollutant constituent and has a hydraulic capacity of 80,000 m^3/week. The treatment cost is $0.1/m^3. Wastewater is discharged into a city sewer with a discharge limit of 80,000 kg/week. There is an effluent discharge fee of $0.1/kg.

(a) What is the best production rate for the factory?
(b) There is a possibility that the effluent charge will be raised to $0.20/kg waste. How does this change the optimal production capacity?
(c) The effluent charge will be raised to $0.20/kg and the treatment plant capacity can be increased to 120,000 m^3/week with a cost of $0.15/m^3 treated. How does this this change the optimal production capacity?
(d) The effluent charge will be raised to $0.20/kg and the efficiency of the existing treatment plant can be increased to 95% at a treatment cost of $0.11/m^3. How does this change the optimal production capacity?

14.11 CHEMICAL MANUFACTURING

A chemical plant uses three raw materials (A, B, and C), which are in limited supply, to make three products (E, F, and G). Each product is produced in a separate process (1, 2, and 3).

Process 1 $\quad 0.6\text{ A} + 0.4\text{ B} \rightarrow \text{E}$
Process 2 $\quad 0.65\text{ A} + 0.35\text{ B} \rightarrow \text{F}$
Process 3 $\quad 0.5\text{ A} + 0.15\text{ B} + 0.35\text{ C} \rightarrow \text{G}$

The available raw materials do not have to be fully consumed. The reactions involving A, B, and C and other process data are in Tables P14.11a and P14.11b. The processing cost includes waste treatment. Maximize the total daily operating profit ($/d).

TABLE P14.11a Raw Material Costs and Availability

Raw Material	Max Available (kg/d)	Cost ($/kg)
A	40,000	0.30
B	30,000	0.40
C	25,000	0.50

TABLE P14.11b Process Chemistry, Costs, and Product Prices

Process	Product (1 kg)	Reactant Used (kg/kg Product)	Processing Cost ($/kg)	Selling Price ($/kg)
1	E	0.6A + 0.4 B	0.30	0.12
2	F	0.65 A + 0.35 B	0.10	0.60
3	G	0.5 A + 0.15 B + 0.35 C	0.20	0.76

14.12 INDUSTRIAL CHEMICALS

A company's primary product is manufactured 10,000 kg/d of P from raw materials A and B. The production of 1 kg of the primary product requires 1 kg of material A and 2 kg of material B. For the upcoming production period, 5,000 kg/d of raw material A and 7,000 kg/d of raw material B will be available. The primary product sells for \$6.50/kg.

For every 1 kg of the primary product, there is also produced 1 kg of liquid waste and 1 kg of solid waste. The solid waste can be given to a local fertilizer plant with no cost for pick up and disposal. Solid waste that is not sent to the fertilizer plant must be converted to P2.

The accounting department has reported these fixed and variable expenses for a typical day of 10,000 kg of P produced.

Total fixed cost = \$16,000 regardless of the production level.

Variable costs

Raw material A = \$1.52/kg
Raw material B = \$1.34/kg
Direct labor = \$0.50/kg P
Treatment of liquid waste = \$0.25/kg

Secondary by-products, P1 and P2, can be produced. Adding 1 kg of raw material A to every 1 kg of liquid waste will produce 2 kg of by-product P1. Adding 1 kg of B to 1 kg of solid waste will produce 2 kg of by-product P2. P1 will net \$0.85/kg and P2 will net \$0.65/kg. Liquid waste that is not converted to P1 must be treated at a cost of \$0.25/kg.

The secondary products will be low in quality and may not be profitable. However, the special treatment alternative also will be a relatively expensive operation. The company's problem is how to satisfy the pollution regulations and still maintain the highest possible profit. How should the liquid waste material be handled? Should the company produce P1, P2, treat the liquid waste, or use some combination of the three alternatives?

14.13 SOLID WASTE DISPOSAL

Two cities are planning a regional solid waste disposal system. There are 40,000 people in city 1 producing 700 T solid waste/week. City 2 has 65,000 people, producing 1,200 T/week. The three disposal sites, described in Table P14.13, use different handling and disposal methods, but the details are not important. The transport costs are \$0.5/T-km. Determine the optimal disposal of solid waste.

TABLE P14.13 Solid Waste Disposal Site Locations, Costs, and Capacities

Site	Distance from City 1 (km)	Distance from City 2 (km)	Fixed Cost (\$/week)	Variable Cost (\$/T)	Processing Capacity (T/week)
1	15	10	3,850	12	1,000
2	5	15	1,150	16	500
3	30	25	1,920	6	1,300

14.14 SEWER CONSTRUCTION

Work is needed at four locations on a sewer construction project, as given in Table P14.14 The company has only 1,500 grader-hours available for this part of the project and a maximum of 1,500 m^3 of concrete can be delivered during the scheduled period of work. Determine the amount of work that can be done in the period to maximize profits.

TABLE P14.14 Work Needs at Locations on a Sewer Construction Project

Location	A	B	C	D
Production target (units)	20	10	20	25
Graders (h/unit)	20	10	20	25
Concrete (m^3/unit)	20	22	24	25
Profit ($/unit)	4,250	4,250	3,000	5,000

14.15 RIVER POLLUTION MODEL

A pollutant that is discharged by three treatment plants will cause nuisance conditions if the concentration in the river exceeds 0.5 mg/L. The treatment plant flows and influent concentrations are in Table P14.15a.

The river will be modeled as 16 completely mixed reactors of equal volume plus one mixing zone for each treatment plant effluent and one mixing zone for a tributary stream. This is 20 segments. Segments 1, 4, and 12 are mixing zones for the treatment plant effluents. Segment 7 is the tributary mixing zone. There is no reaction in the mixing zones.

The segments will be modeled as completely mixed reactors, meaning the concentration in the segment and the effluent from the segment have the same concentration. The material balance on pollutant for segment 2 is

$$\underset{\text{In from 1}}{Q_1C_1} = \underset{\text{Out from 2}}{Q_2C_2} + \underset{\text{Reacted in 2}}{k_2VC_2}$$

TABLE P14.15a Treatment Plant Flows and Concentrations

Plant	Flow (units/d)	Influent Concentration (mg/L)	Cost ($1,000 per mg/L removed)
A	15	15	2.4
B	5	40	2.1
C	14	32	3.4

TABLE P14.15b Stream Segment Reaction Rate Coefficients

Segment	Rate Coefficient *k* (1/d)	Segment	Rate Coefficient *k* (1/d)
1	0	11	0.04
2	0.07	12	0
3	0.07	13	0.04
4	0	14	0.04
5	0.076	15	0.05
6	0.05	16	0.05
7	0	17	0.05
8	0.05	18	0.04
9	0.06	19	0.04
10	0.04	20	0.04

where k_2 = reaction rate coefficient for reactor 2 (1/d). The rate coefficients for each stream segment are given in Table P14.15b

The reactors all have volume $V = 200$ units and detention times of $\theta = V/Q$ days, where Q is the flow entering the segment. Q is either the flow from a mixing zone or the flow from the upstream segment. The flow between mixing zones is constant.

(a) Minimize the total cost of treatment for the three plants.
(b) Is it possible to meet the standard by having all three plants produce the same effluent concentration?
(c) Is it possible to meet the standard by having all three plants remove the same percentage of pollutant?

14.16 LOCATING WASTE PROCESSING STATIONS

A county wants to build the minimum number of waste processing stations to serve six cities such that no city is more than 20 miles from a processing station. Each processing station will be located in a city. The distances are given in Table P14.16. How many stations should be built and where they should be located?

TABLE P14.16 Distance (miles) Between Cities in Service Area

	To					
From	**Site 1**	**Site 2**	**Site 3**	**Site 4**	**Site 5**	**Site 6**
City 1	0	20	25	15	30	20
City 2	10	0	18	28	20	10
City 3	20	15	0	22	34	25
City 4	30	30	20	0	20	25
City 5	30	20	30	10	0	14
City 6	20	10	20	25	14	0

14.17 POLLUTION CONTROL

An industry has \$100 million in capital to invest in pollution prevention. Four projects are desired, if the budget permits. The projects have a capacity of 100,000 or 500,000. Plants of 100,000 capacity cost \$5 million; plants of 500,000 capacity cost \$50 million. Not more than two of the smaller plants can be built. How many plants of each capacity should be built? How will increasing the budget by \$10 million change the solution. Explain your solution with a diagram.

14.18 SEWERAGE PROJECT SELECTION

A sewerage district must decide which capital improvement investments will maximize the net present value. The available capital is \$25 million. The projects are as shown in Table P14.18.

TABLE P14.18 Candidate Projects for Capital Improvements

Project	Net PV ($million)	Capital Cost ($million)
1. Sewer Rehab #1	70	20
2. WWTP Rehab	50	15
3. Sewer Rehab #2	40	12
4. Sewer Rehab #3	30	10
5. Pumping Station	25	6
6. Sludge Dewatering	28	9

14.19 WATER QUALITY MANAGEMENT 1

Five cities discharge waste into a river, as shown in Figure P14.19. Each city will build a treatment plant that provides primary, secondary, or tertiary treatment, which will remove 60%, 80%, or 95% of the influent pollutant load. Tables P14.19a and P14.19b give the influent pollutant loads (kg/d), maximum allowable effluent loads (kg/d), and the treatment costs ($/kg removed) for each plant. Pollutant removal in the river will be ignored. Formulate and solve an integer linear programming model to determine the level of treatment at each plant that minimizes the total treatment cost.

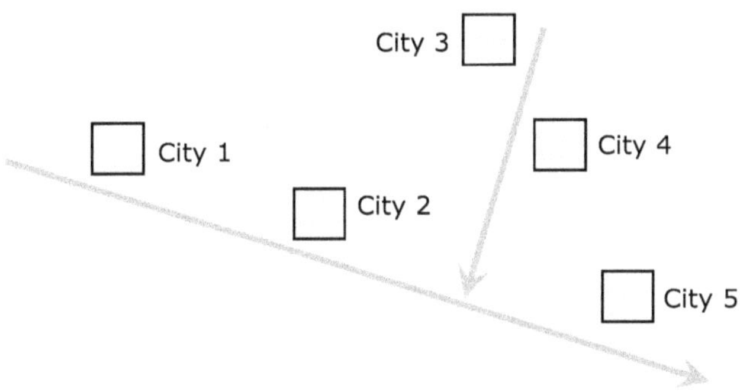

FIGURE P14.19 Location map for cities discharging waste to a river.

TABLE P14.19a Wastewater Data (kg/d)

City	Influent Load (kg/d)	Max Allowed Effluent Load (kg/d)	Mass Removed (kg/d)			Mass Discharged (kg/d)		
						Primary	Secondary	Tertiary
1	1,200	300	720	960	1,140	480	240	60
2	600	350	360	480	570	240	120	30
3	2,000	300	1,200	1,600	1,900	800	400	100
4	500	350	300	400	475	200	100	25
5	1,000	400	600	800	950	400	200	50

TABLE P14.19b Treatment Cost Data

	Unit Cost ($/kg)			Cost ($/d)		
City	Primary	Secondary	Tertiary	Primary	Secondary	Tertiary
1	0.15	0.25	0.6	108	240	576
2	0.2	0.35	0.75	72	168	360
3	0.12	0.18	0.55	144	288	880
4	0.2	0.35	0.75	60	140	300
5	0.16	0.27	0.63	96	216	504

15 Optimization of Nonlinear Problems

Nonlinear problems will have one or more equations (the objective function or the constraints) that contain products of variables, exponents, and other kinds of nonlinear expressions. These problems are not harder to formulate than linear problems. In fact, they may seem easier because engineers frequently use these kinds of equation.

It is convenient to ignore or omit cost items that are constant or nearly constant and to omit minor cost items. For example, an anaerobic sludge digester system has a digestion tank, a tank cover, mixers, sludge pumps, heaters, and piping. The cost of pumps and piping depend on the volume of sludge treated but are independent of the digester volume and the cost of the digester tank and cover. They can be omitted from an optimization model, but they must be included in the final cost estimate.

15.1 FINDING SOLUTIONS FOR NONLINEAR PROBLEMS

Figure 15.1 shows that nonlinear problems may have more than one local minimum. It is possible for an optimization search algorithm to terminate at local minimum instead at the global minimum. Constraints on the range of accepted values for the decision variable may mean that the feasible solution (the constrained minimum) is not the same as the unconstrained minimum. This limits the use of classical calculus.

A simple example is

$$\text{Objective} \quad \text{Maximize} \quad Z = 2 - 6X + X^2$$

Classical calculus sets the derivative equal to zero and solves for X, as follows.

$$dZ/dX = -6 + 2X = 0$$

$$X = 3$$

$$Z = 2 - 6(3) + 3^2 = -7$$

Classical calculus often fails with constrained optimization problems. For example, constrain X to a range of 0–2 and the problem becomes

$$\text{Objective} \quad \text{Maximize} \quad Z = 2 - 6X + X^2$$

$$\text{Constraints} \quad X \leq 2$$

$$X \geq 0$$

This simple constrained problem solves itself. Make X as large as possible, which is $X = 2$, giving

$$Z = 2 - 6(2) + 2^2 = -6$$

The solution of $X = 3$ obtained by classical calculus violates requirement that $X \leq 2$.

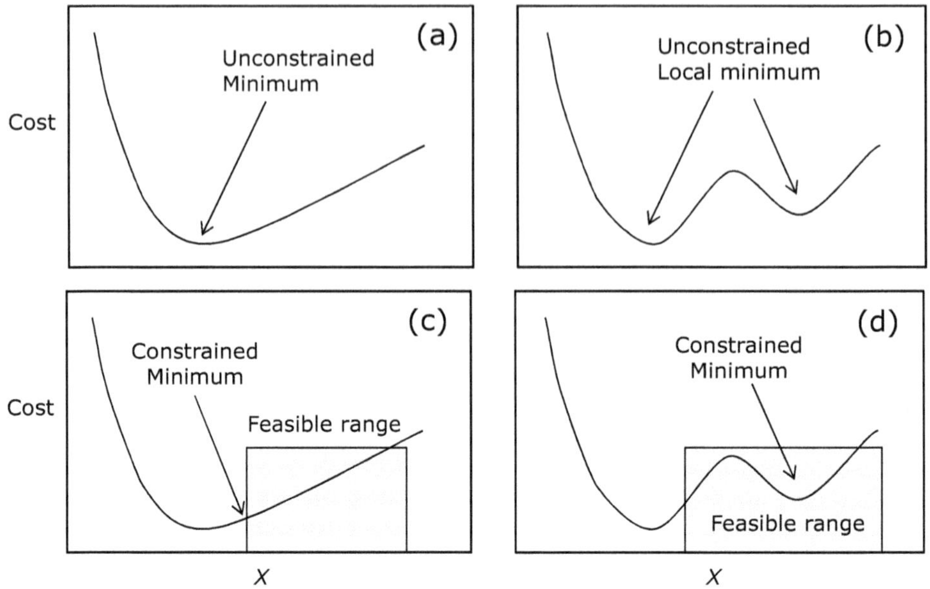

FIGURE 15.1 Four situations of minimizing a nonlinear cost function.

The generalized formulation of a nonlinear optimization problem with n decision variables and m constraint equations is as follows:

Objective function

$$\text{Max}\left(\text{or min}\right) \quad Z = f\left(\alpha,\, x_j\right) \quad j = 1, 2, \ldots n$$

Subject to these constraints

$$\begin{aligned} g_1(\alpha, X_j) &\le (\text{or} \ge)\ b_1 \\ g_2(\alpha, X_j) &\le (\text{or} \ge)\ b_2 \\ \ldots \quad &\le (\text{or} \ge) \ldots \\ g_m(\alpha, X_j) &\le (\text{or} \ge)\ b_m \end{aligned}$$

where

$X_j = X_1, X_2, \ldots, X_n$ are n decision variables
α = coefficients in the constraint equations
$b_1, b_2, \ldots$ = limits on the constraint equations

Figure 15.2 shows an unconstrained search for a two-variable nonlinear problem. A starting point is assigned (guessed) and increasing smaller steps are taken as the search nears the optimum. If the goal is to maximize the objective, steps are always taken in the uphill direction, and the optimum is found when no more uphill steps are possible, even with very small steps. (*very small* must be defined numerically.) The objective function may have more than one optimum, as shown in Figure 15.1b, or it may be twisty and difficult to search. For these reasons, the search may be repeated from different starting positions to verify the solution.

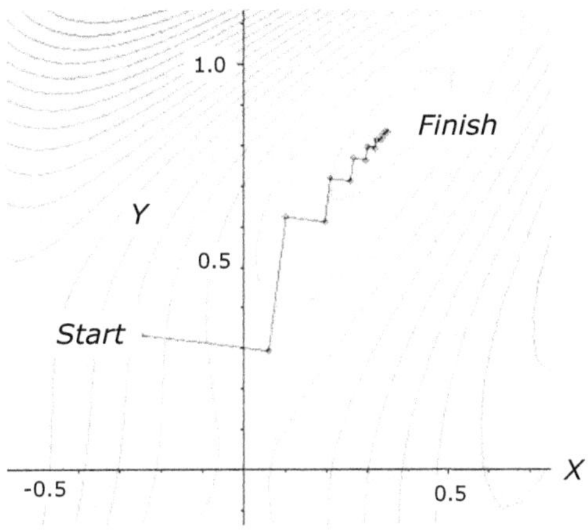

FIGURE 15.2 Two-dimensional search.

There are a number of algorithms for defining the direction and size of the steps. Some are based on calculating slopes (derivatives) and some are entirely arithmetic. Not long ago, it would have been necessary to learn the mathematical details of the search strategies. Using LINGO allows us to omit those details. (Those who wish to learn the mathematics should read Hillier & Lieberman 2001 and Sarkar & Newton 2008 or similar textbooks).

15.2 DESIGN DEGREES OF FREEDOM

The general formulation of a mathematical model consists of M design relations that involve N variables. The design relations may be linear equations, nonlinear equations, inequalities, and graphs or tables.

If $N = M = 2$, there are two equations and two unknown variables. The problem can be solved directly by algebraic manipulation, by iterative (trial-and-error) calculations, or graphically. There is one solution and no optimization to be done.

An optimization problem has more design variables than design relations, $N > M$. The difference between N and M is the design degrees of freedom, F.

$$F = N - M$$

The designer must specify values for $F = N - M$ variables to create a solvable system of equations. The F variables are *design variables* that can be assigned different values and manipulated to search for an optimal solution.

Many problems have a small number of degrees of freedom, sometimes only 2 or 3 even when there are 10 or 20 design equations.

Example 15.1 Blending Problem

Material C is a mixture of materials A and B. The manufacturer has an order for material C that is to contain 15,000 kg of A and to contain A and B in the ratio of A/B = 4. Calculate how much material B to order.

There are $M = 4$ design relations and $N = 4$ variables (A, B, C, and α)

Material balance: C = A + B
Requirement for A: A = 15,000 kg
Quality of the blend: A/B = 4

With $N = M = 4$, there is a unique solution:

B = A/4 = 15,000/4 = 3,750 kg
C = A + B = 15,000 + 3,750 = 18,750 kg

Example 15.2 Separation of Four Materials

Figure 15.3 shows a system that separates 1,000 kg of mixed materials into four products, A, B, C, and D, with X_2 and X_3 as intermediate products. The masses of X_2 and X_3 depend on the split fraction η_1. The split fractions $\eta_2 = 0.7$ and $\eta_3 = 0.95$ are fixed.

There are:

$N = 10$ variables: X_1, X_2, X_3, A, B, C, D, η_1, η_2, and η_3
$M = 7$ design relations:
Material balance equations: $X_1 = 1{,}000$, $X_2 = 1{,}000$, $\eta_1 X_1 = X_2 + X_3$
Outputs: $A = \eta_2 X_2$, $B = X_2 - A$, $C = \eta_3 X_3$, $D = X_3 - C$
$F = N - M = 10 - 7 = 3$

The designer must specify values for three variables. If the split fractions are specified to be $\eta_1 = 0.5$, $\eta_2 = 0.9$, and $\eta_3 = 0.2$, then this is not an optimization problem. The designer must use those values, and the solution is

A = 450, B = 50, C = 100, D = 400

If the designer is free to try different separation efficiencies, this becomes an optimization problem. The efficiencies can be adjusted to make more or less of the four products, and if the products have different selling prices, an optimization problem may be

Max total sales = 2A + 4B + 1C + 3D

where the coefficients are prices per kilogram.

The solution must satisfy the seven design equations, plus constraints or requirements on the amount of each material that is produced, which, for example, would be of this form:

$A \geq 500$, $C \geq 100$, $D \geq 250$
A + B + C + D = 1,000

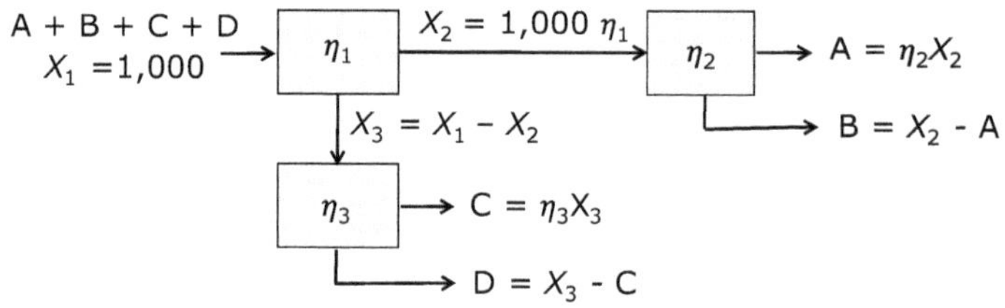

FIGURE 15.3 Separation problem.

15.3 MATHEMATICAL VERSUS ENGINEERING OPTIMIZATION

A mathematical equation is assumed to have known fixed constants, as in $y = 2\,X^{0.6}$. An engineering equation of the same form, Cost = 2(Volume)$^{0.6}$, gives an exact numerical value of *Cost* for a value of *Volume*, but this cost is known to be an estimate. It is not the exact cost because the multiplier 2 and the exponent 0.6 are estimates. The consequence is that the solution of an engineering optimization problem that includes such an estimating equation should be used with a degree of caution. Change 2 to 2.2, or 0.6 to 0.65, and the minimum cost changes. The design variables associated with the calculated minimum cost will also change.

Looking at this in a more optimistic way, the engineer can say that "there is a range of values around the calculated optimum that will be good engineering designs." An optimization always involves trading one cost against other, usually a capital cost against the cost of electricity or some other operating cost. Fortunately, it is often the case that changing coefficients within a margin of reasonable confidence causes a relatively small shift in the location of the optimum solution.

This means that the engineering *optimum* can have some robustness with regard to variations in coefficients that are used in the design equations. The designer should have as much interest in the shape of the cost curve as in the location least cost solution. Some extra work will be required to discover the shape, but the work will be repaid by added information.

We give two illustrations of engineering optima, as contrasted with a purely mathematical exercise.

Example 15.3 Pipe System Design

Pumping system design and operation present opportunities to reduce energy use and operating costs. One aspect of the design is balancing the cost of energy for pumping against the cost of the piping. A larger pipe carries more water at a lower energy cost, but it has a higher capital cost.

A pipe must carry $Q = 0.5\ \text{m}^3/\text{s}$ of water. The annualized fixed cost for pipe with diameter D (m) is

$$C_{\text{Fix}}\left(\$1{,}000/\text{y}\right) = 100\ D^{1.5}$$

The annual operating cost is

$$C_{\text{Op}}\left(\$1{,}000/\text{y}\right) = 20/D^{4.8}$$

The pipe diameter giving the minimum total annual cost is found from:

$$\text{Minimize} \qquad Z = 100D^{1.5} + 20/D^{4.8}$$

This is a single-variable nonlinear optimization; the single design variable is D. The exponents make an algebraic solution awkward, so the easiest solution is to search a range of diameters.

The LINGO code is

```
MIN = TOTALCOST;
      FIXEDCOST =100*D^1.5;
      OPERCOST =20/D^4.8;
      TOTALCOST = FIXEDCOST + OPERCOST;
END
```

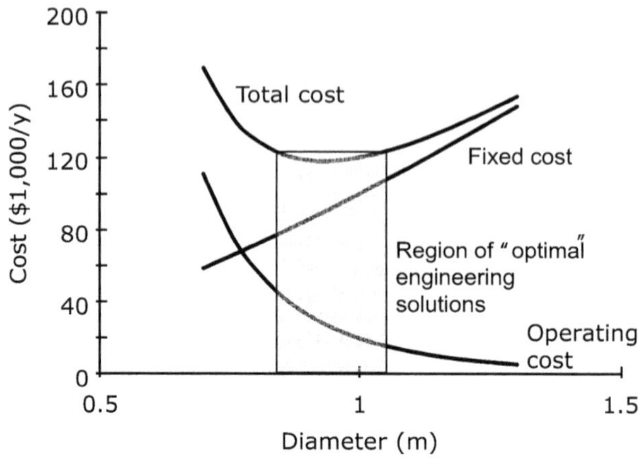

FIGURE 15.4 Annual pumping cost ($1,000/y) as a function of pipe diameter.

TABLE 15.1 Costs for Commercially Available Pipe Sizes

Diameter (m)	Fixed Cost ($/y)	Operating Cost ($/y)	Total Cost ($/y)
0.85	78,400	43,600	122,000
0.9	85,400	33,200	118,500
1.05	107,200	15,800	123,000
1.1	115,400	12,700	128,100
1.2	131,500	8,300	139,800

The solution is

Minimum total cost = $118,000/y
Fixed cost = $89,900/y
Operating cost = $28,100/y
Pipe diameter giving the minimum cost = D = 0.93 m

Figure 15.4 shows how the costs change over a range of pipe diameters. The mathematical minimum cost is $118,000 for a pipe diameter $D = 0.93$. The engineering solution cannot be $D = 0.93$ because pipe is not manufactured in that size.

This illustrates one difficulty with mathematical optimization. The available diameters are given in Table 15.1. The cost difference for 0.85, 0.9, and 1.05 m is about 4%. This gives the designer/owner a choice to trade a higher fixed cost against lower operating costs.

Example 15.4 Consequence of Uncertainty in a Cost Equation

The capital cost of a process is $C = KV^M$, where V is the volume, and the present value of the O&M costs is 10,000/V. This can be formulated as a LINGO program, in this case with $K = 12$ and $M = 0.6$.

TABLE 15.2 "Optimal" Solutions for Four Cost-Estimating Equations

	$K = 10$	$K = 10$	$K = 12$	$K = 12$
	$M = 0.55$	$M = 0.6$	$M = 0.55$	$M = 0.6$
Total cost	222	258	250	290
Capital cost	143	162	161	181
O&M cost	79	97	89	109
Volume, V	127	103	113	92

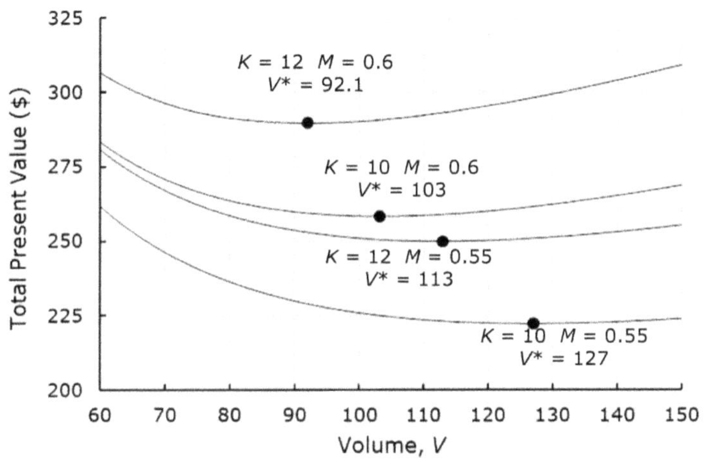

FIGURE 15.5 Cost curves for different values of K and M.

The LINGO code is

```
MIN = CT;
CT = CC + OM;
CC = K*V^M;
OM = 10000/V;
K = 12;
M = 0.6;
END
```

Table 15.2 and Figure 15.5 show four solutions with values of K and M that differ by about 10%. The range of total costs is 67, which is ±20% of the mid-range cost. The optimal volumes range from 92 to 127. Figure P15.5 shows that the present value curves are not changing sharply in the region of the minimum. This indicates that the engineering *optimum* can have some robustness with regard to variations in coefficients that are used in the design equations.

15.4 A NOTE ABOUT LINGO

In our experience LINGO has been flawless with linear optimization problems, but this has not been the case with nonlinear optimization. Models with nonlinear expressions are much more difficult to solve than linear models and LINGO may not find a solution even though one exists, or it may find a solution that appears to be the "best," even though a better one may exist. These results are obviously undesirable. The LINGO 18.0 User's Manual provides some guidance on how to minimize the occurrence of these undesirable results.

Making sure the model is formulated in a way that is most efficient to solve can pay off in terms of solution speed and reliability. Intelligent use of upper and lower bounds on variables will keep LINGO from wasting time searching regions that are unlikely to yield good solutions. For example, suppose you know that, even though the feasible range for a particular variable is between 1 and 100, it is highly improbable that the optimal value is outside the range of 50–75. In this case, using the @BND function to specify a lower bound of 50 and an upper bound of 75 could reduce solution times. Bounding can also help keep the solution search clear of mathematically troublesome areas like undefined regions. For example, if you have a constraint with the term $1/X$, it may be helpful to add a lower bound on X so that it does not get close to 0.

Try to model the problem such that the units involved are of similar orders of magnitude. If the largest number in the model is greater than 1,000 times the smallest number in the model, LINGO may encounter problems when solving the model. This may also affect the accuracy of the solution by introducing rounding problems.

When possible, use linear rather than nonlinear equations. A simple example is a constraint on the ratio of two variables. The constraint $X/Y < 10$ is nonlinear. The equivalent linear constraint is $X < 10*Y$.

Minimize the use of integer restrictions. Solving the model without integer restrictions and then rounding may yield acceptable answers with less chance of the search going astray. (However, rounding a solution will not necessarily yield a feasible or optimal solution.)

These are some additional commands that are needed for nonlinear optimization problems. Notice they are preceded by the @ sign.

@LOG(X) = natural log of X
@EXP(X) = e^X
@LOG10(X) = base 10 logarithm
@SQR(X) = X squared (equivalent to X^2)
@SQRT(X) = square root of X
@ABS(X) = absolute value of X
@SIN(X), @COS(C), @TAN(X) are the trigonometric functions, with X in radians
@PI() = value of pi (3.14159265....)
@FREE(X) makes the variable X free to take on negative values
@BND limits a variable to fall within a given range
@BND(10, X, 20) constrains the variable X to lie in the interval [10, 20]

This next example is not an optimization problem. It is offered to explain the scaling problem and to show how @LOG(X), @EXP(X), and @FREE(X) are used.

Example 15.5 Chemical Equilibrium

The chemical equilibrium equations for the components of phosphoric acid at pH 8.0 are as follows, where the brackets, [ ], indicate molar concentrations.

$$[H^+] = 10^{-8}$$

$$K_1 = \frac{[H^+][H_2PO_4^-]}{[H_3PO_4]} = 6.9\times10^{-3}$$

$$K_2 = \frac{[H^+][HPO_4^{2-}]}{[H_2PO_4^-]} = 6.2 \times 10^{-8}$$

$$K_3 = \frac{[H^+][PO_4^{3-}]}{[HPO_4^{2-}]} = 4.8 \times 10^{-13}$$

This is a computationally difficult problem for LINGO because the values differ by a factor of 10^{10}. The difficulty is greatly reduced if the problem is scaled by taking logarithms.

Define LH = log[H] LP = $\log[PO_4^3]$ LHP = $\log[HPO_4^2]$
LH2P = $\log[H_2PO_4]$ LH3P = $\log[H_3PO_4]$

LINGO Code

```
! Components of phosphorus at pH = 8;
! Total Phosphate concentration = 0.1;
! It helps to scale by taking logs;
      LH = @LOG(10^-8);
      LH + LH2P - LH3P = @LOG(0.0069);
      LH + LHP - LH2P = @LOG(6.2*10^-8);
      LH + LP - LHP = @LOG(4.8*10^-13);
! Convert back to original variables;
      H = @EXP(LH);
      P = @EXP(LP);
      HP = @EXP(LHP);
      H2P = @EXP(LH2P);
      H3P = @EXP(LH3P);
! Total phosphorus;
      H3P+ H2P + HP + P = 0.1;
! Unconstrain the log variables so they can be negative;
      @FREE(LH); @FREE(LP); @FREE(LHP); @FREE(LH2P); @FREE(LH3P);
END
```

LINGO solution

LH = −18.420 LP = −12.396
LHP = −2.452 LH2P = −4.277 LH3P = −17.721
P = 0.413×10^5 HP = 0.0861 H2P = 0.0139 H3P = 0.201×10^7

15.5 EXAMPLES OF NONLINEAR PROGRAMMING PROBLEMS

Example 15.6 Slow Sand Filter Design

Slow sand filters were used in some large US cities in the early days of public water utilities and they are still being used in many countries, especially in villages. This simple system uses no electrical energy, so it can be ideal for rural areas.

The disadvantage of slow sand filtration is the large filter area that is required. The area is determined by the volume of water produced (L/h) and the loading rate per square meter of filter area (L/h-m^2). Loading rate has been optimized by years of research and experience and it is treated as a fixed known quantity. Slow sand filters are effective at removing bacteria and fine particles because a film, called the *schumtzdecke*, forms on the top layer of the sand and this provides excellent purification (some bacteria and most protozoans are removed). At intervals of about one month the filter is cleaned by manually removing the *schumtzdecke* and the top few millimeters of the sand.

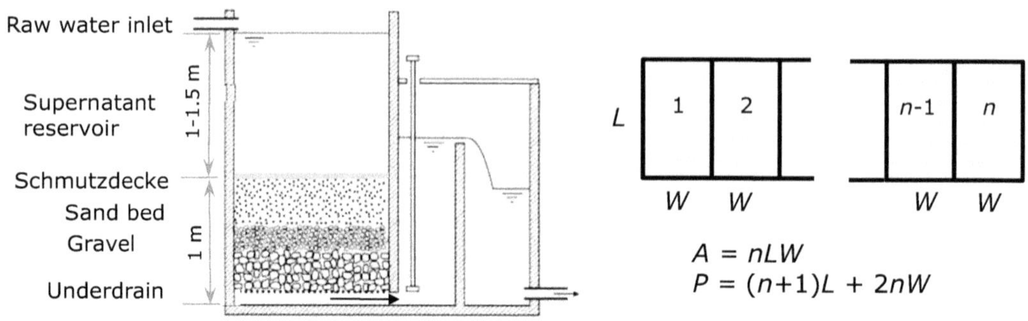

FIGURE 15.6 Slow sand filters built with common walls.

The alternative is rapid gravity filters, which operate at a much higher filtration rate. As a result, the pressure loss through the filter bed increases rapidly and the filters need to be cleaned frequently by pumping clean water upward through the bed to flush away the collected dirt. Chemical pretreatment is always needed.

This example is based on the work of Paramesivam et al. 1981). The filters are usually rectangular with common walls. Figure 15.6 shows n slow sand filters that are built with common walls to minimize the total wall length and volume. Each filter cell has dimensions L and W, so the total surface area is $A = nLW$, where n is the number of filter cells. The cost of building the filter floor is C_a (Rp/m²) and the cost of the wall is C_w (Rp/m) (Rp = rupees).

The filter complex is to treat an average flow of 8,000 L/h at a loading rate of 1 L/h per m² of filter surface area. The depth of the filter bed, underdrains, water above the sand, and freeboard are fixed at 2.5 m. The floor area and volume of sand are the same regardless of the number of cells. The cost of the walls is proportional to the total length of the walls.

The design that minimizes the construction cost will be to build one filter. This is a poor design because it does not provide for operation when that filter is out of service for cleaning or maintenance. Increasing the number of filters will increase the cost, but it will also increase operational flexibility and reliability.

The number of cells, n, to be constructed is a design decision with operational considerations. At least two filters must be provided, because at times one filter is taken out of service for cleaning. The cleaning time is at least one day. If there are only two filters, the one that remains in operation must operate at twice the normal hydraulic loading (a 100% overload), or water production must be reduced. If filtered water storage is available, this is workable for short periods of time.

Minimize	Cost = $C_F A + C_W P$	
Subject to:	$A = (8{,}000\ \text{L/h})/(1\ \text{L/m}^2\text{-h}) = 8{,}000\ \text{m}^2$	Total area required (m²)
	$A = nLW$	Total filter floor area (m²)
	$P = (n+1)L + 2nW$	Total length of walls (m)

This is the cost of the filters that depends directly on n, L, and W.

The area A and the cost of the floor, $C_F A$, is constant, so the objective function can be written as

$$\text{Cost} = K + C_W P$$

The constant K does not change the value of P that gives the minimum cost. The same is true for the constant C_W. The problem can be simplified to minimizing the length of the walls.

Minimize $P = (n+1)L + 2nW$

Subject to: $A = nLW = 8{,}000$

TABLE 15.3 Filter Dimensions and Loadings with One Filter Out of Service

No. of Filters	P (m)	L (m)	W (m)	Filter Area with One Filter Out of Service (m^2)	Loading with One Filter Out of Service (L/m^2)
2	438.2	73.0	54.8	4,000	2.0
3	506.0	63.2	42.2	5,422	1.5
4	565.7	56.7	35.4	6,000	1.33
5	619.7	51.6	31.0	6,400	1.25

This has two equations and three unknowns (n, L, and W) or $F = 3 - 2 = 1$ design degrees of freedom. The designer can specify one variable and solve for the other two. The best choice is to specify the number of filters because this must be an integer.

This gives a series of problems:

For $n = 2$ Minimize $P = (2+1)L + 2(2)W = 4\ W + 3\ L$
Subject to: $A = 2LW = 8{,}000$
P = 438.2 $L = 73.0$ $W = 54.8$

For $n = 3$ Minimize $P = (3+1)L + 2(3)W = 6\ W + 4\ L$
Subject to: $A = 3LW = 8{,}000$
P = 506.0 $L = 63.2$ $W = 42.2$

And so on.

The dimensions would probably be rounded up or otherwise adjusted to whole numbers.

If three filter cells are built, and one is taken out of service, the remaining two operate at a 50% overload. With more filter cells, the overload factor is reduced, as shown by Table 15.3. The design problem is to optimize the physical dimensions for different numbers of cells and then evaluate the feasibility of the resulting operating conditions.

More filter cells require more valves and possibly more piping. The economy of scale for valves and piping is favorable and building five cells instead of three or four has virtually the same hardware cost.

The number of filters can be reduced if there is storage of filtered water. Another interesting problem is to dimension prefabricated filter walls for delivery to many villages. Prefabrication will reduce the loss of materials from construction sites and it should also provide some economy of scale for manufacturing.

Recovery and reclamation of waste materials is a core principle of pollution prevention and control. The balance is always between processing/disposal cost and the value of the recovered material. In especially fortunate cases, there can be a profit, but more often the goal is to reduce disposal costs. This example recovers a useful substance by leaching it from a solid matrix and concentrating the leachate by evaporation.

Example 15.7 Material Recovery by Waste Leaching

A waste material is being hauled to a special landfill at a high cost for transportation and disposal. The contaminant is a substance, P, that has value if it can be separated from the bulk waste material. It is soluble and the bulk material is insoluble, so leaching is a possible recovery process. The leachate can be concentrated to make a salable liquid product. Such a recovery

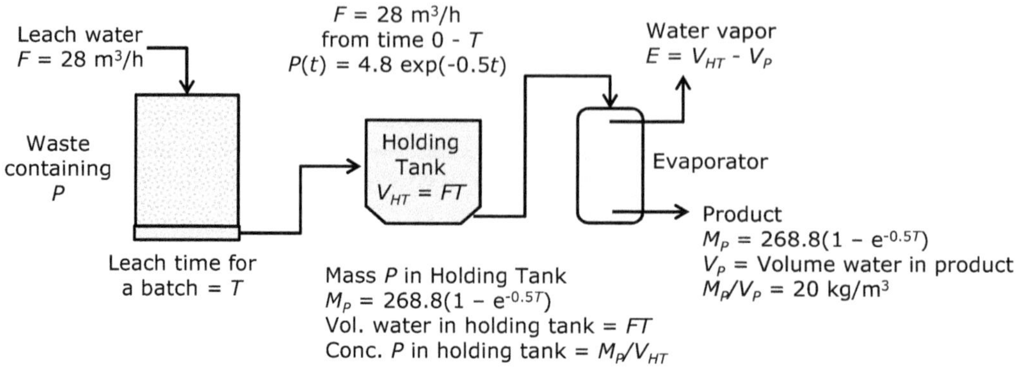

FIGURE 15.7 Material recovery system collects a batch leachate in a holding tank before concentration by evaporation.

process will reduce the cost of waste disposal, and it may earn a profit. The process is shown in Figure 15.7.

A batch of waste is leached with water at a constant rate of $F = 28$ m³/h of water. The initial concentration of P in the leachate is 4.8 kg/m³, but this decreases exponentially over time as leaching progresses. The concentration of P in the leachate at time t after the leaching started is

$$C(t) = 4.8e^{-0.5t}$$

The leaching is stopped at time T which produces a batch of leachate in the holding tank of volume $V_{HT} = (20 \text{ m}^3/\text{h})(T)$. The mass of P in the batch is M_P and the concentration of $P = M_P/V_{HT}$. This concentration will be increased to 20 kg P/m³ by evaporating part of the water.

The 20% concentrate leaving the evaporator can be sold for \$2/kg P. The cost of removing water by evaporation is \$3/m³.

What length of the leaching time, T, will yield maximum profit from each batch?

Let F = flow rate of water used for leaching
T = leaching time
E = volume of water evaporated

The concentration of P in the leachate at any time t is

$$C(t) = 4.8e^{-0.5t}$$

The mass of P leached from the bulk waste in the interval 0 to T is

$$M_P = \int_0^T \left(28\frac{\text{m}^3}{\text{h}}\right)\left(4.8\frac{\text{kg}}{\text{m}^3}\right)e^{-0.5t}dt = \left(\frac{28(4.8)}{0.5}\frac{\text{kg}}{\text{h}}\right)(1-e^{-0.5T}) = \left(268.8\frac{\text{kg}}{\text{h}}\right)(1-e^{-0.5T})$$

The objective is to maximize the net profit per batch

$$\text{Net profit (\$/batch)} = \left(\frac{\$2}{\text{kg } P}\right)\left(\frac{\text{kg } P}{\text{batch}}\right) - \left(\frac{\$3}{\text{m}^3}\right)\left(\frac{\text{m}^3 \text{ evap}}{\text{batch}}\right)$$

Figure 15.8 shows the solution.

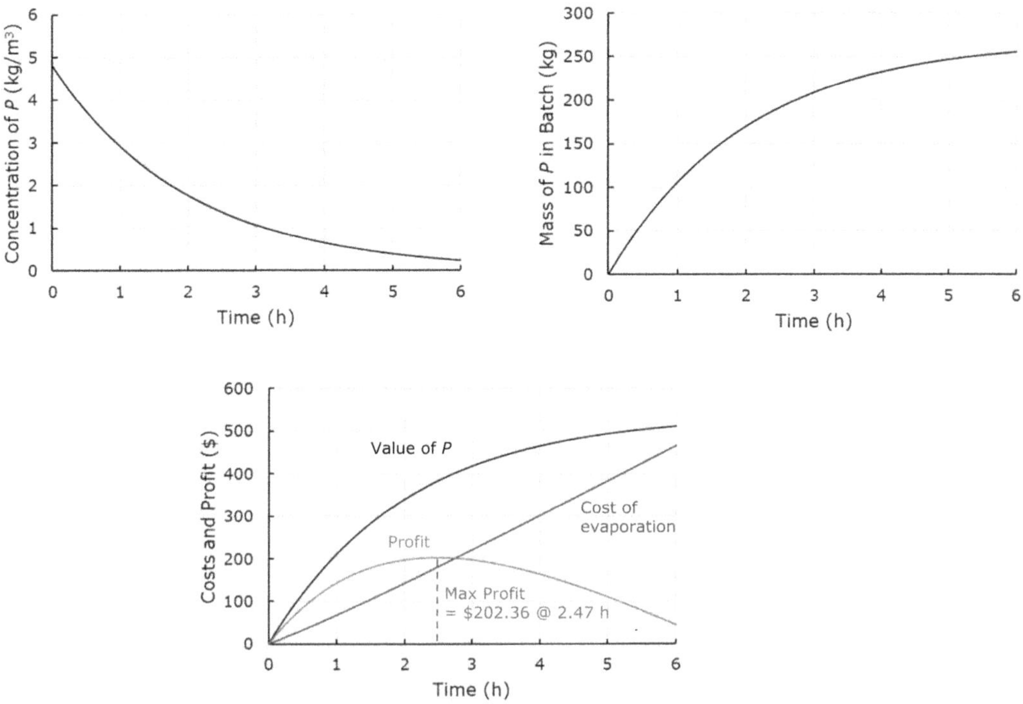

FIGURE 15.8 Solution of waste leaching example.

This is the LINGO Program.

```
! Objective;
      MAX =2*MP - 3*E;
! Define Variables;
! 2*MP = Value of Product P;
! 3E = cost of evaporation;
! Volume of leachate in holding tank;
      VHT = 28*T;
! MP = Mass Product in Leach = Mass into the Concentrator;
      X = @EXP(-0.5*T);
      MP = 268.8*(1 - X);
! Required concentration of Product = 20;
      ! Volume of Product;
      VP = MP/20;
! Volume Evaporated;
      E = VHT - VP;
END
```

Maximum profit = $202.36/batch
Leaching time T = 2.47 h
Batch volume in holding tank V_{HT} = 69.19 m³
Mass of P in batch and product M_P = 190.66 kg
Volume of the product V_P = 9.53 m³
Volume of water evaporated E = 59.65 m³

This seems like a small profit, but it is for one batch of leachate and many batches can be made. More important, it is a profit and there are the additional savings of waste disposal that have been omitted from the analysis.

15.6 CONCLUSION

Engineering optimization problems are not like mathematical optimizations where a precise single value is the correct answer. Engineering problems will have a solution that minimizes the specified objective, but they will also have a range of near optimal solutions. This means that the shape of the cost curve needs to be examined. The curve for an objective function that combines capital and operating costs as a present value tends to have a rather flat region near the calculated optimum. The flatness indicates a trade-off between fixed initial costs and ongoing operating costs.

Another difference is that the cost coefficients, kinetic rate coefficients, and other technical parameters are assigned single values in the problem statement, but many of these estimates are uncertain to some degree. Also, the quality and characteristics of process feed material is variable. For these reasons, near optimal costs may be within the margin of error for the cost estimates and design calculations. One advantage of having a mathematical model is that coefficients and costs can be changed to do a sensitivity analysis. This may show that some values need to be known more precisely than others, in which case work may be needed to refine the estimates, or a management strategy may be needed to accommodate the uncertainties.

15.7 PROBLEMS

15.1 CYLINDRICAL TANK DESIGN

Determine the tank diameter (D) and length (L) that will minimize the cost of a cylindrical tank.

15.2 A LOCATION PROBLEM

An industry is planning to expand into four new cities. To provide good quality service, a new service center must be constructed to support the existing facilities in four current markets. The location coordinates are:

City 1 $x = 10, y = 45$
City 2 $x = 15, y = 25$
City 3 $x = 20, y = 10$
City 4 $x = 55, y = 20$

Locate the new service center to minimize the total distance from the new center to the existing facilities.

15.3 FERTLIZER PRODUCTION

An engineer at a chemical company has synthesized a new fertilizer that is made of just two interchangeable basic raw materials. The company wants to take advantage of this and produce as much as possible of the new fertilizer. Combining amounts A and B of the basic raw materials produces a quantity Q of fertilizer given by

$$Q = 4A + 2B - 0.5A^2 - 0.25B^2$$

The company currently has $40,000 to buy raw materials at a unit price of $8,000 per unit of A and $5,000 per unit of B. What amounts should be purchased to maximize production.

15.4 OPTIMAL TANK DIMENSIONS TO MINIMIZE HEAT LOSS

A cylindrical tank diameter has D and height H. The temperature of the tank contents is T and T^* is the temperature of the surroundings (air, water, or soil). Heat is lost from the tank through the floor, top, and walls:

$$q = UA(T - T^*), \quad T > T^*$$

where

q = heat lost across surface area A, W
A = surface area, m^2
U = heat transfer coefficient, $W/m^2\text{-}°C$

(a) Find the tank dimensions that minimize the heat loss. Assume that heat transfer coefficient and the temperature difference are the same for all surfaces (sides, top, and bottom).
(b) Find the tank dimensions that minimize the heat loss assuming that the bottom half of the tank walls and the floor are insulated with U_B. The top half of the wall and the cover are not insulated with U_T. Assume the temperature differences are the same on all surfaces. Do the dimensions depend on the magnitude of the temperature difference?
(c) Optional: Find a general solution for a fraction f of the walls and floor being insulated, but the top is not insulated. How does the optimal diameter change as f changes?

15.5 PROCESS EXPANSION PLAN

The cost of a process can be estimated by $C = KQ^{0.7}$, where Q is in kg/h. For a process capacity of 15 kg/h, the cost is $180,000. The plan is to increase the capacity from 160 to 400 kg/h. The forecast is for a yearly linear increase of 20 kg/h. What would be the size of two future additions and when would they be made? Use a discount rate of 20%.

15.6 STAGED EXPANSION

A facility (e.g., interceptor sewer or landfill) will be designed for an ultimate demand of Q_{50} = 150,000 units, which will be reached in 50 years. The growth rate is 3,000 units per year. The cost is $C = KQ^{0.75}$. For simplicity, set K= $1.0 million so that the cost to build 150,000 units now is C = $150{,}000^{0.75}$ = $7,622 million.

(a) Evaluate staged construction for i = 3%, 4%, and 5% and recommend an expansion strategy.
(b) There is a proposal to divert 30% of the load. If this can be done, does the expansion strategy change?

15.7 STAGED EXPANSION WITH LOGISTIC GROWTH

The rate of growth in demand can decrease over time for many reasons. One description of this is the logistic growth curve.

$$D_t = \frac{a}{1 + b\exp[-c\,(t + 20)]}$$

where

D_t = demand at time t
a = maximum demand
b and c are shape coefficients

The logistic curve shown in Figure P15.7 has a = 40,400, b = 82, and c = 0.15. There is an unmet initial demand of 8,000 units and the demand will be 40,051 in 40 years. The construction cost is $C = KD_t^{0.6}$. For convenience, set K = 1. Inflation is included in the discount factor, which is i = 6%. Determine the most economical expansion policy for a project having this growth pattern.

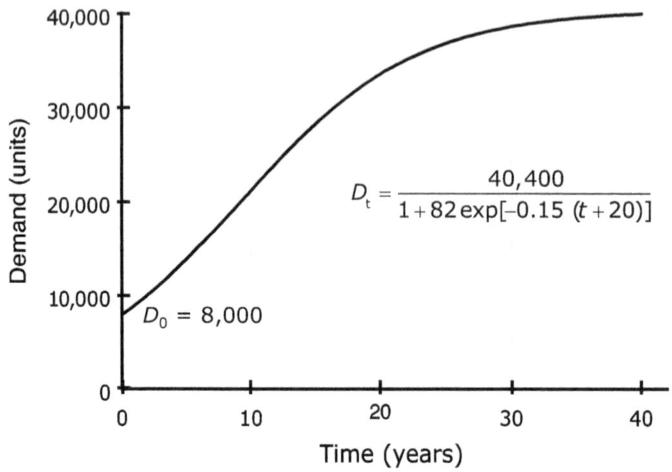

FIGURE P15.7 Logistic growth curve.

15.8 ION EXCHANGE

Water is treated with an adsorbent resin that must be regenerated at regular intervals. A large tank is available at no cost for storing water during regeneration to smooth any variations in demand and supply. Demand for water averages 2.5 m^3/h for all the hours in a year. The time t (h) that the resin can be used before it is regenerated is $t = 5\ V/F$, where V = resin volume (m^3) and F is the flow rate (m^3/h). The fixed annual cost is $C = \$5{,}000\ V^{0.6}$. The cost of each regeneration is \$6 per regeneration over the economic life of the project. What resin volume will minimize the annual cost?

15.9 OXIDATION PROCESS DESIGN

An oxidation process consists of a reactor and chlorine feed equipment. The design flow rate is Q = 5 (m^3/ h) for an eight-year life, after which an expansion will be needed. The process will operate 300 d/y, 20 h/d. The cost of money is 10%/y. The cost of chlorine is \$0.40/kg. The cost of all chemical reagents is 20% of the cost of chlorine. Other annual O&M costs are 10% of the capital cost. The required performance is accomplished if the feed chlorine concentration, C (mg/L), and the reactor time, T (h), meet the condition that

$$T = \frac{5}{C^2}$$

where

$T = V/Q$ = reactor detention time (h)
Q = flow rate (m^3/h)
V = reactor volume (m^3)
C = chlorine dose (mg/L)

The capital costs are

$$\text{Reactor} \quad CR = 2{,}140\ V^{0.7}$$

$$\text{Chlorine feeder} \quad CF = 3{,}700\, M^{0.3}$$

where

$M = CQ$ = Peak demand for chlorine (kg/d)

15.10 FLOW ADJUSTMENT

Adjust the flow network in Figure P15.10 to conform with the law of conservation of mass.

(a) First assume all flows are measured with equal accuracy.
(b) Then recalculate assuming the two metered flows, Q4 and Q7 are measured more accurately and have a variance that is four times smaller than the other flows.

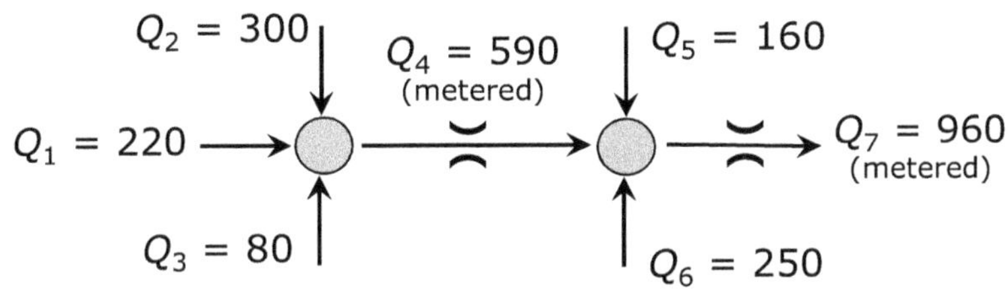

FIGURE P15.10 Flow network.

15.11 MASS BALANCE

Mass flows were measured for a two-stage process, as shown in Figure P15.11. The physical requirement is that for each process, and for the overall system, the mass in must equal the mass out. None of these conditions is satisfied. Assuming the variance is the same for each measured *X*, adjust the data using the method of least squares.

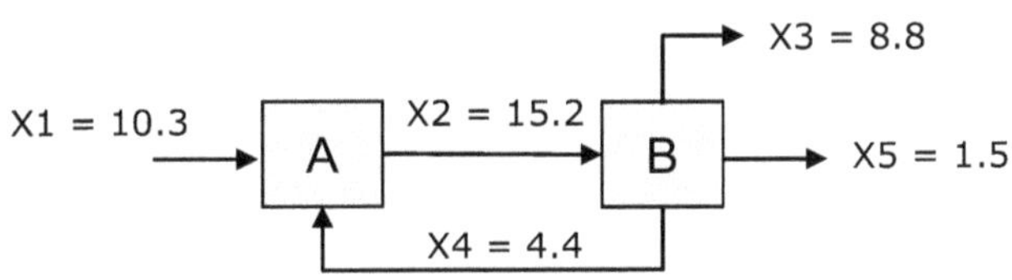

FIGURE P15.11 Unbalanced flows in a two-stage process.

15.12 INDUSTRIAL WASTEWATER TREATMENT

An industrial wastewater has biodegradable COD = 6 kg/m^3 (6,000 mg/L). The average daily wastewater flow is 200 m^3/d. The only nearby wastewater treatment plant treats 4,000 m^3/d that has COD = 0.3 kg/m^3 (300 mg/L). This is a COD loading of (4,000 m^3/d)(0.3 kg/m^3) = 1,200 kg/d. The industry loading is (200 m^3/d)(6 kg/m^3) = 1,200 kg/d. The neighboring plant does not have the capacity to accept this additional load, so the industry will evaluate building a treatment process of its own to produce an effluent with 0.1 kg/m^3 (100 mg/L). This will be accepted by the nearby plant.

The design will be an aerobic biological process with two completely mixed reactors in series, as shown in Figure P15.12. Completely mixed means that the liquid in the reactor is homogeneous, and the reactor contents and the effluent have the same concentration. (If you are not familiar with reactor design, take a moment to think about this.)

The optimal reactor design is defined as the minimum total reactor volume. Calculate the reactor volumes, V_1 and V_2, the total volume $TV = V_1 + V_2$, and the COD concentration leaving reactor and entering reactor 2.

These are the design equations.

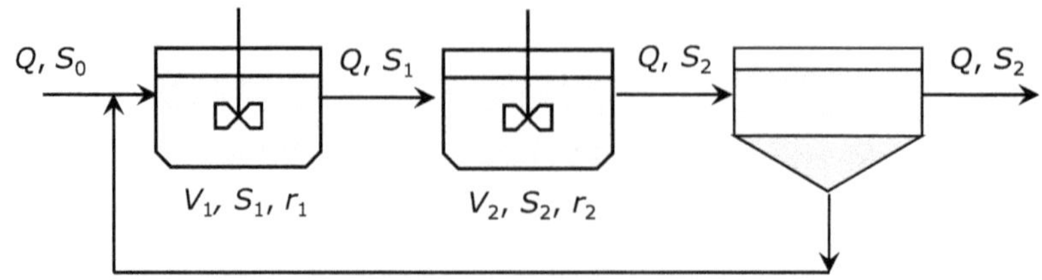

FIGURE P15.12 Two-stage aerobic biological process. Q = volume flow rate (m³/d), V = reactor volume (m³), S = COD concentration (kg/m³), and r = COD removal rate (kg/m³-d).

For steady-state operating conditions, the COD removal rates are the mass of COD removed per unit of time. The rates are different in the two reactors because the volumes are different.

$$r_1 = \frac{Q(S_0 - S_1)}{V_1} \quad \text{and} \quad r_2 = \frac{Q(S_1 - S_2)}{V_2}$$

where

r = COD removal rate of biomass (kg/m³-h)
Q = flow rate (m³/h)
S = biodegradable substrate concentration (mg/L COD)
V = volume of aerobic digester (m³)
X = mass concentration of biomass (kg/m³)

Another definition of the COD removal rate comes from the kinetic reaction rate model. The removal rate typically increases to an asymptotic value as the COD increases. This model is

$$r = \frac{r_{max} S}{K_S + S}$$

The organic compounds in this wastewater inhibit the biological reactions. To account for this, the reaction rate model is modified. The rate equations for the two reactors are

$$r_1 = \frac{r_{max} S_1}{K_S + S_1} \exp(-S_1 / K_I) \quad \text{and} \quad r_2 = K_S \frac{r_{max} S_2}{1 + S_2} \exp(-S_2 / K_I)$$

where

r = *COD* removal rate of biomass (kg/m³-h)
r_{max} = maximum removal rate = 1.5 kg COD/m³-h
K_s = saturation constant = 1 kg COD/m³
K_I = inhibition coefficient = 3.6 kg COD/m³

15.13 CHLORINATION SYSTEM DESIGN

Design a process to disinfect 5,000 m³/d of wastewater by chlorination. The efficiency of disinfection requires CT = 0.5 mg-d/L, where C = applied chlorine dose (mg/L) and $T = V/Q$ = detention time of the chlorination basin (d). V is the volume of the chlorination contact basin (m³). The applied chlorine dose is reduced by building a larger contact tank. Thus, there is a trade-off between capital cost and chlorine cost. Recommend the tank volume for a ten-year life and 8% interest rate. The construction costs ($) are as follows:

Chlorine contact tank $C_{TANK} = \$4,500\, V^{0.78}$

Chlorine feeder $C_{FEEDER} = \$2,000\,M^{0.3}$

where M = kg/d chlorine fed

For simplicity, omit operating costs except for chlorine. Unit cost of chlorine = \$0.35/kg

15.14 PUMPING STATION DESIGN

The pipe diameter that minimizes the cost of a pumping station will be modeled using

$$\text{Cost of pipeline } (\$) = 1.01D^{1.29}L$$

$$\text{Cost of installed pump, motor, and auxiliary equipment } (\$) = 16.14h^{0.642}Q^{0.453}$$

$$\text{Cost of electricity } (\$/\text{y}) = c_e(P)\ (8,760\ \text{h/y})$$

where

D = diameter of pipe (inch)
L = length of pipe (ft)
Q = design flow rate (gal/min)
P = pumping energy (kW) = 0.000189 $Q\,h/\eta$
η = pump/motor efficiency
h = pumping head (ft)

$h = h_S + h_F$ = static lift + friction loss in the pipe

$$h_F\ (\text{ft/ft}) = \frac{0.002083\ Q^{1.852}}{D^{4.8655}}$$

Power (kW) = 0.00019 Qh/η

This pipeline will have no excess capacity for future growth. For operating costs, assume the system will operate at 50% pumping capacity for 12 h/d, 75% for 6 h/d, and 100% for 6 h/d. Design for 2,500 gal/min, pipe length = 6,000 ft, unit cost of electricity = \$0.10 kWh, h_S = 120 m.

Notes:

(a) Cost equations are from Deb, AK 1978, *J Environ. Eng. Div.*, ASCE, pp 127–136.
(b) Friction loss h_F is calculated using the Hazen–Williams equation with a friction factor of 100.
(c) Headloss equation is for pipe inner diameter. For simplicity, use the outer diameter.

15.15 STAGED POWDERED ACTIVATED CARBON ADSORPTION

Countercurrent processing is the most efficient arrangement for equilibrium processes. An equilibrium process means that two phases are in contact and mass moves from one phase to the other. At some point, the two phases are in equilibrium and mass transfer stops. An example is carbon adsorption. When powdered activated carbon (PAC) is mixed with wastewater, COD will adsorb onto the surface of the carbon until an equilibrium is reached. Equilibrium is when the adsorption stops. The adsorptive capacity of carbon in equilibrium with a solution having concentration C_1, measured as kg COD adsorbed/kg PAC, is

$$\frac{C_0 - C_1}{m_{PAC}} = KC_1^n$$

where
C_0 =initial COD concentration (kg/m^3)
C_1 =equilibrium COD concentration of the mixer contents and the effluent (kg/m^3)
m_{PAC} =mass of powdered activated carbon added (kg)

K and n are coefficients that depend on the kind of activated carbon, the nature of the organic compounds that are being adsorbed, pH, and temperature.

The material balance model for a mixer in which the PAC and the wastewater come to equilibrium at concentration C_1 is

$$\underset{\text{Mass removed (kg/d)}}{Q(C_0 - C_1)} = \underset{\text{Mass adsorbed (kg/d)}}{Qm_{PAC}KC_1^n}$$

where
Q = wastewater flow rate (m^3/d)

Figure P15.15 shows a single-stage process and a four-stage countercurrent process. Each stage is a mixing tank and a settler. The adsorption takes place in the mixing tank. The mixer contents are homogeneous. The effluent from a mixer is in equilibrium with the PAC in the mixer. Notice that in the multistage process, the wastewater flows to the right while the PAC flows to the left, hence the countercurrent processing. The fresh PAC is in contact with wastewater that has the lowest COD concentration. The PAC from stage 4 is able to adsorb more COD in stage 3 where the COD concentration is higher, and so on through stage 2 and stage 1. Assume that the PAC and the wastewater comes into equilibrium in each contactor stage. The PAC is removed in the settler, so it can be sent to regeneration or sent to an upstream stage.

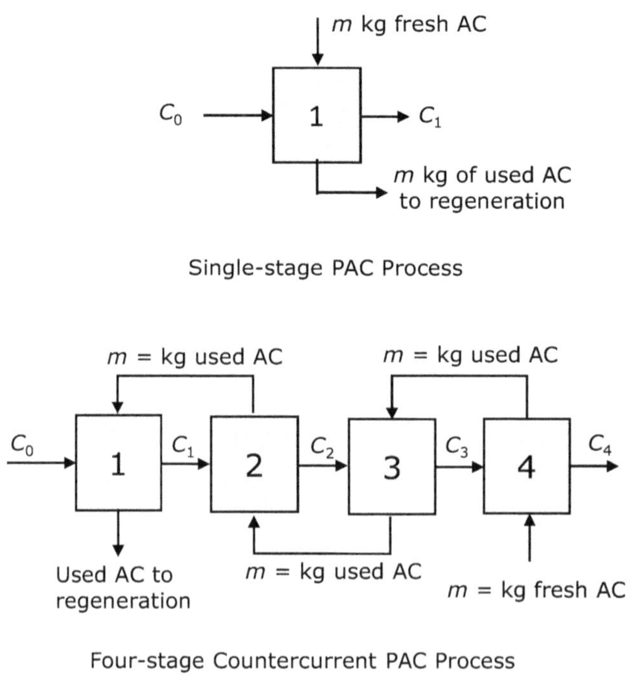

FIGURE P15.15 Single-stage and four-stage countercurrent processing.

Material balance equations for four-stages of mixer/settlers with countercurrent flow of PAC are as follows:

$$C_0 - C_1 = m_{PAC} K C_1^{\,n}$$

$$C_1 - C_2 = m_{PAC} K C_2^{\,n}$$

$$C_2 - C_3 = m_{PAC} K C_3^{\,n}$$

$$C_3 - C_4 = m_{PAC} K C_4^{\,n}$$

Assume that K and n are the same in all stages.

The used PAC is regenerated and reused. About 5% of the mass of PAC is lost in regeneration and there is a cost to replace this. The same mass of PAC, m_{AC}, flows through all four reactors.

The process will treat 25,000 m³/d. Each mixer-settler stage has 2.5 h detention time. This is the mixing time required to reach equilibrium plus the settling time to remove the PAC. Coagulating chemicals are added at a dose of 15 mg/L to each mixer/settler to facilitate settling.

Capital cost related to PAC:

Cost of mixer/settlers = $C = 500\ V^{0.85}$ V = total volume (m³)
Regeneration equipment cost = $8{,}710\ M^{0.43}$ M = kg/d PAC regenerated

Operating costs related to PAC:

Cost of PAC to make up 5% loss in regeneration= \$0.20/kg
Power and fuel for regeneration = \$0.012/kg of PAC regenerated
Polymer to coagulate PAC for settling = \$0.11/kg
Mixer/settler maintenance = \$5,000 $N^{0.8}$, where N = number of stages

Recommend how many stages to build for an influent COD = 0.8 kg/m³ and an effluent concentration of 0.1 kg/m³. Use K = 0.25 kg COD/kg PAC and n = 0.37. Project life = 25 years; discount rate = 6.5%/y

TUTORIAL NOTE

The next problems are Jumbos, our name for larger problems that need more time and thought to solve. Students may need to make some assumptions or approximations, or look up some information. It will be beneficial if students solve one of these. The problem could be selected by the students or assigned. Solutions should be submitted as well-organized and well-written reports. Feedback from the instructor is important. Oral presentations are optional.

15.16 CHLORINATION SYSTEM DESIGN

Chlorine will be used to disinfect sewage effluent before it is discharged into a river. The discharge will increase from 8 to 10 mgd (31,000 to 38,750 m³/d) over the ten-year life of the plant. The process must give an inactivation of at least 99.99% for *Cryptosporidium oocysts*. The design requirement is for CT = 2,280 mg-min/L, or 2 mg-d/L, where C is chlorine concentration in mg/L and T is the chlorination tank detention time in minutes.

Construction costs (\$):

Chlorine contact basins $C_B = 2{,}000{,}000\ V^{0.7}$
Chlorine feed equipment $C_F = 19{,}800\ P^{0.2}$ based on maximum feed in year 10

(includes chlorinator, safety equipment, scale, booster pump) where

V = tank volume in million gal
P = chlorine used (lb Cl_2/d)

Operation and maintenance costs ($/y):

P = lb/d of chlorine applied

Operating labor	$C_{OL} = 319\ P^{0.39}$
Maintenance labor	$C_{ML} = 40\ P^{0.60}$
Materials, other than chlorine	$C_M = 120\ P^{0.29}$
Chlorine	$C_{CL2} = (\$0.10/\text{lb})P$

The discount rate of 8% compensates for inflation of 2% per year.

15.17 ANAEROBIC DIGESTER DESIGN

A wastewater treatment plant with an average daily flow of 37,800 m^3/d produces 175 m^3/d of mixed primary and waste-activated sludge. The mass of total solids entering the digester is 8,165 kg/d, divided into 5,579 kg/d volatile solids (VS) and 2,586 kg/d fixed solids (FS). Design a digester system that has a minimum present value for a 20-year life at 4% discount rate.

The shape of the digester may be assumed as shown in Figure P15.17.

X = height of digester wall
E = depth of excavation = height of digester wall in contact with the ground
D = active depth of sludge in the digester
d = inside diameter of digester
V_S = $(D)(3.1416\ d^2/4)$ = volume of sludge in digester
V_D = $(X)(3.1416\ d^2/4)$ = volume of digester tank
A = $3.1416d^2/4$ = area of roof = area of floor

Excavation depth E is variable and it is possible to have the entire digester underground. The diameter of the hole to be excavated is the inside diameter d plus the thickness of two walls, plus some space for sheeting and forming, say $d + 2$ m.

Simplifications that have been made:

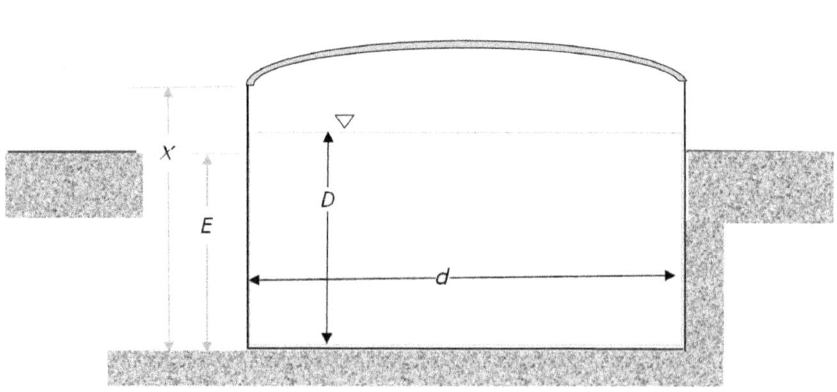

FIGURE P15.17 Generic shape dimensions for an anaerobic digester.

- The cost of excavation will depend on the type of soil/rock, depth, dewatering, sheeting, and the haul to dispose of removed material. There are no unusual excavation problems.
- Many digesters have a conical floor; for simplicity, design a flat floor.
- Assume that the sludge feed volume and composition do not change over the life of the project. Of course, this is not true, but once you have solved this problem, it should be clear how to handle year-to-year changes.
- Certain costs are assumed to be independent of the digester dimensions and will be ignored, including the cost of digester gas cleaning, pumps and pumping costs, process building, instrumentation and controls, electrical, site work other than excavation, etc.

A simplified model for volatile solids removal at a digester temperature of 35°C is

$$\%\ \text{VS destroyed} = 1 - \exp(-k\theta)$$

where

k = VS removal rate coefficient (1/d)
θ = hydraulic retention time = V/Q
Q = sludge flow rate through the digester (m^3/d)
V = active volume of the digester (m^3)

The kinetic coefficients are dependent on temperature of the sludge inside the digester. For temperature other than 35°C,

$$k = 0.058\left(1.02^{(T-35)}\right)$$

Volume of methane produced (m^3/d) at standard conditions (0°C and 1 atm), assuming the biogas is 65% methane.

$$V_{CH4} = 0.65\ \text{m}^3/\text{kg VS destroyed}$$

Annual O&M costs, not including the heating cost, is 5% of the capital cost
Construction costs that depend on the dimensions of the digester are

Cost of wall = \$600/m^2 of reinforced concrete
Cost of floor = \$350/m^2 reinforced concrete
Cost of cover = \$2,000/m^2
Cost of excavation = \$150/m^3
Installed cost of boiler = $C_B = (3{,}100\ \text{kW})^{0.56}$ or $C_B = (49\ \text{J/s})^{0.56}$
where the heat output is measured as kW or J/s
Installed cost of heat exchanger = \$25,700 $A^{0.32}$
where A = heat exchange surface area (m^2)
Required heat transferred

$$H_{\text{req'd}} = UA\Delta T$$

where

$H_{\text{req, d}}$ = heating feed sludge + makeup heat losses from digester (W = J/s)
A = surface area for heat exchange (m^2)

U = heat exchange coefficient = 14.2 W/m^2-°C
ΔT = average temperature difference across the heat exchange surface = 65°C

Design the digester heating system for January, which is the coldest month of the year.

Feed sludge = 11°C
Air = 8.8°C
Soil abutting the wall = 7°C (averaged from surface-to-floor level)
Soil under floor = 2°C

Heat requirement for sludge heating:

$$H_D = Q\rho\, c_p\left(T_D - T_S\right)$$

where
Q = volume of sludge to be heated (m^3/d)
ρ = density of sludge = 1,000 kg/m^3
C_p = heat capacity of sludge = 4,200 J/kg-°C
T_D = temperature of sludge in the digester = 35°C
T_S = temperature of the unheated sludge fed to the digester (11°C)

Heat required to makeup loss from the digester through walls, floor, and roof:

$$H_L = UA\left(T_E - T_D\right)$$

where
U = overall heat transfer coefficient (W/m^2-°C)
A = surface area of wall and floor through which heat is lost (m^2)
T_D = temperature in the digester (°C)
T_E = temperature of air or soil outside the digester (°C)

Heat transfer coefficients, U:

Wall above the ground	U = 0.70 W/m^2-°C
Wall exposed to dry earth	U = 0.62 W/m^2-°C
Moist earth below floor	U = 0.85 W/m^2-°C
Insulated roof	U = 0.95 W/m^2-°C

15.18 RECOVERY OF VOC BY CONDENSATION

A granulated pharmaceutical intermediate is dried to evaporate 500 kg/h of VOC from the product. The VOC is carried in 1,000 m^3/h of exhaust gas from the dryer. There is a proposal to build a condensation process to recover and reuse 90% of the VOC, or 450 kg/h. This is the volume at standard temperature and pressure. The temperature of the exhaust gas is 45°C. The VOC recovery process operates 2,000 h/y. Is 90% VOC recovery the best design?

The condenser is a heat exchanger that cools the feed gas to change VOC vapor to a liquid. Figure P15.18a shows how solvent recovery by condensation is integrated with the manufacturing process. The cooling fluid and the solvent-laden air are separated by a metal surface. Countercurrent flow of the liquid and the gas gives the minimum condenser surface area and coolant flow.

The installed cost of the heat exchanger is

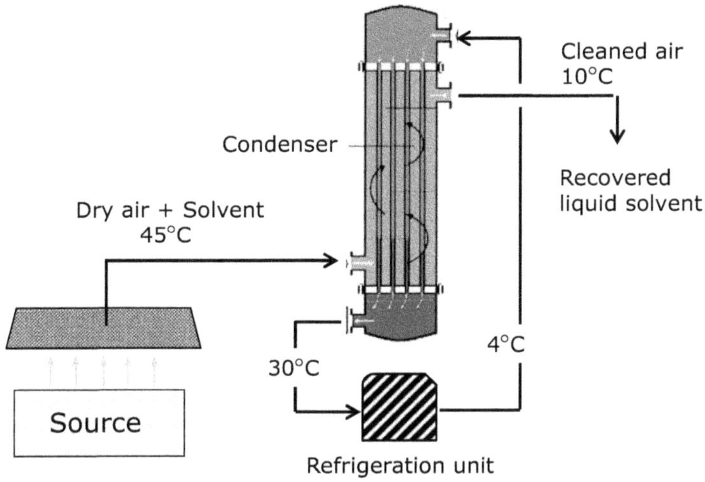

FIGURE P15.18a Refrigerated countercurrent surface condenser for VOC recovery.

$$C = 105{,}000\ A^{0.4}, \quad A = \text{area (m}^2\text{)}$$

Net cost of cooling \$0.00008/kJ
Price of VOC = \$0.2/kg

The Basics of Condenser Design

Heat exchanger design requires attention to many details that are in the realm of chemical and mechanical engineering. The design is done in the most elementary way in this problem, which is not about heat exchanger design. There are other simplifying assumptions:

- The condenser feed is dry air plus VOC vapor. The calculations are more difficult for moist air.
- The condenser coolant is cold water.
- The condensed VOC separates easily from the air and leaves the condenser as a liquid.
- The condensation is done in one stage. It could be done in two stages (two heat exchangers), the first using ambient cooling water at 20°C and the second using refrigerated water. The refrigerated water would cost more than the ambient water.
- VOC that is not removed by condensation be removed by another method (adsorption or catalytic incineration). The cost of this will not be included. Adsorbed VOC can be recovered and incineration provides recoverable heat energy. Assume the net cost of this treatment is zero.

Condensation will occur if the vapor pressure of the VOC in the air exceeds the saturation vapor pressure at the temperature of the exiting air. The vapor pressure depends on temperature. Lowering the temperature of the gas lowers the saturation concentration and enhances condensation. The vapor pressure, p, at temperature, T, is calculated from the Antoine equation:

$$\log(p) = A - \frac{B}{C + T}$$

where

T = temperature (°C)

A, B, and C = Antoine coefficients for the VOC

The Antoine coefficients for volatile chemicals can be found in the literature or online. Here is a basic design procedure for condensing a VOC from air.

(1) Convert the volume flows (m³/h) of air and VOC to mass flow rates (kg/h).
(2) Convert mass to moles: moles = (mass flow, kg/h)/(kg/mol)
   1 kg-mol = 22.4 m³ at 1 atm and 0°C.
(3) Calculate the volume fraction of the VOC = (volume VOC/(Volume gas)
(4) Calculate the partial pressure p (mmHg) = (volume VOC)(760 mmHg)
(5) Specify the exit temperature of cooled air and calculate the partial pressure of VOC at that temperature.
(6) Calculate the heat that must be removed from the gas to reach the desired exit temperature of the gas. This is to cool the VOCs and the air, and to remove the latent heat of the VOC.

$$Q_{\text{cool}} = m\,c_p \Delta T$$

$$Q_{\text{latent}} = \left(840 \text{ kJ/kg}\right) m$$

$$Q = Q_{\text{cool}} + Q_{\text{latent}}$$

where

m = mass of gas to be cooled (kg/h)

c_p = heat capacity of gas = 1 kJ/kg-°C for air and VOC vapor

ΔT (°C) = initial temperature (°C) – final temperature (°C)

(7) Calculate the surface area of the condenser using the general heat transfer equation

$$Q = UA\Delta T_m$$

where

Q = amount of heat to be removed (kJ/h)

U = overall heat transfer coefficient

Use U = 400 W/m²-°C = 1,440 kJ/ m²-°C

ΔT_m = log-mean temperature difference (°C)

The log-mean temperature is explained in Figure P15.18b.

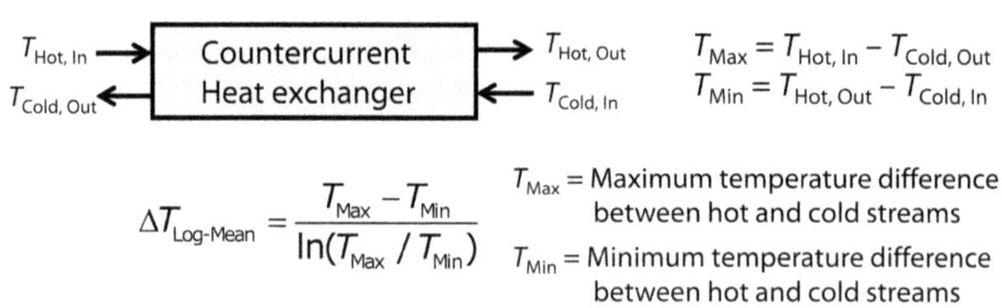

FIGURE P15.18b Definition of log-mean temperature for countercurrent condensers (heat exchangers).

15.19 IRON REMOVAL

An industry needs 3,750 m^3 of high-quality water. A well of sufficient capacity yields water that will be suitable for manufacturing if the iron concentration is reduced by 95%. Design an iron removal process for options 1 and 2, as shown in Figure P15.19. Use a ten-year life, and a discount rate of 8%. The water quality of the well water is

pH = 6.3
Total alkalinity = 50 mg/L as $CaCO_3$ = 0.0005 mol/L = 1.0 x 10^3 equivalents/L
CO_2 concentration = 44 mg/L as CO_2 = 10^3 mol/L = 0.001 equivalents/L
Ferrous iron concentration (Fe^{2+}) = 1.6 mg/L

Capital Cost

Aeration tower and blower cost = C_A = \$30,000 $H^{0.4}$
Contact tank cost = C_T = \$24,000 $T^{0.8}$

where
H = tower height (m)
T = V/Q = detention time (h)

Filter cost is the same for all options and is ignored.
Annual electric cost for pump and blower (\$/y) = \$450 $H^{0.8}$
Annual cost of lime (\$/y) = (\$0.11/kg)(365 d/y)(3.75 B kg/d) = 150 B

where
B = lime added (mg/L); kg/m^3 = (mg/L)/1,000
Lime added (kg/d) = (3,750 m^3/d) (B/1,000) = 3.75 B
Unit cost of lime ($Ca(OH)_2$) (\$/kg) = \$0.11

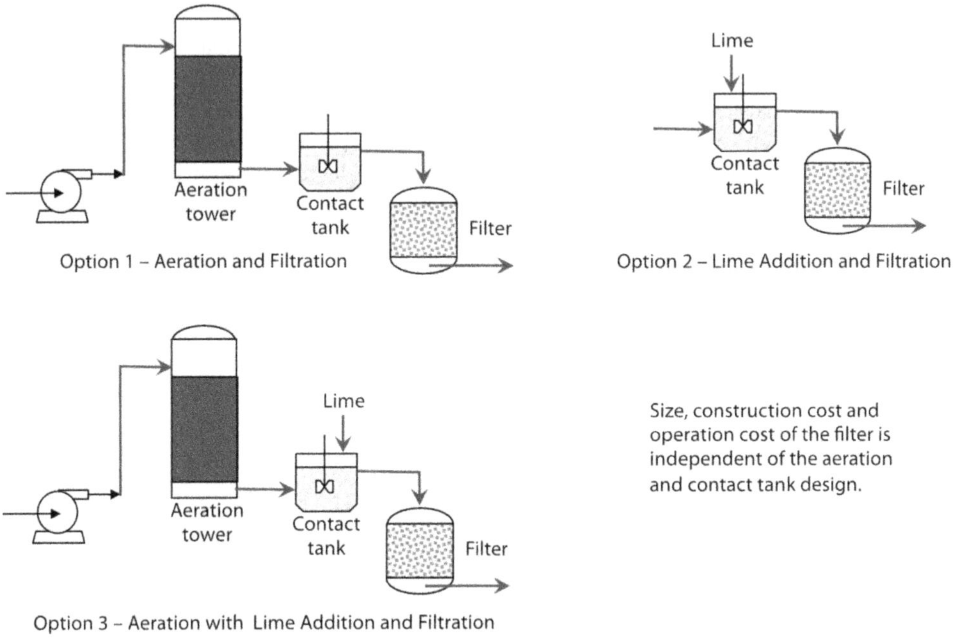

FIGURE P15.19 Three iron removal processes. (a) Aeration to raise the pH by removing carbon dioxide. (b) Lime addition to increase the pH. (c) Combination of aeration and lime addition.

The pH can be changed by (1) removing carbon dioxide by air stripping, (2) adding a basic chemical such as lime or sodium hydroxide, or (3) a combination of stripping and chemical addition can be used. All chemical and process constants are given for 20°C, so assume 20°C for the temperature of ambient water and air. The cost of building and operating the filter is the same regardless of the design details of the iron oxidation system and can be ignored. The design should minimize the present value of the cost over ten years at 8% interest.

The rate of iron oxidation will determine the detention time on the contact tank. It is assumed that the ferrous iron (Fe^{2+}) that disappears from solution is converted to an insoluble precipitate that can be removed by filtration. The rate of oxidation depends on the pH and the oxygen concentration (Stumm & Lee 1961). The rate model is

$$\frac{d[Fe^{2+}]}{dt} = -k[Fe^{2+}]$$

where

k = reaction rate coefficient

= $(3.8 \times 10^{13})[OH^-]^2$ mol Fe^2/L-min (at 2°C and 0.21 atm O_2)

Small changes in pH cause large changes in k:

For pH = 7, $[H^+] = 10^7$, $[OH^-] = 10^7$ $k = 0.38$

pH = 7.5, $[H^+] = 10^{7.5}$, $[OH^-] = 10^{6.5}$ $k = 3.80$

pH = 8, $[H^+] = 10^8$, $[OH^-] = 10^6$ $k = 38.0$

Aeration supplies the oxygen to oxidize the iron from the soluble ferrous form (Fe^{2+}) to the insoluble ferric form (Fe^{3+}). Aeration also changes the pH by stripping dissolved carbon dioxide (CO_2) from the water. Dissolved CO_2 acts as an acid, so removing it favors iron precipitation. The pH can also be increased by adding a basic chemical, which is usually hydrated lime, $Ca(OH)_2$, but could be sodium hydroxide (NaOH). Increasing the pH will increase $[OH^-]$ and k. For example, a 1 unit increase in pH corresponds to a tenfold change in $[OH^-]$ and k. Increasing the pH also will reduce the solubility of $Fe(OH)_3$ that is formed as a precipitate, since $K_{sp} = [Fe^{3+}][OH^-]^3$. We do not take this into account because the iron removal will be satisfactory at any reasonable pH.

A taller aeration tower will remove more carbon dioxide and reduce the need for adding lime. After pH adjustment by aeration or chemical addition, the water is held for some time in a contact tank where ferric oxide is precipitated. The iron oxidation reactions occur in contact tank. The time required depends on the pH, that is, it depends on the amount of CO_2 that is removed and/or the amount of lime that is added. A larger contact tank will allow greater completion of the oxidation reaction.

Assume that the contact tank is a completely mixed reactor; the liquid in the tank is homogeneous with respect to iron concentration and pH. The flow rate and other operating conditions are constant. The material balance on ferrous iron for the contact tank is

$$Fe = \frac{Fe_0}{1+kt} = \frac{Fe_0}{1+kV/Q}$$

where

Fe_0 = influent ferrous iron concentration (mg/L)

Fe = effluent ferrous iron concentration (mg/L)

t = V/Q = hydraulic detention time of the reactor (min)

k = reaction rate coefficient = $(3.8 \times 10^{13})[OH^-]^2$ (1/min)

The treatment requirement is to reduce the iron concentration by 95%, from 1.6 to 0.08 mg/L.

$0.08\ \text{mg/L} = (1.6\ \text{mg/L})/(1 + kt) \rightarrow 1 + kt = 20$
$t = 19/k$ min

Aeration Tower Design

Aeration removes carbon dioxide which increases the pH. At pH less than 8.3,

$$[\text{OH}^-] = 2\times10^{-8}(m/p)$$

where
m = initial total alkalinity, mol/L
p = concentration of CO_2 remaining after aeration stripping, mol/L

With total alkalinity = 50 mg/L as $CaCO_3$ = 0.0005 mol/L

$$[\text{OH}^-] = \frac{(2\times10^{-8})(0.5\times10^{-3})}{p} = \frac{1\times10^{-11}}{p}$$

The CO_2 concentration, p, depends on the degree of aeration, and this depends on the height of the aeration tower. A taller tower will remove more CO_2 and bring about a greater change in pH. A gravity-type tray aeration tower drops the liquid through slotted trays. In contrast, a forced-air system blows air up through the liquid as it falls. The change in CO_2 for a gravity type tower is

$$p = j\exp[-K_La\left(2h/g\right)^{0.5}$$

where
j = initial CO_2 concentration (mol/L)
p = concentration of CO_2 remaining after aeration stripping (mol/L)
K_La = overall mass transfer coefficient for aeration (1/min)
g = gravitation constant (9.81 m/s^2)
h = height of tray aerator (m)
C_s = saturation concentration of carbon dioxide at 20°C

Lime Addition

The concentration $[OH^-]$ can be increased by adding hydrated lime, $Ca(OH)_2$. The lime will increase the alkalinity and decrease the carbon dioxide concentration. For pH levels below 8.3,

$$[\text{OH}^-] = 2\times10^{-8}\left(\frac{m+2B}{j-2B}\right)$$

where
m = initial total alkalinity (mol/L)
j = initial CO_2 concentration (mol/L)
B = lime dose (mol/L)

15.20 IRON REMOVAL BY AERATION COMBINED WITH LIME ADDITION

For the conditions in Problem 15.19, it may be that a combination of aeration and lime (or NaOH) addition will cost less than aeration alone or chemical treatment alone. Evaluate this option.

16 Building and Fitting Statistical Models

Most design equations come from material and energy balances on separation and reaction processes. These are *mechanistic models*. There are times when experiments are needed to estimate parameters in these models, such as mass transfer coefficients and reaction rate coefficients.

When there is no mechanistic model, experiments are designed to discover the important predictor variables and build an *empirical model* that describes the process performance.

The following are the four kinds of statistical problems that are useful in solving pollution prevention and control problems:

Identifying the predictor variables
Building and fitting empirical models
Estimating parameters in mechanistic models
Judging the goodness of fit of models (empirical and mechanistic)

Solving these problems requires having a good experimental strategy and efficient experiment designs. *Experimental design* means specifying the independent variables to be studied and the experimental settings of those variables. *Efficient* means getting maximum information from a minimum number of test runs.

The experimenter's *state of knowledge* about the process increases as experiments are designed, data are collected, and models are fitted. Obviously, the kind of experiment that should be done depends on the initial state of knowledge.

A low state of knowledge means there is uncertainty about which variables are important and the kind of model that can be developed with available resources. An experimenter working from the low state usually wants an *empirical model*. This will usually be a smooth interpolating function that describes (in two dimensions) a plane, a warped plane, or a quadratic surface. Sometimes preliminary screening experiments are needed to determine which of several *potentially important factors* (independent or predictor variables) actually change the performance of the process. Then additional experiments are run using only the important variables.

A higher state of knowledge is when the mathematical form of the model is known, or the experimenter has a good idea of what it may be. These are often *mechanistic models* that describe material and energy balances, process chemistry, mass transfer, and reaction kinetics. Typical predictor variables are concentration, temperature, pH, and other factors. These models contain parameters (e.g., kinetic and mass transfer coefficients) that must be estimated from data that is collected under specific conditions. This is especially true for pollution prevention and control systems because the process inputs and operating conditions are site-specific.

16.1 LINEAR AND NONLINEAR MODELS

What is commonly known as *fitting the model to the data* is calculating the values of the unknown parameters. The distinction of *linear* and *nonlinear* models is important because of the way the parameters are estimated. Parameters in linear models can be calculated directly using linear algebra in a process called *linear regression*. There is no algebraic solution for the parameters in nonlinear models; they must be estimated by a numerical search. The method is called *nonlinear regression*, or *nonlinear least squares*.

The terms *linear* and *nonlinear* as used in model building refer to the parameters and not to the independent variables (the x's). Once the data have been collected, the x and y values change from symbolic variables to numbers and the model equations contain only the unknown parameters that are to be estimated.

Data from two experiments [$x_1 = 1.0$, $y_1 = 2.3$; $x_2 = 2.0$, $y_2 = 4.1$] are used to fit a model that is nonlinear in x but linear in the parameters β_1 and β_2.

$$y_i = \beta_1 x_i + \beta_2 x_i^2$$

Substituting the data into the model gives

$$2.3 = \beta_1(1) + \beta_2(1)^2 = \beta_1 + \beta_2$$

$$4.1 = \beta_1(2) + \beta_2(2)^2 = 2\beta_1 + 4\beta_2$$

The quadratic terms are gone and the result is two linear equations with the parameters β_1 and β_2 as the unknowns. There is an algebraic solution for simultaneous linear equations.

In contrast, the nonlinear model $y_i = \theta_1 \exp(-x_i \theta_2)$ for the same data gives

$$2.3 = \theta_1 \exp(-1\theta_2)$$

$$4.1 = \theta_1 \exp(-2\theta_2)$$

with θ_1 and θ_2 as the unknowns. There is no algebraic solution and the estimated parameter values are found by a numerical search. Statistical software such as Minitab, Systat, StatEase, and others can be used to estimate the parameters and report other statistics that are useful for evaluating the model. The solver function in an Excel spreadsheet program can estimate parameters, but it will not provide the supplementary information.

To maintain a distinction between linear and nonlinear models, the parameters in empirical models are identified by β; those in mechanistic models by θ. The estimated parameter values will be indicated by b and k, respectively.

In general, there will be n observations of y, but not necessarily n different settings for x because tests at some settings *might* be replicated. After substituting the data into the model, there will be n equations with p unknown parameters, with $n \geq p$. If the n resulting equations are linear, the parameters can be estimated directly by using linear algebra.

Linear models do not have to describe a straight line (in one dimension) or a plane (in two dimensions). Terms can be added to create polynomial models that can describe straight lines, planes, warped planes, and quadratic surfaces. A quadratic model can define a hill, a valley, or a saddle by including quadratic terms and cross-products (x_1x_2, x_1x_3). The goal is an adequate model that has the fewest number of parameters (terms). Diagnostic checks of the fitted models are used to decide which terms are needed.

If there are two independent variables, x_1 and x_2, the common models are planes, warped planes, or quadratic curves.

Plane $\quad y = \beta_0 + \beta_1 x_1 + \beta_2 x_2$

Warped Plane $\quad y = \beta_0 + \beta_1 x_1 + \beta_2 x_2 + \beta_{12} x_1 x_2$

Quadratic $\quad y = \beta_0 + \beta_1 x_1 + \beta_2 x_2 + \beta_{12} x_1 x_2 + \beta_{11} x_1^2 + \beta_{22} x_2^2$

A linear equation has one basic form, but nonlinear equations have many. Examples of nonlinear equations are

$$y = \theta_1 e^{-\theta_2 x}, \quad y = \theta_1 x^{\theta_2}, \quad y = \frac{1}{\theta_1 + \theta_2 x}, \quad y = \frac{\theta_1 x}{\theta_2 + x}$$

These nonlinear equations can be transformed to get linear forms.

$$\ln y = \ln \theta_1 - \theta_2 x, \qquad \ln y = \ln \theta_1 - \theta_2 \ln x, \qquad \frac{1}{y} = \theta_1 - \theta_2 x, \qquad \frac{1}{y} = \frac{\theta_2}{\theta_1 x} + \frac{1}{\theta_1}$$

It is tempting to use linear regression to fit linearized models, but fitting ln(y) or $1/y$ will give results that are different from fitting y, and parameters estimated from the transformed version are often biased (Berthouex and Brown 2002).

In most cases, it is best to fit nonlinear models using nonlinear regression methods; then check the adequacy of the model and make modifications (including transforming variables if that is indicated). Diagnosing the goodness of fit of models will be explained in Section 16.4.

16.2 THE METHOD OF LEAST SQUARES

Suppose that the true model for a process is

$$\eta = f(\beta, x)$$

where

η is the true value of dependent variable that is calculated using the true values of the β's
x = a vector of independent variables, or predictor variables
β = a vector of parameters or coefficients in the model

The true values of the parameters, β's, are unknown and the best we can do is to estimate them from experimental data.

The observed values of the dependent variable, indicated by y_i, are the true but unknown values of η plus experimental errors:

$$y_i = f(\beta, x_i) + e_i$$

where the e_i are random experimental errors (i = 1 to n), also known as the *residual errors*, or simply *residuals*.

According to the least squares criterion, the best estimates of the β's minimize the sum of the squared residuals. This quantity is called the *residual sum of squares*, *RSS*.

$$\text{RSS} = \min \sum_{i=1}^{n} e_i^2 = \min \sum_{i=1}^{n} [y_i - f(\beta, x_i)]^2$$

Put another way

$$\text{RSS} = \min \sum_{i=1}^{n} e_i^2 = \min \sum_{i=1}^{n} (y_{\text{obs},i} - y_{\text{calc},i})^2$$

where

$y_{\text{obs},\, i}$ = value obtained from an experiment
$y_{\text{calc},\, i}$ = value calculated from the model for an assumed value of β

Figure 16.1 shows the residuals (e_i) and the squared residuals (e_i^2) for an arbitrary set of data (t_i, y_i) and model $y_{\text{calc},\, i} = C_0 \exp(-kt_i)$. The RSS is found by determining the values of the parameters, C_0 and k, such that the total area of the squares in the right panel is minimized.

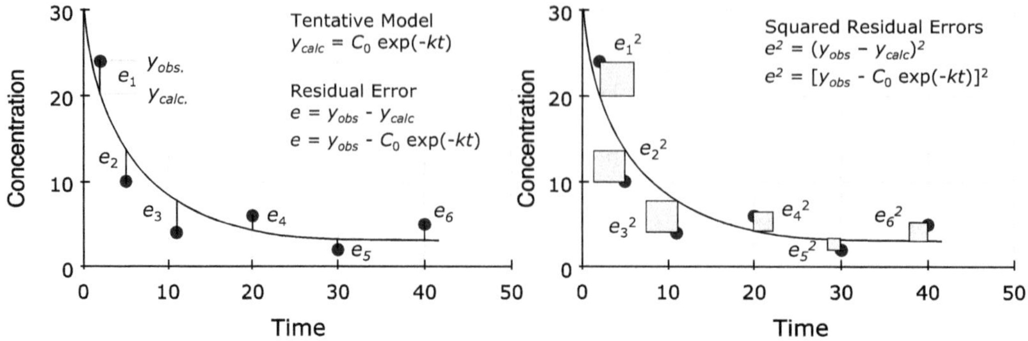

FIGURE 16.1 The sum of squares concept. On the left are the residuals (e_i) and on the right are the squared residuals (e_i^2). The RSS is found when the total sum of squares – that is, the total area of the squares – is minimized.

Example 16.1 Least Squares Parameter Estimates for Two Simple Models

Data were collected to estimate the parameters in two models, one linear and one nonlinear, as identified by the subscript L or NL on the dependent, or response, vartiable, η.

Linear model $\eta_L = \beta x$
Nonlinear model $\eta_{NL} = 20[1 - \exp(-\theta x)]$

where η is the true but unknown value of the response for the true but unknown value of the parameter β or θ.

Note: In most real experiments, the *y* intercept and the asymptote are not known and the models to be fitted would have two parameters.

$$\eta_L = \beta_0 + \beta_1 x \quad \text{and} \quad \eta_{NL} = \theta_2[1 - \exp(-\theta_1 x)]$$

Using one-parameter models makes it possible to show the least squares calculations in a concise format. Both models will be fitted by calculating the RSS for different parameter values. The parameter value that minimizes the RSS gives the best fit.

Table 16.1 shows a sample calculation, using a trial value of $\beta = 1.8$, of the predicted response, the residual error, and squared residuals. The residual sum of squares (RSS) is the sum of the squared residuals.

$$\text{RSS} = e_1^2 + e_2^2 + e_3^2 + e_4^2 = 0.64 + 1.00 + 0.16 + 10.89 = 12.69$$

By trying other values of β, as shown in Table 16.2, the least squares estimate for β is $b = 1.95$, with minimum RSS = 9.87. The fitted least squares model is

TABLE 16.1 Example Calculation of Squared Residuals for $\eta_L = \beta x$ using $\beta = 1.8$

Data	Calculated *y*	Residuals (e)	Squared Residuals (e^2)
$x = 2$, $y_{obs} = 2.8$	$y_{calc} = 1.8(2) = 3.6$	−0.8	$e_1^2 = (-0.8)^2 = 0.64$
$x = 4$, $y_{obs} = 6.2$	$y_{calc} = 1.8(4) = 7.2$	−1	$e_2^2 = (-1.0)^2 = 1.00$
$x = 6$, $y_{obs} = 10.4$	$y_{calc} = 1.8(6) = 10.8$	−0.4	$e_3^2 = (-0.4)^2 = 0.16$
$x = 8$, $y_{obs} = 17.7$	$y_{calc} = 1.8(8) = 14.4$	3.3	$e_4^2 = (3.3)^2 = 10.89$

TABLE 16.2 Sum of Squares Calculations for the Linear Model $\eta_L = \beta x$ and the Nonlinear Model $\eta_{NL} = 20[1 - \exp(-\theta x)]$

Data		Linear Model $y_{calc} = \beta x$						
x	y_{obs}	Squared Residuals for Trial Values of β						
		1.8	**1.85**	**1.9**	1.95	**2.0**	**2.05**	**2.1**
2	2.8	0.64	0.81	1.00	1.21	1.44	1.69	1.96
4	6.2	1.00	1.44	1.96	2.56	3.24	4.00	4.84
6	10.4	0.16	0.49	1.00	1.69	2.56	3.61	4.84
8	17.7	10.89	8.41	6.25	4.41	2.89	1.69	0.81
	RSS =	12.69	11.15	10.21	**9.87**	10.13	10.99	12.45
Data		**Nonlinear Model $y_{calc} = 20[1 - \exp(-\theta x)]$**						
x	y_{obs}	**Squared Residuals for Trial Values of θ**						
		0.27	**0.28**	**0.29**	**0.3**	**0.31**	**0.32**	**0.33**
2	9.6	1.575	1.049	0.637	0.332	0.129	0.021	0.004
4	13.6	0.154	0.016	0.017	0.141	0.375	0.704	1.118
6	16.2	0.025	0.005	0.084	0.244	0.471	0.753	1.079
8	18.6	0.822	0.532	0.320	0.172	0.076	0.021	0.001
	RSS =	2.575	1.602	1.057	**0.889**	1.051	1.500	2.201

$$y = 1.95x$$

The bottom section of Table 16.2 gives the similar search calculations for an exponential nonlinear model

$$y = 20[1 - \exp(-\theta x)]$$

The least squares parameter estimate for θ is $k = 0.3$ at the minimum RSS = 0.889

$$y = 20[1 - \exp(-0.3x)]$$

Figure 16.2 shows the data and the fitted models. Figure 16.3 plots the RSS values for a range of parameter values. The sum of squares curve for the linear model is symmetrical. This is always true for linear models. This is not true for nonlinear models. The range of the diagram is expanded from the table in an attempt to show this.

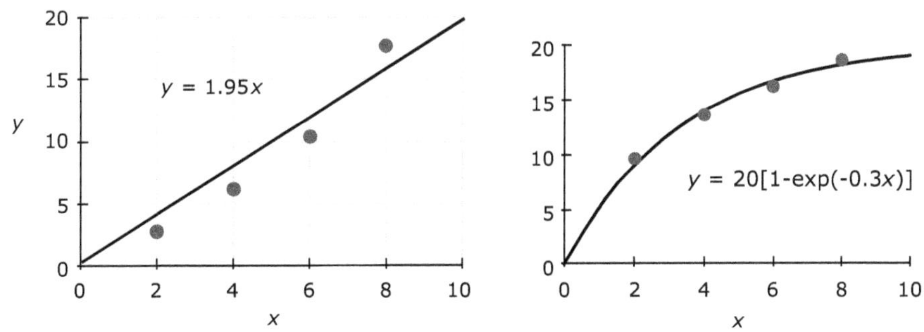

FIGURE 16.2 Fitted linear and nonlinear models.

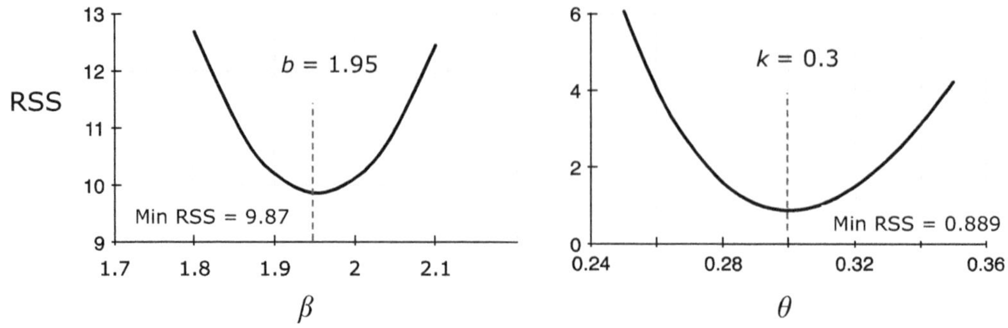

FIGURE 16.3 Sum of squares function for the one-parameter model.

In summary, the method of least squares is the statistical procedure that will give the best estimates of the parameters. These *best* estimates are those that give the minimum residual sum of squares (RSS). Example 16.1 demonstrated this with one-parameter models. The interesting problems have two or more parameters and this requires a better estimation process. Models that are linear in the parameters have an exact algebraic solution for these parameter estimates. Models that are nonlinear in the parameters require a methodical search for the minimum. The details of these calculation methods maybe found in any statistic reference book, and a number of excellent software applications can be used to execute the calculations.

16.3 JOINT CONFIDENCE REGION

A 95% confidence interval of a parameter has an upper and a lower bound which is expected to contain the true parameter value with 95% confidence. For example, if the least squares estimate of a parameter is 8.1 with a 95% confidence interval of 7.7–8.5, then we are 95% confident that the true value of the parameter is within that interval; and 5% confident that it is not.

It is tempting to interpret this to mean that there is a 95% probability that the true parameter value is in this interval. This is not strictly correct because the true parameter value is an unknown constant, and either is in the interval, or it is not. We simply don't know. So, we ascribe to the interval our level of confidence that it contains the unknown value of the parameter.

A *joint confidence region* defines a region that applies to two or more parameters that have been estimated jointly. For example, a straight line has two parameters, the slope and the intercept. The least squares parameter estimates are calculated from a single set of algebraic equations, so the precision of those estimates should be evaluated jointly, and they should be described by a joint confidence region. The 95% joint confidence region for the slope and intercept of a straight line is the range of paired slope and intercept values in which we are 95% confident that the true parameter values lie.

Figure 16.4 is the sum-of-squares surface for a hypothetical two-parameter model. The vertical axis is the residual sum of squares (RSS) for pairs of parameter values.

The contour map of the sum of squares surface for a linear model will be ellipses on the β_1, β_2 plane a for two-parameter model, and symmetrical higher-order surfaces for three or more parameters. The shapes are symmetrical which makes it possible to calculate precise confidence intervals for the estimated parameters.

The sum-of-squares surface for nonlinear models is not symmetrical. The contours can have interesting shapes and are frequently elongated and curved (a banana shape). The asymmetry makes it impossible to calculate exact confidence limits for the parameters in nonlinear models, but approximate confidence regions can be defined. An example is given in Example 16.2 (Figure 16.9).

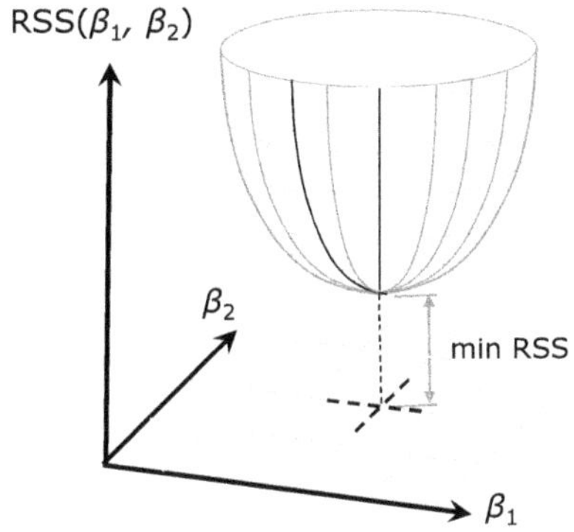

FIGURE 16.4 Sum of squares surface for a two-parameter linear model.

The *joint confidence region* is best understood graphically and this will be illustrated for fitting a straight line. The model is

$$y_i = \beta_0 + \beta_1 x_i + e_i$$

The parameters β_0 and β_1 are estimated using the method of least squares. The least squares estimates are b_0 and b_1, which are obtained by minimizing the residual sum of square.

$$\text{min RSS} = \sum_{i=1}^{n} e_i^2 = \sum_{i=1}^{n} [y_{\text{obs},i} - (\beta_0 + \beta_1 x_i)]^2$$

Table 16.3 describes a hypothetical experiment. The true model $\eta = 1.0 + 2.5x$ was used with the settings of $x = 2$, 4, 6, 8, and 10 to get the true values of $y = 6$, 11, 16, 21, and 26. Random errors were added to the true values to create the five observed values, y_{obs}, in Table 16.3. These represent the outcome of one random experiment with five observations. The fitted model is (with RSS = 1.5946) (Figure 16.5)

$$y_{\text{calc}} = 0.349 + 2.583x$$

The experiment was repeated 1,000 times by creating 1,000 data sets of five observations (y's) for the same settings of x. Random errors from a normal distribution with mean zero and standard deviation of 0.8 were added to the true values of x to create the observed values of y. Each of the

TABLE 16.3 Observed and Calculated Values for a Fitted Linear Model

x	y_{obs}	y_{calc}	e^2
2	5.95	5.518	0.1900
4	9.70	10.684	0.9746
6	16.23	15.850	0.1475
8	21.47	21.016	0.2025
10	25.90	26.182	0.0800
		RSS =	1.5946

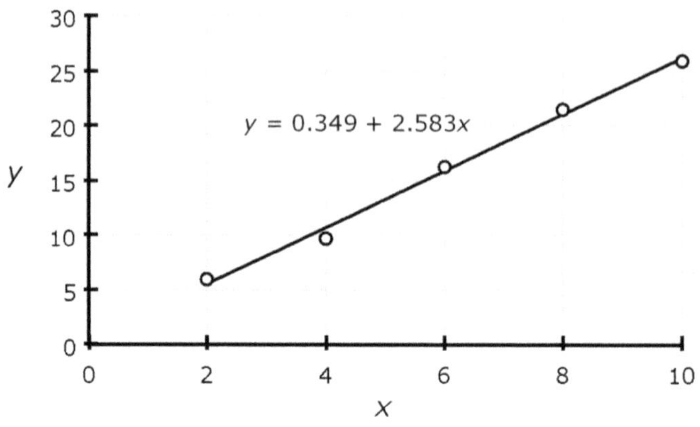

FIGURE 16.5 Fitting the example set of five observations gives $y = 0.349 + 2.853x$.

1,000 data sets was used to estimate a slope (b_0) and intercept (b_1) of the line. Each of the 1,000 dots in Figure 16.6 locates one pair of estimated values of b_0 and b_1.

There are two notable characteristics of this plot. One is the variation in the estimated parameter values. The estimated intercepts (b_0) range between −1.2 and 3.5; some of the estimated values are negative. The slopes (b_1) range from 2.05 to 2.85.

The other is the NW to SE orientation of the plotted points. This shows that a larger estimated value of the slope corresponds to a smaller estimate of the intercept. Parameter estimates that show this kind of corresponding changes are said to be *correlated.* Correlation means that the estimated parameter values are not independent – a smaller value of b_0 pairs with a larger value of b_1. Mild correlation, as shown in this example, is common and is not a problem.

A joint confidence region could have been drawn for each of the 1,000 fitted lines. In theory, 95% of these confidence regions will capture the true parameter values. This was done and 950 paired

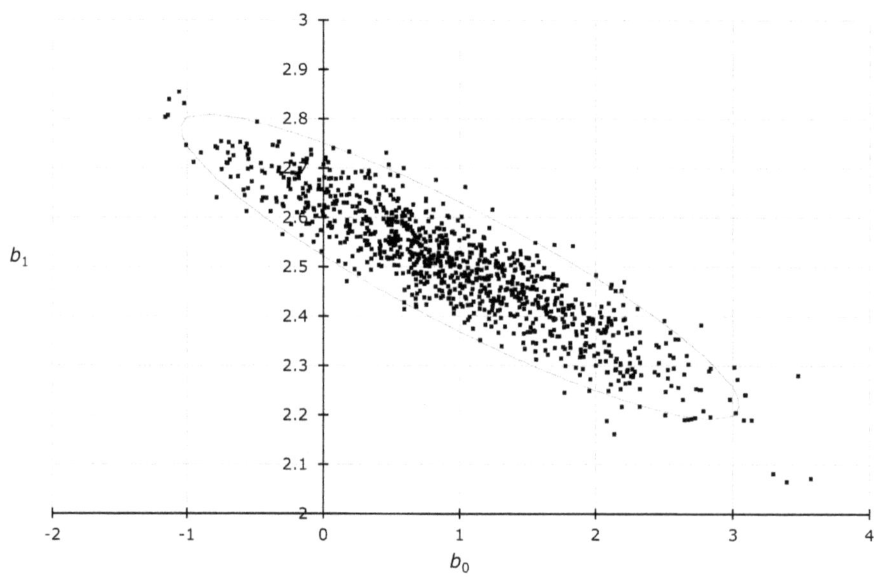

FIGURE 16.6 Pairs of estimated slopes and intercepts from fitting 1,000 simulated sets of data with five observations.

estimates did capture the true values of $\beta_0 = 1.00$ and $\beta_1 = 2.50$. (This was luck; we cannot expect exactly 95%, but with 1,000 simulations we expect something close to 95%.)

Showing the ellipses from1,000 hypothetical experiments would be messy and difficult to interpret. Instead, imagine that one of the simulated experiments happened to predict $b_0 = 1.00$ and $b_1 = 2.50$, which are the true values of β_0 and β_1. The ellipse in Figure 16.6 is the 95% joint confidence interval for that hypothetical experiment. The number of dots inside the ellipse is 954, which is very close to 95% of all the pairs of parameter estimates.

16.4 GRAPHICAL DIAGNOSIS OF THE GOODNESS OF FIT OF A MODEL

A statistical condition for a fitted model is that the residual errors are random, have constant variance, and are normally distributed. The first two conditions are the most important. They are often ignored when checking a model, even though they are the easiest to evaluate and the most informative. Graphical diagnosis is simple, legitimate, and effective.

The first step in checking a fitted model is to plot the calculated values over the observed values. This gives a useful impression of the fit. A second and clearer visualization comes from *plots of the residual errors* against the independent variables and the calculated values. If a model fits the data, the residuals will be randomly distributed about a mean of zero. Random means no trends or patterns, and it means that roughly half the residuals are positive and half are negative.

Example 16.2 Model for The BOD of Soybean Oil Wastewater

Treating soybean oil wastewater is an unusual problem. There is no historical data, so long-term BOD tests were done to estimate θ_1 and θ_2 in this model:

$$y_i = \theta_1(1 - e^{-\theta_2 t_i})$$

where

y_i = BOD concentration observed at time t_i
t_i = time in days for observation i
θ_1 = ultimate BOD (g BOD/g soybean oil)
θ_2 = BOD reaction rate coefficient (1/d)

BOD observations were made in quadruplicate at seven different times, as given in Table 16.4. BOD is usually measured as mg/L, but in this example the measure is grams of BOD per gram of soybean oil (g BOD/g oil).

TABLE 16.4 BOD Measurement for Soybean Oil

BOD	Time (d)						
(g/g oil)	1	2	3	5	7	12	15
Replicate 1	0.40	0.99	0.95	1.53	1.25	2.12	2.42
Replicate 2	0.55	0.95	1.00	1.77	1.35	2.21	2.28
Replicate 3	0.61	0.98	1.05	1.75	1.90	2.34	1.96
Replicate 4	0.66	0.95	1.20	1.95	1.95	1.95	1.92
Average	0.555	0.968	1.050	1.750	1.612	2.155	2.145

The residual error for the first observation is

$$e_1 = 0.40 - \theta_1(1 - e^{-\theta_2(1)})$$

This abbreviated sum of squares function shows the first two and the last terms.

$$\text{RSS} = \left[0.40 - \theta_1\left(1 - e^{-\theta_2(1)}\right)\right]^2 + \left[0.99 - \theta_1\left(1 - e^{-\theta_2(2)}\right)\right]^2 + \cdots + \left[1.92 - \theta_1\left(1 - e^{-\theta_2(15)}\right)\right]^2$$

Many software applications (e.g., Minitab, Systat, StatEase, and LINGO can be used to do the nonlinear regression and estimate the two parameters.

The parameter estimates for θ_1 and θ_2 are

$$k_1 = 2.16 \text{ g BOD/g oil} \quad \text{and} \quad k_2 = 0.263 \text{ d}^{-1}$$

This also can be written using the "hat" notation, where the hat indicates a predicted value.

$$\hat{\theta}_1 = 2.16 \text{ g BOD/g oil} \quad \text{and} \quad \hat{\theta}_2 = 0.263 \text{ d}^{-1}$$

The minimum RSS is 1.2402.

The most obvious test of the model is to plot the fitted curve over the data. This is Figure 16.7. The curve passes nicely through the data and the asymptote is well defined. (The asymptotic BOD is known as the ultimate BOD.)

A plot of the residuals, Figure 16.8, is an even better diagnostic tool. The residuals should be plotted against the predicted BOD and the independent variables. If the model has deficiencies, this will be revealed by the residual plots. *Flattening* the data makes it easier to see whether the residuals are random and whether there are trends that indicate lack of fit.

Both residual plots in Figure 16.8 show that the residuals appear to satisfy the requirement of being randomly scattered about zero.

Figure 16.9 shows the approximate 95% joint confidence region for the parameter estimates (k_1 = 2.16 and k_2 = 0.263). We are 95% confident that the true parameter values fall within this region. The true parameter values are likely to be in the range from about 1.9–2.4 for θ_1, and in the range from about 0.19 to 0.36 for θ_2, with a special condition that is shown by the shape of the confidence region. The shape of the region is not an ellipse as it would be for a linear model. It looks more like a bean or sausage. The NW–SE orientation to the region that indicates a negative correlation between the parameter estimates. Large estimated values of k_1 pair with small values of k_2. The pair k_1 = 2.3 and k_2 = 0.33 is not within the joint confidence region and is not plausible.

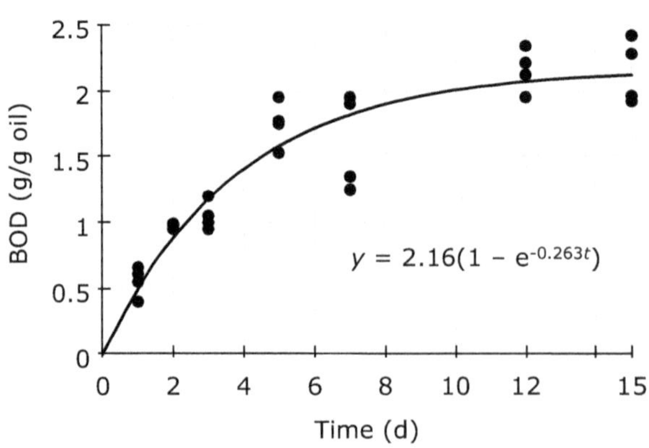

FIGURE 16.7 Soybean oil BOD data with the curve fitted by nonlinear least squares.

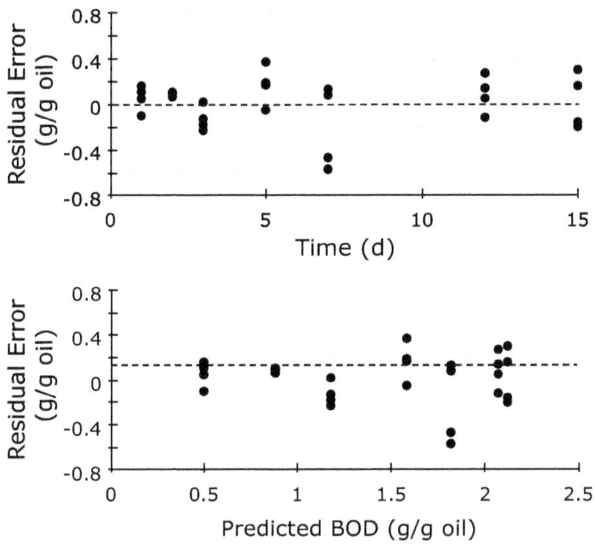

FIGURE 16.8 A plot of the residual errors is an excellent diagnostic tool for judging the goodness of fit.

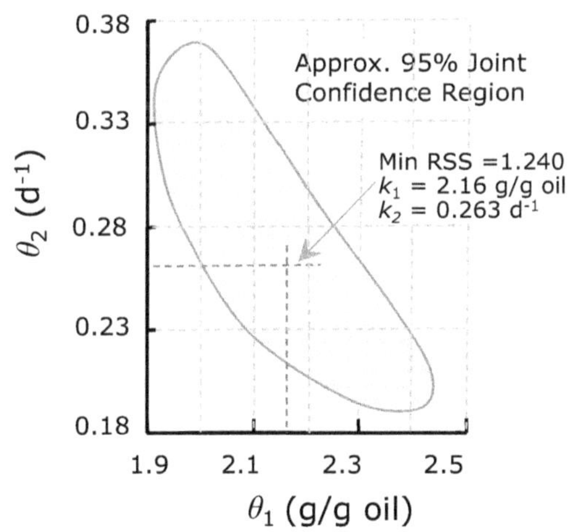

FIGURE 16.9 Sum of squares surface and the approximate 95% joint confidence region.

Different parameter estimates would be obtained if the experiment were repeated (or if each of the replicate experiments was fitted independently). This is because of random experimental error in the measured values. We expect that 95% of all joint confidence regions from many repeated experiments will contain the true (but unknown) values of the parameters.

Statistical software produces an array of diagnostic statistics, such as R^2 and t values. R^2 and t are introduced in Section 16.5, and R^2 is explained in Section 16.6 using data from a case study on modeling the settling of suspended solids in wastewater. For those who are uncertain about the interpretation of these statistics, or who want a refresher, we recommend Box, Hunter & Hunter (2005) and Berthouex and Brown (2002).

16.5 CASE STUDY: EMPIRICAL MODEL FOR SUSPENDED SOLIDS REMOVAL

Solid particles can be removed from a liquid by settling. An ideal settling tank can be created in the laboratory in the form of a batch column, shown in Figure 16.10, that is filled with industrial wastewater or sewage. Samples are taken over time from sampling ports located at several depths along the column. The measure of sedimentation performance will be total suspended solids (TSS) that are removed, or the fraction of TSS removed, both measured as a function of time and depth. The goal is an empirical statistical model of the removal efficiency of a real sedimentation process (Berthouex & Stevens 1982, Berthouex & Brown 2002).

A column settling test was done on a suspension with initial total suspended solids (TSS) concentration, C_0, that is uniform throughout the test column. C_0 is measured as mg/L. The total volume of the column is $V = AZ$ (in liters) and the initial mass of suspended solids is $M_0 = C_0AZ$ (mg). Small samples of wastewater were taken from the column at several times and depths. The sample volumes are small in comparison with the water depth and the total mass of solids in the column is constant.

At the beginning of the test the concentration is uniform over the depth of the test settling column. After settling has progressed for time t, the concentration near the bottom of the column has increased relative to the concentration at the top to give a solids concentration profile that is a simple function of depth at any time t. The model for the concentration profile, $C(z, t)$, will be a polynomial that is derived from the experimental data.

The mass of solids remaining above depth z at time t is

$$M(t) = \int_0^z C(z,t)dz$$

The fraction of solids removed in an ideal settling tank at any depth z, that has a detention time t, is estimated as

$$R(z,t) = \frac{AZC_0 - A\int_0^z C(z,t)dz}{AZC_0} = 1 - \frac{1}{ZC_0}\int_0^z C(z,t)dz$$

The initial concentration was C_0 = 560 mg/L. Table 16.5.shows the TSS concentrations that were measured at depths of 0.8, 1.6, and 2.4 m (measured from the water surface) at 20, 40, 60, and 120 min after the start of the test.

The model building problem starts by fitting a plausible model and continues by examining the residuals for inadequacies in the model and adding or removing terms (predictor variables) to

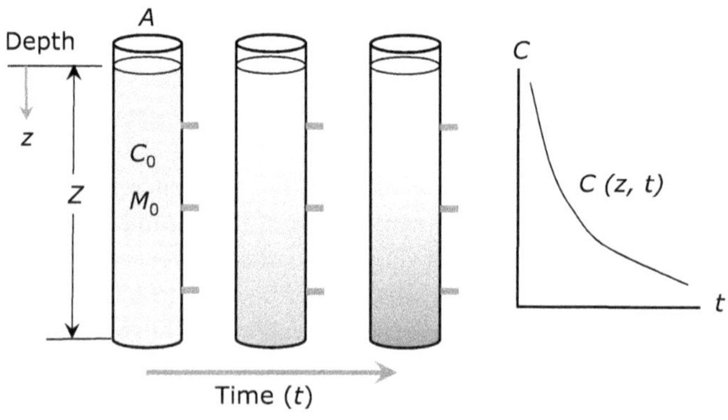

FIGURE 16.10 Settling test columns at three times.

TABLE 16.5 Total Suspended Solids Concentrations from a Settling Test Column

Depth (m)	Time (min)			
	20	40	60	120
0.8	135	90	75	48
1.6	170	110	90	53
2.4	180	126	96	60

improve the model. This process can start with a simple model (with few terms) or a larger model. The best model will adequately describe the data with the fewest number of terms (parameters).

Define $C(z,t)$ as the TSS concentration measured at depth z and time t. The simplest possible model with z and t.

$$C(z,t) = b_0 + b_1 z + b_2 t$$

The full quadratic model is

$$C(z,t) = b_0 + b_1 z + b_2 t + b_3 zt + b_4 z^2 + b_5 t^2$$

where the bs are model coefficients (parameters) to be estimated from the data.

Fitting the simple model, where y is the predicted TSS concentration, gives

$$y = 132.3 + 17.8z - 0.97t$$

Figure 16.11 shows the diagnostic residual plots for this model. The residuals plotted against predicted solids concentration, time, and depth are not random. There is a parabolic, or quadratic, pattern. This model is not adequate.

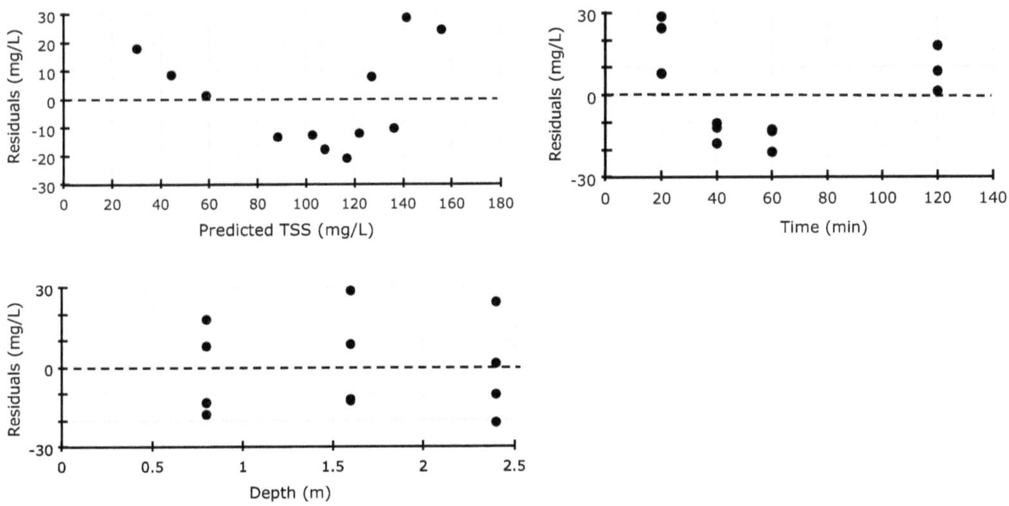

FIGURE 16.11 Diagnostic plots of the residuals for $y = 132.3 + 17.8\ z - 0.97\ t$.

The quadratic shape of residuals versus time suggests adding t^2 to the model. Doing this gives

$$y = 186 + 17.8z - 3.06t + 0.0143t^2$$

The residual plot of TSS concentration for this augmented model, Figure 16.12, shows no inadequacies. The residuals appear random and are evenly scattered above and below zero (constant variance). This model is adequate to describe the data.

The model building process can also start by fitting the most complicated model and then checking to determine which terms can be dropped from the model (Berthouex & Brown 2002). Terms can be dropped if they do not improve the model in a statistically significant way. The process will be outlined without explaining the statistical calculations that are behind the decision-making.

Table 16.6 summarizes RSS and R^2 for the eight possible linear models. (The *t*-ratios are shown for the estimated parameters of Models A and D. These will be ignored for now.)

The fitted six-parameter Model A is

$$y = 151.7 + 52.37z - 2.735t - 0.201zt - 7.031z^2 + 0.0143t^2$$

Simplifying this by dropping the z^2 and zt terms (Models B and C) hardly changes the value of R^2 or the residual sum of squares (RSS). Dropping t^2 (Models E–H) makes a notable difference in RSS

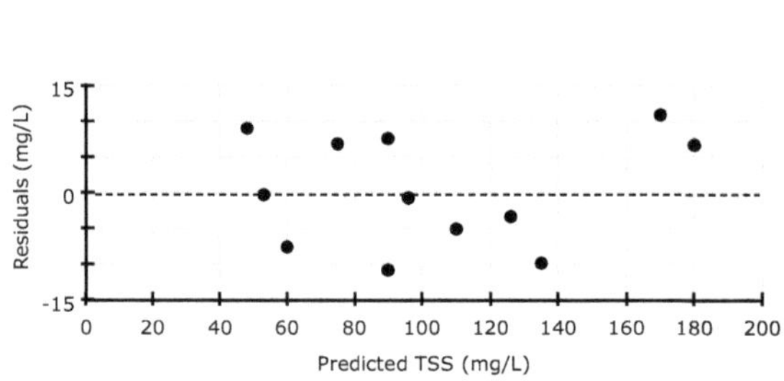

FIGURE 16.12 Diagnostic residuals plot for $y = 186 + 17.8\,z - 3.06\,t + 0.143\,t^2$.

TABLE 16.6 Summary of Possible Models for Solids Removal by Settling

Model	Model with Estimated Parameters	RSS	R^2
A	$y = 151.7 + 52.37z - 2.735t - 0.201zt - 7.031z^2 + 0.0143t^2$	309	0.985
t-ratio	2.3 8.3 2.4 1.0 7.0		
B	$y = 166.7 + 29.88z - 2.735t - 0.201zt + 0.014t^2$	362	0.982
C	$y = 171.0 + 40.31z - 3.057t - 7.031z^2 + 0.014t^2$	598	0.971
D	$y = 186.0 + 17.81z - 3.057t + 0.0143t^2$	*652*	*0.968*
t-ratio	4.46 8.0 5.6		
E	$y = 98.0 + 52.37z - 0.646t - 0.201zt - 7.03z^2$	2,859	0.864
F	$y = 113.0 + 29.88z - 0.646t - 0.20zt$	2,681	0.858
G	$y = 117.3 + 40.31z - 0.969t - 7.03z^2$	3,309	0.849
H	$y = 132.3 + 17.81z - 0.968t$	3,388	0.844

and these models are not satisfactory. Adding z^2 and zt decreases the residual sum of squares (and increases R^2) but the change is not statistically significant. There are formal ways to test whether the change in RSS is significant, but an explanation is beyond the scope of this chapter.

Model D is again identified as the simplest adequate model. It fits the data as well as Models A, B, and C with fewer terms. The residual plots support this conclusion.

R^2 decreases as terms are dropped from the model, although $R^2 = 0.86$ for Model E is still "large." It is natural to be fascinated by high R^2 values and this tempts us to think that the goal is to make R^2 as high as possible by putting more high-order terms into a model. Increasing R^2 is the wrong goal. Instead, prefer the most parsimonious model – the model that adequately describes the data with the fewest parameters.

The t-ratios are diagnostic statistics that give guidance about whether a term in a linear model provides significant information about the response. By significant, we mean information above that associated with the random error in the data. Roughly speaking, if the t ratio of a parameter is less than 2.5, the true value of the parameter could be zero, and that term can be dropped from the model. As parameters are dropped, the smaller (fewer parameters) model should be refitted and checked. An equivalent approach that is more satisfying is to calculate confidence intervals for the estimated parameters. If the interval includes zero, the true value of the parameter could be zero and the term associated with the parameter can be dropped from the model (Box, Hunter and Hunter 2005; Berthouex and Brown 2002).

Notice that the predicted concentration at $z = 0$, $t = 0$ is not the measured initial concentration of 560 mg/L. The reason is that this empirical model will be used to predict TSS concentrations at times and depths that are relevant to real settling tanks. These predictions are not improved by including data near the surface and at short settling times. A model to fit those conditions would need cubic terms and that added complexity is not justified.

Note: Settling tests liked the ones described here are part of most environmental engineering laboratory courses, and they have been used frequently for industrial wastewater treatability studies, so many examples can be found in the literature. Most of these will start making measurements after five or ten min, and then the frequency of sampling decreases. It should be done in the opposite way. Data in the first 20 min give little useful information. It is true that conditions are changing rapidly, but the times of interest for design are usually between 30 and 90 min, and this is where most sampling should be done. Samples in the early minutes actually make modeling more difficult because more terms (e.g., cubic terms) will be needed to fit all the data. If such data are reported, delete them and fit a simpler model. The best model describes the conditions of greatest interest to the designer with the fewest number of parameters.

16.6 WHAT IS R^2?

The *coefficient of determination*, R^2, has been called "a measure many statistician's love to hate." The usual definition says that R^2 is that proportion of the *total variability* in the dependent variable that is *explained by the regression equation.* The usual interpretation is that high R^2 means that a model is *good* because it *explains* a lot, and, conversely, that low R^2 means that a model is poor or not useful. These interpretations are inadequate. R^2 says nothing about the goodness of a model in any absolute sense. A model can have serious lack of fit and still have large R^2, and a model with low R^2 can be valid and useful.

The *total sum of squares* (Total SS) is the *residual sum of squares* (RSS) plus the *regression sum of squares* (Reg SS):

$$\text{Total SS} = \text{RSS} + \text{Reg SS}$$

$$\text{Reg SS} = \text{Total SS} - \text{RSS}$$

Reg SS is the part of Total SS that is "explained" by the regression model. RSS, the random leftovers of the least squares estimation, is the part of Total SS that is not explained by the regression. The definition of R^2 is

$$R^2 = \frac{\text{Reg SS}}{\text{Total SS}} = \frac{\text{Total SS} - \text{RSS}}{\text{Total SS}} = 1 - \frac{\text{RSS}}{\text{Total SS}}$$

Total SS is calculated from the *null model.* The null model is the simplest possible model that can be fitted to any set of data. It is a straight horizontal line at the level of the average value of the y's, indicated by

$$\text{Total SS} = \sum (y_i - \bar{y})^2$$

So R^2 measures how much the regression model improves the null model. Substituting the algebraic expressions for RSS and Total SS gives a computational definition of R^2.

$$R^2 = 1 - \frac{\text{RSS}}{\text{Total SS}} = 1 - \frac{\sum (y_{\text{obs}} - y_{\text{calc}})^2}{\sum (y - \bar{y})^2}$$

R^2 will be explained using the settling test data and Model D, the preferred model. Table 16.7 lists the measured solids concentrations, y_{obs}. The average solids concentration is 102.75 so the

Null model $\bar{y} = 102.75$

The residuals for the null model are (135–102.75), (90–102.75), and so on. The sum of the squared null residuals is the *total sum of squares*; Total SS = 20,564.

Also given are the concentrations predicted from the fitted model, y_{calc}, the residual errors for the fitted model, $e = (y_{\text{obs}} - y_{\text{calc}})$, and the squared residuals, e^2. The residual sum of squares, RSS = 652. This is the part that is not explained by the fitted model.

Fitting the model reduces the unexplained variance from 20,564 to 652. The variance *explained* by the model is

$$\text{Reg SS} = \text{Total SS} - \text{RSS} = 20{,}564 - 652 = 19{,}910$$

TABLE 16.7 Calculation of R^2 for the Settling Test Model D

Observed Concentration (y_{obs})	Null Residual (e_{null})	Null Residual Squared (e_{null})^2	Calculated Concentration (y_{calc})	Model D Residual (e)	Residual Squared (e^2)
135	32.25	1040.06	141.38	−9.83	96.54
90	−12.75	162.56	97.34	−10.89	118.69
75	−27.75	770.06	64.74	6.57	43.15
⋮	⋮	⋮	⋮	⋮	⋮
⋮	⋮	⋮	⋮	⋮	⋮
126	23.25	540.56	119.05	−3.39	11.52
96	−6.75	45.56	86.45	−0.93	0.87
60	−42.75	1827.56	57.29	−8.35	69.70
Average = 102.75		Total SS = 20,564			RSS = 652

R^2 is the proportion of the total variance that is explained by the model, and

$$R^2 = \frac{19,910}{20,564} = 0.968$$

16.7 CASE STUDY: CHICK'S LAW OF DISINFECTION

Drinking water and wastewater effluents are disinfected using chlorine, UV radiation, and ozone to inactivate pathogenic organisms. Coliforms, like most water-borne pathogens, are excreted from the gut of warm-blooded animals, so they serve as a marker for possible fecal contamination. An absence of coliform bacteria is taken to mean an absence of pathogenic microorganisms.

The classical disinfection model, Chick's law, says the inactivation of coliform bacteria is directly proportional to the product of chlorine dose, C, and contact time, t.

$$\ln(F) = -K\,t\,C$$

where

$F = N/N_0$ = fraction of coliform bacteria surviving for a given contact time
N_0 = initial bacterial count
N = bacterial count at time t and chlorine dose C
K = proportionality coefficient (L/mg-min)
t = contact time (minutes)
C = chlorine dose (mg/L)

A modification of Chick's law changes the proportionality of concentration by adding an exponent α to give

$$\ln(F) = -K\,t\,C^{\alpha}$$

These models can be tested with the data for disinfection of wastewater with chlorine that are given in Table 16.8 and plotted in Figure 16.13.

TABLE 16.8 The Fraction (*F*) of Coliforms Surviving Given Contact Time and Chlorine Dose

Contact Time	Chlorine Dose (mg/L)			
(min)	1	2	4	8
0	1.000	1.000	1.000	1.000
15	0.690	0.280	0.400	0.200
30	0.400	0.110	0.090	0.039
45	0.250	0.090	0.032	0.0075
60	0.150	0.0320	0.008	0.0012
75	0.080	0.0080	0.0016	0.0003
90	0.050	0.0045	0.0009	0.00005

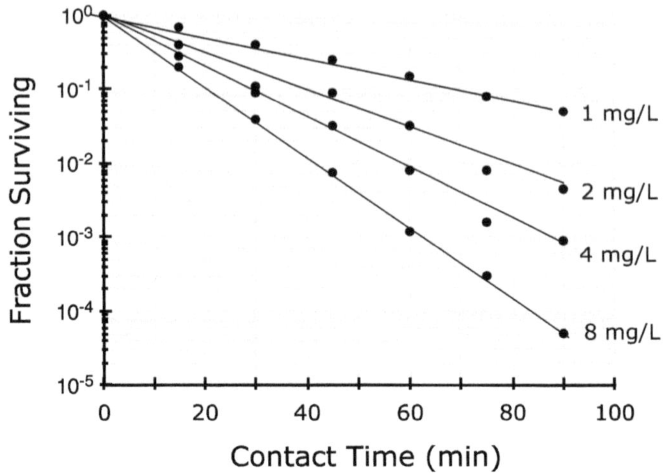

FIGURE 16.13 Disinfection of wastewater with chlorine.

Fitting Chick's law to the data for each chlorine dose gives

1 mg/L Cl_2	$K = 0.0326$ L/mg-min	$\ln(F) = 0.0326\, t$ (1 mg/L) = 0.0326 min^{-1} (t)
2 mg/L Cl_2	$K = 0.0304$ L/mg-min	$\ln(F) = 0.0304\, t$ (2 mg/L) = 0.0608 min^{-1} (t)
4 mg/L Cl_2	$K = 0.0200$ L/mg-min	$\ln(F) = 0.0200\, t$ (4 mg/L) = 0.0800 min^{-1} (t)
8 mg/L Cl_2	$K = 0.0138$ L/mg-min	$\ln(F) = 0.0138\, t$ (8 mg/L) = 0.1104 min^{-1} (t)

The models very closely fit the data in Figure 16.13 but they are inadequate for two reasons: one engineering and one statistical. The engineering reason is that the model should give good predictions for any chlorine dose. These four equations provide no easy way to do that because the K values are a function of chlorine dose (see Figure 16.14) and Chick's law does not account for this.

Fitting Chick's law to all the data simultaneously gives

$$\ln(F) = -0.016\, t\, C$$

The residual plots in Figure 16.15 show definite patterns and non-constant variance. Both are signs of an inadequate model. The residuals for the 8 mg/L chlorine dose are all positive; the other doses

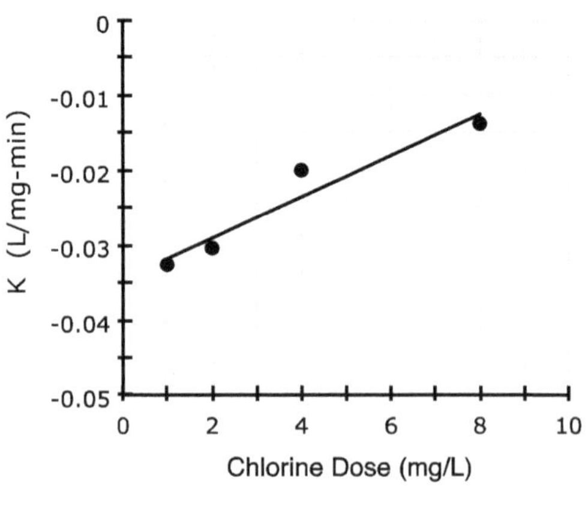

FIGURE 16.14 Rate coefficient in Chick's law is a function of chlorine concentration.

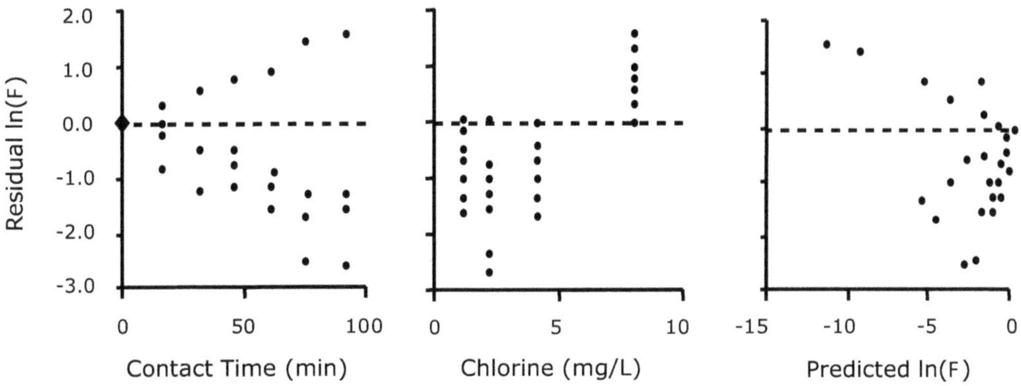

FIGURE 16.15 Residuals for the Chick's law model.

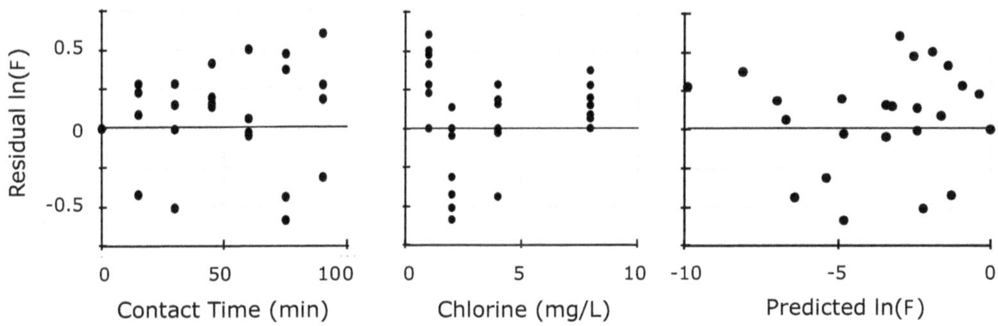

FIGURE 16.16 Residuals for the modified chlorination models.

have negative residuals. Non-constant variance means that the magnitude of the residuals is different at levels of the variables.

Fitting the modified Chick's law gives

$$\ln(F) = -0.04\,t\,C^{0.5}$$

This model fits all the data and interpolates between the four discrete chlorine doses used in the experiment. The residual plots, Figure 16.16, show some weakness related to chlorine dose, but the modified model is better than the original Chick's law.

16.8 DESIGN OF EXPERIMENTS FOR PARAMETER ESTIMATION

There are two competing experimental desires: (1) to minimize the number of measurements (n) and (2) to reduce the duration of the test program. This is especially true when each test run takes hours or days. The good news is that precise estimates can be obtained with a small number of experiments *if* the observations are made at the right settings.

Figure 16.17 shows four possible experimental designs to estimate θ_1 and θ_2 in this model

$$y = \theta_1[1 - \exp(\theta_2 t)]$$

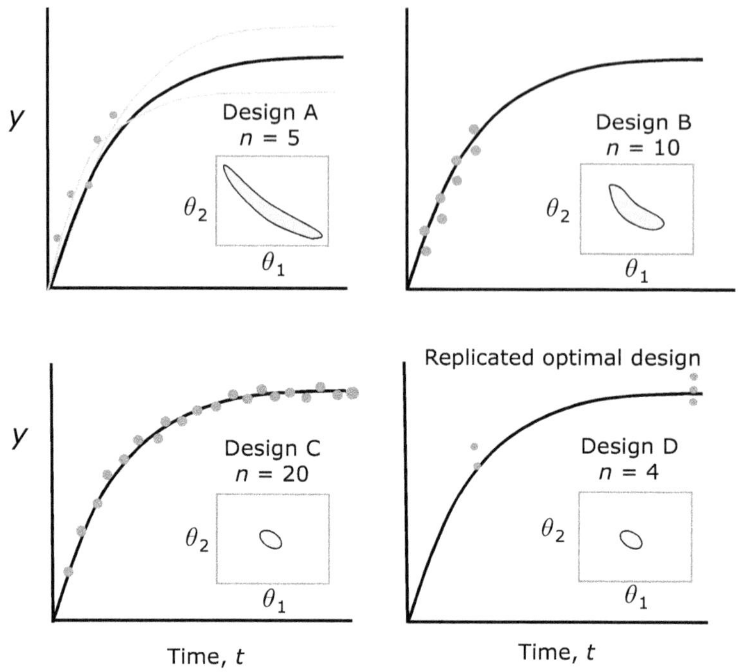

FIGURE 16.17 The location of measurements determines the precision of the estimated parameters. Shaded approximate joint confidence regions are not to scale, but the shapes and relative sizes are correct.

where

θ_1 = a saturation value (the asymptote) that is reached at some large value of t
θ_2 = rate coefficient that determines how rapidly the asymptote will be reached
t = time (d)

The size (area) of the joint confidence region, shown in the insert, depends mainly on the number of observations and the magnitude of the random experimental error. The shape depends mainly on the settings of the independent variable. A good experimental design will give an elliptical or nearly circular confidence region that has a small area.

Design A has five observations, spaced one day apart. Design B replicates design A. Design C has 20 observations spaced one day apart. Design D is the most efficient of the four.

Design A may seem attractive because it takes 5 days instead of the 20 days for designs C and D. But design A is five days of wasted work because neither θ_1 nor θ_2 has been estimated with any precision. The banana-shaped (hyperbolic) joint confidence region shows that the parameter estimates are strongly correlated with each other. The gray lines show that curves with larger θ_1 and smaller θ_2 will reasonably fit the data, and the reverse is also true (smaller θ_1 and larger θ_2). This is *parameter correlation* – the parameter estimates are not independent. A hyperbola describes the product of two variables, in this case the product of $\theta_1\theta_2$. What is being estimated is the initial slope of the curve:

$$\left.\frac{dy}{dx}\right|_{x=0} = \theta_1\theta_2$$

The data are sufficient to estimate the product of the parameters, but they cannot give independent estimates of the individual parameters and this is a serious weakness in the design.

Design B, with ten observations (replicates at times of 1–5 days), is slightly better, but the parameter estimates are not precise and the parameter correlation is not eliminated. Adding more observations at the wrong locations is wasteful.

It should be clear that getting a precise estimate of the asymptotic value, θ_1 requires making observations on the asymptotic part of the curve, that is, at large values of t. Design C, with $n = 20$ observations, an expansion of design A, will accomplish the goal. The confidence region is small (approximately 10% of design A) and the estimated values show little correlation.

The design in the bottom right panel (call it the optimal design) gives an excellent result with just four observations because the observations are *in the right location.* For an asymptotic model, measurements are needed at the largest feasible setting of t. This gives independent information about θ_1. The second important location is $t = 1/\theta_2$, which is at about 60% of θ_1 on the *y axis*. The informative regions are shown in Figure 16.18.

There may seem to be a dilemma. We are doing the experiment to find the value of θ_2, but to design the optimal experiment we need to know θ_2. This is not a serious problem because the experimenter will have some idea of the value, for example, θ_2 is between 0.2 and 0.4. The design and the parameter estimates are robust with respect to the exact value of t. Any setting within this range will give a good result. Assume a value (your best guess), make some measurements (say at $t = 4$ and $t = 20$), and estimate the parameters. To improve the estimates, make a few more measurements at the same or slightly shifted times. Replicated measurements in the informative regions will be the most cost-effective design.

The same concept applies to exponential decay models, as shown in Figure 16.19. An important question in groundwater and soil remediation is "Will the curve go to zero or will it flatten out at some non-zero level?"

The statistical problem is not so much to estimate θ_2, but to predict the trend toward some near-zero level. A common approach is frequent sampling in the early days of a cleanup program and less sampling as time goes on. Early samples are needed to check that remediation is happening (i.e., concentrations are going down) and to estimate the mass of pollutant being removed. But data from the early times contribute no useful information about the ultimate asymptotic concentration. Estimating the slope of the curve as it becomes flat is statistically difficult. Failure to understand this will result in undersampling and poor estimates.

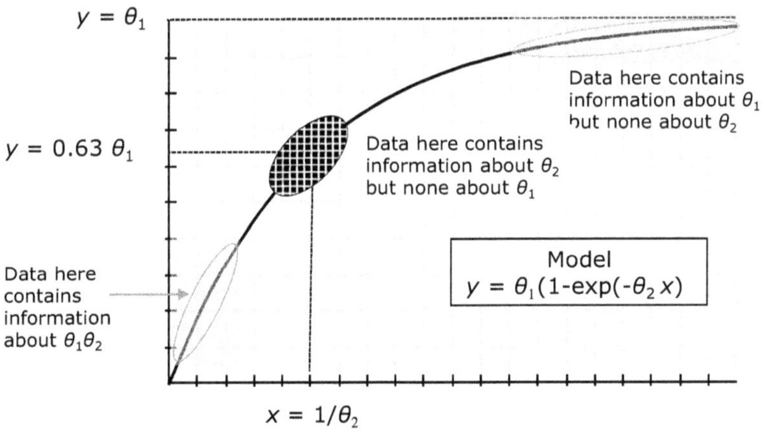

FIGURE 16.18 Most informative regions for parameter estimation of the first-order model.

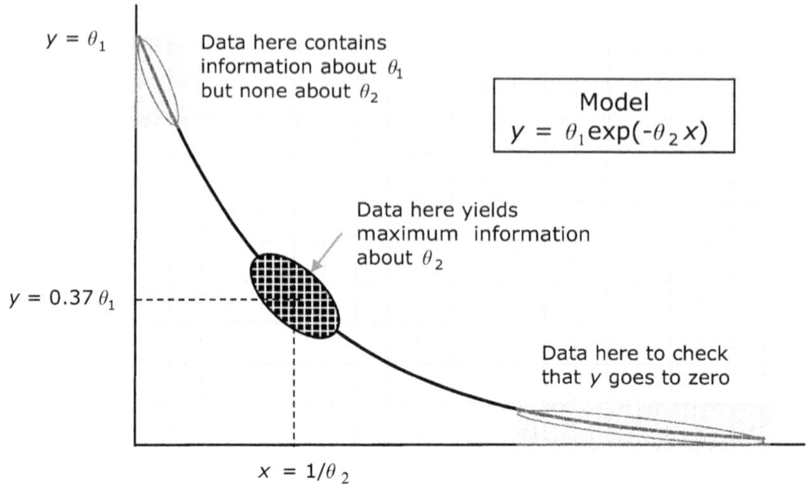

FIGURE 16.19 Most informative regions for exponential decay models.

16.9 REPLICATION AND RANDOMIZATION

Two fundamentals of experimental design are replication and randomization. *Replication* improves the precision of the estimated parameters and increases the information that can be used to evaluate the goodness of fit of a model. *Randomization* removes the effect of systematic errors in the measurements and, by doing so, increases the validity of all statements made about the fitted model.

Consider an experimental design to estimate the slope of a straight line:

$$y = \beta_0 + \beta_1 x + e$$

Suppose the range of x is from 5 to 50 and the budget allows making six runs (a run is to set x and measure y). Figure 16.20 shows three ways the runs could be arranged. The numbers indicate the order in which the tests will be done. The design on the left is weak because there is no randomization and no replication. The designs improve going from left to right.

How the six runs are used should depend on the experimenter's *a priori* knowledge of the system. If the response is known to be a straight line (for example, a calibration curve for most instruments),

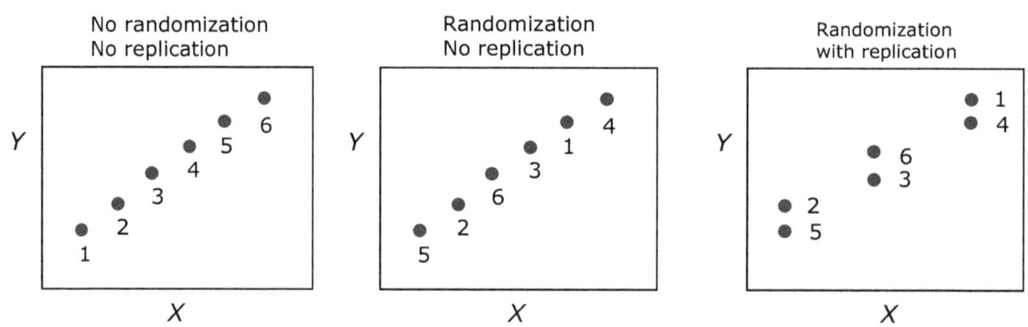

FIGURE 16.20 Three ways to organize and experiment with six runs when the model is expected to be a simple curve (straight-line or quadratic function).

the most efficient design locates three runs at each end of the range to be covered in the experiments. This will give the most precise estimates of the slope and intercept of the line.

A design to check for curvature needs runs within the range, say two runs at each extreme and two points in the center of the range. Obviously, there are other designs with runs spread over the range of x. Designs with replication will always be superior, and replication at all chosen settings of x is preferred.

True replication means observations (measurements of y) on an independent set of experiments. The goal is to measure the experimental error, including all the experimental factors that might introduce variation in the observed values of y. Setting x once and measuring y twice or three times is not true replication because it measures only the repeatability of the analytical measurements. The variance of repeated measurements will be smaller than the variance of replicate measurements. A model must be evaluated using a valid measure of the true experimental error, which has components of variation due to experimental setup, sample collection, sample handling and storage, and instrumental analysis. This measure is obtained by true repliction.

16.10 CONCLUSION

Good experimental designs are needed to produce data that contain the information needed to estimate the parameters in a model and test its goodness of fit. This is common sense, but experimental design is often overlooked in a rush to start collecting data.

The collection of statistical methods known as experimental design, model building, and model discrimination should be used to plan ways to gather the best data about a design through experiments and pilot plant studies. Good experimental designs will be rewarded by savings in time and money.

The parameter values that give the minimum residual sum of squares are called "estimates" because their values are affected by the experimental error in measurements of the response (independent) variable, y. Experimental error will cause new or repeat experiments to have slightly different values of y, even if performed at the same levels of the dependent variable, x. The residual errors and the residual sum of squares will change and so will the parameter estimates.

The accuracy and precision of the measurements, the degree of replication, *and* the location of the observations in the experimental space (that is, the experimental design) all contribute to the quality of the fitted model. All observations are not equal in terms of providing useful information for building and diagnosing models. It is not always true that a poor design can be overcome by making a large number of observations. A common mistake, especially with exponential models, is making too many measurements at the beginning and not enough near the asymptote.

The goodness of models should be based on an estimate of variance that is obtained from true replicated experiments.

16.11 PROBLEMS

16.1 LEAST SQUARES ESTIMATES OF RATE COEFFICIENTS

Table P16.1 is the data for degradation of a pesticide treated by oxidation. The initial concentration is known to be 32.2 mg/L. Estimate the reaction rate coefficients for these two models by calculating and plotting the residual sum of squares function for a range of values of k_1 and k_2. Which model best fits the data?

First order: $$C = C_0 \exp(-k_1 t)$$

Second order: $$\frac{1}{C} = \frac{1}{C_0} + k_2 t$$

TABLE P16.1 Kinetic Data

Time (min)	Concentration (mg/L)
0	32.2
10	23.4
20	20.6
30	14.9
40	13.8
50	11.6
60	10.8

16.2 WASTEWATER TREATMENT IN INDONESIAN MALLS

Use the historical cost data in Table P16.2 to derive a model for estimating the capital cost of wastewater treatment plants for shopping malls in Indonesia.

TABLE P16.2 Cost of Wastewater Treatment in Surabaya, Indonesia, shopping Malls (Razif 2015)

Mall	Design Flow (L/s)	Capital Cost (Million Rupiah)
A	0.88	361
B	1.31	474
C	1.41	503
D	1.44	518
E	2.08	688
F	2.46	785
G	2.70	839
H	4.06	1,097
I	6.03	1,653
J	6.72	1,803

16.3 LINEAR PROGRAMMING COST MODEL

An industry can process waste in two ways, both earn a profit on reclaimed materials. The company can shift the waste load from process 1 to process 2. The units of waste treated annually (units/y) by processes 1 and 2 are X1 and X2. A linear optimization problem is being formulated to calculate the best waste treatment schedule. The objective function is

MAX = NETCOST
NETCOST = REVENUE − PROCESSING COSTS
REVENUE = C1*X1 + C2*X2

C1 and C2 are the coefficients for revenue earned by each process ($/unit). These coefficients need to be estimated using the last ten years of revenue data given in Table P16.3. Fit a linear model to the data to estimate C1 and C2.

TABLE P16.3 Revenue Data for Two Industrial Processes

	Process 1		Process 2	
Year	Capacity (units)	Revenue ($)	Capacity (units)	Revenue ($)
2010	16,860	240,200	21,250	250,500
2011	19,450	250,600	21,450	255,800
2012	21,660	265,100	21,650	259,600
2013	22,250	266,400	21,850	265,400
2014	23,350	268,600	22,050	270,400
2015	24,522	275,700	22,250	274,700
2016	24,453	276,400	22,450	278,300
2017	25,520	280,900	22,650	282,300
2018	26,550	282,600	22,850	286,300
2019	27,888	288,400	23,050	289,100

16.4 URBANIZATON AND THE WATER QUALITY OF STREAMS

Urbanization replaces cropland and forestland with impervious surfaces and this changes watershed hydrology and the quantity and quality of stormwater runoff. One measure of water quality, total suspended solids (TSS), is a good indicator of other pollutants, particularly nutrients and metals carried on the surfaces of sediment in suspension. Turbidity (T) is a less expensive analytical method than TSS, and it can be measured in the field. TSS is a measure of the mass of solids captured by filtering a water sample. Turbidity is a measurement of the scatter in a light beam as it passes through a sample of water. High turbidity can be caused by colloidal particles that will pass through a filter. Both are related to particles in the water, and both increase as the particle count increases, but there is the question of whether the correspondence is strong enough to be a reliable predictor.

Table P16.4 gives the data from 20 representative measurements of TSS and turbidity that can be used to investigate and evaluate the validity of using turbidity to estimate TSS in urbanized streams.

TABLE P16.4 Turbidity and TSS Data for 20 Samples from Urbanized Streams

Sample	Turbidity (NTU)	TSS (mg/L)	Sample	Turbidity (NTU)	TSS (mg/L)
1	50	85	11	5	10
2	52	96	12	22	65
3	27	42	13	10	18
4	69	120	14	40	98
5	11	20	15	7	16
6	52	140	16	12	27
7	18	35	17	46	97
8	6	9	18	84	135
9	26	55	19	4	6
10	39	72	20	63	166

(a) Plot the data (TSS vs. turbidity).
(b) Fit a straight-line model, TSS = $\beta_0 + \beta_1$(turbidity), using the method of least squares and examine the residuals to assess the goodness of fit of the model.
(c) If the straight-line mode is inadequate, propose and fit an alternative model to improve the fit.

16.5 BOILER STEAM PRODUCTION MODEL

The amount of boiler steam that must be produced depends on the steam and electricity that are required by the industry. Fit this proposed model to the data in Table P16.5.

$$S_b = \beta_0 + \beta_1 S_d + \beta_2 E$$

where

S_b = boiler steam required (1,000 lb)
S_d = delivered steam (1,000 lb)
E = delivered electricity (MWh)
β_0 = average internal steam usage (1,000 lb)
β_1 = (lb boiler steam/lb delivered steam)
β_2 = (lb boiler steam/MWh delivered electricity)

TABLE P16.5 Typical 24-Month Cogeneration Plant Output

Year 1	Delivered Steam (1,000 lb)	Delivered Electricity (MWh)	Boiler Steam (1,000 lb)	Year 2	Delivered Steam (1,000 lb)	Delivered Electricity (MWh)	Boiler Steam (1,000 lb)
Jan.	220,517	19,584	361,785	Jan.	270,359	22,080	402,159
Feb.	216,566	18,168	338,720	Feb.	233,287	19,896	359,106
Mar.	203,342	18,648	332,196	Mar.	200,758	19,824	345,032
April	147,034	17,040	279,985	April	146,173	19,512	306,099
May	118,384	19,224	286,422	May	133,234	19,512	312,789
June	142,052	18,312	291,783	June	148,862	21,120	312,789
July	170,273	16,896	287,527	July	153,431	23,256	344,057
Aug.	162,038	21,336	340,535	Aug.	148,245	23,112	331,096
Sept.	140,468	20,184	296,051	Sept.	137,128	20,472	311,740
Oct.	152,206	21,216	326,868	Oct.	144,794	21,768	315,770
Nov.	176,210	20,112	331,705	Nov.	151,721	22,632	325,949
Dec.	199,742	20,184	350,670	Dec.	197,626	21,072	344,927

Source: McMasters, RL 2002.

16.6 SETTLING TEST MODEL

The removal of suspended solids (TSS) from wastewater by settling was measured in a quiescent laboratory settling column. The initial TSS concentration was 195±2 mg/L. A coagulant, $FeCl_3$, was mixed into the wastewater to enhance solids removal. Samples were removed from the settling column at intervals over a 90-min test at depths ranging from 0.25 to 2.5 m. The depth of the testing column was 4 m. The results from an optimum dose of 10 mg/L $FeCl_3$ are given in Table P16.6. Derive an empirical statistical model to describe suspended solids removal in the settling column.

TABLE P16.6 Suspended Solids Concentration Data from Settling Column Test at Initial TSS Concentration = 195±2 mg/L and Coagulant Dose = 10 mg/L $FeCl_2$

Depth, *z* (m)	Suspended Solids Concentration at Different Sampling Times *t* (min)						
	0	**20**	**30**	**45**	**60**	**75**	**90**
0.25	189	176	160	132	115	113	111
0.50	181	179	161	151	143	133	122
0.75	184	178	170	163	148	137	125
1.00	182	185	173	159	150	143	133
1.25	191	191	187	171	151	141	135
1.50	192	182	175	166	154	141	136
1.75	195	184	180	174	155	144	137
2.00	196	188	185	181	158	148	140
2.50	194	190	187	180	163	155	143

16.7 COEFFICIENT OF DETERMINATION, R^2

Regression software produces R^2 as a routine part of the output. Without using linear regression software, calculate the total sum of squares, residual sum of squares, and R^2 for this straight-line model that fits the data in Table P16.7. The fitted model is

$$y = 0.8731 + 0.5624x$$

TABLE P16.7 Sample Data for R^2 Calculation

$x =$	2	5	8	12	15	18	20
$y =$	1.7	4	5.1	8.1	9.2	11.3	11.7

16.8 IS A LOW R^2 VALID?

Figure P16.8 shows 200 weekly pH measurements in a stream. These data support the hypothesis that the stream has been affected by acidic deposition from regional coal-fired power plants. The linear regression equation that fits the data is

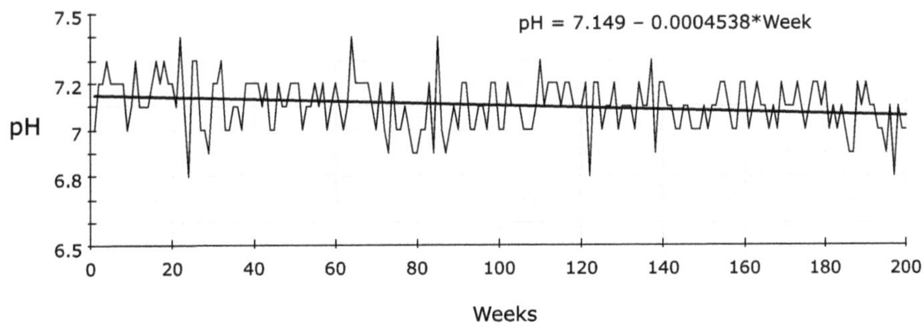

FIGURE P16.8 200 weekly measurements of stream pH.

$$pH = 7.149 - 0.0004538 * Week$$

The slope (−0.0004538) is small, but statistical tests confirm that it is, with a high degree of confidence, larger than zero. This is a valid model.

The average of the 200 pH readings is 7.103. The total sum of squares (Tot SS) = 2.41820 and residual sum of squares (RSS) = 2.28096. Calculate the Regression Sum of Squares (Reg SS) and R^2.

16.9 FITTING REPLICATE DATA

The data in Table P16.9 comes from an experiment having five settings of x and four replicate measurements of y at each x. A replicate is an independent repeat of the experiment. The setting of x is made anew for each replicate in random order. It is not setting x once and taking four samples. It is not making four measurements on the same sample. It has been suggested that fitting the five averages of the replicates is better than fitting all of 20 observations. Compare the two methods. Which should be preferred? Why?

TABLE P16.9 Replicate Experimental Data

x	y_1	y_2	y_3	y_4	Average y
10	140	110	180	80	127.5
15	250	210	190	160	202.5
20	350	320	280	240	297.5
25	370	400	430	340	385.0
30	530	510	460	440	485.0

16.10 PUMP STATION COSTS

Use the cost data in Table P16.10 to derive a cost-estimating equation for pumping stations.

TABLE P16.10 Pump Station Costs (US$ 2015)

Installed Horsepower, Q	Cost ($)	Installed Horsepower, Q	Cost ($)
100	620,000	2,000	3,500,000
200	930,000	3,000	4,200,000
300	1,200,000	4,000	5,100,000
400	1,500,000	5,000	5,800,000
500	1,700,000	6,000	6,600,000
600	1,800,000	7,000	7,200,000
700	1,900,000	8,000	7,800,000
800	2,100,000	9,000	8,500,000
900	2,200,000	10,000	9,000,000
1,000	2,400,000	20,000	14,000,000
		30,000	17,000,000

TUTORIAL NOTE

The following are nonlinear least squares problems and the parameters are to be estimated by minimizing the sum of the squared residuals. This calculation is a numerical search best done using a statistical software package like Minitab, SYSTAT, StatEase, or SPSS, but it can be done using Excel or LINGO. Use the method of calculation that is most convenient for you.

16.11 BIOCHEMICAL OXYGEN DEMAND (BOD)

Biochemical oxygen demand (BOD) measures the amount of oxygen consumed by microorganisms as they consume organic compounds in wastewater. The standard test runs five days (at 20°C) and the result is called the 5-day BOD. Table P16.11 shows the 5-day BOD is 11.3 mg/L. Incubation times other than five days yield a higher or lower BOD. The ultimate BOD is the value that would be obtained if the test were extended until all the organic matter has been consumed. This is not practical, so the ultimate BOD is estimated from measurements at times that are practical, which is usually not longer than 20 days. The rate at which oxygen is consumed is proportional to the concentration of organic matter, which decreases over time. The BOD reaction rate coefficient is also estimated from BOD measurements made over time. The model for biochemical oxygen demand (BOD) that is used to estimate the ultimate BOD and the reaction rate coefficient is

$$y = \theta_1[1 - \exp(-\theta_2 t)]$$

where

y = BOD at time t (mg/L)
θ_1 = ultimate BOD (mg/L)
θ_2 = reaction rate coefficient, d^{-1}

(a) Fit the model to the data in Table P16.11 to estimate θ_1 and θ_2.
(b) Plot the data with the fitted model.
(c) Plot the residuals to check whether the model adequately describes the data.

TABLE P16.11
Experimental BOD Data

Time, t (d)	0	2	5	7	10	12	15	20
BOD, y (mg/L)	0	3.9	11.3	11.7	16	18.1	18.5	20.8

16.12 KINETIC MODELS

The data in Table P16.12 were fitted in Problem 16.1 using the initial value, $C_0 = 32.2$, as a known value. Perhaps it had been strictly controlled to start the experiment or it had been measured repeatedly, so it was known much more precisely than the other measurements. In this problem the initial concentration carries the same weight – the same experimental error – as the other concentration data. A hand-drawn plot of the data (not shown) suggests the model might be a first-order reaction.

$$\frac{dC}{dt} = -k_1 C, \qquad C = C_0 \exp(-k_1 t)$$

TABLE P16.12 Data for Fitting Two-Parameter First- and Second-Order Kinetic Models

Time (min)	Concentration (mg/L)
0	32.2
10	23.4
20	20.6
30	14.9
40	13.8
50	11.6
60	10.8

A second-order model is also worth checking.

$$\frac{dC}{dt} = -k_2C^2, \qquad \frac{1}{C} = \frac{1}{C_0} + k_2t$$

Fit the two models and select the one that is best.

16.13 TANNIN REMOVAL

Tannins are water-soluble polyphenolic compounds of varying molecular weight. They have traditionally been used for converting animal hides into leather, due to their ability to interact and precipitate proteins found in animal skin. Tannins are known to inhibit the activity of many microorganisms.

The strength of wastewater from the tannin manufacturing process is measured as chemical oxygen demand, COD (mg/L). The carbonaceous organic matter is tannins and other organic compounds. The tannins inhibit the biological process; the other organics do not, but the tannins tend to be a constant fraction (31%), the COD, and inhibition increases as COD increases. Inhibition means that the COD removal rate increases and then, at some critical concentration, it decreases.

A model of the COD removal rate comes from an inhibition kinetic model for tannin removal.

$$r = \frac{r_{\max}S}{K_S + S}\exp(-S / K_I)$$

where

r = COD removal rate of biomass (kg/m³ d)
S = COD concentration (kg/m³)
$r_{\max}$ = maximum removal rate (kg COD/m³ d)
K_s = saturation constant (kg COD/m³)
K_I = inhibition coefficient (kg COD/m³)

Seven lab-scale replicate steady-state experiments at six different loading conditions were performed and gave the data in Table P16.13. Fit the inhibition model to these data to get estimates of the kinetic parameters $r_{\max}$, K_S, and K_I. The values in the table are the averages of seven replicate tests at each condition.

TABLE P16.13 Results of Tannin Removal Experiments (Gorsek & Tramsek 2008)

Experiment	Effluent S (kg/m^3)	r (kg/m^3-d)
1	0.150	0.544
2	0.240	0.826
3	0.750	0.957
4	1.240	1.081
5	2.040	1.010
6	3.600	0.570

16.14 GROUNDWATER DATA

Remedial pumping of a Superfund site has accomplished a steady decrease in the groundwater contamination level. Pumping can be stopped when the concentration is "stable" and significant further change is not expected. That is, will the contaminate concentration stabilize at a non-zero level, C^*, or will it slowly decline toward zero? Use the data in Table P16.14 to answer this question by fitting these two models.

Model 1: $C = C_0 \exp(-kt)$

Model 2: $C = C^* + (C_0 - C^*)\exp(-kt)$

TABLE P16.14 Groundwater Data (Berthouex and Brown 2002)

Time (days)	Concentration (μg/L)	Time (days)	Concentration (μg/L)
0	2,200	1,000	180
30	2,200	1,100	230
100	1,400	1,200	200
200	850	1,300	190
300	380	1,400	280
350	500	1,500	200
400	600	1,600	250
500	800	1,700	290
600	230	1,800	200
700	130	1,900	140
800	120	2,000	180
900	200		

16.15 REACTIONS IN SERIES

Material A reacts to form product B which in turn forms product C. Both reactions are first-order with respect to the reactant.

$$A \xrightarrow{k_1} B \xrightarrow{k_2} C$$

The model for the concentration of B, with initial concentrations ($t = 0$) of $C_{A0} = 1$, $C_{B0} = 0$, and $C_{C0} = 0$ is

$$C_B = \frac{\theta_1}{\theta_2 - \theta_1} C_{A0} \left[\exp(-\theta_1 t) - \exp(-\theta_2 t) \right]$$

where θ_1 is the reaction rate coefficient for the transformation of A to B and θ_2 is the reaction rate coefficient for the transformation of B to C. The data are in Table P16.15.

(a) Fit the model to the data to estimate the two parameters in the model
(b) Plot the predicted and observed concentrations of B versus time
(c) Plot the residuals versus time and versus the predicted concentration of B
(d) Evaluate on the adequacy of the model.

TABLE P16.15 Data from Two Consecutive First-Order Reactions

Time (min)	C_B (mg/L)
0	0
2.5	0.08
5	0.22
10	0.29
20	0.41
40	0.38
50	0.28
60	0.23
70	0.2

16.16 BACTERIAL GROWTH RATE

Experimental data for fitting the two models was collected in a laboratory-scale biological reactor. Each experimental setting of the reactor's dilution rate (flow rate divided by reactor volume) produces a pair of values of bacterial growth rate, μ, and substrate concentration S, as given in Table P16.16. Two possible mechanistic models are by Monod (1942) and Teissier (1936). Fit the models and plot the residuals to evaluate the goodness of fit:

Monod model: $$\mu = \frac{\theta_1 S}{\theta_2 + S}$$

Teissier model: $$\mu = \theta_3 (1 - \exp^{-\theta_4 S})$$

where μ is the growth rate observed at substrate concentration S

TABLE P16.16 Bacterial Growth Rate Data (Schulze and Lipe 1964)

Test No.	Experimental Data	
	S (mg/L)	μ (1/h)
1	5.1	0.059
2	8.3	0.091
3	13.3	0.124
4	20.3	0.177
5	30.4	0.241
6	37.0	0.302
7	43.1	0.358
8	58.0	0.425
9	74.5	0.485
10	96.5	0.546
11	112	0.610
12	161	0.662
13	195	0.725
14	266	0.792
15	386	0.852

16.17 OXYGEN TRANSFER

Oxygen transfer in rivers and streams is an important process for a healthy aquatic environment. Forty experiments were performed to model the oxygen mass transfer coefficient, k_2, as a function of stream hydraulic conditions. Two competing models are as follows:

$$\text{Model 1:} \quad k_2 = a\frac{V^b}{H^c}$$

$$\text{Model 2:} \quad k_2 = a\frac{S^b}{H^c}$$

where

k_2 = mass transfer coefficient, d^{-1}
V = mean stream velocity, ft/s
H = mean stream depth, ft
S = slope of stream channel, ft/ft

Fit the two models to the data in Table P16.17 to determine the better model.

TABLE P16.17 Oxygen Mass Transfer Data (Thackston 1966)

Exp No.	H (ft)	V (ft/s)	S (ft/ft)	k_2 (1/d)	Exp No.	H (ft)	V (ft/s)	S (ft/ft)	k_2 (1/d)
1	0.125	1.323	0.00596	18.72	21	0.078	0.848	0.00524	39.45
2	0.109	1.449	0.00894	20.75	22	0.082	0.802	0.00438	16.74
3	0.124	1.878	0.01161	25.45	23	0.085	0.365	0.00315	16.32
4	0.092	1.707	0.01446	34.85	24	0.070	0.389	0.00600	36.45
5	0.098	1.858	0.01623	40.6	25	0.101	0.645	0.00659	14.62
6	0.122	2.230	0.01785	44.9	26	0.119	0.865	0.00794	21.95
7	0.117	2.029	0.01418	52.5	27	0.146	1.100	0.00791	25.87
8	0.170	2.079	0.00843	24.35	28	0.082	0.636	0.01076	24.25
9	0.112	1.225	0.00557	19.57	29	0.129	1.110	0.01074	24.72
10	0.104	0.994	0.00430	18.38	30	0.057	0.558	0.02012	64.05
11	0.107	0.921	0.00380	21.10	31	0.141	0.476	0.00144	10.88
12	0.094	0.802	0.00236	15.39	32	0.175	0.955	0.00426	11.38
13	0.107	0.940	0.00423	16.69	33	0.232	0.484	0.00065	5.780
14	0.113	0.773	0.00245	9.980	34	0.154	1.471	0.01332	44.55
15	0.099	0.776	0.00282	13.19	35	0.15	1.743	0.02038	31.55
16	0.065	0.754	0.00571	26.15	36	0.189	2.015	0.01688	39.30
17	0.065	1.116	0.01124	39.55	37	0.089	1.333	0.00241	23.80
18	0.063	0.961	0.00894	40.55	38	0.067	0.978	0.00174	18.45
19	0.072	1.069	0.00897	43.80	39	0.037	0.592	0.00182	28.80
20	0.052	0.678	0.00674	31.90	40	0.080	0.843	0.00103	14.36

17 Experimental Methods for Engineering Optimization

Engineers are often asked:

> "How would the process work if we change the operating conditions?"
>
> "What will happen if we do something different?"

To learn what happens when a process will be changed, *you have to change it.*

The *experiment* is running tests (i.e., making observations) at specified conditions to collect the data. This is technical work. *Data analysis*, or model building, is interpreting how those changes are reflected in performance. This is statistical work and the methods described in Chapter 16 are tools for doing this work.

Experimental design is a strategy for solving engineering design problems. It is deciding what changes should be investigated, which factors should be changed and by how much, how many test conditions should be investigated? These are engineering considerations more than statistical ones. The common sense and thoughtful problem-solving skills of good engineers should control the statistical experimental design problem.

The engineering comes first. An elaborate experimental design cannot compensate for weak engineering, just as no amount of data analysis can overcome an inefficient experimental design. Many designs are weak because they try to investigate too many variables at too many settings. Use the first experiments to explore. Test the potential independent variables to see if they are really important. Use early experiments as a basis for changing the settings at which responses are measured, and even for adding or dropping independent variables.

In some books, this chapter would be Statistical Experimental Design. We prefer to think of it as the engineering design of experiments based on statistical concepts that will produce a lot of useful information with a small number of experiments.

17.1 ITERATIVE EXPERIMENTS

Figure 17.1 is the *Design-Test-Analyze* cycle of investigation. It is rare that one experiment yields all the needed information. For example, if there is uncertainty about which factors actively change the process response, experiments are needed to identify those factors before building empirical or mechanistic models of process performance.

Whatever the research budget may be, do not spend it all on one grand experiment. Design and run a modest exploratory experiment, analyze the data, and use what was learned to plan additional experiments. These might replicate the initial design to increase confidence in the results, or change the settings of the predictor variables to explore more promising conditions. The investment in exploratory tests will be repaid.

Consider the problem of locating the optimal operating conditions for a process. The process behavior is described by a *response surface* of unknown shape, which may be a smooth slowly changing surface, a peaked surface, a valley, or a ridge. The goal is to find to an optimum, if one exists. This is to be done with the minimum number of tests and every test must provide useful information. Two-level factorial experiments are economical and efficient designs for doing this.

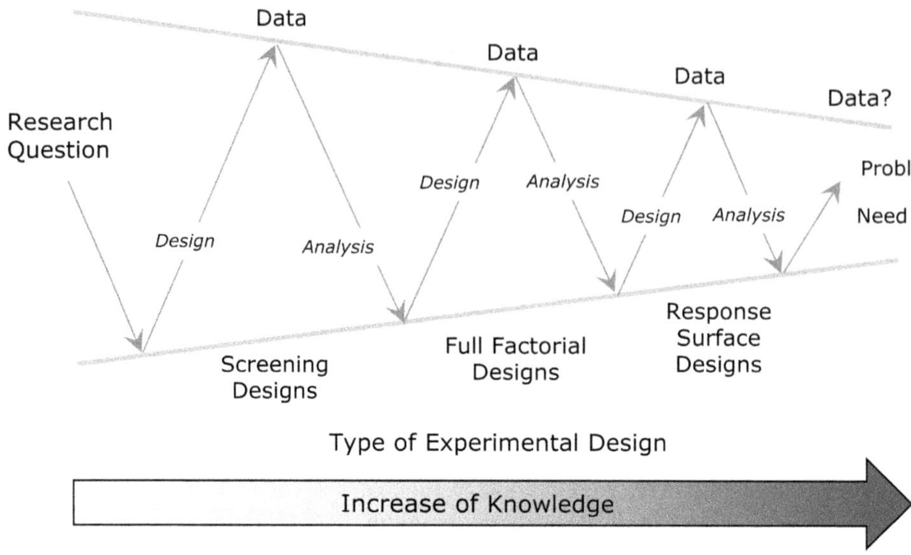

FIGURE 17.1 The iterative cycle of model building.

Figure 17.2 shows how a simple exploratory design with four test conditions (two-factors, each set at two levels) can identify a direction of steepest ascent (or descent) toward a region of potentially better results. This region is explored with a second two-factor, two-level design (gray dots on square, upper left). The data here may not show a clear gradient in any direction because the tests are within a region of stationarity. In this case, a third iteration could add the *center point* and the *star points*. This will allow a quadratic model to be fitted if indeed there is an optimum.

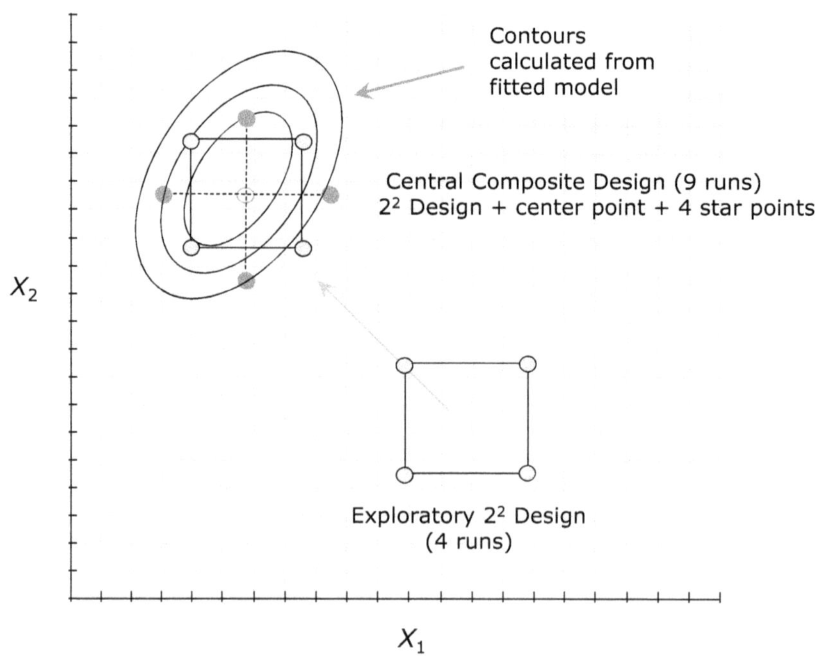

FIGURE 17.2 Exploratory design guides experiments toward the optimal region.

The example used four exploratory tests, a second design with four more tests, followed by adding a center point and then four star points. This is a total of 13 test conditions. Contrast that with an experiment, for example, that used 36 tests, arranged in a 6 × 6 grid, to cover a large region. This would be a wasteful experiment because most tests would provide little useful information. Unfortunately, experiments like this are far too common.

Two-level experiments give the maximum possible information for a unit of testing effort. They show the interaction of variables, which is what gives the direction for moving the design, and they are easily shifted, resized, or augmented to explore for a maximum or minimum.

17.2 ORTHOGONAL AND BALANCED EXPERIMENTAL DESIGNS

The most efficient experimental designs are *balanced* and *orthogonal*. *Orthogonal* in simple geometric terms means at right angles.

A *balanced design* has greater statistical power because there is an equal number of observations at all levels of testing. A factorial design is balanced because each level of the factors is run the same number of times.

Experimenters will at times need to work with unbalanced designs. Balanced designs may produce unbalanced data when something goes wrong: test animals die, machinery breaks, test materials don't arrive when needed, operators or subjects get sick. Design for balance and hope for best.

Figure 17.3 shows some possible experimental designs in two and three variables. The designs shown in the top and middle frames are orthogonal. Those in the bottom frames are not.

Each observation in an orthogonal design does double or triple duty. For example, the top left design for two variables, each at two levels, giving four experiments that are located at the corners of the box. This is a two-factor, two-level design, or $2^2 = 4$ test conditions (runs). To estimate the effect of changing X_2 from the low level to the high level, subtract the low-level value from the high

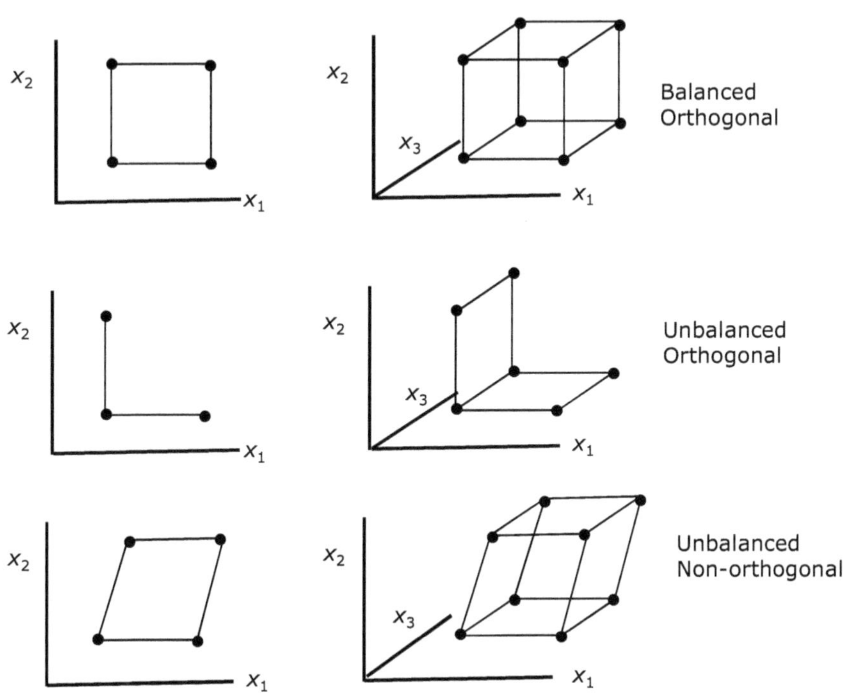

FIGURE 17.3 (Top) Balanced orthogonal design, (middle) unbalanced orthogonal, (bottom) unbalanced non-orthogonal design.

level. The two measurements at the high level and two at the low level gives two independent estimates of the effect of X_2. And, the same four measurements will give two independent estimates of X_1. Each observation does double duty. This is an efficient design.

As a bonus, the interaction of X_1 and X_2 can be estimated by comparing the diagonal corners. The interaction is what happens when two factors are changed at the same time. Interactions are common in biological, chemical, and physical experiments. In some processes, the interactions are the most interesting effects.

The four experiments make it possible to fit a model of the form

$$y = b_0 + b_1X_1 + b_2X_2 + b_{12}X_1X_2$$

That is a lot of information from just four tests. That is the power of a good experimental design. It does not rely on brute force or large numbers of tests. It gets a lot of information from each observation.

The middle frame (the L-shaped layout) is orthogonal, but not balanced. With these three test conditions, it is possible to estimate the effect of X_1 and the effect of X_2. The model that can be fitted is

$$y = b_0 + b_1X_1 + b_2X_2$$

The estimates b_1 and b_2 will be less precise than those from the square (or rectangular) design, and the interaction term ($b_{12}X_1X_2$) cannot be estimated. Losing the interaction ($b_{12}X_1X_2$) is a serious flaw in the experiment. This experiment is three-fourths the work of the 2^2 design, but it yields less than half the information. It is not a good experimental design.

The unbalanced non-orthogonal design wastes the chance to make each observation do double duty and the results are more difficult to interpret.

17.3 TWO-LEVEL FACTORIAL EXPERIMENTS

The number of test conditions in a two-level factorial design is 2^N, where N is the number of factors. A test at one condition is called a *run*. A three-factor design has $2^3 = 8$ runs; a four-factor design has $2^4 = 16$ runs.

Figure 17.4 shows the arrangement of a two-factor and a three-factor design. The combinations of +1 and −1 at the corners are the coded variables that indicate high level (+1) and low level (−1).

The number of runs will be larger than the number of test conditions if replicate measurements are made. A replicated 2^3 design has 8 + 8 = 16 runs. If half the test conditions are replicated, the number of runs is 8 + 4 = 12.

Two-level factorial designs can be quickly written in the form of the design matrix, which has columns for each factor (independent variable) that show the high-low combinations coded as high (+1) or low (−1).

Table 17.1 is the design matrix for a two-level experimental design with two, three, and four factors. The two-factor design has $2^2 = 4$ runs that are described by four rows in the design matrix; each column has two −1's in two columns and two +1's in each column. The three-factor design has $2^3 = 8$ runs and 8 rows, with four −1's and four +1's in each column. The first two columns of the three-factor design repeat the 2^2 design. A third column is added for the third factor to complete the eight combinations. These are balanced orthogonal designs.

A five-factor experiment would have $2^5 = 32$ runs, which may be an unreasonable experimental burden. In such a case, the experimenter may choose to run 16 of the 32 runs. This is called a half-fraction, or $(2^5)/2 = 2^{5-1} = 16$ runs. This kind of design is called a *fractional factorial design*.

Fractional designs will give independent estimates of the main effects but estimates of the interactions may be compromised. This is not a serious problem because the purpose of fractional

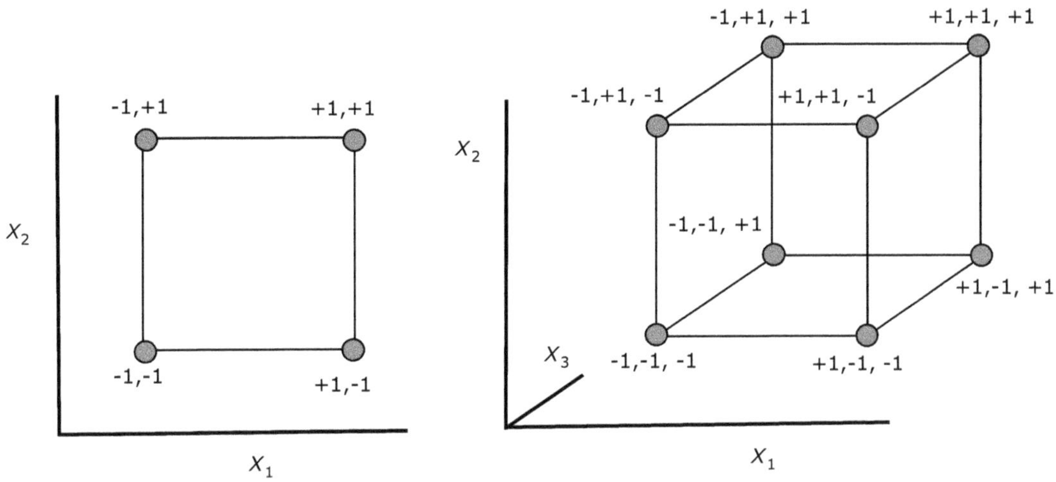

FIGURE 17.4 Arrangements of test conditions for 2^2 and 2^3 factorial experimental designs.

TABLE 17.1 Design Matrix for Two-Level Experimental Designs in Two, Three, and Four Factors

Two Factors			Three Factors				Four Factors				
Run	X_1	X_2	Run	X_1	X_2	X_3	Run	X_1	X_2	X_3	X_4
1	−1	−1	1	−1	−1	−1	1	−1	−1	−1	−1
2	+1	−1	2	+1	−1	−1	2	+1	−1	−1	−1
3	−1	+1	3	−1	+1	−1	3	−1	+1	−1	−1
4	+1	+1	4	+1	+1	−1	4	+1	+1	−1	−1
			5	−1	−1	+1	5	−1	−1	+1	−1
			6	+1	−1	+1	6	+1	−1	+1	−1
			7	−1	+1	+1	7	−1	+1	+1	−1
			8	+1	+1	+1	8	+1	+1	+1	−1
							9	−1	−1	−1	+1
							10	+1	−1	−1	+1
							11	−1	+1	−1	+1
							12	+1	+1	−1	+1
							13	−1	−1	+1	+1
							14	+1	−1	+1	+1
							15	−1	+1	+1	+1
							16	+1	+1	+1	+1

designs, with five or more factors, is usually to identify factors that are not important, so subsequent experiments can be done with fewer factors. See Box, Hunter, and Hunter (2005) to learn how fractional factorial designs are constructed.

17.4 CASE STUDY: USING FLY ASH TO CONSTRUCT LAGOON LINERS

Managing the disposal or containment of fly ash from coal-fired power plants is a difficult and expensive problem. Some fly ashes contain minerals that make them pozzolanic; they have properties similar to a low-grade cement. When mixed with water, they will set into a rock-like material and this makes them interesting as construction materials. Pozzolanic fly ash is sold and mixed with Portland cement. Power plants would like to use this material to build impermeable liners for ash storage ponds and lagoons.

This experiment investigated the effect on the density of lagoon liners constructed with the pozzolanic fly ash caused by changing

W = water content (% water in the wet mix)
C = compression (lb/inch^2 or psi)
T = reaction time before compression (minutes)

Fly ash was mixed with the specified amount of water, measured as percent water in the total mixture, to form a moist solid material. The mixture was placed in a cylindrical mold where it was allowed to cure for the specified reaction time, and then compressed to the specified pressure. The density (lb/ft^3) of the material after compression was measured. The data are in Table 17.2.

The *main effects* are calculated from the differences in the response, in this case the density of solidified liner material, as an experimental factor is changed from its low to high level. There are three main effects to be calculated: water (W), compression (C), and time (T). The same eight values of density are used in each main effect calculation, which means that each test run is doing triple duty in the calculation of the effects. (They will also do triple duty in calculating the interactions.) This is why factorial experiments are so efficient.

The *main effect of water* is the average difference of the densities at 10% water minus the densities at 4% water. Figure 17.5 shows that there are four independent estimates of the effect of water. Calculate these differences and divide by four to get the average, which is called the *main effect*.

The *main effect of reaction time* is the average difference of the densities at 20 min minus the densities at 5 min. The main effect of compression is the average difference between the high and low pressures.

The main effect of changing the water content from 4% to 10% is

$$\text{Main effect of water} = \frac{(121-108)+(118-100)+(126-119)+(119-108)}{4}$$

$$= \frac{13+18+7+11}{4} = 12.25 \text{ lb/ft}^3$$

TABLE 17.2 Experimental Design Matrix and Density of Lagoon Liners Constructed with Fly Ash

Run	W (%)	C (psi)	T (min)	W	C	T	Density (lb/ft^3)
1	4	60	5	−1	−1	−1	108
2	10	60	5	+1	−1	−1	121
3	4	260	5	−1	+1	−1	119
4	10	260	5	+1	+1	−1	126
5	4	60	20	−1	−1	+1	100
6	10	60	20	+1	−1	+1	118
7	4	260	20	−1	+1	+1	108
8	10	260	20	+1	+1	+1	119

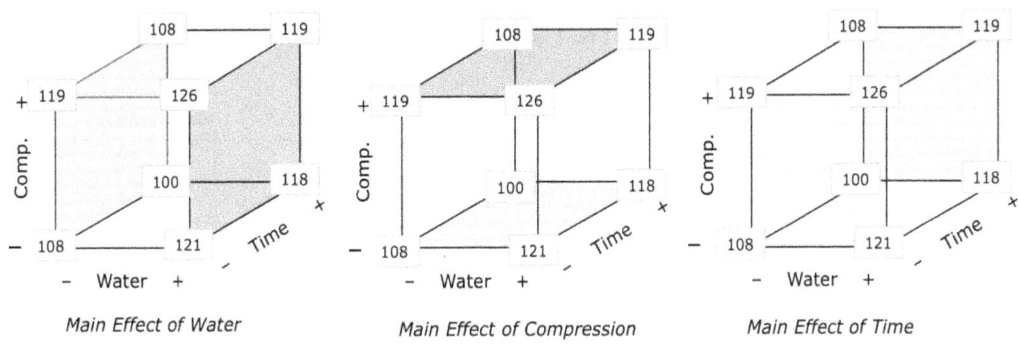

FIGURE 17.5 Graphical interpretation of lagoon liner density experiment.

The main effects of compression pressure and reaction time are calculated in the same way:

$$\text{Main effect of compression} = \frac{(119-108)+(126-121)+(108-100)+(119-118)}{4}$$

$$= \frac{11+5+8+1}{4} = 6.25\ \text{lb/ft}^3$$

$$\text{Main effect of reaction time} = \frac{(100-108)+(118-121)+(108-119)+(119-126)}{4}$$

$$= \frac{(-8)+(-3)+(-11)+(-7)}{4} = -7.25\ \text{lb/ft}^3$$

The interpretation of the main effects is as follows:

- Changing the water content from 4% to 10% increases the density by an average of 12.25 lb/ft^3.
- Changing the compression pressure from 60 to 250 psi increases the density by an average of 6.25 lb/ft^3.
- Changing the reaction time before compression from 5 to 20 min decreases the density by an average of 7.25 lb/ft^3. This is because the pozzolanic fly ash starts to harden very quickly and it cannot be compacted effectively if too much time elapses.

An easier way to calculate the effects uses the coded design matrix. Simply multiply the column vectors in the design matrix times the densities, *y*, add the columns, and divide by 4. The plus and minus signs take care of the arithmetic, as shown in Table 17.3.

The analysis can be extended to calculate two-factor interactions between water and compression, water and time, compression and time. Two-factor interactions indicate whether the factors in the experiment act independently. Negligible or near-zero interaction means the factor main effects are independent. Positive interaction means they are synergistic, while negative interaction means they are antagonistic. The interactions are shown in Table 17.4 as *WC*, *WT*, and *CT*. The column for *WC* is the product of the + and – values for *W* and *C*; likewise, for *WT* and *CT*. Multiply the density by the *WC* column and do the addition to calculate the *WC* interaction.

An equation that describes the response as a function of the coded variables (the *X*'s), is

$$y = 114.9 + 12.25X_1 + 6.25X_2 - 7.25X_3 + 12.25X_1X_3$$

TABLE 17.3 Calculation of the Main Effects of Water (*W*), Compression (*C*), and Reaction Time (*T*)

Run	Factors			Coded Factors			Density	Main Effects		
				X_1	X_2	X_3	Y	y^*X_1	y^*X_2	y^*X_3
	W (%)	C (psi)	T (min)				(lb/ft³)	(lb/ft³)	(lb/ft³)	(lb/ft³)
1	4	60	5	−1	−1	−1	108	−108	−108	−108
2	10	60	5	+1	−1	−1	121	121	−121	−121
3	4	260	5	−1	+1	−1	119	−119	119	−119
4	10	260	5	+1	+1	−1	126	126	126	−126
5	4	60	20	−1	−1	+1	100	−100	−100	100
6	10	60	20	+1	−1	+1	118	118	−118	118
7	4	260	20	−1	+1	+1	108	−108	108	108
8	10	260	20	+1	+1	+1	119	119	119	119
Average =							114.9			
Sum =								49	25	29
Main effect = Sum/4 =								12.25	6.25	7.25

TABLE 17.4 Calculation of the Main Effects and Interactions

Coded Factors and Interactions							Density	Main Effects			Interactions			
W	*C*	*T*	*WC*	*WT*	*CT*	*WCT*	(lb/ft³)	*W*	*C*	*T*	*WC*	*WT*	*CT*	*WCT*
−1	−1	−1	+1	+1	+1	−1	108	−108	−108	−108	108	108	108	−108
+1	−1	−1	−1	−1	+1	+1	121	121	−121	−121	−121	−121	121	121
−1	+1	−1	−1	+1	−1	+1	119	−119	119	−119	−119	119	−119	119
+1	+1	−1	+1	−1	−1	−1	126	126	126	−126	126	−126	−126	−126
−1	−1	+1	+1	−1	−1	+1	100	−100	−100	100	100	−100	−100	100
+1	−1	+1	−1	+1	−1	−1	118	118	−118	118	−118	118	−118	−118
−1	+1	+1	−1	−1	+1	−1	108	−108	108	108	−108	−108	108	−108
+1	+1	+1	+1	+1	+1	+1	119	119	119	119	119	119	119	119
Average =							114.9							
							Sum=	49	25	−29	−13	49	−7	−1
Effect of interaction = Sum/4 =								**12.25**	**6.25**	**−7.25**	−3.25	**2.25**	−1.75	−0.25
									Main Effects			**Interactions**		

The two-factor interaction of water content and reaction time shows that the best results are with high water content and a short reaction time.

The two-factor interactions for *WC* and *CT* too are small to be included (they are within the realm of experimental error), and the same is true for the three-factor interaction *WCT*. Three-factor interactions are expected to be small and statistically insignificant. For more instruction on assessing the significance of factor effects, consult the texts by Box, Hunter, and Hunter 2005; and Berthouex and Brown 2002.

17.5 CASE STUDY: BOD REDUCTION FOR POLYESTER MANUFACTURING

This is an example of how a problem might be identified, defined, analyzed, and implemented (Akesrisakul & Jiraprayuklert 2007).

The cost of treating wastewater from a polyester manufacturing process had to be reduced. The high treatment cost was due to ethylene glycol (EG), a biodegradable organic chemical, being lost from manufacturing into the wastewater. Higher EG losses increase the BOD and the treatment cost. The goal was to minimize the loss of EG to the wastewater (and thus lower wastewater treatment cost), while maintaining polyester quality.

The source of the EG in the wastewater was the esterification reaction in the manufacturing process. Process experts believed that four factors might affect the concentration (mg/L) of EG in the wastewater. The factors must be controlled within the stated ranges.

T = Reaction temperature (288°C–292°C)
S = Slurry level in the heat exchanger (67%–70%)
F = Feed rate of EG (200–300 kg/h)
T_{VS} =Temperature at vapor separator (103°C–109°C)

These factors were investigated in preliminary experiments under actual manufacturing conditions and the level and range of the experimental factors were held within the process specifications. It was quickly discovered that T_{VS} had a negligible effect on EG losses over the range of 103–109°C and it was dropped from further studies.

The other three factors were "active" over the specified range and they were studied in a two-level, three-factor (2^3) factorial design, giving eight test conditions. The settings of the factors, in the actual levels and the coded −1 and +1 levels are given in Table 17.5, along with the measured ethylene glycol concentrations for each test condition.

Figure 17.6 shows the eight test conditions and the measured ethylene glycol concentrations. The average level of EG in the wastewater is 67.9 mg/L. The lowest EG concentration was for run 3:

Low reaction temperature = 288°C
High slurry level in the heat exchanger = 70%
Low feed rate of EG = 200 kg/h

Calculations for the factor effects from the factorial experiment using the coded levels from the design matrix are shown in Table 17.6.

Temperature had the largest effect, and flow rate had the smallest effect, for the factor ranges that were tested.

TABLE 17.5 The 2^3 Factorial Experimental Design to Investigate Ethylene Glycol (EG) Levels in Polyester Manufacturing Wastewater

	Operating Conditions			Coded Factors			Response
Run	Temperature (°C)	Slurry Level (%)	Feed Rate (kg/h)	X_1	X_2	X_3	EG Concentration (mg/L)
1	288	67	200	−1	−1	−1	65.0
2	292	67	200	+1	−1	−1	73.6
3	288	70	200	−1	+1	−1	61.5
4	292	70	200	+1	+1	−1	68.1
5	288	67	300	−1	−1	+1	67.0
6	292	67	300	+1	−1	+1	75.2
7	288	70	300	−1	+1	+1	62.9
8	292	70	300	+1	+1	+1	70.1

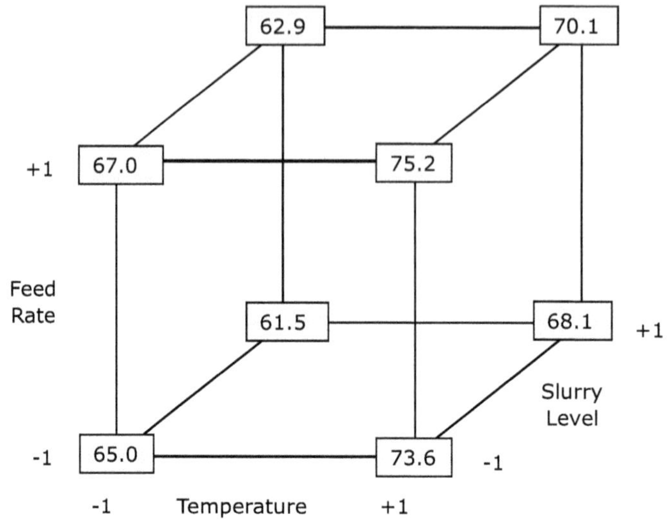

FIGURE 17.6 Cube plot showing the eight test conditions and the measured ethylene glycol concentrations.

TABLE 17.6 Calculation of Main Effects and Interactions for EG Case Study

Run	Coded Factors			EG	Main Effects		
	X_1	X_2	X_3	Concentration (mg/L)	X_1	X_2	X_3
1	−1	−1	−1	65.0	−65.0	−65.0	−65.0
2	+1	−1	−1	73.6	73.6	−73.6	−73.6
3	−1	+1	−1	61.5	−61.5	61.5	−61.5
4	+1	+1	−1	68.1	68.1	68.1	−68.1
5	−1	−1	+1	67.0	−67.0	−67.0	67.0
6	+1	−1	+1	75.2	75.2	−75.2	75.2
7	−1	+1	+1	62.9	−62.9	62.9	62.9
8	+1	+1	+1	70.1	70.1	70.1	70.1
Average =				67.9			
Sum =					30.6	−18.2	7
Factor effect = Sum/4 =					**7.65**	**−4.55**	**1.75**

Coded Factors and Interactions							EG	Interactions			
X_1	X_2	X_3	X_1X_2	X_1X_3	X_2X_3	$X_1X_2X_3$	Concentration (mg/L)	X_1X_2	X_1X_3	X_2X_3	$X_1X_2X_3$
−1	−1	−1	+1	+1	+1	−1	65.0	65.0	65.0	65.0	−65.0
+1	−1	−1	−1	−1	+1	+1	73.6	−73.6	−73.6	73.6	73.6
−1	+1	−1	−1	+1	−1	+1	61.5	−61.5	61.5	−61.5	61.5
+1	+1	−1	+1	−1	−1	−1	68.1	68.1	−68.1	−68.1	−68.1
−1	−1	+1	+1	−1	−1	+1	67.0	67.0	−67.0	−67.0	67.0
+1	−1	+1	−1	+1	−1	−1	75.2	−75.2	75.2	−75.2	−75.2
−1	+1	+1	−1	−1	1	−1	62.9	−62.9	−62.9	62.9	−62.9
+1	+1	+1	+1	+1	+1	+1	70.1	70.1	70.1	70.1	70.1
Average =							67.9				
Sum =								−3.0	0.2	−0.2	1.0
Factor effect = Sum/4 =								**−0.75**	**0.05**	**−0.05**	**0.25**

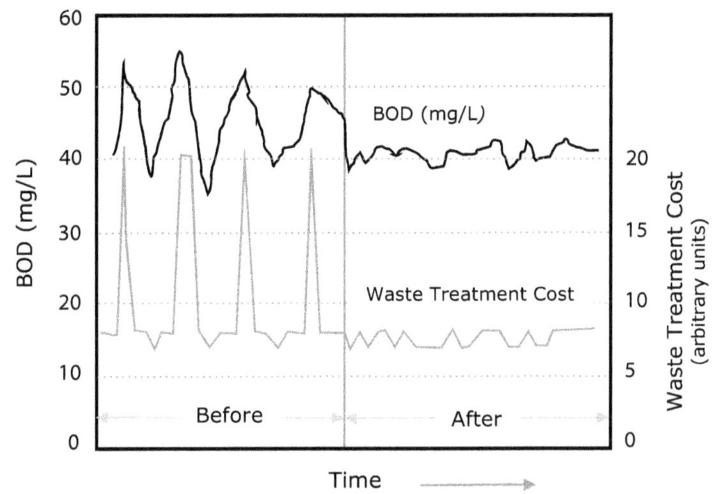

FIGURE 17.7 The before and after results for wastewater treatment at the polyester manufacturing plant. BOD concentrations were reduced and the high variability was removed. The average wastewater treatment cost has been reduced.

Reducing the reaction temperature (X_1) from the high level (292°F) to the low level (268°F) decreases the EG concentration in the wastewater by an average of 7.65 mg/L.

Changing the slurry level (X_2) from the low level to the high level decreases the EG concentration in the wastewater by 4.55 mg/L. Changing the EG feed rate (X_3) from the low level to the high level increases the EG concentration in the wastewater by 1.75 mg/L.

A model for EG concentration in the wastewater (y), written in terms of the coded variables (the X's), is

$$y = 67.92 + 7.65X_1 - 4.55X_2 + 1.75X_3$$

A solution is worthless if it is not implemented. Implementation is largely a matter of communication and persuasion. Figure 17.7 shows the variable BOD concentrations and high treatment costs that correspond to the higher BOD concentrations. Run 3 gave the best response. Operating at temperature = 288°C, slurry level = 70%, and feed rate = 200 kg/h reduced the EG concentration from 70.5 to 61.5 mg/L while maintaining the same level of polyester quality. The wastewater treatment cost was reduced 27.8%.

17.6 FACTORIAL DESIGNS VERSUS ONE-FACTOR-AT-A-TIME DESIGNS

A popular (unfortunately) experimental approach is to vary one factor at a time. That is, hold all factors constant except variable X_1 and determine the best level of Y. Then in a subsequent experiment, hold X_1 at its "best" level, change one more variable while holding all other factors constant, measure Y, and so on.

One-factor-at-a-time experiments (OFAT) are usually disappointing and inefficient. A large number of test runs may not be rewarded with a large amount of information. The greatest weakness is that they give no indication of interactions between variables and no indication about how factors should be changed to improve the system. The experimenter who fails to discover interactions wastes time and money.

This case study shows an OFAT for treating an industrial wastewater with initial suspended solids concentration of 5,000 mg/L. The treatment was to add a combination of ferric chloride ($FeCl_3$)

and sulfuric acid (H_2SO_4) to a 1-L sample of the wastewater, mix briefly to promote solids flocculation, allow some time for solids to settle, and sample the supernatant.

Tests with 100 mg/L H_2SO_4 and various levels of $FeCl_3$, given in Table 17.7, indicate that the best dosage for $FeCl_3$ is about 130 mg/L, and the minimum suspended solids concentration is 400 mg/L. A second series of tests with ferric chloride at 130 mg/L and different acid doses supports this conclusion. Data from the two experiments are shown in Figure 17.8. The curves were drawn by eye and the minimums could be shifted, but these tests seem to indicate that the best conditions are 130–135 mg/L $FeCl_3$ and 5–7 mg/L H_2SO_4.

The flaw in one-factor-at-a-time designs is that they fail to capture the interaction between factors. In chemical systems, there is almost always an interaction, and to see this you must simultaneously vary both factors.

A much different result is found when this problem is investigated using two-level factorial experiments, as shown in Figure 17.9. Experiment 1 shows that better results might be obtained with less $FeCl_3$ and more H_2SO_4. Experiment 2 locates three conditions that are better than the OFAT "optimum." A third two-level uses less $FeCl_3$ and more acid to achieve excellent suspended solids removal. This is how the strategy of *design-test-analyze* shown in Figure 17.1 works.

The OFAT approach used nine tests and did not find the best operating conditions. Even worse, it did not discover that there are better treatments along a valley. To move down the valley, increase the ferric chloride dose and lower the acid dose.

The best economic conditions will depend on the relative cost of the two chemicals, how much sludge is produced, and the cost of sludge disposal. A few simple two-level experiments provide information that will let the designer or operator optimize the process, not only in terms of solids removal, but also for cost.

TABLE 17.7 Treatment Results for 100 mg/L H_2SO_4

$FeCl_3$ (mg/L)	120	130	140	150
Suspended solids (mg/L)	1,000	400	650	1,000

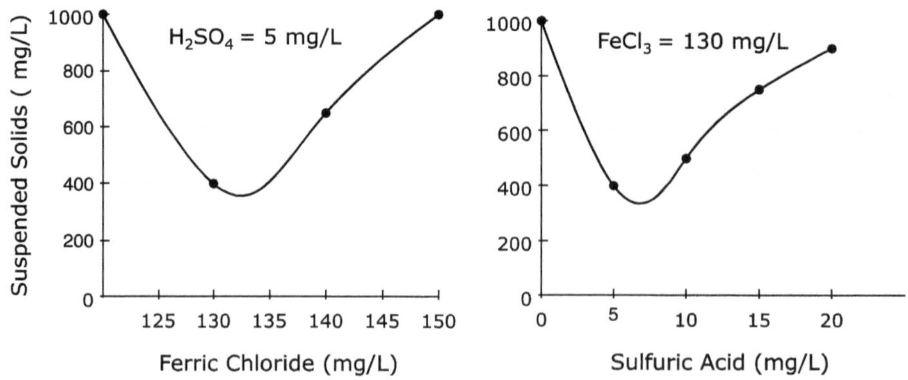

FIGURE 17.8 One-factor-at-a-time experiment seems to confirm that 130 mg/L $FeCl_3$ and 5 mg/L H_2SO_4 are the best conditions.

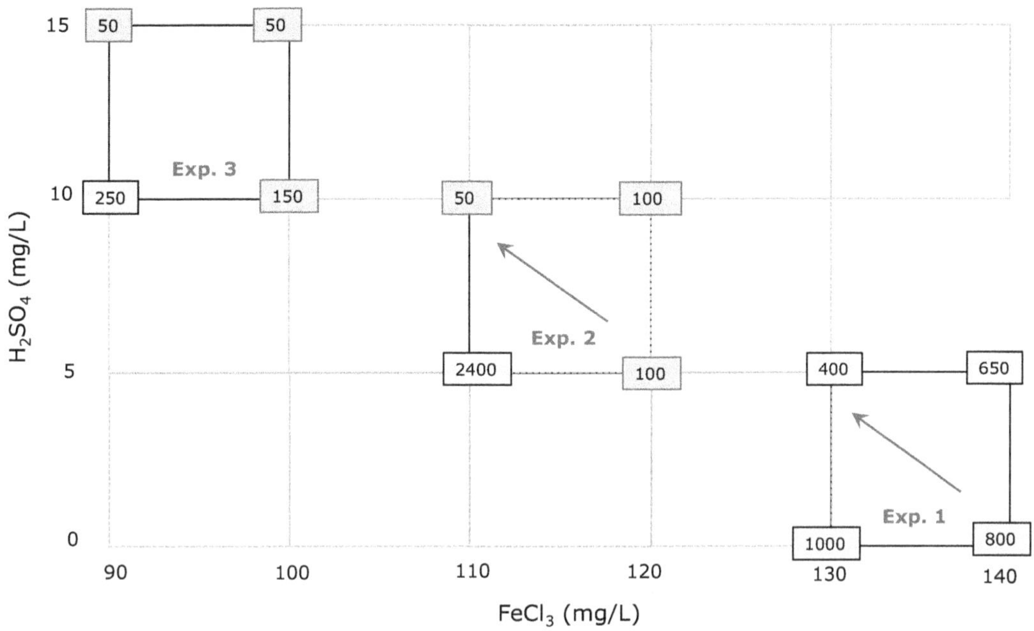

FIGURE 17.9 Initial two-level, two-factor experimental design.

17.7 RESPONSE SURFACE EXPERIMENTS

Response surface methodology is an experimental approach to optimizing the performance of systems for which no detailed process model is known. It is the ultimate application of the iterative approach to experimentation, first demonstrated by Box and Wilson (1951), uses sequential exploratory designs.

The strategy is to explore, analyze what has been learned, design a new experiment, and repeat the learning cycle shown in Figure 17.1. The new experiment may replicate the previous experiment, or move toward more promising conditions.

We visualize the response surface as a hill that will be climbed by stopping periodically to resurvey the most promising path to the summit. At the beginning, the hillside has a relatively smooth upward slope, which we can describe approximately with a smooth surface (a plane or a warped plane). The interesting feature is the steepness of the slope. The shortest path to the top will be the steepest path, and this path can change directions and steepness. The search can stop in two ways: (1) the settings of the factors reach the boundary of feasible operation or (2) an optimum operating condition (a peak) may be located. If a peak is located or suspected, the design can be modified to fit the curvature at the summit.

17.8 CASE STUDY: BIOLOGICAL DEGRADATION OF PHENOL

Wastewater from a coke plant contains phenol, which is known to be biodegradable at low concentrations and inhibitory at high concentrations. A laboratory-scale treatment system was used to determine the operating condition at which the removal rate is a maximum (Hobson & Mills 1990; Berthouex & Brown 2002).

The biological oxidation rate of phenol depends on the concentration of phenol (C) in the reactor and the dilution rate (D). Dilution rate is the reactor volume divided by the flow rate of wastewater through the reactor. Other factors, such as temperature, oxygen level, and pH were held constant.

The iterative approach to experimentation, as embodied in response surface methodology, will be illustrated. The steps in each iteration are *design*, *data collection*, and *data analysis*. Here only design and data analysis are discussed. Each experimental run takes several days to complete, so the experiments were performed in small sequential stages.

The first exploratory experiment was a 2^2 factorial design with dilution rate (D) and phenol concentration (C) as the factors. The response is phenol oxidation rate (R). Each factor was investigated at two levels and the observed phenol removal rates given in Table 17.8. Figure 17.10 shows the first and second exploratory designs.

The four observations from this design were used to fit a regression model that describes a warped plane.

$$R = -0.022 - 0.018C + 0.200D + 0.300CD$$

The direction of steepest ascent toward potentially higher removal rates is indicated by the arrow. The experimenter must decide how far to move in that direction. Published experience indicates that inhibitory effects will probably become evident in the range of 1–2 g/L phenol. This suggests exploring a modest increase of 0.5 g/L in phenol concentration. Thus, the second iteration of testing

TABLE 17.8 Experimental Design and Results for the Exploratory Design

Run	C (g/L)	D (1/h)	R (g/h)
1	0.5	0.14	0.018
2	0.5	0.16	0.025
3	1.0	0.14	0.030
4	1.0	0.16	0.040

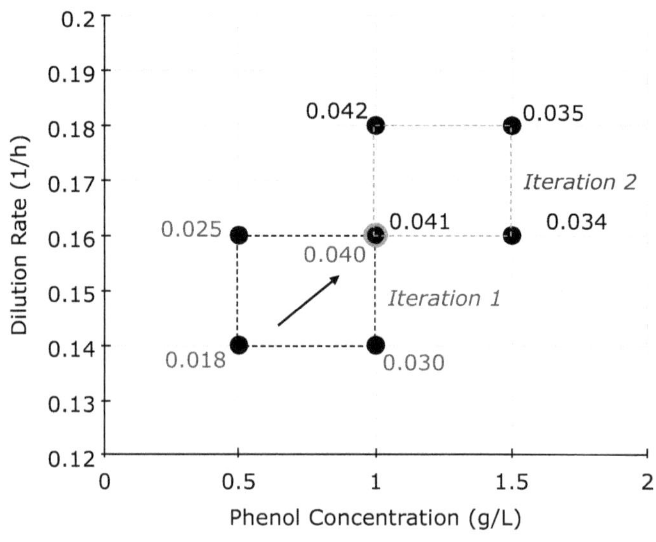

FIGURE 17.10 First and second iterations of 2^2 exploratory designs.

TABLE 17.9 Experimental Design and Results for Iteration 2

Run	C (g/L)	D (1/h)	R (g/h)
1	1.0	0.16	0.041
2	1.0	0.18	0.042
3	1.5	0.16	0.034
4	1.5	0.18	0.035

used $C = 1.0$ and 1.5 mg/L as the low and high settings. Going roughly along the line of steepest ascent, this gives dilution rates of $D = 0.16$/h and 0.18/h as the low and high settings of D. This leads to the second 2^2 factorial experiment shown in Table 17.9. Notice that the setting ($C = 1.0$, $D = 0.16$) is the same in iterations 1 and 2. The observed phenol oxidation rates (0.040 and 0.041 g/h) give us information about the experimental error.

The average performance has improved and two of the response values are larger than the maximum observed in the first iteration. The estimated coefficient for the CD interaction was zero, indicating that the factors C and D act independently in affecting the phenol oxidation rate. The fitted model

$$R = 0.047 - 0.014C + 0.05D$$

describes a plane that is almost horizontal, as indicated by the small coefficients of both C and D.

One way to get a small effect for both variables is if the four corners of the 2^2 experimental design straddle the peak of the response surface. Also, the direction of steepest ascent has changed from iteration 1 to iteration 2. An experimental design is needed that can detect quadratic effects that will define an optimum. Fortunately, the second two-level design can be augmented to do this.

The third iteration of the experiment anticipates fitting a model that contains some quadratic terms, such as

$$R = b_0 + b_1C + b_2D + b_{12}CD + b_{11}C^2 + b_{22}D^2$$

where the b's are model coefficients to be estimated from the data.

TABLE 17.10 Experimental Results for the Third Iteration

Run	C (g/L)	D (1/h)	R (g/h)	Notes
5	1.0	0.16	0.041	Iteration 2 design
6	1.0	0.18	0.042	Iteration 2 design
7	1.5	0.16	0.034	Iteration 2 design
8	1.5	0.18	0.035	Iteration 2 design
9	0.9	0.17	0.038	Augmented "star" point
10	1.25	0.156	0.043	Augmented "star" point
11	1.25	0.17	0.047	Center point
12	1.25	0.184	0.041	Augmented "star" point
13	1.6	0.17	0.026	Augmented "star" point

The second exploratory 2^2 factorial design was augmented by adding "star" points to make a *composite design* (Box 1999). Rather than move the experimental region, use the four points from iteration 2 and add four more in a way that maintains the symmetry of the original design. The augmented design has eight points. Adding one more point at the center of the design will provide a better estimate of the curvature while maintaining the symmetric design. The nine experimental settings and the results are shown in Table 17.10 and Figure 17.11.

These data were fitted to a quadratic model to get

$$R = -0.76 + 0.28C + 7.54D - 0.12C^2 - 22.2D^2$$

The *CD* interaction term had a very small coefficient and was omitted. This model could be used to create a contour plot in the region of the augmented experimental region (Figure 17.12).

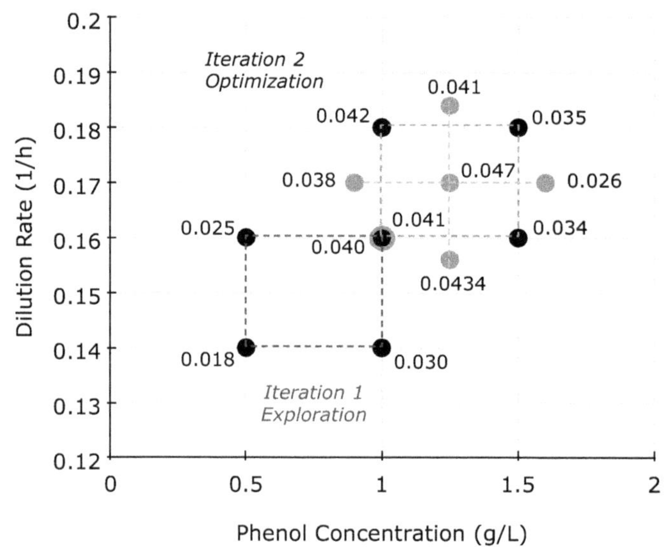

FIGURE 17.11 Recapitulation of the three iterations of experimentation.

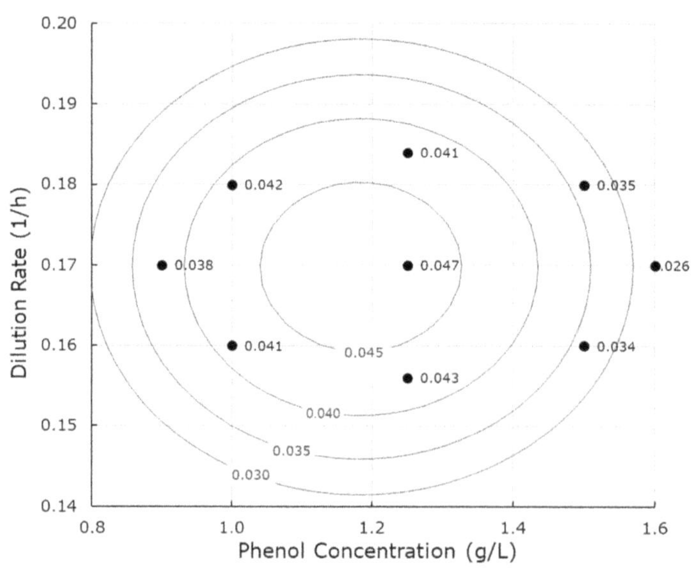

FIGURE 17.12 Experimental data and response surface for the fitted phenol bio-oxidation model.

The maximum predicted phenol oxidation rate is 0.047 g/h at $C = 1.17$ g/L and $D = 0.17$ h^{-1}.

Is a fourth iteration needed? We have learned that the dilution rate should be in the range of 0.16–0.18 h^{-1} and that the process seems to be inhibited if the phenol concentration is higher than about 1.1 or 1.2 mg/L. This was accomplished with only 13 strategic settings of the factor levels. As a practical matter, more precise estimates may not be important. If they are, the experiment could be replicated.

17.9 CASE STUDY: COAGULATION TREATMENT OF LANDFILL LEACHATE

Stabilized leachate from a landfill that is high in chemical oxygen demand (COD), ammonia–nitrogen, color, and alkalinity is discharged to a municipal wastewater treatment plant. The landfill pays a substantial surcharge based on flow and an average COD concentration of 1,925 mg/L. The flow cannot be changed or controlled, so the only way to reduce the surcharge is to remove COD by pretreatment at the landfill. The only feasible treatment is chemical coagulation and settling.

The potential cost savings, or COD reduction, can be estimated before the experiments are done. The current surcharge is \$0.44/kg of COD. At an average daily flow of 1,000 m^3/d and 1,925 mg COD/L, this is 1,925 kg COD/d, or \$847/d and \$309,155/y. Removing 40% of the COD will reduce the surcharge to \$185,493/y, a savings of \$123,662/y. The cost of chemicals must be less than this or pretreatment is not cost-effective.

Some of the COD is particulate, some is colloidal, and some is soluble. The distribution is unknown, but low TSS (80 mg/L) indicates low particulate COD. Experience suggests that 50% or more will be soluble and will not be removed by coagulation. Nitrogen associated with the solids (not measured) will be removed, but soluble nitrogen, mainly in the form of ammonia, will not.

COD removal by coagulation–flocculation with poly-aluminum chloride (PAC) was investigated as a function of pH and PAC dose. PAC works well at near-neutral pH and an ideal result would be to operate within the ambient pH range of 8.2–8.5. The first experiments are at the ambient raw leachate pH of 8.5 and a more neutral level of pH 7.5. The pH is the level before PAC is added.

The budget will allow 20 runs, but management hopes a good experimental strategy will save part of this money. An iterative experimental strategy will minimize the number of tests. The preliminary exploration evaluates two factors, each set at two levels, for a total of $2^2 = 4$ runs.

The test settings and the percent COD removal are shown in Table 17.11 and Figure 17.13. The percentages in the diagrams have been rounded.

The best COD removal was at 2 g PAC/L and pH 7.5. The second 2^2 factorial design, shown in Figure 17.14, was performed at PAC doses of 2 and 2.5 g/L and pH 7.5 and 8.0. The test at pH 7.5 and PAC dose of 2 g/L duplicates a test from the first iteration. Duplicate tests are a good investment because they provide information on the precision of the measurements (magnitude of experimental error).

There is no stable "optimum" in practice because leachate composition and operating conditions change from day to day. What is pleasing to the engineer and the operator is a region of robust operating conditions around a PAC dose of 2 mg/L and a pH of 7.5.

TABLE 17.11 Experimental Design and Results for First Iteration

Run	PAC Dose (g/L)	pH	COD Removal (%)
1	1.0	7.5	29.8
2	2.0	7.5	43.4
3	1.0	8.5	17.8
4	2.0	8.5	30.1

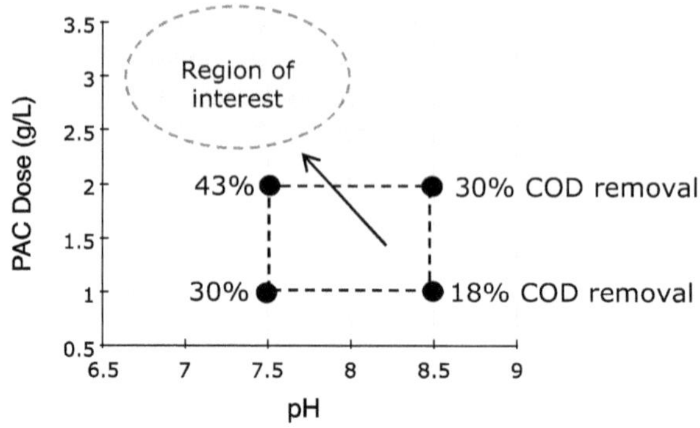

FIGURE 17.13 Percent COD removal for first iteration (values in the diagram have been rounded).

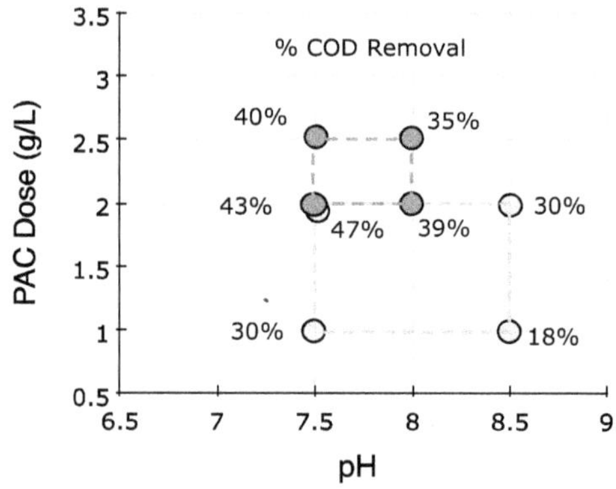

FIGURE 17.14 First and second experimental iterations.

The 2019 price of PAC (powdered form) is about US$0.40/kg. Forty percent COD removal will cost $800/d. This is a dose of 2 kg/m^3 PAC applied to 1,000 m^3/d for a daily PAC consumption of 2,000 kg/d. This is $47/d less than the COD surcharge, not including the cost of the treatment facility and sludge disposal. The cost of pH adjustment, if needed, also needs to be considered. This project is financially marginal.

17.10 THE MISUSE OF RESPONSE SURFACE DESIGNS

Many recent papers that claim to be response surface studies do not use the iterative approach that is the strength of the method. These studies use central composite designs (CCD) similar to the one shown in Figure 17.15.

The CCD is a two-level factorial design augmented with center and "star" points. An augmented 2^3 design has 15 test conditions. The center points are often replicated. The distance from the center to the star points can be calculated in different ways. One is to make every point the same distance from the center points, which is already the case for the eight corners of the factorial design. Another way is to set them two units from the center, as done in this example.

This design is orthogonal and balanced. It was intended to be used to check quadratic effects as the final stage of an experiment (Box and Wilson 1951, Box 1954, Margolin 1985, Box 1999). Too many studies skip the initial exploration and design one experiment to cover a large experimental space, usually with numerous replications of the center point. This is the antithesis of the iterative approach recommended in the previous sections.

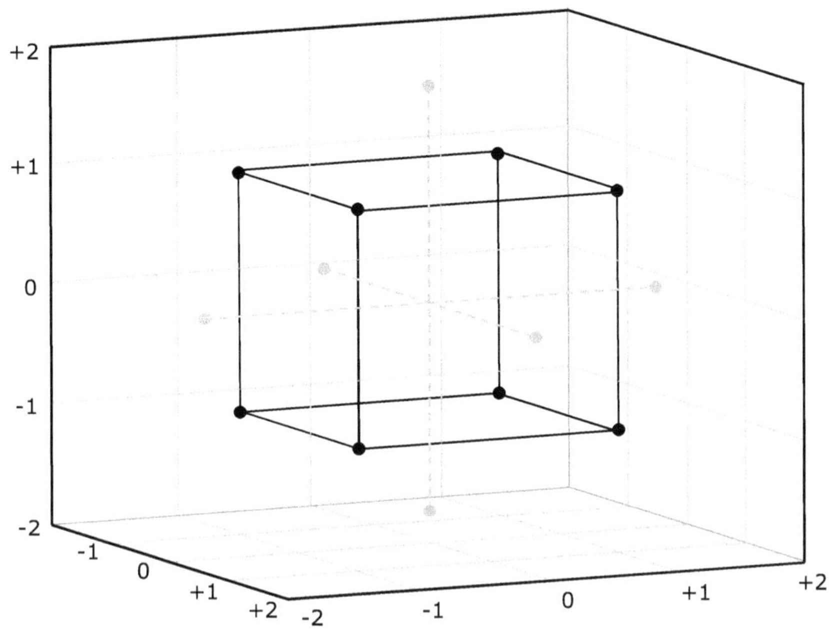

FIGURE 17.15 Central composite experimental design is an augmented two-level factorial design.

The CCD experiment described below was used to determine the best (optimal) settings of three factors for removing COD in printer's ink wastewater. The three factors were coagulant dose (mg/L), flocculant dose (mg/L), and pH. The experimental design shown in Table 17.12 has 23 test runs, including seven replicates of the center point. The coded levels and the actual levels are given.

The average wastewater pH is 7.8, with fluctuations between 7.0 and 8.6. The average COD = 38,595 mg/L (60% soluble), total suspended solids = 4,200 mg/L, and total solids is 27,600 mg/L (60% volatile). All of these fluctuate by up to 50% of the average value. It is not stated whether all tests were done on samples from the same batch of process wastewater, but we shall assume that was the case.

The tests were done in 0.5 L beakers (jars). The coagulant was added, the pH was adjusted, and the flocculant was rapidly mixed into the wastewater for 2 min, followed by 10 min of slow mixing, and 5 min of settling. The COD of the supernatant was measured.

Think about this experiment as an engineer with no knowledge of experimental design. Some questions might be as follows:

- The average COD removal is 97% at the nine center points (pH = 7.5, coagulant = 7500 mg/L, and flocculant = 80 mg/L). This is excellent COD removal. Should one expect that conditions distant from the center will be much better? What is the experimenter looking for?
- It is not possible to adjust the pH to precisely 6.61 and 8.39. The actual settings will be pH $\approx$ 6.6 and $\approx$ 8.4, where the$\approx$ means some error in the settings. A more workable design would control the amount of acid (HCl) or base (NAOH or lime) added and measure pH as a response.
- Is the initial state of knowledge about pH so low that testing pH from 6.61 to 9.0 is necessary? Coagulation–flocculation processes typically are sensitive to changes in pH; shifting the pH by 0.2 pH units can make a dramatic change in the removal of color, which might be important in treating printing ink waste. pH control may be less critical when removing

other pollutants, such as solids. Some preliminary testing would be worthwhile to narrow the range.

- Committing to 23 runs without some preliminary testing is not an efficient experimental strategy.
- The center point is replicated nine times. Replication is always beneficial, and replication at other settings (not all at the center) would provide more information about the experimental error.
- The standard deviation of the nine center point CODs observations is 32 mg/L or 3.3%. The COD test typically has a standard deviation of 10% of the measured value due to random measurement errors, so the reported precision of the replicate measurements is suspicious. They may not be true replicates.
- Furthermore, the variances of the calculated percent removal will be greater than the variance of the individual measurements. This is shown in Chapter 18.
- An interesting observation is that runs 1–4, 6–9, and 10 had essentially the same COD removal (about 96%) as the center points.

It is always helpful to plot the data before doing any statistical calculations. Figure 17.16 shows percent COD removed as a function of the three independent variables. pH shows no effect in the range of 6.5–8.5. COD removal is 95% or more except for five runs that are at extreme settings (pH 6 or 9) or tests with low coagulant dose.

TABLE 17.12 Response Surface Design for Treatment of Printer's Ink Wastewater by Coagulation–Flocculation

Run	X_1	X_2	X_3	Coagulant (mg/L)	Flocculant (mg/L)	pH	% COD Removal	Final COD (mg/L)
1	−1	−1	−1	6,910	70	6.61	96.9	994
2	1	−1	−1	8,090	70	6.61	97.1	917
3	−1	+1	−1	6,910	90	6.61	96.8	1,033
4	+1	+1	−1	8,090	90	6.61	96.2	1,265
5	−1	−1	1	6,910	70	8.39	88.5	4,236
6	+1	−1	1	8,090	70	8.39	96.2	1,265
7	−1	1	1	6,910	90	8.39	83.9	6,012
8	+1	+1	+1	8,090	90	8.39	96.9	994
9	−2	0	0	6,510	80	7.5	95.5	1,535
10	+2	0	0	8,490	80	7.5	96.0	1,342
11	0	−2	0	7,500	60	7.5	90.3	3,542
12	0	+2	0	7,500	90	7.5	95.3	1,612
13	0	0	−2	7,500	80	6.0	86.0	5,201
14	0	0	2	7,500	80	9.0	77.1	8,636
15	0	0	0	7,500	80	7.5	96.9	994
16	0	0	0	7,500	80	7.5	97.1	917
17	0	0	0	7,500	80	7.5	97.0	956
18	0	0	0	7,500	80	7.5	96.9	994
19	0	0	0	7,500	80	7.5	97.0	956
20	0	0	0	7,500	80	7.5	97.1	917
21	0	0	0	7,500	80	7.5	96.9	994
22	0	0	0	7,500	80	7.5	96.9	994
23	0	0	0	7,500	80	7.5	97.0	956

Final COD was calculated from an assumed initial COD = 38,595 mg/L.

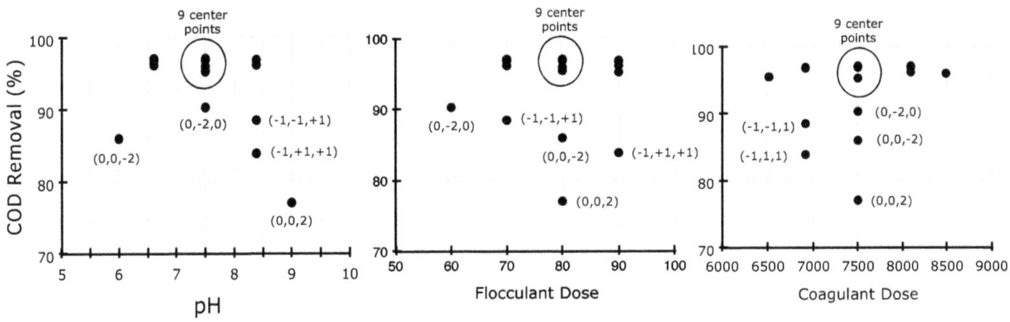

FIGURE 17.16 Results of the central composite design plotted as a function of the three independent variables.

The curvature in the response surface comes from the extreme conditions. These runs are not helpful for modeling the behavior at the region of best performance. There is some curvature for coagulant dose, but this is uninteresting because COD removal decreases on both sides of the central dose of 7,500 mg/L. The plots show no curvature for flocculant and pH within the range of best performance. There is an interaction between coagulant and flocculant. Designing a large experiment to fit a quadratic model with no preliminary exploration was inefficient.

The reported result was a quadratic model with ten parameters (4 main effects, 3 two-factor interactions, and 3 quadratic terms).

$$y = 96.88 + 1.54X_1 + 0.25X_2 - 2.67X_3 + 0.48X_1^2 - 0.55X_2^2 - 4.53X_3^2$$
$$+0.56X_1X_2 + 2.63X_1X_3 - 0.36X_2X_3$$

Every term was reported as statistically significant. Finding ten significant parameters in a model is highly unusual. This is an artifact of using nine COD measurements at the center point to estimate the experimental error. The percent COD removals at the center points are 97.1, 97.0, 96.9, 97.0, 97.1, 96.9, 96.9, and 97.0. The lack of variation in the replicates is remarkable. The high and low values are 97.1% and 96.9%. The average of the replicates is 96.99, with standard deviation 0.08997. Using this as an estimate of the experimental error will make all terms in the model, even very small ones, appear to be statistically significant.

One suspects that the center points are not true replicates. They may be repeated COD measurements on a single run. A true replicate is an entirely independent setup of the test, so it includes all the components of variance, from measuring the volume of the wastewater and the chemical additions to adjusting the pH, controlling mixing times, collecting the sample of treated wastewater, and measuring the COD. Every step introduces random measurement errors and collectively they should accumulate to more variance than is seen in the nine center points.

There are two major weaknesses in this experiment. One is asking the design to cover too large an experimental space. Setting the factor values so widely apart will make every term in the fitted model seem to be statistically significant, especially if all replicate runs are concentrated at a single experimental condition. This results in overfitting the model and is not helpful. It provides a poor description of the optimal region because the model is distorted to fit the extreme data points, which are of no interest. For example, the fitted model predicts a COD removal in excess of 100% for test run 10.

The second weakness is the premature commitment of all resources to one experiment. The greater the cost of each test run, the greater the reward of some preliminary exploration. Another consideration is that many engineering experiments have to run over an extended period of time. Even if the experiments are simple jar tests, most experimenters cannot do more than four to six runs in a half day. The time per run will be extended if physical equipment needs to be altered or adjusted between runs. To do 23 runs will take several days. Reagents, temperature, and other tests conditions change from day to day. It is better to organize the experiment into smaller blocks, so conditions are similar within each block.

What could have been learned with a 2^3 experiment? Table 17.13 and Figure 17.16 show the eight-run 2^3 experiment extracted from the larger central composite design. The COD removals in runs 5 and 7, at the high pH and low coagulant settings, stand out as being oddly low. Figure 17.17 shows that changing the pH had no effect at the high coagulant dose.

Figure 17.18 shows the main effects and two-factor interactions for the 2^3 design. Flocculant dose has no effect. The main effect of coagulant dose (C) is 5.08% COD removal. The main effect of pH is–5.38% COD removal. There is a strong positive interaction between coagulant dose and pH because the COD removal efficiency decreases by about 10% when coagulant is low and pH is high. The fitted model from this 2^3 experiment is

$$\%\text{COD Removal} = 94.06 + 5.08X_1 - 5.38X_3 + 5.28X_1X_3$$

TABLE 17.13 Two-Level Three-Factor (2^3) Design

Run	Coded Factors			Factors			
	X_1	X_2	X_3	Coagulant (mg/L)	Flocculant (mg/L)	pH	% COD Removal
1	−1	−1	−1	6,910	70	6.61	96.9
2	+1	−1	−1	8,090	70	6.61	97.1
3	−1	+1	−1	6,910	90	6.61	96.8
4	+1	+1	−1	8,090	90	6.61	96.2
5	−1	−1	+1	6,910	70	8.39	88.5
6	+1	−1	+1	8,090	70	8.39	96.2
7	−1	+1	+1	6,910	90	8.39	83.9
8	+1	+1	+1	6,090	90	8.39	96.9

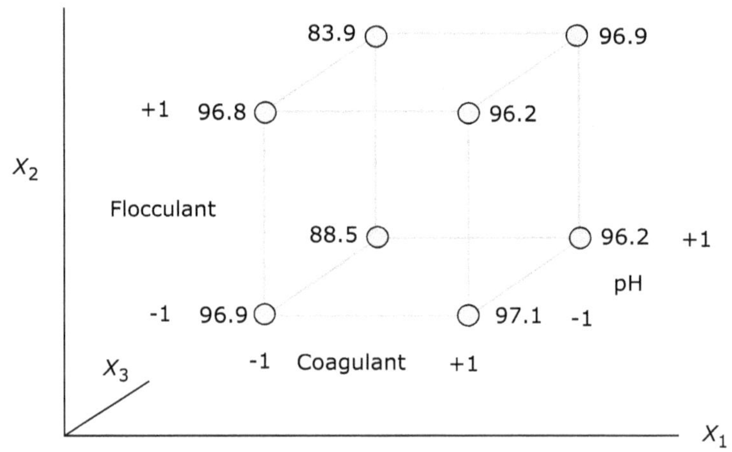

FIGURE 17.17 Interpretation of a 2^3 design extracted from the central composite design.

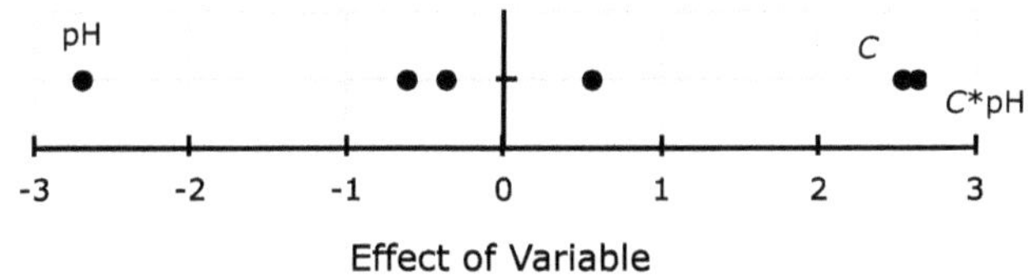

FIGURE 17.18 Main effects and two-factor interactions for removing printer's ink COD.

This is a case where replication would be helpful. If the study had started with these eight runs, the investigator could have checked the measurement process and improved the experiment.

If runs 5 and 7 are correct, the experiment should drop flocculant as a factor and focus on the region centered on pH 7.0 or 7.5 and a coagulant dose of 7,500 mg/L. The range of the variables should be reduced. A center point might be added.

17.11 CONCLUSIONS

The iterative experimental approach using factorial experimental designs is an efficient way to study engineering problems when the important factors need to be identified, the effects of changing the factor settings needs to be quantified, the interactions between factors need to be investigated, and when we can work with a smooth approximation of the response surface.

Response surface methodology should have a natural appeal to environmental engineers because their experiments often take a long time to complete and are conducted a few at a time. For these reasons, it is attractive to use a few early results to guide the design of additional experiments.

Response surface methods are not designed to faithfully describe large regions of the possible experimental space. This is acceptable, and perhaps desirable, because large parts of the experimental space are not near the optimum process conditions and testing in those regions is wasteful. The goal is to explore and to locate the promising regions as efficiently as possible. Early tests – the exploration phase – are used to approximate the response surface with a simple model and use that model to locate the direction of greatest improvement. If the experiments suggest a peak in the response surface, the experimental design can be adapted to explore that possibility and a smooth quadratic model fitted to the data.

17.12 PROBLEMS

TUTORIAL NOTE

The first three problems can be assigned when study of this chapter begins, but should not be due until some progress has been made through the chapter. Problem 17.1 is designed as an in-class game which will take about 15 minutes. Problem 17.2 is suited for a short class discussion. Problem 17.3 should be assigned when study of this chapter begins, but is due after the material has been studied.

17.1 EXPERIMENTAL DESIGN TO OPTIMIZE CHEMICAL TREATMENT

This class exercise is a fun way to start Chapter 17. It will take 15–20 min.

Students will work in teams of three. Two team members will be process engineers who set the experimental conditions and the third is the "chemist" who produces test results from a table that is provided in the study guide. Only the chemist sees the table until the process engineers terminate their testing.

Background of the project: Free oil can be removed from wastewater by gravity flotation or dissolved air flotation. Emulsified oil does not separate unless the emulsion is broken, which can be done by adding chemicals. There are reasons to believe that effective chemical doses are in the range of 0.1–0.4 g/L of acid and 0.6–1.4 g/L of $FeCl_3$. Your task is to locate the best treatment conditions. You do this by proposing a test condition, for example, acid = 0.2 g/L and $FeCl_3$ = 0.8 g/L. A run is one test condition. An experiment is a collection of runs in any number and arrangement you specify. The team that finds the best treatment conditions with the fewest runs wins the contract to design the process.

There are six rules: (1) The team may discuss their strategy before testing starts. (2) One member of the team is designated the "chemist." The chemist is given the table of data which the process engineers are not allowed to see. The data are the oil concentration (mg/L) remaining in the wastewater after treatment has been applied. (3) The process engineers may freely discuss the experimental design and the results obtained. (4) All communication between the chemist and the team is in writing. The experimental settings are given to the chemist in writing and the test results are reported back in writing. (5) When your team has finished testing, write down your "best" treatment conditions and the oil concentration achieved at that condition and give the solution to the instructor. (6) Violation of any rule causes disqualification.

17.2 LIMITATIONS OF OFAT EXPERIMENTS

One-factor-at-a-time (OFAT) experimentation is often used and taught in lab courses, even though it is an inefficient way to perform engineering experiments. Identify an experiment that you may have performed that used this method, or take an example in the professional literature, and explain how it would be improved by varying more than one factor at a time. Explain the limitations of OFAT. Explain what is to gain by varying factors simultaneously.

17.3 GUIDELINES FOR A GOOD EXPERIMENTAL DESIGN

You have been assigned to give a talk about the design of experiments to a group of managers who will not enjoy or understand mathematical explanations. Assume they may have heard about design of experiments (DOE), but have no clear notion of what it is or how it is done. Outline the main concepts of doing efficient experiments. Explain how an engineer should plan an experiment. Try to formulate some general principles and useful guidelines.

17.4 OPTIMIZE EFFICIENCY OF A CO_2 SCRUBBER 1

Industrial scrubbers absorb air pollutants such as SO_2 or CO_2 from flue gas. The flow of liquid and gas is countercurrent. Water enters at the top of the column and flows down over a packing material. The CO_2-laden gas is blown up from the bottom. The packing creates a large surface area for contact between the liquid and the gas. Adding caustic (NaOH) to the water will change the pH and also the absorption of $CO_{2.}$ The CO_2 is converted to carbonate which is insoluble at the high operating pH of the process.

A laboratory absorption column is available for students to do some testing. A 9-cm diameter column that is 1.7 m tall contains a depth of 0.55 m packing material. The packing used provided 21.3 m^2 of surface area per cubic meter of column volume. An air–CO_2 mixture was supplied at a constant flow rate. The CO_2 was 10% by volume (compared to a typical value of 16% CO_2 in dry exhaust gas from coal-fired power plants). The goal is a low volume percent of CO_2 in the exit.

Table P17.4 describes a 2^2 factorial experiment that evaluated the effects of water flow rate (F) and the caustic concentration in the scrubbing liquid (C) on the CO_2 concentration of the exiting

TABLE P17.4 Data from CO_2 Scrubbing Experiments (Smart, JL, University of Kentucky)

	Actual Values of Factors		Coded Factor Values		Response
Run	*F* Flow Rate (L/min)	*C* Caustic (mass %)	X_1	X_2	CO_2 in Exit Gas (vol%)
1	2.5	0	−1	−1	9.87
2	4.5	0	+1	−1	8.97
3	2.5	1	−1	+1	7.45
4	4.5	1	+1	+1	7.28

gas. The type of packing to use is an important design decisions, but all tests in this experiment use the same packing material.

(a) Interpret the test results. Which settings give the best results? Calculate the main and interaction effects of the factors.
(b) If you were to do another set of four experiments to find values of *F* and *C* that give better results (lower CO_2 responses), which would you use? Why?

17.5 OPTIMIZE EFFICIENCY OF A CO_2 SCRUBBER 2

The wet scrubber column used in experiment described in Problem 17.4 contained packing material of type P1. The liquid trickles over the packing and this increases the exposed surface area between the air and the liquid. The experiment in Table P17.5 adds a second type of packing expanding the design to make a 2^3 factorial. Each experiment was replicated four times and the response is the average of the four replicates.

(a) Display the data in a box plot. Which settings give the best results?
(b) Calculate the main and interaction effects of the factors. Interpret the results.

TABLE P17.5 Results of Testing of Eight Conditions, Replicated Four Times (Response Is the Average of Four Runs at Each Condition)

	Actual Values of Factors			Coded Factors			Response
Run	*A* Flow Rate (L/min)	*B* Caustic (mass %)	*C* Packing Type	X_1	X_2	X_3	CO_2 in Exit Gas (vol%)
1	2.5	0	P1	−1	−1	−1	9.87
2	4.5	0	P1	1	−1	−1	8.97
3	2.5	1	P1	−1	1	−1	7.45
4	4.5	1	P1	1	1	−1	7.28
5	2.5	0	P2	−1	−1	1	9.90
6	4.5	0	P2	1	−1	1	9.93
7	2.5	1	P2	−1	1	1	6.10
8	4.5	1	P2	1	1	1	5.18

17.6 STORMWATER SCREENING

The Portland Sewerage District was testing a rotating screen (Figure P17.6) to treat storm sewer overflows. The planned project duration was 18 months. At 14 months, no useful conclusions had been reached. Progress was made after the experimental approach was changed to use factorial experimental designs. One design is shown in Table P17.6a.

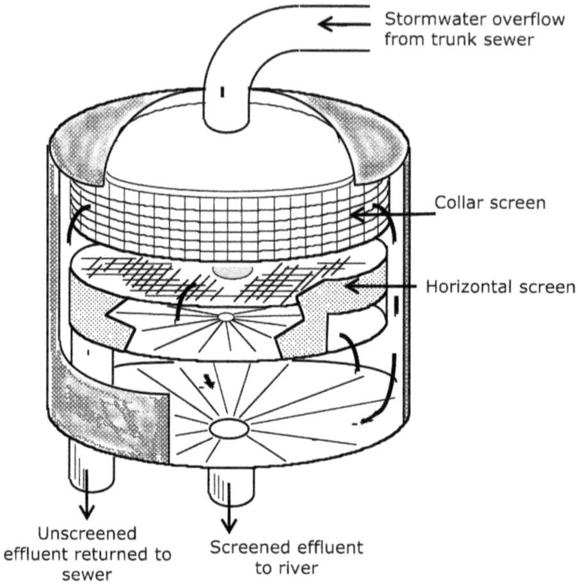

FIGURE P17.6 Stormwater screen with horizontal and collar screens.

TABLE P17.6a Experimental Design with Three Factors, Each Set at Two Levels (Marske 1970)

Factor	Symbol	Settings	
Size opening of horizontal screen	*H*	fine = 110 mm	coarse = 175 mm
Size opening of collar screen	*C*	fine = 110 mm	coarse = 175 mm
Flow rate	*F*	low = 700 gpm	high = 1,200 gpm

TABLE P17.6b Two-Level Experimental Design for Improving a Stormwater Screen

Factors			Factors			Coded Factors			% Solids
C	*H*	*F*	*C*	*H*	*F*	X_1	X_2	X_3	Removed
Fine	Fine	Low	110	110	700	−1	−1	−1	81
Coarse	Fine	Low	175	110	700	1	−1	−1	92
Fine	Coarse	Low	110	175	700	−1	1	−1	79
Coarse	Coarse	Low	175	175	700	1	1	−1	75
Fine	Fine	High	110	110	1,200	−1	−1	1	85
Coarse	Fine	High	175	110	1,200	1	−1	1	93
Fine	Coarse	High	110	175	1,200	−1	1	1	86
Coarse	Coarse	High	175	175	1,200	1	1	1	62

There are eight possible combinations of the high and low settings of three factors. Table P17.6b shows three ways these can be identified. The coded factors are a convenient way to show the two levels as either high (+) or low –). This is a generic description that is independent of the actual levels that are assigned to the factors.

Analyze these data and make recommendations for improving the design of the rotating screen.

17.7 OPTIMIZING METHANE PRODUCTION

An experiment was conducted to optimize operating conditions for methane production (ft^3/lb feed VS) in anaerobic digesters using a factorial design approach. Three operating parameters were of interest: (1) total solids loadings (TS), (2) mean cell residence time (MCRT), and (3) mixing intensity (MI). The design and data are in Table P17.7. The responses are the averages of those measurements. In each experiment, the digester was allowed to come to equilibrium and the response was measured three to five times over a period of about a week.

(a) Display the data in a box plot. Which settings give the best results?
(b) Calculate the main and interaction effects of the factors. Interpret the results.

TABLE P17.7 Anaerobic Digester Design Data (Park C, et. al., 2009 *Residuals and Biosolids*, pp, 350–361.)

	Actual Factor Levels			Coded Factor Level			Response
Run	Feed TS (%)	MCRT (d)	Mixing Intensity	X_1	X_2	X_3	Methane Production (ft^3/lb Feed VS)
1	3.5	15	low	–1	–1	–1	6.65
2	7	15	low	1	–1	–1	6.04
3	3.5	25	low	–1	1	–1	6.22
4	7	25	low	1	1	–1	5.89
5	3.5	15	high	–1	–1	1	6.39
6	7	15	high	1	–1	1	5.47
7	3.5	25	high	–1	1	1	6.86
8	7	25	high	1	1	1	6.56

17.8 FISH CANNING WASTEWATER

Biological treatment was applied to previously settled fish canning wastewater to evaluate the organic matter removal efficiency by the activated sludge process. A sample of suspended biomass from a municipal wastewater treatment plant was used as inoculum (Cristóvão 2015).

A 3^2 factorial design was used to analyze the influence of the hydraulic retention time (HRT) and the feed stream dissolved organic carbon (DOC) on the removal of DOC. The operating ranges for each factor were typical for this type of biological treatment. The wastewater DOC value is highly variable, with a minimum value of 153 mg/L, a maximum value of 984 mg/L, and an average of 500 mg/L. Based on this, upper and lower limits of 800 and 200 mg/L were chosen in order to encompass the majority (74%) of the original values recorded. This way, the initial DOC range was established as 200–800 mg/L and the HRT between 4.2 and 8.1 h. Each independent variable was coded at three levels between–1 (low level), 0 (middle point), and +1 (high level).

Fit an empirical model to the data in Table P17.8. Use the actual values of the factors, not the coded values. Under what conditions is the response optimal?

TABLE P17.8 Factorial Design Matrix (3^2 Experiment) with Experimental Results for Fish Canning Wastewater DOC Removal by Activated Sludge

Run	Factors		Coded Factors		Response
	HRT (h)	DOC (mg/L)	X_1	X_2	DOC Removal (%)
1	4.20	200	−1	−1	39.7
2	4.20	500	−1	0	38.7
3	4.20	800	−1	1	31.5
4	6.15	200	0	−1	82.2
5	6.15	500	0	0	86.2
6	6.15	800	0	1	73.3
7	8.10	200	1	−1	57.2
8	8.10	500	1	0	64.9
9	8.10	800	1	1	42.3

17.9 EXPLORATION OF A BIOLOGICAL PROCESS

The biological oxidation rate of a compound depends on its concentration (C) in the reactor and the dilution rate (D). Dilution rate is the reactor volume divided by the flow rate of wastewater through the reactor.

The first exploratory experiment was a 2^2 factorial design with concentration (C) and dilution rate (D) as the factors. Other factors, such as temperature, oxygen level, and pH were held constant. The response was the reaction rate (R). Each factor was investigated at two levels and the observed oxidation rates are given in Table P17.9a.

(a) Fit the data by least squares and determine the path of steepest ascent.
(b) Plan a second iteration of experiments to explore the region where higher reaction rates might be expected.
(c) The second cycle of experiments that actually was done gave the results in Table P17.9b. The location of the experiments and the direction moved from the first-cycle design may be different than you proposed in part (b). Note this design has one point in common with the initial design ($C = 1.6$, $D = 0.4$). Plot and interpret the data. What would you suggest for the third experimental iteration?
(d) The results of the second iteration experiment showed a decrease in reaction rate at high concentrations. A third iteration was designed to capture the curvature in the response surface. It consisted of adding the center and star points to the second iteration design. The results are in Table P17.9c. Fit a full quadratic model to the data from iterations 2 and 3, and estimate the location of the maximum reaction rate.

TABLE P17.9a Exploration Phase

C (g/L)	D (1/h)	R (g/h)
0.8	0.3	0.10
1.6	0.3	0.17
0.8	0.4	0.14
1.6	0.4	0.22

TABLE P17.9b Second Exploratory Iteration

C (g/L)	D (1/h)	R (g/h)
1.6	0.4	0.22
2.4	0.4	0.19
1.6	0.5	0.23
2.4	0.5	0.20

TABLE P17.9c Optimization Phase

C (g/L)	D (1/h)	R (g/h)	Notes
1.4	0.45	0.21	Augmented "star" point
2.0	0.38	0.24	Augmented "star" point
2.0	0.45	0.26	Center point
2.0	0.52	0.23	Augmented "star" point
2.6	0.45	0.14	Augmented "star" point

17.10 SLUDGE CONDITIONING

Polymer (P) and fly ash (F) were mixed with sludge to improve the performance of a dewatering filter. Performance was measured as the yield of the dewatered sludge, Y (kg/m^2-h). The first cycle of experimentation used a 2^2 factorial design with three replicated points near the center of the design. The design and data are in Table P17.10a.

(a) Fit the data by least squares and determine the path of steepest ascent.
(b) Plan a second cycle of experiments, assuming that second-order effects might be important.
(c) The second cycle of experiments that actually was done gave the results in Table P17.10b. The location of the experiments and the direction moved from the first-cycle design may be different than you proposed in part (a). Fit a full quadratic model to the data and find the location of the maximum yield.

TABLE P17.10a Exploratory Design

Polymer (P) (g/L)	Fly Ash (F) (% by wt)	Yield (Y) (kg/m^2-h)
40	114	18
40	176	24
50	132	21
50	132	23
50	132	17
60	114	18
60	176	35

Note: The "center points" ($P = 50$, $F = 132$) are not exactly centered in the experimental space.

TABLE P17.10b Composite Design

P (g/L)	*F* (% by wt.)	Yield (kg/m²-h)	Notes
55	140	29	Second iteration 2^2
55	160	100	Second iteration 2^2
65	140	105	Second iteration 2^2
65	160	108	Second iteration 2^2
60	150	120	Center point
60	150	121	Center point
60	150	118	Center point
60	150	120	Center point
60	150	118	Center point
53	150	77	Star point
60	165	99	Star point
60	135	102	Star point
67	150	97	Star point

17.11 PRINT INK WASTEWATER

The manager of a printing plant is searching for a pollution control treatment method. The average wastewater pH is 7.8, with fluctuations between 7.0 and 8.6. The average COD = 40,000 mg/L (90% soluble), total suspended solids = 2,000 mg/L (70% volatile), and total solids = 28,000 mg/L (60% volatile). The normal color of the wastewater is black, but at times it has other hues.

The manager believes that COD removal is the issue. He has an interest in designed experiments and proposes to copy the augmented two-level three-factor factorial design (a central composite design) described in Section 17.9. On that test 90+% of COD removal was achieved by coagulation–flocculation. A process engineer who believes that color removal is more important than COD removal has discovered Figure P17.11. The data are from a different printing company and the engineer does not expect the same result at his company, but the abrupt change in color removal over a narrow range of pH needs to be investigated.

Discuss how these data should influence the design of experiments. Propose a design, or a series of designs, that you believe would be efficient.

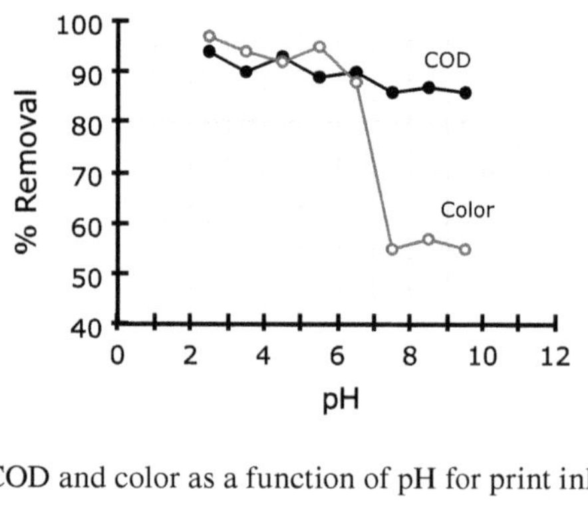

FIGURE P17.11 Removal of COD and color as a function of pH for print ink wastewater.

17.12 ARSENIC REMOVAL FROM WATER

Arsenic (As) can cause skin lesions, cancer, cardiovascular, respiratory, and neurological disorders. This is a serious risk in West Bengal, India, and Bangladesh where many natural water sources contain 20–3,000 μg/L of arsenic. The WHO and the US EPA limit is 10 μg/L, which gives an estimated excess lifetime risk of skin cancer of 6×10^4.

The characteristics of the tap water used in the experiments are given in Table P17.12a. This is excellent quality water except for the Arsenic(V), or As^{5+}. As(V) species are negatively charged in the pH range of 4–10 and can be adsorbed or coprecipitated with other solids that are formed by coagulation. The solids can be removed by filtration.

Arsenic removal was investigated using a three-factor Box–Behnken experimental design, as shown in Figure P17.12. The Box–Behnken design is orthogonal and balanced. It can capture quadratic effects (because it has three levels). A full 3^3 factorial design has 27 runs, or 28 if a center point is added. The basic Box–Behnken design has 12 runs, plus center points.

This study used a Box–Behnken Design with 17 runs, including 5 replicates at the center point. (*Note:* This design was not explained in the text, but it is an efficient orthogonal design. There is no need to critique or comment on the arrangement of the geometry of runs on the runs, but you are expected to comment on the settings of the factor levels.)

TABLE P17.12a Tap Water Characteristics

Water Quality Variable	Concentration
pH	7.9
Arsenic(V) (μg/L)	500
Turbidity (NTU)	0.1
Chloride (mg/L)	44
Nitrate (mg/L)	4
Iron (mg/L)	0.040
Aluminum (mg/L)	0.012
Manganese (mg/L)	0.001
Sodium (mg/L)	22.8
Conductivity (μS)	462

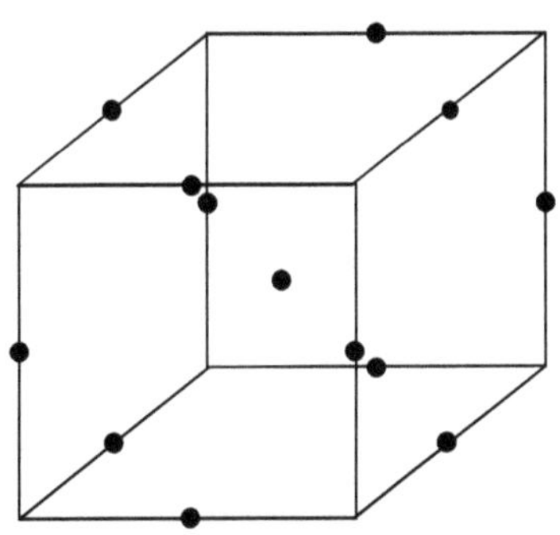

FIGURE P17.12 Box–Behnken design in three factors with 13 runs, including a center point. The runs are located at the mid-range values and not at the corners as is the case with a two-level factorial design.

The three factors were initial As(V) concentration, coagulant dose, and pH. The design matrix with the experimental results is given in Table P17.12b. The stated goal was to maximize arsenic removal.

Evaluate the experimental design as an engineer. No statistical calculations are needed. You may wish to do some reading about arsenic removal.

TABLE P17.12b Box–Behnken Design with 17 runs, Including 5 Center Points

	Coded Factors			Actual Factors			Response
Run	Initial As(V) X_1	Coagulant dose X_2	pH X_3	Initial As(V) (μg/L)	Coagulant (mg/L)	pH	As(V) Removal (%)
1	1	0	−1	1,000	40.25	4.0	37.30
2	−1	1	0	10	80	6.5	90.00
3	0	0	0	505	40.25	6.5	95.25
4	1	0	1	1,000	40.25	9.0	84.40
5	0	0	0	505	40.25	6.5	95.45
6	0	0	0	505	40.25	6.5	95.64
7	0	0	0	505	40.25	6.5	94.65
8	1	−1	0	1,000	0.5	6.5	14.70
9	0	−1	1	505	0.5	9	13.07
10	−1	0	1	10	40.25	9	70.00
11	0	0	0	505	40.25	6.5	96.44
12	−1	−1	0	10	0.5	6.5	0.00
13	0	−1	−1	505	0.5	4	11.68
14	−1	0	−1	10	40.25	4	30.00
15	0	1	1	505	80	9	96.63
16	0	1	−1	505	80	4	37.03
17	1	1	0	1,000	80	6.5	99.10

18 Designing under Uncertainty

Uncertainty is an uncomfortable position but certainty is an absurd one – Voltaire

Uncertainty is caused by a lack of information and data, having to use imperfect or approximate models, or not having precise values for model parameters. There are limits on uncertainty; the flow rate for next month is not known and not knowable, but it can be forecasted within workable limits, and this allows plans to be made. Using safety factors in design provides a hedge against uncertainty.

Figure 18.1 shows how the effluent concentration from a treatment process depends on the magnitude of the reaction rate coefficient, k. This is a simple example of how one can look at the effect of uncertainty and shift the design to accommodate the uncertainty. The construction cost of the process is proportional to the hydraulic detention time t. There is uncertainty in k. It is known to be within the range of 0.25–0.35, with a most likely value of 0.27, as shown by the triangle. The effluent concentration must be less than 20, so a detention time of t = 15 is satisfactory so long as $k \geq 0.27$. To be sure of 20 mg/L or less requires a detention time longer than 15. To provide a margin for error the design might use t = 17 or 18. Trying to save money by designing a smaller tank increases the risk of violating the effluent limit. Economy of scale favors the safer design. If the uncertainty can be reduced, perhaps by running some tests, the design can be modified.

Variability refers to changes in physical phenomena and processes over time or space. Temperatures, atmospheric pressure, mass and volumetric flow rates, and concentrations change. Short-term variation can be handled with process control, blending, and equalization. Long-term variability, such as peak flows, are handled in design by applying peaking factors.

We will not deal with catastrophic events, such as floods, hurricanes, and tsunamis.

18.1 THE EXPECTED VALUE CRITERION

The *expected value* of a random variable X is the weighted average of the possible values of X, where the weights are the respective probabilities $P(X)$.

$$E(X) = P(X)X$$

The possible outcomes X and probabilities $P(X)$ are subjective estimates (or guesses) made by someone who has experience or special knowledge of the system can make with the information at hand.

Some problems have a few discrete outcomes and probabilities. Others that have continuous distributions come later in this chapter.

Example 18.1 Expected Payoff

Based on available information, it is expected that an investment will earn at least \$40,000 but not more than \$80,000, with \$50,000 being the most likely value. The probabilities, which must add to 100%, for the three possible outcomes are

X_1 = Payoff = \$80,000 Probability = $P(X_1)$ =15%
X_2 = Payoff = \$40,000 Probability = $P(X_2)$ = 25%
X_3 = Payoff = \$50,000 Probability = $P(X_3)$ = 60%

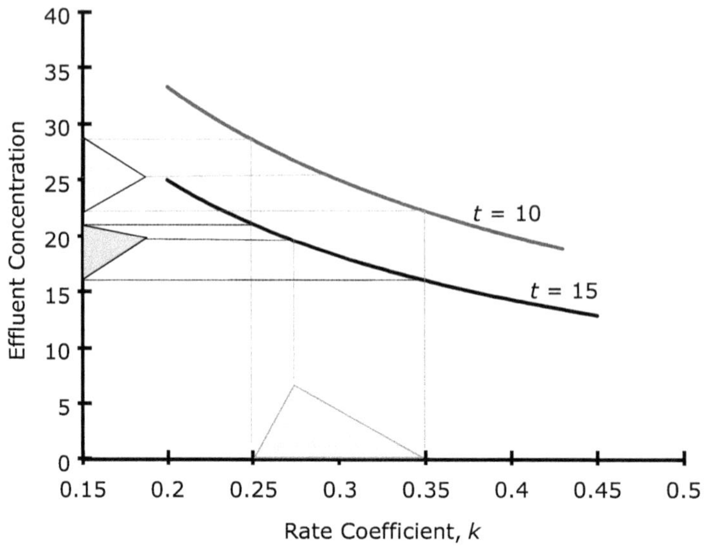

FIGURE 18.1 Accommodating uncertainty in design.

The possible outcomes are not equally likely, so they must be weighted by the probability that each will occur. The expected payoff is

$$E(\text{Payoff}) = P(X_1)X_1 + P(X_2)X_2 + P(X_3)X_3$$
$$= 0.15(\$80{,}000) + 0.25(\$40{,}000) + 0.6(\$50{,}000)$$
$$= \$12{,}000 + \$10{,}000 + \$30{,}000 = \$52{,}000$$

Example 18.2 Payoff for Multiple Events

Two projects might be undertaken with uncertain costs. There are only two possible outcomes for each project, as given in Table 18.1

$$E(\text{Project A}) = 0.4(\$2{,}000{,}000) + 0.6(\$3{,}000{,}000) = \$2{,}600{,}000$$
$$E(\text{Project B}) = 0.3(\$3{,}800{,}000) + 0.7(\$2{,}500{,}000) = \$2{,}890{,}000$$

The expected value analysis favors project B.
The best possible outcome of $3,800,000 also favors project B.
The worst outcome of project B is only 4% less than the expected payoff for project A.

TABLE 18.1 Project Outcomes and Probabilities

Project	Outcome	Probability
A	A_1 = \$2,000,000	$P(A_1) = 40\%$
	A_2 = \$3,000,000	$P(A_2) = 60\%$
B	B_1 = \$3,800,000	$P(B_1) = 30\%$
	B_2 = \$2,500,000	$P(B_2) = 70\%$

Example 18.3 Risk of a Hazardous Waste Leak

Four years ago, a special sealer was applied to the floor of the hazardous waste storage building to prevent spills from seeping into the concrete floor and through the floor into the ground. The contractor gave a five-year warranty on the sealant protection, and that warranty expires at the end of this year. The question is whether to replace the sealer or to gamble on having to clean up a spill.

The appearance of the floor suggests that the integrity of the sealer may be compromised. The cost for a new application of the sealer at the end of the year, which gets a new five-year warranty, is \$187,000.

If the floor is not resealed with a new warranty, any spill damage or cleanup will be the financial liability of the company. If the floor sealer fails, cleanup would require tearing out and replacing the floor of the building, and excavating and disposing of the contaminated soil, at a cost of \$350,000.

From past experience, the number of spills per year has varied from one to four, with the probabilities given in Table 18.2.

The expected number of spills per year is two:

$$E(n) = 0(0.06) + 1(0.25) + 2(0.38) + 3(0.25) + 4(0.06) = 2$$

Fortunately, spills have been easy and inexpensive to clean up because the floor sealer was effective. Nevertheless, an average of two spills per year should not be tolerated. The reason for the spills should be identified, and measures should be taken to reduce the number.

Once beyond the 5-year warranty period, there is an increasing possibility that the sealer will fail whenever there is a spill. In year 6, the estimated probability of a sealer failure in case of a spill is 10% per spill. In year 7, this increases to 20%, then to 25% in years 8 and 9, and 30% thereafter.

The expected value of the cost of cleanup in any year is given by

$$E(\text{Cost}) = E(n)P(\text{Sealant failure})(\text{Cost of cleanup})$$

For example, in year 6

$$\begin{aligned} E(\text{Cost}) &= (2 \text{ spills})(0.1 \text{ sealant failures/spill})(\$350{,}000/\text{sealant failure}) \\ &= \$70{,}000 \end{aligned}$$

The expected costs per year (probability of a sealer failure times the cost per failure) are given in Table 18.3.

The cumulative expected cost of cleanup due to sealant failure exceeds the cost of replacing the sealer (\$187,000) in year 7. The smart decision is not to gamble. Reseal the floor and get a new five-year warranty.

TABLE 18.2 Probability of Chemical Spills onto the Floor

Number of Spills per Year (n)	Probability (%)
0	6
1	25
2	38
3	25
4	6

TABLE 18.3 Probabilities of Sealant Failures and the Expected Cost of Cleanup

Year	Expected Number of Spills	Probability Failure per Spill	Expected Sealant Failures	Cost ($)	*E*(Cost) ($)	*PV* of *E*(Cost) at 6%	Cumulative *PV* of *E*(Cost)
6	2	0.1	0.2	350,000	70,000	66,038	66,038
7	2	0.2	0.4	350,000	140,000	124,600	190,638
8	2	0.25	0.5	350,000	175,000	146,935	337,574
9	2	0.25	0.5	350,000	175,000	138,619	476,192
10	2	0.3	0.6	350,000	210,000	156,927	633,120

Example 18.4 Probability of Profit and Loss

An initial investment of $11,000 will earn a profit if the annual cash flow is large enough, but a profit is not guaranteed. There are three discrete levels for cash flow, project lifetime, and interest rate.

Cash flow is either $1,500, $2,000, or $3,000 with probabilities $P(C) = 0.2$, 0.5, and 0.3.
Project lifetime is 8, 10, or 11 years with probabilities $P(L) = 0.2$, 0.6, and 0.2.
Interest rate is $i = 7$, 8, or 9% with probabilities of $P(i) = 0.3$, 0.4, and 0.3.
The expected values are

$$E(\text{Cash flow}) = (\$1{,}500/\text{y})(0.2) + (\$2{,}000/\text{y})(0.5) + (\$3{,}000/\text{y})(0.3)$$

$$= \$300/\text{y} + \$1{,}000/\text{y} + \$900/\text{y} = \$2{,}200/\text{y}$$

$$E(\text{Project life}) = 8\,\text{y}(0.2) + 10\,\text{y}(0.6) + 11\text{y}(0.2) = 1.6 + 6.0 + 2.2 = 9.8\,\text{y}$$

$$E(\text{Interest rate}) = 7\%(0.3) + 8\%(0.4) + 9\%(0.3) = 2.1 + 3.2 + 2.7 = 8.0\%$$

A case is one realization of a combination of these three factors. There are 27 possible discrete cases (3^3). The probability of a case occurring

$$P(\text{Case}) = P(C)\,P(L)\,P(i)$$

There are 27 discrete possible outcomes listed in Table 18.4. The Profit(Loss) column is the *PV* of cash flow less the initial investment of $11,000. Most of the cases with low cash flow lose money. The NPV values for each case are ranked from smallest to largest and plotted against the cumulative probability, as shown in Figure 18.2. There is a 30% chance of losing money (NPV ≤ 0), a 50% chance of earning more than $2,400/y, and 20% chance of earning more than $8,250/y. The project will lose money if the cash flow is $1,500/y or less. A cash flow of $3,000/y will make a profit even if the business lasts only eight years.

The expected value for a continuous variable is

$$E(X) = \int P(X)\,f(X)\,\mathrm{d}X$$

where

f(*X*) is a function of the uncertain variable *X*
P(*X*) is the probability distribution of the uncertain variable *X*.

TABLE 18.4 Enumeration of 27 Cases for the Investment of $11,000

Case	Cash Flow ($/y)	Life (y)	*i* (%)	Probabilities *P*(*C*)	*P*(*L*)	*P*(*i*)	*P*(Case)	*PV* Factor	*PV* of Cash Flow ($)	NPV ($)
1	1,500	8	7	0.2	0.2	0.3	0.012	5.9713	8,957	−2,043
2	2,000	8	7	0.5	0.2	0.3	0.030	5.9713	11,943	943
3	3,000	8	7	0.3	0.2	0.3	0.018	5.9713	17,914	6,914
4	1,500	10	7	0.2	0.6	0.3	0.036	7.0236	10,535	−465
5	2,000	10	7	0.5	0.6	0.3	0.090	7.0236	14,047	3,047
6	3,000	10	7	0.3	0.6	0.3	0.054	7.0236	21,071	10,071
7	1,500	11	7	0.2	0.2	0.3	0.012	7.4987	11,248	248
8	2,000	11	7	0.5	0.2	0.3	0.030	7.4987	14,997	3,997
9	3,000	11	7	0.3	0.2	0.3	0.018	7.4987	22,496	11,496
10	1,500	8	8	0.2	0.2	0.4	0.016	5.7466	8,620	−2,380
11	2,000	8	8	0.5	0.2	0.4	0.040	5.7466	11,493	493
12	3,000	8	8	0.3	0.2	0.4	0.024	5.7466	17,240	6,240
13	1,500	10	8	0.2	0.6	0.4	0.048	6.7101	10,065	−935
14	2,000	10	8	0.5	0.6	0.4	0.120	6.7101	13,420	2,420
15	3,000	10	8	0.3	0.6	0.4	0.072	6.7101	20,130	9,130
16	1,500	11	8	0.2	0.2	0.4	0.016	7.1390	10,708	−292
17	2,000	11	8	0.5	0.2	0.4	0.040	7.1390	14,278	3,278
18	3,000	11	8	0.3	0.2	0.4	0.024	7.1390	21,417	10,417
19	1,500	8	9	0.2	0.2	0.3	0.012	5.5348	8,302	−2,698
20	2,000	8	9	0.5	0.2	0.3	0.030	5.5348	11,070	70
21	3,000	8	9	0.3	0.2	0.3	0.018	5.5348	16,604	5,604
22	1,500	10	9	0.2	0.6	0.3	0.036	6.4177	9,626	−1,374
23	2,000	10	9	0.5	0.6	0.3	0.090	6.4177	12,835	1,835
24	3,000	10	9	0.3	0.6	0.3	0.054	6.4177	19,253	8,253
25	1,500	11	9	0.2	0.2	0.3	0.012	6.8052	10,208	−792
26	2,000	11	9	0.5	0.2	0.3	0.030	6.8052	13,610	2,610
27	3,000	11	9	0.3	0.2	0.3	0.018	6.8052	20,416	9,416

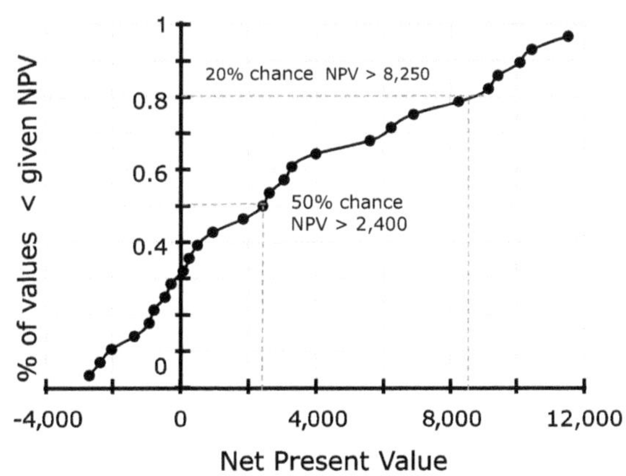

FIGURE 18.2 Ranked NPV values show the probability of various outcomes.

This will be illustrated for a uniform distribution, which means that every value of X within a given interval is equally likely to occur. For example, if the range of X is from a to b, then

$$P(X) = \frac{1}{b-a}$$

Example 18.5 Expected Cost of Equipment

The cost-capacity estimating model for an item of equipment is

$$C = KQ^{0.7}$$

The value of K is uncertain, but it will be between 4 and 6 and each value within that range is equally likely. What is the expected cost?

$$P(K) = \frac{1}{\Delta K} = \frac{1}{6-4} = 0.5 \quad \text{for} \quad 4 \le K \le 6$$

$$E(C) = \int_4^6 KQ^M P(K)dK = \int_4^6 KQ^M \left(\frac{1}{6-4}\right) dK$$

$$E(C) = \frac{1}{2}\left(\frac{K^2}{2}\right)Q^M \Big|_{K=4}^{K=6} = \frac{1}{2}\left(\frac{6^2}{2} - \frac{4^2}{2}\right)Q^M = 5Q^M$$

18.2 THE PROPAGATION OF ERROR

Some of the variables and coefficients in the equations engineers use to calculate the behavior of systems are measurements that are subject to error or estimates are uncertain. Most of the coefficients are estimated from an experiment or taken from a handbook. Some of the design variables are forecasts or historical values, and some are extrapolated from experience with similar equipment or processes.

Values calculated from equations are uncertain if the inputs values are uncertain. If the precision of each measured or estimated quantity is known, then simple mathematical rules can be used to estimate the precision of the calculated value. This is called *propagation of errors* or *propagation of variance*.

The best-known expressions of variability in measured values are the variance $V(x)$ and the standard deviation s_x. The variance is the standard deviation squared, $V(x) = s_x^2$.

The general model for the variance of y calculated from $y = f(x_i)$ is

$$V(y) = \sum_{i=1}^{n}\left(\frac{\partial y}{\partial x_i}\right)^2 V(x_i) + 2\sum_{i=1}^{n}\sum_{j=i+1}^{n}\left(\frac{\partial^2 y}{\partial x_i \partial x_j}\right) Cov(x_i, x_j)$$

The first term accounts for the variance contributed by each individual x_i. The partial derivatives are evaluated at the expected values of the x_i and are known as the sensitivities of y to each x_i.

The second term accounts for any covariance effects among the x_i. Covariance is when a change in one variable causes a predictable change in another. If all the x_i and x_j are uncorrelated (plots of x_i vs. x_j show no relation between the two variables), the Cov terms are all zero. For correlated variables, they can be positive or negative. The correlations are rarely known, and these terms are routinely ignored.

TABLE 18.5 Variances from Algebraically Combined Data

$y = x$	$V(y) = V(x)$	$y = x_1 x_2$	$V(y) = x_2^2 V(x_1) + x_1^2 V(x_2)$
$y = x_1 \pm x_2$	$V(y) = V(x_1) + V(x_2)$	$y = \frac{1}{x}$	$V(y) = \frac{V(x)}{x^2}$
$y = \frac{x_1}{x_2}$	$V(y) = \frac{x_1}{x_2}\left[\frac{V(x_1)}{x_2^2} + \frac{V(x_2)}{x_1^2}\right]$	$y = x^2$	$V(y) = 2xV(x)$
$y = e^{ax}$	$V(y) = a^2 e^{ax}$	$y = \ln(x)$	$V(y) = \frac{V(x)}{x}$
$y = \sqrt{x}$	$V(y) = \frac{V(x)}{2\sqrt{x}}$	$y = \log_{10}(x)$	$V(y) = \frac{V(x)}{x\ln(10)}$

Consider these two linear models. The θ's are known constants and the errors in x_1, x_2, and x_3 are assumed to be random:

$$y = \theta_0 + \theta_1 x_1 + \theta_2 x_2 + \theta_3 x_3$$

$$y = \theta_0 - \theta_1 x_1 - \theta_2 x_2 - \theta_3 x_3$$

The variance of y is the same for both models. The signs are irrelevant because the derivative terms are squared and the variances are additive:

$$V(y) = \theta_1^2 V(x_1) + \theta_2^2 V(x_2) + \theta_3^2 V(x_3)$$

$$+ 2\theta_1\theta_2 \,\mathrm{Cov}(x_1 x_2) + 2\theta_1\theta_3 \,\mathrm{Cov}(x_1 x_3) + 2\theta_2\theta_3 \,\mathrm{Cov}(x_2 x_3)$$

Assuming the covariance terms are zero, gives

$$V(y) = \theta_1^2 V(x_1) + \theta_2^2 V(x_2) + \theta_3^2 V(x_3)$$

Each term represents the separate contribution of an x variable to the overall variance of the calculated variable y. The derived $V(y)$ estimates the true variance of y, and can be used to compute a confidence interval for y:

$$y \pm t\, s_y$$

where t is the value of the t-statistic which has a value of approximately 2 for a 95% confidence interval.

Expressions for the variance of some common models are in Table 18.5.

Example 18.6 Exponential Decay

Evaluate the uncertainty in y for the exponential decay model $y = \exp(-kt)$ if the value of k is uncertain with an expected value = 0.2 and a standard deviation = 0.02. Calculate y, $V(y)$, and s_y for times $t = 2$ and $t = 5$.

For $t = 2$,

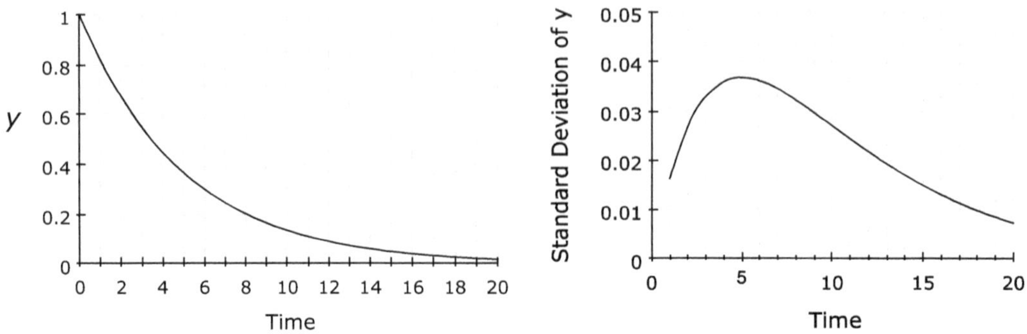

FIGURE 18.3 Propagation of variance in k for the model $y = \exp(-0.2t)$.

$$y = \exp[(-0.2)(2)] = 0.670$$

$$V(y) = \left(\frac{\partial y}{\partial k}\right)^2 V(k) = \left[-t\exp(-kt)\right]^2 V(k) = \left[-t\exp(-kt)\right]^2 (0.0004)$$

$$= [(-2)\exp[(-0.2)(2)]]^2(0.0004) = 1.797(0.0004) = 0.0007189$$

$$s_y = 0.0268$$

For $t = 5$,

$$y = \exp[(-0.2)(5)] = 0.368$$

$$V(y) = [(-5)\exp[(-0.2)(5)]]^2(0.0004) = 3.383(0.0004) = 0.0001353$$

$$s_y = 0.0368$$

Figure 18.3 shows the exponential decay curve and the standard deviation of y for t = 1–20. The maximum standard deviation (and variance) is at t = 5.

Example 18.7 Reactor Kinetics

The material balance for a lab-scale reactor is

$$SQ = S_0Q - kXS_0V \quad \Rightarrow \quad S = \frac{S_0}{1 + kXV/Q}$$

where S_0 = influent pollutant concentration, S = effluent pollutant concentration, X = concentration of active biomass in the reactor, V = reactor volume, Q = volumetric flow rate, and k = reaction rate coefficient. The values of Q and V are fixed at Q = 1.0 L/h and V = 1.0 L.

The purpose of the experiment is to estimate the value of k and its precision. The precision can be estimated from multiple estimates of k, but that can only be done after the experimental work has been completed. It is desirable to make an estimate before the research starts to understand how random errors in the measurements will translate into the estimates of k.

Both X and S are difficult to measure precisely, so replicate measurements will be beneficial. How many replicate samples of X and S are needed if k is to be estimated within plus or minus 5%? Must both X and S be measured precisely or is one more critical than the other? Measuring S is expensive. Can money be saved if errors in S are not strongly transmitted into the calculated value of k?

Many experiments would be improved if questions like these were answered as the experiment is planned rather than in the data analysis phase of the experiments.

The expected values of the variables are

$$S_0 = 200 \text{ mg/L} \qquad S = 20 \text{ mg/L} \qquad X = 2{,}000 \text{ mg/L}$$

The expected value of k is

$$k = \frac{(200 \text{ mg S/L} - 20 \text{ mg S/L})(1 \text{ L})}{(20 \text{ mg S/L})(2{,}000 \text{ mg X/L})(1 \text{ L/h})} = 0.0045 \frac{\text{h}}{\text{mg X/L}}$$

Values calculated from experimental data should vary about this expected value if the model is correct. Roughly 95% of the observed values should fall within the interval 0.0045 ± 2*s*, where *s*, is the standard deviation of k. The standard deviation is the square root of the variance, which will be estimated using

$$V(k) = \left(\frac{\partial k}{\partial S_0}\right)^2 V(S_0) + \left(\frac{\partial k}{\partial S}\right)^2 V(S) + \left(\frac{\partial k}{\partial X}\right)^2 V(X)$$

$$\frac{\partial k}{\partial S_0} = \frac{Q}{SXV} = \frac{1}{20(2{,}000)(1)} = 2.5 \times 10^{-5}$$

$$\frac{\partial k}{\partial S} = -\frac{Q}{SXV} = -\frac{1}{20(2{,}000)(1)} = -2.5 \times 10^{-5}$$

$$\frac{\partial k}{\partial X} = -\frac{S_0 - S}{SX^2V} = -\frac{200-20}{20(2{,}000)^2(1)} = -2.25 \times 10^{-6}$$

The measurement errors expressed as standard deviations are $s_{S0} = 20$, $s_S = 2$, and $s_X = 100$. The variances are

$$V(S_0) = 20^2 = 400 \qquad V(S) = 2^2 = 4 \qquad V(X) = 100^2 = 10{,}000$$

$$V(Q) = 0 \quad \text{and} \quad V(V) = 0 \quad \text{because they are fixed.}$$

The estimated variance of k is computed from the separate contributions of the three uncertain variables,

$$V(k) = \left(\frac{\partial k}{\partial S_0}\right)^2 V(S_0) + \left(\frac{\partial k}{\partial S}\right)^2 V(S) + \left(\frac{\partial k}{\partial X}\right)^2 V(X)$$

$$= (2.5 \text{ x } 10^{-5})^2(400) + (-2.5 \text{ x } 10^{-5})^2(4) + (-2.25 \text{ x } 10^{-6})^2(10{,}000)$$

$$= 2.5 \text{ x } 10^{-7} + 0.025 \text{ x } 10^{-7} + 0.5062 \text{ x} 10^{-7} = 3.031 \text{ x } 10^{-7}$$

The contribution of each term to the variance is

$$S_0 = 82.5\%, \rightarrow \qquad S = 0.8\%, \rightarrow \qquad X = 16.7\%$$

This may seem surprising since $V(X)$ is larger than $V(S_0)$. The reason is that the weight applied to the $V(X)$ is smaller than the weight applied to $V(S_0)$. That is $(\partial k/\partial X)^2 < (\partial k/\partial S_0)^2$.

The estimated standard deviation of k is

$$\text{Standard deviation} = \sqrt{\text{Var}(k)} = \sqrt{3.03 \times 10^{-7}} = 0.00055$$

This is 12.25% of the expected value.

As a useful and reasonable approximation, assume that the distribution of k values will be symmetrical about the expected value and can be described by a normal distribution. These numbers are small, but the expected value of $k = 0.0045$ is also small. Based on this approximation, roughly two-thirds of the experimental estimates will fall within ±1 standard deviation:

$$k = 0.0045 \pm 0.00055$$

$$0.00395 \le k \le 0.00505$$

An approximate 95% confidence interval for k is the expected value plus and minus two standard deviations:

$$k = 0.0045 \pm 2(0.00055) = 0.0045 \pm 0.110$$

$$0.0034 \le k \le 0.0056$$

We may be disappointed that the precision is not better, say something like $0.0043 \le k \le 0.0047$. It is better to learn the expected precision before spending the money so that the experimental plan can be modified to get a more precise estimate. One such modification might be to make replicate measurements of S_0 to reduce the variance. Duplicate measurements (two replicates) will reduce the variance by 30%; four replicates will reduce it by 50%. An alternative strategy is to run the experiment many times and average the multiple estimates of k that are obtained.

18.3 MEASUREMENTS AND MEASUREMENT ERRORS

All measurements have a measurement error. There are three kinds of errors. *Gross errors* are caused by mistakes or blunders (mislabeling a sample, recording the wrong value, etc.). These are eliminated by working more carefully. *Bias* is a consistent deviation of the measurement from the true value; the measurement is always too high or too low. Bias is corrected by better calibration and quality control.

Measurements that have no gross errors or bias will have *random errors*. Some measurements are too high, and some are too low, and these occur in roughly equal proportions. The sizes of the errors are also random, and they often fall into the familiar bell-shape of a normal distribution (Gaussian distribution). Random errors cannot be eliminated, but they can be quantified and managed, and sometimes made smaller. Averaging is one method of dealing with random errors.

Measurements can be adjusted to make them conform to known conditions or constraints. The three angles of a triangle must add to 180°, but the measured values almost never do, even when measured with great care with a precision protractor or surveyor's transit. Good practice is to measure all three angles and check the closure. Any discrepancy in the measurements can be adjusted (corrected) to conform with the correct total. Adjustment is only possible when closure is checked by measuring all the angles. If only two angles are measured, errors in those measurements will be propagated into the calculation of the third angle.

The same problem exists with measured volumetric or mass flows in a treatment plant or volumetric flows in a hydraulic system. The mass leaving any part of the system must equal the mass that entered (assuming no matter is stored within the system). There is always some discrepancy, or error, in the balance of measured values. Most often these errors are ignored in wastewater treatment systems, because the natural variability overwhelms the need for precision. Also, it is common that not all inflows and outflows are measured; some are calculated from measurements on others. This means that errors in the measured values are translated into errors in the calculated values.

The next example uses the *method of least squares* to adjust measurement errors in a network so that the material flowing into a node or junction equals the mass flowing out. The method of least squares minimizes the sum of all the squared deviations (errors). This method gives best estimates of the true values. It is valid only for random errors.

Example 18.8 Adjusting Flows

Figure 18.4 shows seven flows that were measured in a collection system. Flows Q_1, Q_2, and Q_6 are from industries, and the extra measurements were to resolve some disputes over wastewater charges. One criterion for fairness in billing customers is that conservation of mass is satisfied by the measured flows.

Q_4 and Q_7 are measured by flow meters that have good accuracy and precision, but the measurements, like all measurements, are subject to random error. Flows 1, 2, 3, 5, and 6 were measured by injecting a dye at a known rate and measuring the concentration after the dye was well mixed with the flow. There is no reason to believe that one measurement is more correct than any other.

According to the conservation of mass, the volume (mass) of water entering a junction (node) must equal the volume that leaves. Thus, it should be true that

$$Q_1 + Q_2 + Q_3 - Q_4 = 0 \quad \text{and} \quad Q_4 + Q_5 + Q_6 - Q_7 = 0$$

but in this case the measured flows give

$$35 + 21 + 10 - 60 = 6 \quad \text{and} \quad 60 + 16 + 20 - 102 = -6$$

Define the adjusted flow as q_1, q_2, and so on. The difference between the measured and adjusted flows are *measurement errors* that are deemed to be random: error = $q - Q$. The adjusted flows will be calculated using the method of least squares (which is explained more carefully in Chapter 16).

The objective is to

$$\text{Minimize } (q_1 - Q_1)^2 + (q_2 - Q_2)^2 + \cdots + (q_7 - Q_7)^2$$

$$\text{Minimize } (q_1 - 21)^2 + (q_2 - 35)^2 + \cdots + (q_7 - 102)^2$$

This is the LINGO program that solves this problem.

```
! Least squares criterion for adjusting flows;
   MIN = (q1 - 21)^2 + (q2 - 35)^2 + (q3 - 10)^2 + (q4 - 60)^2
   + (q5 - 16)^2 + (q6 - 20)^2 + (q7 - 102)^2;
!Material Balance Constraints;
   q1 + q2 + q3 - q4 = 0;
   q4 + q5 + q6 - q7 = 0;
END
```

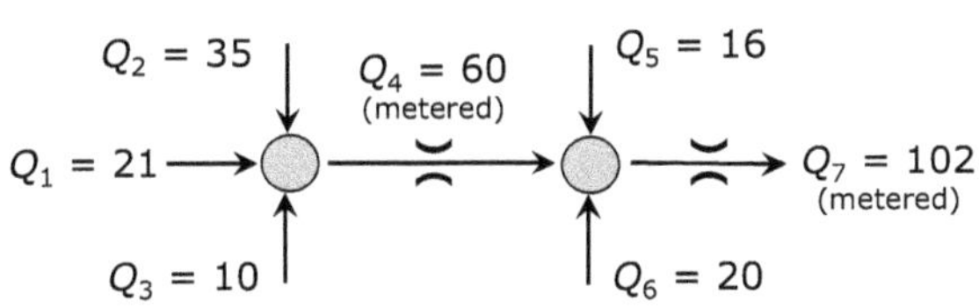

FIGURE 18.4 Flows measured in a collection system.

TABLE 18.6 Data Adjustment Solution

Flow	Measured	Adjusted	Adjustment
q_1	21	19.8	−1.2
q_2	35	33.8	−1.2
q_3	10	8.8	−1.2
q_4	60	62.4	+2.4
q_5	16	17.2	+1.2
q_6	20	21.2	+1.2
q_7	102	100.8	−1.2

The solution is in Table 18.6. Each flow has been adjusted by a small amount, the smallest amount that brings all the flows into balance.

This calculation has assumed that the variance of the errors is the same in all measurements. If this is not the case (e.g., errors may be proportional to the flow), the least squares adjustment can be modified to improve the solution (Berthouex and Brown, 2002).

18.4 CONCLUSION

Variation and uncertainty are common in environmental engineering. There is a natural tendency to ignore factors that are not well understood and to avoid decisions based on partial or conflicting information. The empirical art of overdesign evolved to provide some measure of security. Now statistical decision theory helps to remove some of the empiricism from the design process.

The expected value criterion is a useful measure of desirability in an uncertain environment. This introduction to the topic has been enriched by additional experience and research.

Error propagation tools are useful in identifying the major sources of uncertainty in process modeling and design.

18.5 PROBLEMS

18.1 EXPECTED PAYOFF

The values in Table P18.1 are payoffs for the indicted options and outcomes. Compute the weighted payoff by multiplying the payoff of each state by the probability of occurrence then sum the results. The maximum expected payoff is the best choice.

TABLE P18.1 Probabilities and Payoffs for Options and Outcomes

Alternatives	Level of Demand		
	Low	Moderate	High
Replace all equipment	353,500	736,000	1,456,000
Rehab and replace some equipment	176,750	365,000	728,000
Do nothing	88,375	184,000	364,000
Probability	0.4	0.3	0.3

18.2 PUMP-AND-TREAT GROUNDWATER REMDIATION

A contaminated groundwater is being remediated using pump-and-treat technology. The most likely time to completion is 30 years for the existing treatment system. The most optimistic completion time for the existing system is 20 years. The expected cost for the existing system is $2,302,570.

An improved version of the existing treatment system can do the cleanup in 10 years at best and 22 years at worst with completion expected in 15 years. A completely new system is expected to complete the cleanup in 5 years, with 3 years being possible and 12 years being the longest foreseeable duration. The worst case for the new technology shows a $135,000 expense at year 10 because some equipment must be replaced if the project runs that long. Table P18.2 shows the annual total present values for the improved technology and the new technology. *Year* is time measured from today. Generally, the present value decreases from year to year because of the discounting, but some years have a higher present value because some special work is needed.

The scenarios are not equally likely. The probabilities are given in row 4 of the table. Analyze the possibilities and recommend which system should be used for the duration of the project.

TABLE P18.2 Total Present Values of Annual Treatment Costs ($) for the New and Improved Treatment Systems

Year	Improve the Existing System			Replace the Existing System		
	Best Case	Most Likely	Worst Case	Best Case	Most Likely	Worst Case
	Likelihood Scenario Will Occur					
	30%	40%	30%	25%	50%	25%
1	117,000	117,000	117,000	300,000	300,000	300,000
2	115,000	115,000	115,000	500,000	500,000	500,000
3	111,000	111,000	111,000	78,000	78,000	533,416
4	108,000	108,000	108,000	–	76,000	76,000
5	105,000	105,000	105,000	–	74,000	74,000
6	102,000	102,000	102,000	–	–	72,000
7	99,000	99,000	99,000	–	–	69,000
8	96,000	96,000	96,000	–	–	135,000
9	93,000	93,000	93,000	–	–	92,000
10	90,000	90,000	90,000	–	–	89,000
11	–	87,000	87,000	–	–	86,000
12	–	85,000	85,000	–	–	83,000
13	–	82,000	82,000	–	–	–
14	–	80,000	80,000	–	–	–
15	–	78,000	78,000	–	–	-
16	–	76,000	76,000	–	–	–
17	–	74,000	74,000	–	–	–
18	–		72,000	–	–	–
19	–		72,000	–	–	–
20	–		72,000	–	–	–
21	–	–	72,000	–	–	–
22	–	–	72,000	–	–	–
Total	1,036,000	1,598,000	1,958,000	878,000	1,028,000	2,109,416

18.3 SAN FRANCISCO BAY, PIER 70, REMEDIATION PROJECT

A remediation project at Pier 70, San Francisco, California, will cost about $70,000,000. Table P18.3 lists the key project activities, the estimated low, likely, and maximum costs, and the probability associated with each cost. Calculate the likely cost and the expected cost of the project.

TABLE P18.3 Summary of Scenarios and Cost Estimates for Environmental Remediation of Pier 70, San Francisco, California (Treadwell & Rollo, 2009)

Parcel/Description	Estimated Cost ($1,000)			Probability (%)		
	Low	Likely	High	Low	Likely	High
Crane Cove Park and Slipway Park						
1. Impacted soils	2,795	4,887	14,477	35	50	15
2. Mass excavation, removal of fill, regrading	1,114	1,872	3,084	35	35	30
3. Stabilization or treatment of near-shore sediments	532	1,900	2,933	50	35	15
Parcels 1 and 3						
1. Impacted soils	349	1,714	3,852	35	50	15
2. Vapor intrusion mitigation		334	964	45	35	20
Parcels 2 and 4						
1. Impacted soils	751	4,491	7,714	35	50	15
2. Vapor intrusion mitigation	279	537	616	15	50	35
3. Steam and fuel line mitigation	198	297	396	20	50	30
Parcels 5, 6, 7, and 8						
1. Impacted soils	910	6,064	15,065	35	50	15
2. Vapor intrusion mitigation	763	1,473	3,285	25	50	25
Groundwater						
1. Parcels 1 and 3 – UST residuals management	56	146	269	25	50	25
2. Parcels 2 and 4 – UST residuals management	201	414	537	15	50	35
3. Parcels 5, 6, 7 and 8 – UST residuals management	71	201	414	15	50	35
4. All parcels – groundwater remediation	1,015	1,595	3,335	15	50	35
Site-wide and programmatic costs						
A. RMP compliance (security, dust control, etc.)	200	400	600	25	50	25
B. IC implementation (monitoring and enforcement)	500	750	1,000	25	50	25
C. Regulatory oversight	450	750	1,200	25	50	25

18.4 UNCERTAINTY IN PRESENT VALUE

The conventional calculation of the present value treats all costs as though they are known, as in

$$\text{Initial cost:} \quad P = \$1,000$$

$$\text{Annual costs:} \quad R = \$100/\text{y}$$

Assuming all costs are known

$$PV = P + \frac{R}{1+i} + \frac{R}{(1+i)^2} + \frac{R}{(1+i)^3} + \cdots + \frac{R}{(1+i)^n} = P + R\sum_{j=1}^{n} \frac{1}{(1+i)^j}$$

The initial cost may be known with certainty if a price has been quoted (a quotation is a contracted price). The annual costs cannot be known exactly, but this is usually ignored. This is an analysis of uncertainty in the ongoing annual costs.

A project has annual payments with an average value of \$100 and standard deviation of \$20, giving a variance of $\sigma_R^2 = 400$. For P = \$1,000 and i = 6% and n = 10 years, the PVF = 7.36009. Calculate the expected value and standard deviation of PV. Estimate the approximate 50% and 95% confidence interval for PV.

18.5 INFLATION RATE

The conventional calculation of the present value with inflation treats all costs as though they are known, as in

$$PV = P + \frac{R(1+r)}{(1+i)} + \frac{R(1+r)^2}{(1+i)^2} + \frac{R(1+r)^3}{(1+i)^3} + \cdots + \frac{R(1+r)^n}{(1+i)^n} = P + R\sum_{j=1}^{n} \frac{(1+r)^j}{(1+i)^j}$$

A project has annual payments with an average value of \$100 and standard deviation of \$20, giving a variance of $\sigma_R^2 = 400$. For P = \$1,000, i = 8%, r = 2%, and n = 10 years, calculate the expected value and standard deviation of PV. Estimate the approximate 50% and 95% confidence interval for PV.

18.6 EXPECTED COST OF EQUIPMENT 1

The cost-capacity estimating model for an item of equipment is

$$C = KQ^M$$

The value of K is uncertain but it will be between 4.5 and 6.2 and each value within that range is equally likely. Show that the expected cost is $C = K_{\text{Avg}}Q^M$.

18.7 EXPECTED COST OF EQUIPMENT 2

The cost-capacity estimating model for an item of equipment is

$$C = KQ^M$$

Each value of M within the range [0.5, 0.7] is equally likely. What is the expected cost for Q = 100?

18.8 LOAD PROJECTIONS

The load to a wastewater treatment plant has been projected to grow from the current load of 1,000 units/d at an expected rate of r = 5%/y. This rate is uncertain. It could be 4%/y or 6%/y, all rates between these values being equally likely. What is the expected load after 10 y? After 20 y?

The projected load model is L = 1,000 exp(rt). For r = uniformly distributed over the range from a to b, the expected value of L is

$$E(L) = \int_r (L)f(r)dr = \int_{r=a}^{r=b} (1{,}000\, e^{rt})\left(\frac{1}{b-a}\right)dr = \frac{1{,}000}{(b-a)\, t}(e^{bt} - e^{at})$$

where $f(r) = \dfrac{1}{b-a}$ = uniform probability distribution of r.

Optional: Estimate the standard deviation of the expected loads.

18.9 OPTIMIZING REACTOR VOLUME UNDER UNCERTAINTY

A pollutant is destroyed in a completely mixed reactor according to a first-order reaction. The cost of the reactor, including installation, auxiliary equipment, instrumentation, overhead, maintenance, etc., is \$0.01/L-h. The chemical and operating cost is \$50/unit of pollutant removed. The cost (\$/h) of the treatment process is

$$C = \$0.01\ V + \frac{\$50\ kV}{\left(kV - 1{,}000\right)}$$

where

C = cost (\$/h)
V = reactor volume (L)
k = reaction rate coefficient

(a) Determine the optimum volume and the optimized cost if the expected value of $k = 0.2$.
(b) Determine the volume that gives the minimum expected cost when k is equally likely to have any value between 0.15 and 0.25.

18.10 EXPONENTIAL DECAY PROCESS MODEL

There is a legal requirement that the effluent concentration from a process shall be $y \le 20$ mg/L. The process detention time will be calculated using this exponential decay model:

$$y = (100\ \text{mg/L})\exp(-kt)$$

where

y = effluent concentration (mg/L)
k = reaction rate coefficient (1/h)
t = reactor detention time (h)

The true value of k is unknown, but it has an expected value of $E(k) = 0.25$ and standard deviation $s = 0.03$, giving $\text{Var}(k) = (0.03)^2 = 0.0009$. There will be a redesign penalty of \$200,000 if $k = 0.2$ when the process goes into operation. A 5% probability of $y > 20$ mg/L for the designer's expected penalty cost is

$$E[\text{Penalty cost}] = (0.05)(\$200{,}000) = \$10{,}000$$

What design detention time will control the risk of paying the penalty to less than 5%?

18.11 FLOW MEASUREMENT

Two flows, Q_1 and Q_2, are combined to give the total flow Q. The expected values of Q_1 and Q_2 are 3.6 and 1.4. The standard deviations of flows 1 and 2 are $\sigma_1 = 0.2$ and $\sigma_2 = 0.3$, respectively.

(a) What is the expected value and standard deviation of Q?
(b) Does this standard deviation change if the upstream flows change?

(c) The cost of providing Q to a process is

$$C = KQ^M$$

For K = \$4,000 and M = 0.67, calculate the expected value and standard deviation of C.

18.12 FLOW MEASUREMENT BY DILUTION OF A TRACER

An industry pays a monthly fee to the city of \$1/m^3 based on volume of wastewater discharged from a drain into a city sewer. The flow in the drain is unmetered. From time to time the flow is estimated by mixing a solution of inert tracer (dye or salt) with the wastewater in the drain. The concentration of the tracer solution is 40,000 mg/L, and it is added at a rate of 1 L/min. This volume is a negligible portion of the total flow. The upstream flow, Q_U, contains none of the tracer material and the upstream tracer concentration is zero. The downstream flow, Q_D, has a tracer concentration of C_D.

C_D was measured five times while the flow rate was constant to give concentrations of 200, 230, 192, 224, and 207 mg/L. What is the estimated discharge to the city sewer? What is the average monthly discharge fee? What is the standard deviation of the estimated flow and of the discharge fee? If this measurement is made every month, what is the probability the industry will overpay by 10%?

18.13 BOD OXYGEN UTILIZATION MODEL

Evaluate the uncertainty in y for the BOD oxygen utilization model $y = L(1 - \exp(-kt))$ if the values of L and k are uncertain. The expected value of L = 16 with standard deviation = 2, and the expected value of k = 0.12 with standard deviation = 0.025. Calculate y, $V(y)$, and s_y for times t = 5 and t = 10. The derivatives are

$$\frac{\partial y}{\partial L} = 1 - \exp(-kt) \quad \text{and} \quad \frac{\partial y}{\partial k} = kL\exp(-kt)$$

18.14 UNCERTAINTY IN OXYGEN TRANSFER RATES

The oxygen transfer rate in clean water, OTR_{CW}, will be accepted as known based on manufacturer's test. The oxygen transfer rate in wastewater, OTR_{WW}, is less than OTR_{CW} for several reasons. Manufacturers report their equipment's efficiency as though it is operating in 20°C water at zero dissolved oxygen concentration. The transfer efficiency depends on the difference between the saturation concentration of oxygen in the liquid and the actual concentration in the liquid. For clean water at 20°C, this difference is $\Delta\text{DO}_{\text{CW}}$ = (9.18 mg/L – 0 mg/L) = 9.18 mg/L. The saturation concentration in wastewater is less than 9.18, perhaps only 8.5 mg/L, and the actual DO concentration is at least 1 mg/L, giving $\Delta\text{DO}_{\text{WW}}$ = (8.5 mg/L – 1.0 mg/L) = 7.5 mg/L or even lower. The OTR also must be adjusted for temperature. A major correction factor is the empirical alpha factor (α), which is less than 1.0, typically in the range of 0.6–0.8, depending on the characteristics of the wastewater and the aeration equipment. The corrections combined give

$$\text{OTR}_{\text{WW}} = \text{OTR}_{\text{CW}}\,\alpha\left(\frac{C^*_{\text{WW}} - C_{\text{WW}}}{C^*_{\text{CW}}}\right)1.02^{T-20}$$

where

C_{WW} = DO concentration in the wastewater at operating conditions, mg/L
C^* = DO saturation concentration in wastewater (WW) and clean water (CW), mg/L
α = alpha factor (dimensionless)
T = temperature of wastewater, °C

Assume that the variances OTR_{CW}, T, and C_{CW}^* are small enough to be neglected. Further assume that C_{WW}^* and α have a large variance.

Show that the variance in OTR_{WW}, $Var(OTR_{WW})$, is directly proportional to $Var(\alpha)$ and that $Var(OTR_{WW})$ is between $50Var(\alpha)$ and $70Var(\alpha)$.

18.15 F/M LOADING

Wastewater treatment plant operators often calculate food to microorganism ratio (F/M) for an activated sludge process.

$$\frac{F}{M} = \frac{QS_0}{XV}$$

where

Q = influent flow rate (m^3/d)
S_0 = BOD concentration entering the aeration tank (mg/L)
X = suspended solids concentration in the aeration tank (mg/L)
V = aeration tank volume (m^3)

V is fixed, Q is variable and can be measured with acceptable accuracy. X and S_0 are measured daily and they are uncertain due to measurement error. If the value of F/M falls outside a certain range, the activated sludge floc and the removal of suspended solids in the final clarifier will be poor and the effluent quality will be poor. The operator will try to control the F/M by adjusting the recycle of sludge from the final clarifier to the aeration tank. This is difficult because the true value of F/M is changing, but the difficulty is increased because the calculated F/M is even more variable than the true value. This raises the possibility that the operators overcontrol the process because they are reacting to measurement error. F/M greater than 0.5 is undesirable. The values in Table P18.6 give $F/M = 0.44$. Use the standard deviations in Table P18.15 to calculate the probability that the true value will be more than 0.5.

TABLE P18.15 Data for *F/M* Calculations

Variable	Average	Std. Dev.	Variance
Q (m^3/d)	3,500	200	40,000
S_0 (mg/L)	300	15	225
X (mg/L)	1,825	150	22,500
V (m^3)	13,000	–	–

18.16 A POLLUTION PREVENTION GAME

Games have fixed rules that are known to both sides and do not change during the game. This game, like all games, is not about reality. It is about testing strategies and improving them by gaining experience.

This is a game between a state pollution control inspector and two polluters who are located on rivers A and B. The available treatment strategies are in Table P18.16. They cannot start and stop their waste discharge, so whatever treatment strategy they select will operate for one year. Both polluters now operate with strategy 5. They will change strategy, but they are not responsive to moral arguments. They will operate to minimize their waste disposal costs.

The state agency has decided that polluters should be fined at the rate of \$1,000/d. They hope that vigorous enforcement to maximize the total amount of fines collected will persuade the polluters

TABLE P18.16 Cost of Waste Treatment and Expected Number of Days with Violations

Operating Strategy	Polluter A		Polluter B	
	Cost ($1,000)	Days	Cost ($1,000)	Days
1. Avoid all pollution	100	0	60	0
2. Good	95	4	55	4
3. Fair	80	18	45	10
4. Poor	70	30	35	30
5. Very poor	60	80	30	50

to adopt a treatment strategy that is less polluting. One stream inspection is made each day and the inspection will always correctly detect pollution in a river. The inspector can proportion inspections between rivers A and B in any way he thinks will maximize total fines collected.

For example, if polluter A uses operating plan 3 and the inspector chooses to spend all their time on stream A, the expected number of violations is 18 days. Every day of pollution will be detected and the total fines collected will $18,000. The total cost to polluter A is $18,000 + $80,000 = $98,000. If both A and B select strategy 3 and the inspector works 180 days on each river, the expected violations in 180 days are 9 days for A and 5 days for B, resulting in fines of $9,000 and $5,000, and total costs of $89,000 and $50,000, respectively.

The inspector and the polluters have the same information. The inspector has the data in Table P18.16. The polluters know that the goal of the state is to maximize the collection of fines.

Develop a matrix of expected costs for each polluter. Their expected costs are based on the treatment strategy adopted and the probability that the inspector will survey river A or river B. What strategy should the polluters choose? How much improvement will be made by each? What inspection strategy should the inspector choose? Does maximizing fines minimize the days of pollution? All in all, how effective is this enforcement plan? If you think it is not effective, recommend a better plan.

19 Monte Carlo Simulation

Monte Carlo simulation is a method for obtaining information about the consequences of variability and uncertainty on the performance of complex systems. The method consists of numerically *building* or *operating* a system many times in order to empirically build a distribution of possible outputs. The distribution will show the range of possible output values and the frequency with which each may be expected to occur.

The inputs are *random values* that are drawn from probability distributions with known properties, such as the mean and standard deviation for a normal distribution. *Random* means that a selected value cannot be predicted from values that were selected in the past.

The distribution of values might represent actual variation in a system, for example, the concentration of a pollutant, a flow rate, a processing temperature, or the reaction rate of a biological process. Or, it might reflect uncertainty about a factor that is estimated or cannot be measured accurately. Many interesting problems have some factors that vary (sometimes randomly) and others that are uncertain.

The designer must specify the problem structure, decide which variables will be treated as variable or uncertain, specify the range to be considered for each variable, and select statistical distributions to describe the variability.

Consider a system that has one input, x, that determines the output $y = f(x)$. Figure 19.1 shows how a distribution of inputs, the x values, is translated into a distribution of outputs, the y values. Notice how the range of the outputs will change if the inputs are shifted to smaller x values, and how the distribution would change if the range of x is changed. In practice, such changes might be made by better process control. Also, the output distribution can be changed by a redesign to shift the position or shape of the transforming function.

Many problems have more than one input variable ($x_1, x_2, \ldots, x_n$), so a graphical interpretation like Figure 19.1 is impossible. The procedure then is to define the probability distributions for each independent variable (x_i) and draw random values for each x_i to construct the probability distribution for the dependent variable(s), such as shown in Figure 19.2. This random drawing of x and predicting outputs continues through many trials (typically 1,000 or more) until the distribution of output values becomes clear.

19.1 DESCRIBING UNCERTAINTY WITH PROBABILITY DISTRIBUTIONS

Probability distributions are used to describe variability and uncertainty. The probability distribution can only capture the engineer's state of knowledge. When variability is the issue, there is usually historical data that will help to identify the best distribution. Uncertainty is a status of having little or no data. The distribution is unknown, but a statement is needed about the range of possible values and how they are distributed about some most likely value. This statement will be based on limited available data, and assumptions based on experience with similar problems.

Six probability distributions are described and developed in more detail in examples.

- *Uniform or rectangular distribution:* Only the upper and lower bounds can be estimated for a design variable or parameter, and there is no reason to prefer any single value between these bounds.
- *Triangular distribution:* The upper and lower bounds can be estimated and there is one intermediate value that is more likely than any other. The triangle does not need to be symmetrical.

- *Normal distribution:* This symmetric bell-shaped distribution is used when the mean value (or median) can be estimated and it is believed that values near the mean are more likely than values far from the mean.
- *Truncated normal distribution:* A bell-shaped distribution is reasonable, but the tail of one side needs to be removed because those values are not allowed. For example, the amount of water people drink or the amount of air they breathe must exceed some finite value for them to survive.
- *Log-normal distribution:* The log-normal distribution is not symmetrical; it has tail of high values. Most values are less than the average and the peak of the distribution (the mode) is not the average as it is in the normal distribution. Environmental data often have this distribution (the size of particles in the air, for example). If taking logarithms makes the data normally distributed, the data have a lognormal distribution.
- *Exponential distribution:* The probability decreases exponentially according to a single parameter that is the average rate at which random events occur.

Appendix E explains the Excel commands that can be used to generate random numbers with these distributions.

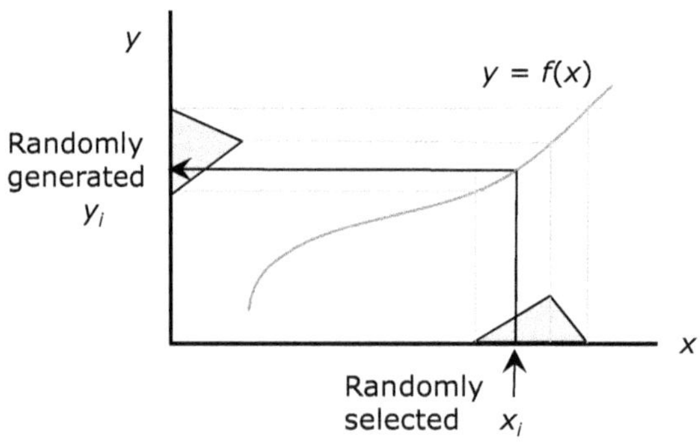

FIGURE 19.1 The Monte Carlo simulation method. A design model, $y = f(x)$, transforms a distribution of input values into a distribution of output models.

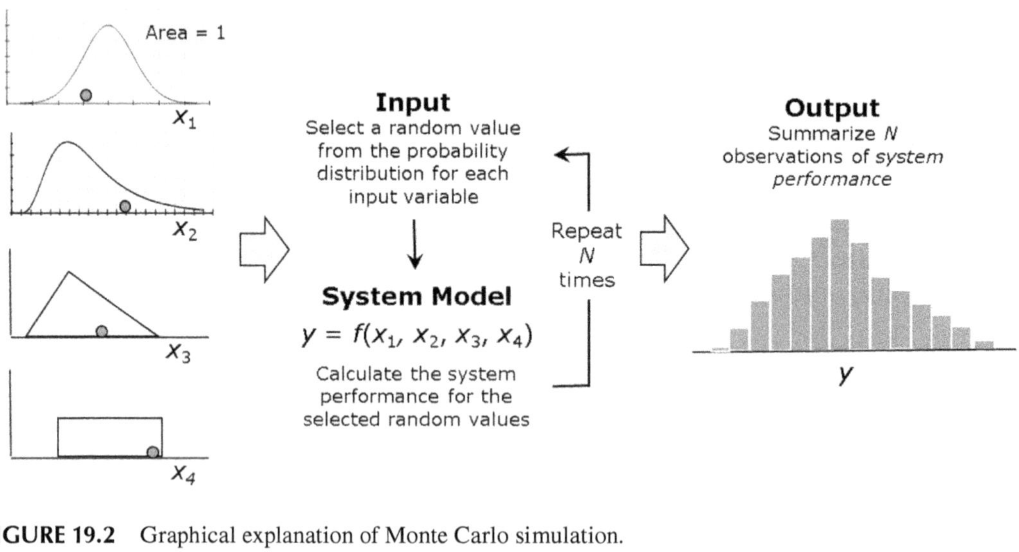

FIGURE 19.2 Graphical explanation of Monte Carlo simulation.

19.2 THE UNIFORM AND NORMAL PROBABILITY DENSITY FUNCTIONS (PDF)

A *probability density function* (PDF), denoted by $f(x)$, can be defined for any continuous distribution, where continuous means that the random variable can assume any value within the bounds of the distribution. The total area under the PDF is unity (1.0). The PDF is also called the *probability distribution.*

The common and convenient interpretation is that the height of the distribution is the probability that x has the given value. However, the function, $f(x)$, actually gives the density of probability that x has the given value. Thus, the probability that a value of x falls within a window or range of values is the area under the PDF between those values. For a small (infinitesimal) interval, x, around x, the probability that x lies in the interval is $f(x)\Delta x$.

Figure 19.3 shows the PDFs for the standard uniform distribution and the standard normal distributions.

The most basic PDF is the *standard uniform distribution*, $R_U(0, 1)$, which shows that any value of x between 0 and 1 is equally likely. The mean of the uniform distribution is $\mu = 0.5$ and the variance is $\sigma^2 = 1/12$. Random numbers with a standard uniform distribution are denoted by $R_U(0, 1)$.

The mean and variance of a uniform distribution with range from a to b are

$$\mu_U = \frac{a+b}{2} \quad \text{and} \quad \sigma_U^2 = \frac{(b-a)^2}{12}$$

The best known PDF is the bell-shaped *standard normal distribution* that has a mean of zero ($\mu = 0$) and a standard deviation of one ($\sigma = 1$), and a variance of one ($\sigma^2 = 1$). Random numbers with a standard uniform distribution are denoted by $R_N(0, 1)$.

Notice that the (0, 1) descriptors have entirely different meanings. For the standard uniform distribution, (0, 1) means a range from zero to one. For the standard normal distribution, zero is the mean ($\mu = 0$) and one ($\sigma = 1$) is the standard deviation. For convenience, the descriptor *standard* will be omitted in most of the following discussion.

These standard distributions can be shifted and scaled to create uniform and normal distributions with other bounds, means, and standard deviations, as shown in Figure 19.3. The location is shifted by adding a constant. The spread is scaled by multiplication of the range or standard deviation.

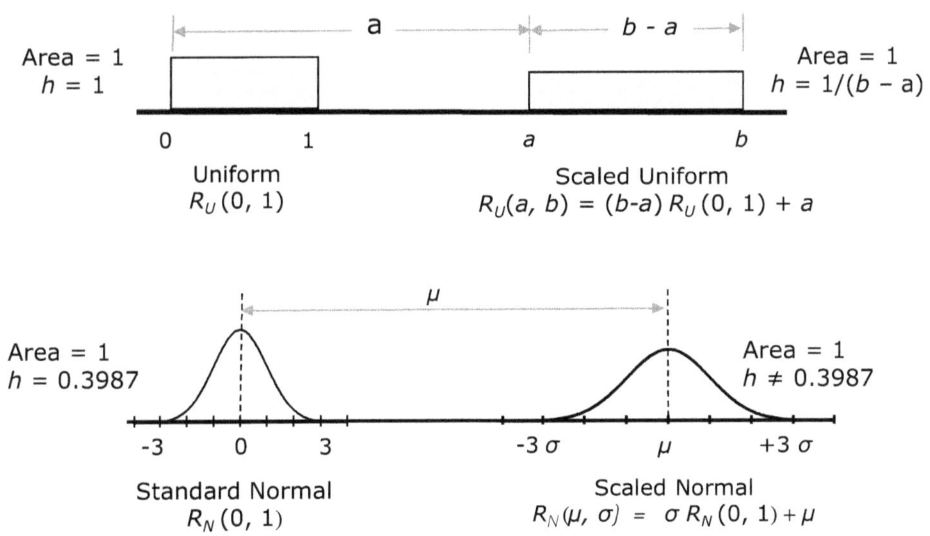

FIGURE 19.3 Shifting and scaling the standard normal and uniform distributions.

The PDF of the general uniform distribution is

$$y = f(x) = \frac{1}{b-a}, \quad \text{for} \quad a \le x \le b \text{ and } 0 \text{ elsewhere}$$

The PDF of the standard normal density function is

$$y = f(x) = \frac{1}{\sqrt{2\pi}} \exp\left(-\frac{x^2}{2}\right)$$

The general normal distribution, with mean μ and standard deviation σ is

$$y = f(x) = \frac{1}{\sigma\sqrt{2\pi}} \exp\left(-\frac{(x-\mu)^2}{2\sigma^2}\right)$$

It can be useful to make a quick sketch of the normal distribution. The peak of the bell-shaped curve is located at $x = \mu$, which gives exp(0) = 1, and the peak height

$$h = \frac{1}{\sigma\sqrt{2\pi}}$$

For the standard normal curve, $\sigma = 1.0$ and $h = 0.3987$. The area under a normal distribution is always 1.0, so when the distribution is scaled (stretched or squeezed) to a new value of σ, the peak height will change. It will be higher for $\sigma < 1.0$ and lower for $\sigma > 1.0$.

Any statistical software package can generate uniformly and normally distributed random numbers for any preferred location and spread. LINGO has excellent simulation capabilities.

Example 19.1 Selecting The Distribution

Table 19.1 lists the average size of truck loads (T/d) of municipal solid waste delivered to a processing center. On 95% of operating days, the number of truck loads delivered is 18–22. The size of the load and the number of loads per day are independent.

TABLE 19.1 Distribution of Municipal Waste Hauling Data

Day	Average Load (T/d)	Day	Average Load (T/d)
1	14.0	12	12.0
2	11.6	13	10.8
3	11.5	14	9.1
4	9.7	15	14.5
5	10.0	16	11.1
6	12.1	17	12.4
7	11.1	18	13.3
8	11.6	19	13.4
9	10.0	20	11.1
10	9.6	21	10.4
11	14.6		

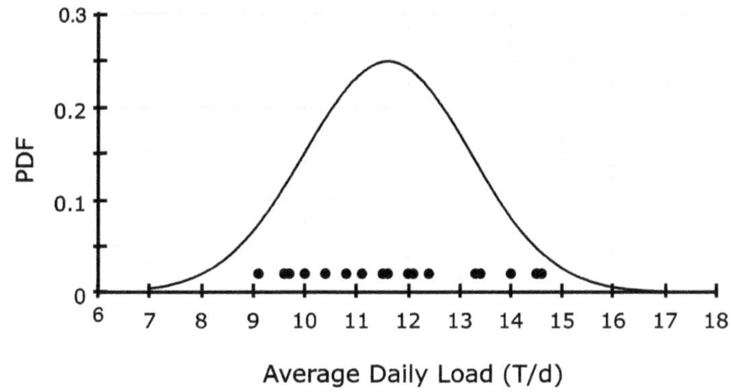

FIGURE 19.4 Distribution of the average truck load of municipal refuse delivered to a processing center.

Each dot in Figure 19.4 show the average load for one of the 21 days, which range from 9 to 14.6 T/d. There is no tendency for load size in the middle of this range to occur more often than any other value. This indicates that a uniform distribution will describe the data. This can be simulated as

$$\text{Random load (T/d)} = (b - a)R_U + a = (15 - 9)R_U + 9$$

The upper bound is set as $b = 15$ instead of the maximum observed value of 14.6. The sample size is relatively small ($n = 21$), so those bounds serve as guides to set limits based on all that is known about the system. One adjustment might be for time of year. These data were for November and there may be evidence that other months or seasons are different. A given fleet of trucks will have a distribution of known maximum capacities, and there will be a limit on how many loads per day each truck can deliver. This kind of information can be used to modify this empirical distribution.

For comparison, the plot also shows the PDF for a normal distribution with mean of 11.6 T/d and standard deviation of 1.6 T/d calculated from 21 days; that is $N(11.6, 1.6)$. Clearly, for this sample, the density of the dots (the data) does not follow the pattern of a normal distribution. The dots are not more densely spaced in the mid-range and less so in the tails of the distribution.

19.3 THE CUMULATIVE DENSITY FUNCTION

The *cumulative density function* (CDF), denoted by $F(x)$, is the integral of the PDF. The CDF gives the probability that x is less than or equal to the given value. Because the area under the PDF is 1.0, the values of the CDF run from 0 to 1.

$$F(x) = \int f(x)dx \quad \text{and} \quad f(x) = \frac{dF(x)}{dx}$$

The standard uniform distribution can be used to create random values for any arbitrary PDF using the CDF.

Example 19.2 CDF for Discrete Values

For example, suppose that random events can have only a few *discrete* values. Table 19.2 and Figure 19.5 show the PDF and CDF for a situation with four possible discrete outcomes.

A small simulation could be done by drawing cards (with replacement), numbered 2, 3, 4, and 5 from a bowl. The number of cards with each random value must be in proportion to the PDF, that is 10 cards numbered 2, 15 cards numbered 3, etc. Alternatively, making a large number of random trials can be done converting standard uniform random numbers, $R_U(0,1)$, to x, according to

TABLE 19.2 Empirical PDF and CDF, e.g., Discrete Random Variable

Value of Random Variable x	Probability PDF = $f(x)$	Cumulative Probability CDF = $F(x)$
2	0.10	0.10
3	0.15	0.25
4	0.50	0.75
5	0.25	1.00

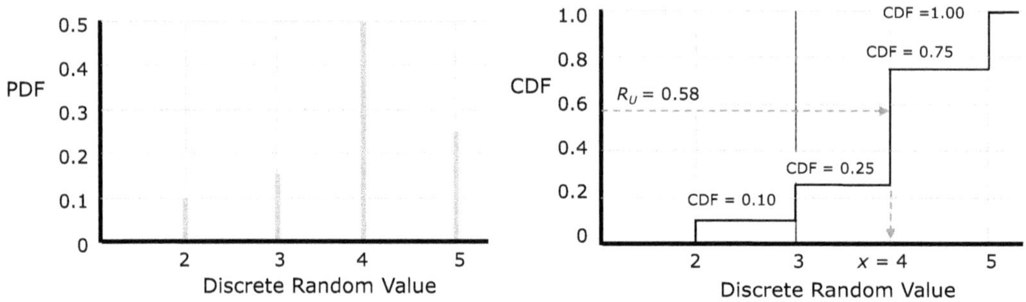

FIGURE 19.5 Empirical probability distribution and cumulative distribution for discrete values.

$$\text{if } 0.00 \le R_U \le 0.10, \quad \text{then} \quad x = 2$$

$$\text{if } 0.10 < R_U \le 0.25, \quad \text{then} \quad x = 3$$

$$\text{if } 0.25 < R_U \le 0.75, \quad \text{then} \quad x = 4$$

$$\text{if } 0.75 < R_U \le 1.00, \quad \text{then} \quad x = 5$$

19.4 TRIANGULAR DISTRIBUTION

A triangular distribution is used when an upper and a lower bound and a most probable value (peak) can be estimated, but there is not enough data to estimate a mean and standard deviation. The triangular distribution can serve as reasonable approximation of the normal or log-normal, shown in Figure 19.6. If the peak falls equidistant between the bounds, as in the left-hand panel, the triangular distributions can approximate a truncated normal distribution. (The tails are truncated.) If

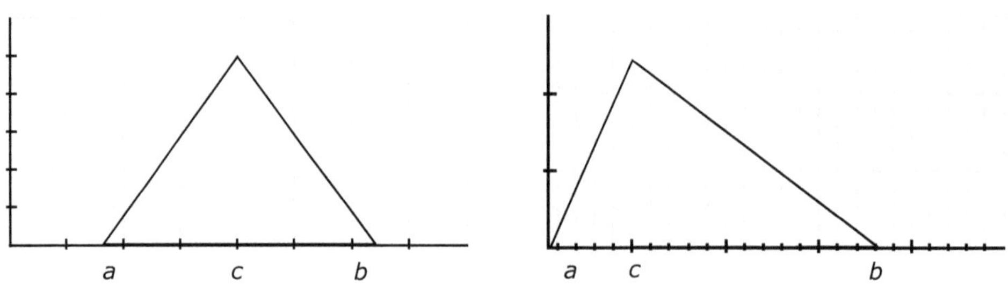

FIGURE 19.6 Triangular approximation of normal and log-normal distributions.

the triangle is not symmetrical, as in the right-hand panel, it can be a reasonable approximation of a truncated log-normal distribution. This demonstrates that the triangular distribution is legitimate – it captures the essential information.

A triangular distribution is specified by upper and lower bounds, denoted by a and b in Figure 19.6, and a peak or most probable value, denoted by c. The mean and variance of the distribution are

$$\mu = \frac{a+b+c}{3} \quad \text{and} \quad \sigma^2 = \frac{a^2+b^2+c^2-ab-ac-bc}{18}$$

The geometry is shown in Figure 19.7. The area under the PDF is 1.00, so

$$\frac{h(b-a)}{2} = 1.00$$

where h is the peak height. The cumulative distribution, which is derived by integrating the PDF, has an inflection point at c, which is the value of x at the peak of the triangle.

The PDF for the triangular distribution is

$$f(x) = \begin{cases} \dfrac{2(x-a)}{(b-a)(b-c)} & \text{for} \quad a \le x \le c \\ \dfrac{2(b-x)}{(b-a)(b-c)} & \text{for} \quad c \le x \le b \end{cases}$$

The CDF of the triangular distribution is

$$F(x) = \begin{cases} \dfrac{(x-a)^2}{(b-a)(b-c)}, & a < x < c \\ 1 - \dfrac{(b-x)^2}{(b-a)(b-c)}, & c < x < b \end{cases}$$

The inflection point in the CDF curve is at $x = c$, where the

$$\text{CDF} = \frac{c-a}{b-a}$$

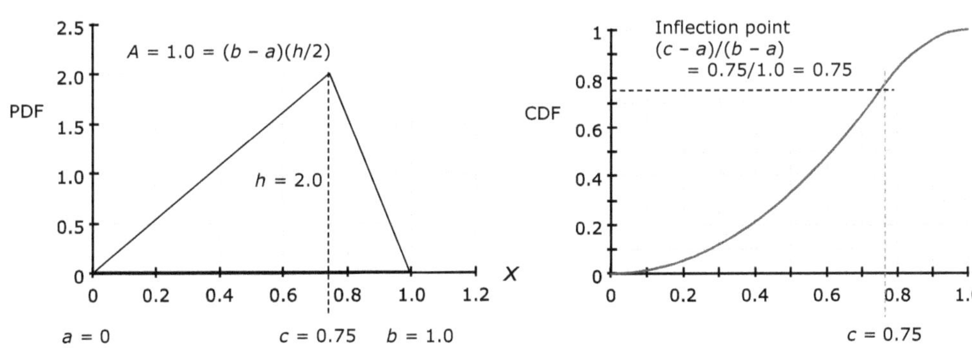

FIGURE 19.7 Geometry of the triangular distribution.

19.5 LOGNORMAL DISTRIBUTION

Many environmental measurements have a distribution with a tail that skews toward high values and is bounded by zero on the lower end. This shape is common for the size of soil grains and particulates in air, and for the concentration of most chemicals in water and wastewater.

The logarithms of such measurements are often normally distributed. In this case, the distribution is called log-normal. A random variable x is log-normally distributed if $\ln(x)$ has a normal distribution. (Usually natural logarithms are used, but $\log_{10}(x)$ also will give a normal distribution.)

The left-hand panel of Figure 19.8 shows a log-normal distribution in the natural variable x. The right-hand panel replots the distribution on a log-scale, which is equivalent to plotting the log-transformed values, as shown by the lower right-hand panel. The normal distribution has mean $\mu = 5$ and standard deviation $\sigma = 1.0$.

The PDF for the log-normal distribution is

$$f(x) = \frac{1}{x\sigma\sqrt{2\pi}} \exp\left[-0.5\left(\frac{\ln x - \mu}{\sigma}\right)^2\right]$$

where μ and σ are the mean and standard deviations of the log-transformed values; μ is also the median of the log-normal distribution.

One can look at a normal probability distribution and estimate the mean and standard deviation. The mean, median, and mode are at the peak and the standard deviation is, roughly, the range of values divided by six.

This is not possible for a log-normal distribution. The mean and the median are not at the peak. And the mean and median are not equal, as they are in the normal distribution. The most likely value is the mode, which is at the peak. The standard deviation is even more obscure.

If data are available, plot the histogram of the log-transformed data. If the resulting distribution appears to be a normal distribution, the mean and standard deviation of the transformed data can be estimated. These can be used to calculate the mean and standard deviation of the log-normal distribution, but these statistics are not needed to simulate log-normal data.

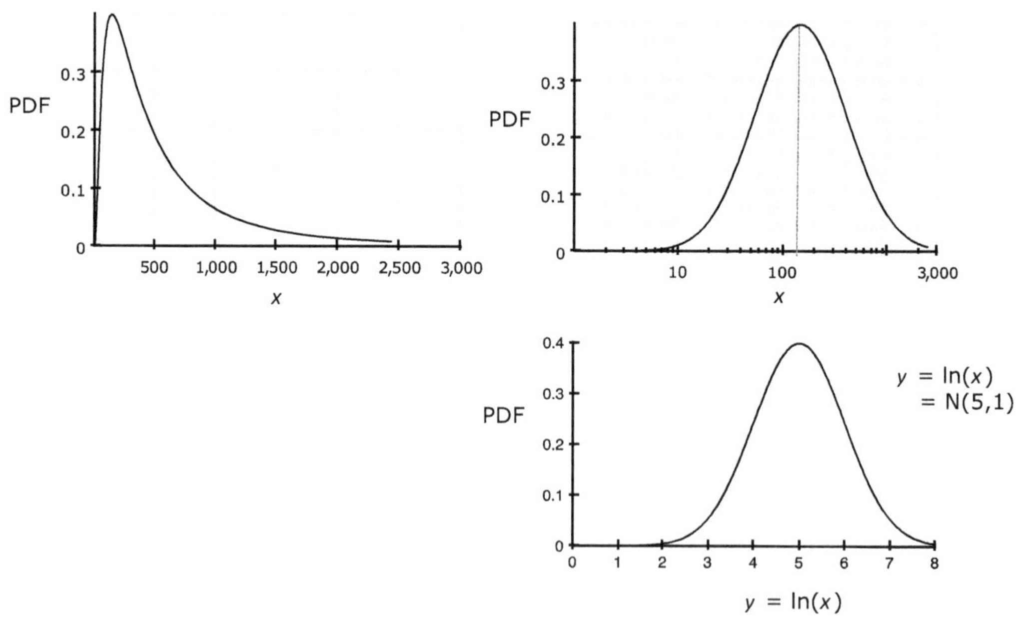

FIGURE 19.8 Log-normal distributions.

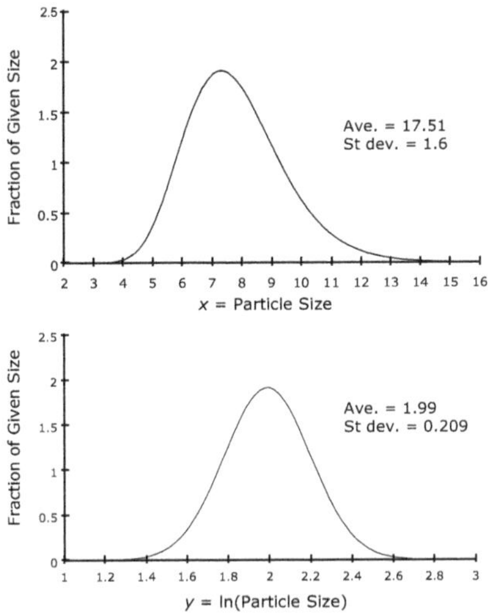

FIGURE 19.9 Log-transformation of particle size data.

Random normally distributed values can be used to generate random log-normally distributed values. Define the log-normal values as x and log-transformed values as $y = \ln x$, which are distributed $N(\mu, \sigma)$. Generate a series of normally distributed values and convert them to x using

$x = \exp(y) = e^y$.

Appendix E gives more details.

Example 19.3 Log Transformation of Particle Size Measurements

Many environmental data have a log-normal distribution. Particles of soil or air particulates are examples. Figure 19.9 (top) shows the particle size distribution as measured and the bottom panel shows the log-transformed data, which are normally distributed.

19.6 REACTOR KINETICS AND THE REACTION RATE COEFFICIENT

Example 18.7 used propagation of variance to estimate the precision of a biological reaction rate coefficient k. The model was

$$S = \frac{S_0}{1 + kXV/Q}$$

The volume and flow rate were fixed at $V = 1$ and $Q = 1$. The measured influent and effluent pollutant concentrations (S_0 and S) and the concentration of active biomass (X) in the reactor were uncertain because they are subject to random measurement errors. The estimated expected values and standard deviations were

$$\begin{array}{lll} S_0 = 200 & S = 20 & X = 2{,}000 \\ s_{S0} = 20 & s_S = 2 & s_X = 100 \end{array}$$

The estimated expected value was $k = 0.0045$ with standard deviation 0.00055. The calculation is based on linear approximations and this will predict a normal distribution for k and a symmetrical confidence interval. The approximate 95% confidence interval was reported as

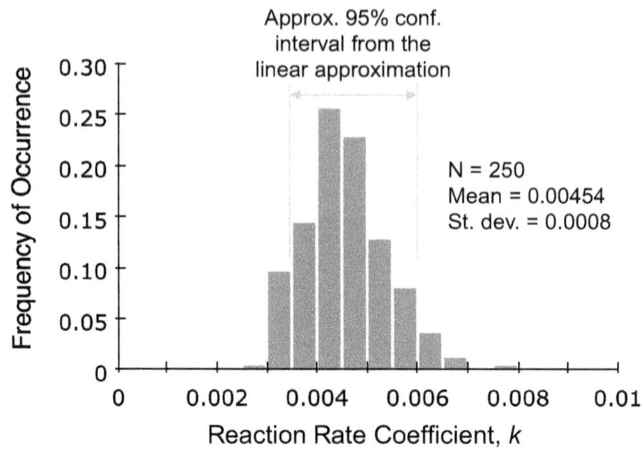

FIGURE 19.10 Histogram of simulated reaction rate coefficients, k.

$$k = 0.0045 \pm 2(0.00055) = 0.0045 \pm 0.110$$

$$0.0034 \le k \le 0.0056$$

This result was checked by Monte Carlo simulation with $N = 250$ runs. Figure 19.10 is a histogram of the simulated k values. It deviates from the assumed symmetrical normal distribution by having a slight skewness toward higher values of k. The 95% confidence interval should contain 95% of the k values. For the simulated values, this range is $k = 0.00323$–0.0064.

The linear approximation of the propagation of variance was useful, but the results from a Monte Carlo simulation may be more informative.

19.7 CASE STUDY: COST OF CONTAMINATED SOIL EXCAVATION AND TREATMENT

The owner of a contaminated property wants to know the expected cleanup cost and an upper bound for a contingency fund. There is uncertainty about the cost because there has been limited sampling and site characterization. The area and volume of soil that must be treated are unknown, and so are contamination levels. The pollutant concentration varies from place to place; hot spots will be discovered, and this will increase the cost. The needed work will adapt to the reality in the ground, and the cost will follow.

The processing capacity, in T/d, of the soil treatment equipment has been fixed, so the total cleanup time will depend on the volume of soil that must be handled. The contaminants vary in type and concentration. Lightly contaminated loads will cost less to treat because less processing is required. Some kinds of contamination will require disposal in hazardous waste landfills, and this is expensive.

Only an upper and a lower bound can be estimated for each variable. There is not enough data to make a reliable estimate of an average value, so rectangular (uniform) distributions will be used.

The total operating cost is the product of the three factors.

$$\text{Total cleanup cost} = MCD$$

where

M = mass of soil to be cleaned = 300–600 T (1 T = 1,000 kg)
C = level of contamination = 75–125 units/T
D = cost of excavation, treatment, and disposal = \$2–\$3.5/T per unit of contamination

Uniform distributions will be assigned to each variable because only the ranges are known. The uniform random numbers, R_U, between 0 and 1 are scaled and shifted to produce uniform distributions for M, C, and D.

$$M = (600 - 300)R_{U,M} + 300$$

$$C = (125 - 75)R_{U,C} + 75$$

$$D = (3.5 - 2)R_{U,D} + 2$$

The first few simulated values are shown in Table 19.3. The distribution of the cost estimates based on 1,000 simulations is in Figure 19.11. The histogram is slightly skewed toward higher costs, but this does not affect the central values. The mean (the expected cost) is \$123,000, the median (50% value) is \$128,000, and the mode (most frequent value) = \$120,000.

The financial risk is shown by the spread of the distribution. The minimum estimated cost is \$56,000; the maximum is \$248,000. The middle 50% of the distribution (the mid-range) is from \$100,000 to \$145,000. There is a 50% probability that the cost will fall within this range. The 90th percentile cost is \$170,000, which is much lower than the maximum of \$248,000. There is only a 2% chance the cost will exceed \$200,000.

TABLE 19.3 Simulated Costs for the Excavation, Treatment, and Disposal of the Soil Calculated from Random Values for Soil Mass (*M*), Level of Contamination of Soil (*C*), and Unit Disposal Cost per Cubic Meter (*D*)

Trial	$R_{U,M}$(0.1)	$R_{U,C}$(0,1)	$R_{U,D}$(0,1)	Mass (T)	Contaminant (units/T)	Disposal (\$/T-unit)	Cost (\$)
1	0.3189	0.7333	0.0901	396	112	2.14	94,338
2	0.5077	0.6747	0.7846	452	109	3.18	156,255
3	0.6901	0.4082	0.0563	507	95	2.08	100,831
4	0.9406	0.3615	0.0106	582	93	2.02	109,231
5	0.5257	0.9675	0.8649	458	123	3.30	186,198
6	0.2955	0.0032	0.2660	389	75	2.40	70,081
7	0.1988	0.8192	0.7286	360	116	3.09	128,988
8	0.9492	0.3856	0.4216	585	94	2.63	145,123
9	0.4420	0.9588	0.4535	433	123	2.68	142,547
10	0.4304	0.7860	0.5586	429	114	2.84	139,202

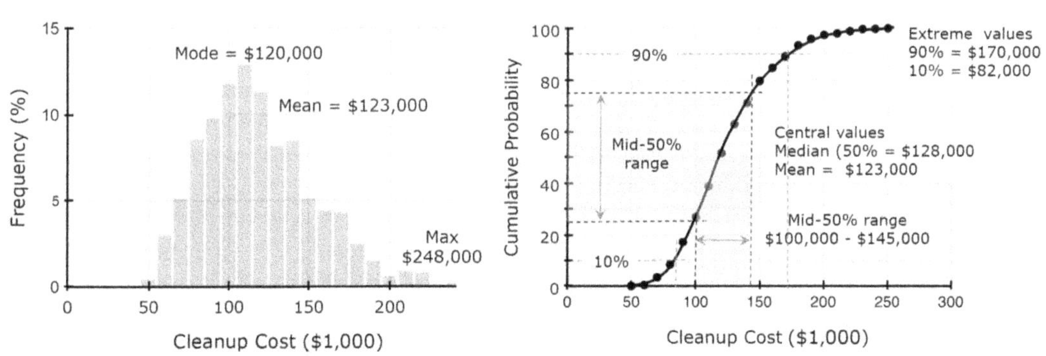

FIGURE 19.11 Distribution of simulated cleanup costs (n = 1,000 runs).

19.8 CASE STUDY: EXPOSURE TO CHROMIUM IN DRINKING WATER

An important problem is risk assessment to estimate the probability of humans being harmed by a toxic chemical that has been released into the environment. The information required to make this estimate includes (1) the dose–response curve for the chemical, (2) estimates of transport of the chemical from its source to the human population (groundwater, vaporization, etc.), and (3) the intake of the chemical by the exposed human population (diet, water consumption, adsorption through skin, etc.). None of these factors are not precisely known, but reasonable estimates usually can be made.

Many toxic chemicals have a threshold value of exposure below which the substance is harmless. This threshold value, known as the *reference dose* (RfD), is based on the *no observed adverse effect level* (NOAEL) from bioassay tests. NOAELs are extrapolated from animals to humans, which introduces an element of uncertainty. The NOAEL is divided by a safety factor to deal with this uncertainty. Other *safety* factors may be applied based on the quantity and quality of the toxicity data. The result is a numerical value (published by the US EPA or other international environmental agency) that is not precisely known even though it is precisely specified in regulations.

The reference dose (RfD) for the oral intake of chromium (Cr^{6+}), according to the US EPA, is 0.003 mg/kg-d. The RfD will be treated as a fixed known quantity because it is established in environmental regulations.

The daily intake rate (IR) is the mass per day of potentially harmful material that enters the body. The *chronic daily intake* (CDI) is the IR adjusted for body weight, with units of mass/day-kg. *Chronic* means repeated low-level exposure. In the context of drinking water, mass ingested is proportional to the chemical concentration and the volume of water consumed. The definition is

$$\text{CDI (mg/kg-d)} = \frac{C \text{ (mg/L)} \times \text{IR (L/d)}}{\text{BW (kg)}}$$

where

C = metal concentration in the drinking water (mg/L)
IR = intake rate (L/d)
BW = body weight (kg)

The *hazard quotient* (HQ), a simple measure of risk, compares the CDI and the RfD:

$$\text{HQ} = \frac{\text{CDI}}{\text{RfD}}$$

HQ < 1 means the chronic daily intake is less than the RfD, so there should be no risk. HQ > 1 indicates a potential health risk. Assuming BW = 70 kg and IR = 2 L, a chromium concentration of 0.105 mg/L gives HQ = 1.

The HQ is not a fixed known quantity because the CDI varies depending on water intake, water contamination level, and body weight, and these quantities are also variable and uncertain to some extent. Simulation is one way to investigate how variation in these quantities translates into uncertainty in the risk as measured by the HQ.

There is not enough information to fit a normal, log-normal, or other formal distribution, but enough data is available to make a useful estimate of an expected Cr^{6+} concentration and the maximum and minimum values. The empirical triangular distribution for chromium, shown in Figure 19.12, captures the essence of the available information.

The water intake rate was modeled with normal distribution N ($\mu = 2$ L/d, $\sigma = 0.5$ L/d) that is truncated to exclude values less than 1 L/d; humans need at least 1 L/d to survive. Body weight

(BW) was modeled with normal distribution N (μ = 70 kg, σ = 20 kg) that is truncated to exclude BW less than 40 kg. The analysis does not apply to small children. These distributions are shown in Figure 19.13.

The results of 1,500 simulations are summarized in Figure 19.14. Of the 1,500 values, 142 values were censored because IR rate or BW were truncated. Eighty percent of the theoretically exposed

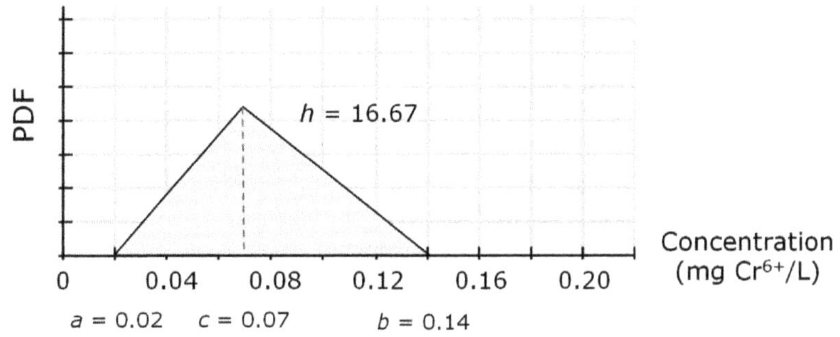

FIGURE 19.12 Triangular distribution for chromium (Cr^{6+}) concentrations.

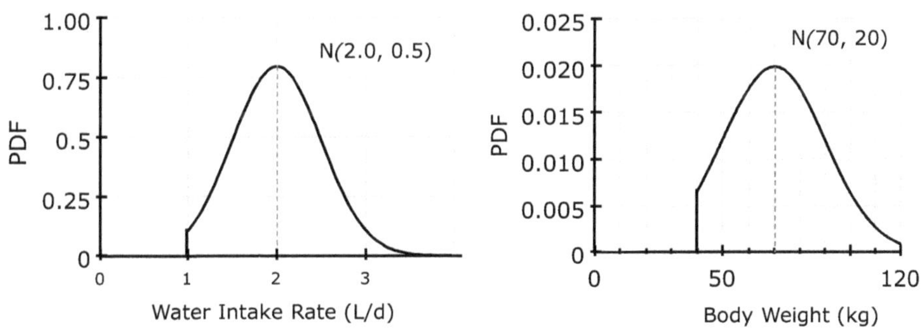

FIGURE 19.13 Truncated normal distributions for water consumption and body weight.

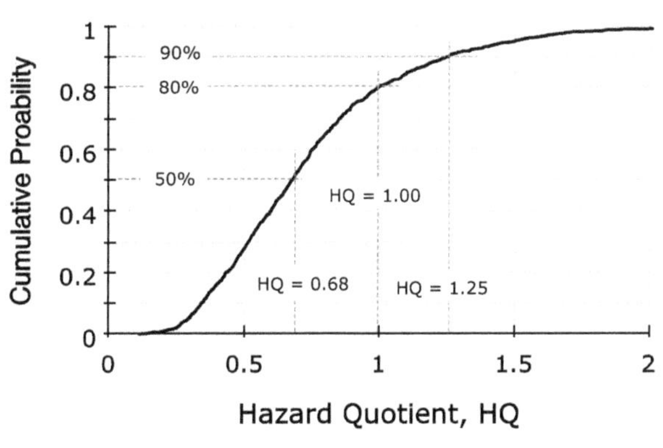

FIGURE 19.14 Cumulative distribution of chromium hazard quotient.

individuals have a hazard quotient less than 1 and are not at risk. Ninety percent of the individuals have HQ less than 1.25.

The only way to reduce the risk is to reduce the contaminant level in the water.

19.9 CASE STUDY: TRUCK FUELING STATION ANALYSIS

City trucks arrive between 7 am and 4 pm for fueling at time intervals described by an exponential probability distribution. The mean rate of arrival is six trucks per hour for an average of one truck every ten min.

Fueling takes eight minutes once a truck reaches the fuel pump. If a truck arrives at the end of a queue, it must wait in line. The time to refuel is a normal business expense, but waiting in a queue wastes money. The cost of a truck driver is \$55/h (\$26/h plus benefits and indirect costs).

Management wants to know if a second fuel pump will be justified. This depends on the amount of time being wasted at the pumps.

Specifically they want to know the:

(1) Average waiting time in the queue
(2) Average queue length
(3) Amount of time the fuel pumps are idle
(4) Percent of time the fueling system is being used

A Monte Carlo experiment to simulate the arrival of trucks at the fueling using

$$\text{Arrival of truck}(n) = \text{Arrival of truck}(n-1) + \text{Time to next arrival}$$

The time to the next truck arrival was simulated as a random variable having an exponential probability distribution with mean λ = 10 min. The random intervals between arrivals is calculated

$$t = (-\lambda)\ln(1 - R_U)$$

where

t = time between arrival of two trucks
λ = average time between arrivals
R_U = uniformly distributed random number over the range of 0–1.

As an example, the time between arrivals is t = 4.26 min for λ = 10 min and R_U = 0.347.

Figure 19.15 shows four days of simulated operation with 50 trucks per day arriving randomly with an average interval of 10 min (λ = 10). Table 19.4 shows the simulated operation of the fueling station between 7:00 and 8:28 a.m. for the first 10 of 50 trucks that were fueled that day. The random times between arrivals have been rounded to the nearest minute. The average time waiting in the queue was 3.5 min. The maximum queue length was two trucks. The fuel pump was idle for 22 min. The percent pump utilization was (88 min – 22 min)/88 min = 75%.

This small simulation does not give a dependable prediction of the average values. Furthermore, it gives no useful information about the extreme conditions, such as what is the probability there will be five trucks in the queue (important if there is holding space only for three) or of a truck having to wait more than 20 min in the queue (important when truck operation is \$50 and more per hour). These questions can be answered by running the simulation

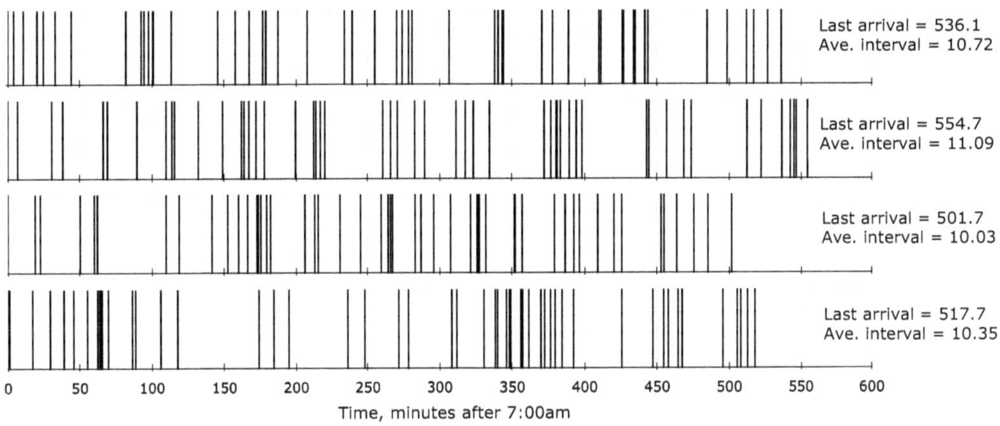

FIGURE 19.15 Four simulations of 50 trucks arriving the fueling station for $\lambda = 10$ and random times between arrivals generated using $t = (\lambda)\ \ln(1\ x)$.

TABLE 19.4 Simulated Results for the First 10 of 50 Trucks That Were Fueled in on a Particular Day

Truck	Time Arrive at Queue	Wait at Pump?	Wait Time (min)	Arrive at Pump	Leave Pump	Total Time = Wait + Fueling (min)	Queue Length	Idle Time of Pump (min)	Random Number	Time to Next Arrival (min)
1	7:00	NO	0	7:00	7:08	8	0	0	0.5613	8
2	7:08	NO	0	7:08	7:16	8	0	0	0.3954	5
3	7:13	YES	3	7:16	7:24	11	1	0	0.2011	2
4	7:15	YES	9	7:24	7:32	17	2	0	0.5830	9
5	7:24	YES	8	7:32	7:40	16	1	0	0.4994	7
6	7:31	YES	9	7:40	7:48	17	2	0	0.5975	9
7	7:40	YES	8	7:48	7:56	16	1	0	0.9714	36
8	8:16	NO	0	8:16	8:24	8	0	20	0.6454	10
9	8:26	NO	0	8:26	8:34	8	0	2	0.1465	2
10	8:28	YES	6	8:34	8:42	14	1	0	0.8916	22

long enough to plot histograms of the waiting time, fuel pump idle time, queue length, and other factors.

This simulation was repeated for 50 trucks for each of 10 days of operation in order to get an idea of the day-to-day variability of refueling operations. The results are in Table 19.5.

Some of these results are not intuitive, e.g., that the average wait time can vary by as much as a factor of 6 from one day to another, or that the fuel pump utilization rate is fairly stable at 80%.

This number of simulations is still a small sample. Several months of simulated operations are necessary to get good estimates of day-to-day variation. This may be worth doing if large amounts of money are at stake.

TABLE 19.5 Summary of Simulated Results for 10 Days of Truck Refueling Operations with 50 Trucks per Day

Day	Time for 50 Trucks (min)	Average Time between Trucks (min)	Average Time in Queue (min)	Total Wait Time (min)	Fraction with No Wait (%)	Average Queue Length (Trucks)	Max Queue Length (Trucks)	Pump Idle Time (min)	Pump Utilization Time (%)
1	569	11.4	4.5	227	46	0.8	3	156	72
2	524	10.5	8.9	445	36	1.4	5	141	74
3	497	9.9	6.8	338	36	1.1	4	103	80
4	517	10.3	24.5	1223	20	3.5	8	130	75
5	577	11.5	7.8	388	28	1.3	4	183	69
6	485	9.7	10.3	516	24	1.6	3	86	82
7	458	9.2	14.2	712	22	2.1	6	62	87
8	500	10.0	7.9	394	20	1.4	4	109	79
9	429	8.6	26.1	1306	12	3.7	7	46	90
10	508	10.2	4.2	211	34	0.8	3	117	77
Mean	506	10.1	12.0	576	28	1.8	5	113	78
Std. Dev	45	1.0	8.0	390	10	1.0	2	42	6

19.10 CONCLUSION

In most design problems, calculating a few well-chosen design conditions will produce good design decisions. When inputs are variable or design factors are uncertain, it may be prudent to learn the frequency of extreme conditions before making final design decisions. Monte Carlo simulation is a useful and easy method for making these investigations.

This leaves the design engineer to define the problem structure and specify which variables will be simulated and which will be fixed. Decisions are also needed about which statistical distributions will be used to generate the random numbers, and what range of numbers will be considered.

Simulation should not be done on every problem where there is some degree of variability or uncertainty. The danger in having software that makes the calculations easy is that the original uncertainty may be buried in an avalanche of simulated conditions that will obscure rather than illuminate.

19.11 PROBLEMS

19.1 SIMULATING AN EMPIRICAL PROBABILITY DISTRIBUTION

Simulate 100 random numbers using the empirical distribution shown in Figure P19.1.

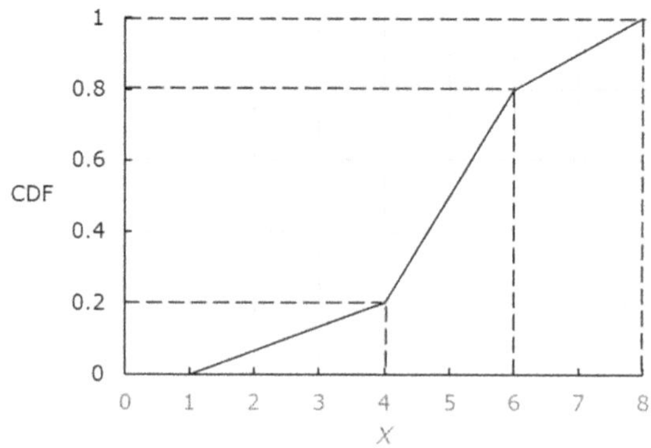

FIGURE P19.1 CDF for an empirical probability distribution.

19.2 A GAME OF CHANCE

You are invited to play a game of chance, where an unbiased coin is flipped repeatedly until a run of three tails (T) or three heads (H) ends the game. You pay \$1 for each flip of the coin and you receive \$8 at the end of each game. Table P19.2 shows five games. Winning two games out of five produced net winnings of \$2, or \$0.40 per game. There is one more rule. If you start the game, you must play for at least 100 trials. Assuming that you have \$100, do you want to play the game? Calculating the theoretical probabilities is difficult so investigate the game by simulation. Devise your own method and play the game. All that is needed is a binary random number generator. Summarize your findings and report your conclusion.

TABLE P19.2 Outcomes for Five Games of Chance

Game	H/T Outcome	Cost (\$)	Payoff (\$)	Net Gain/Loss (\$)
1	THHTHTH TTT	10	8	-2
2	TT HHH	5	8	3
3	HH TTT	5	8	3
4	HTHHTH TTT	9	8	−1
5	HTHTTH TTT	9	8	−1
	Totals	38	40	2

19.3 LOG-NORMALLY DISTRIBUTED DATA

Confirm that the data in Table P19.3 are skewed to the right, i.e., they have a tail of high values. Devise a method to simulate random values that will have approximately the same distribution.

TABLE P19.3 Skewed Experimental Data

144	1,145	1,038	422	36	172	840	23	99	181	52	136
140	155	83	69	107	793	142	82	252	132	336	42
179	250	90	447	890	372	1,417	29	551	223	229	79
594	96	200	210	115	361	135	148	65	870	949	91
69	233	16	56	23	232	116	175	240	111	378	66
840	23	99	181	52	136	840	23	99	181	52	136

19.4 MODELING AND VARIABILITY

Air pollutants are reduced by reaction, adsorption, or settlement. An exponential decay model can represent all of these mechanisms:

$$C = C_0 e^{-kt} \quad \Rightarrow \quad C = C_0 e^{-kx/v}$$

where

C_0 = concentration at the source at $t = 0$
C = concentration after travel downwind for time t
t = x/v = time of travel from the source (d)
x = distance from the source (mile)
v = velocity (mile/d)
k = rate coefficient for pollutant disappearance (1/d)

The average source concentration is C_0 =100, but it varies between 20% higher and 10% lower than this. We also know that, because of variations in temperature and other physical–chemical factors,

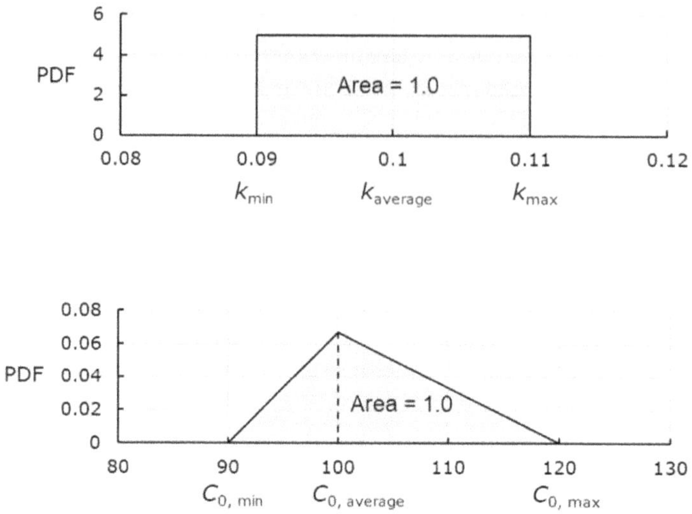

FIGURE P19.4 PDFs for rate coefficient, k, and source concentration, C_0.

the value of k varies between ±10% of its average value of 0.1. Develop a method to determine the probability distribution of C at a distance equal to five days travel time from the source. The units of k are d^{-1}. Figure P19.4 shows a uniform distribution for k and a triangular distribution for C_0. What are the values of C with cumulative probability 50%, 90%, and 95%.

19.5 HAZARDOUS WASTE CLEANUP

A rough estimate of a hazardous waste cleanup site reveals that the quantity of soil that needs to be handled (excavated, cleaned, etc.) is between 10,000 and 12,500 m^3, and the cost is \$16–\$19/m^3. Your preliminary state of knowledge is such that you have no reason to prefer any value within the ranges given. Prepare a probability distribution of the cost of the remediation. Plot the cumulative remediation cost distribution and estimate the expected cost, and a 95% confidence interval for that cost.

19.6 INDUSTRIAL POLLUTION CONTROL

A proposed industry expects to produce between 3,000 and 5,000 m^3/d of wastewater, containing a concentration between 0.6 and 0.8 kg/m^3 of a toxic substance. The costs of treatment will be in the range of \$1.00–\$2.25/kg of toxic substance. Prepare a probability distribution showing the possible costs of treatment. Plot the cumulative treatment cost distribution and estimate the expected cost, and a 50% and 95% confidence interval for that cost. State all assumptions.

19.7 INDUSTRIAL POLLUTION PREVENTION

An industry expects that a new process will produce 130–150 m^3/min of contaminated exhaust gas, having a solvent concentration of 500–750 ppmv. The density of the VOC vapor is 10 kg/m^3. The cost of treatment (net present value) will be in the range of \$1,000–\$1,400/m^3/min of gas treated. Recovered solvent will have a value of \$0.80/kg. Prepare a probability diagram showing the possible costs of treatment, the possible value of recovered solvent, and the net cost (or savings) of recovery. State all assumptions.

19.8 DISSOLVED OXYGEN MODEL

There is a critical aquatic habitat between 1.5 and 3 days flow downstream of a point source of pollution. The Streeter–Phelps oxygen sag model will be used to calculate the quantity of BOD that can be put into a stream without driving the dissolved oxygen concentration below 6 mg/L during low flow and when the stream temperature is 20C. The model is

$$D = \frac{k_1 L_a}{k_2 - k_1}\left[\exp(-k_1 t) - \exp(-k_2 t\right] + D_a \exp(-k_2 t)$$

where

D	= dissolved oxygen (DO) deficit (mg/L)
L_a	= initial concentration of BOD in the stream (mg/L)
D_a	= initial dissolved oxygen deficit (mg/L)
k_1 and k_2	= deaeration and reaeration coefficients (1/d)

The DO deficit D is the difference between the DO saturation concentration and in-stream DO concentration, C. The DO saturation concentration at 20C is 9.18 mg/L.

$$D\text{ (mg/L)} = 9.18\text{ mg/L} - C\text{ (mg/L)} \quad \text{or} \quad C\text{ (mg/L)} = 9.18\text{ mg/L} - D\text{ (mg/L)}$$

(a) Is it possible, or likely, that the DO at day 1.5 could be less than the critical concentration of $C = 6.0$ mg/L ($D = 3.18$ mg/L)?
(b) How does the standard deviation of the DO change as the DO concentration changes?

Answer these questions by simulating the dissolved oxygen deficit and estimating its standard deviation at travel times (t) of 1.5 days and 3.0 days for the conditions in Table P19.8.

TABLE P19.8 Average and Standard Deviations for the Dissolved Oxygen Sag Model Parameters

Parameter	Average	Standard Deviation
L_a (mg/L)	15	2.0
D_a (mg/L)	0.47	0.05
k_1 (d^1)	0.52	0.1
k_2 (d^1)	1.5	0.2

19.9 TRUCK FUELING STATION

The case study in Section 19.9 showed a simulation for a truck fueling station. Management wants to know if a second fuel pump will be justified. This depends on the amount of time being wasted at the pumps, specifically they want to know the:

(1) Average waiting time in the queue
(2) Average queue length
(3) Amount of time the fuel pumps are idle
(4) Percent of time the fueling system is being used

This can be done manually as a team exercise or individually as a computer simulation. Explain clearly how truck arrivals are generated. Simulate enough truck arrivals to demonstrate that you know how to evaluate the four system performance statistics given above.

19.10 TRUCK FUELING

City trucks arrive for fueling between 7 am and 12 pm at discrete intervals of 5, 10, 15, or 20 min with these probabilities:

40% of trucks 5 min between arrivals
30% of trucks 10 min between arrivals
20% of trucks 15 min between arrivals
10% of trucks 20 min between arrivals

The discrete arrivals times (5, 10, 15, and 20 min) are not realistic, because the time interval between arrivals could be fractions of minutes, but this simplifies some concepts and calculations.

A sequence of truck arrivals can be simulated by drawing uniform random numbers, $R_U(0, 1)$ that fall between 0 and 1.0. Table P19.10a shows how random numbers are associated with the interval between arrivals.

Fueling takes 10 min once a truck reaches the fuel pump. If a truck arrives at the end of a queue, it must wait in line. The time to refuel is a normal business expense, but waiting in a queue wastes money. The cost of a truck driver is $55/h.

The numbers of trucks arriving to refuel in October, a typical month, are given in Table P19.10b.

Management wants to know the following:

(1) Average waiting time
(2) Average queue length (there has been congestion in the drive that serves the fuel pumps)
(3) Amount of idle time at the fuel pumps
(4) Percent of time the fueling system is delivering fuel to trucks

A second fuel pump will be justified if the waiting time cost is high enough.

Simulate one day's operation at the fuel pump and calculate preliminary estimates of the items of interest to management.

TABLE P19.10a Distribution of Truck Arrival Times

If	Time to Next Arrival
$0 \le R_U < 0.4$	5 min
$0.4 \le R_U < 0.7$	10 min
$0.7 \le R_U < 0.9$	15 min
$0.9 \le R_U \le 1.0$	20 min

TABLE P19.10b Total Number of Trucks Fueled during October

Week	Mon	Tue	Wed	Thu	Fri
1	31	28	26	32	29
2	30	28	25	30	27
3	30	26	26	29	29
4	29	26	26	34	32
5	29	26	–	–	–

19.11 BLENDING SIMULATION

A process receiving material of variable equality should be insensitive to the variations. A simple means of protecting the process from input variations is a simple uncontrolled blender. This may be the only choice when rapid analysis of the composition of the incoming material is impractical. The blender is a vessel in which feed material is well mixed with material that has been retained from previous batches. The quality of the blend is a weighted time average of the material that has arrived.

For convenience imagine that the feed arrives in batches of equal volume. The quality of the batches varies according to no well-defined pattern; the variation is random and uncorrelated. The long-range average quality of the incoming batches is A, and the long-term average of the outgoing batches will be the same, assuming there is no reaction or loss in the blender. At any time the blender contains C batch volumes. When a new batch comes, an equivalent volume of the mixer contents is discharged. All batches have the same volume but the concentration varies.

For this special case, Rudd and Watson (1968) showed that the variance of the input to a bender, σ_A^2, and the variance of the output, σ_B^2, are related by

$$\frac{\sigma_B^2}{\sigma_A^2} = \frac{1}{2C - 1}$$

where C = blender volume measured in batches.

Notice that when $C = 1$, $\sigma_B^2 = \sigma_A^2$ and the blender does nothing. As C becomes large, σ_B^2/σ_A^2 tends to 0; a very large blender can smooth away all the variations. At more reasonable values, $C = 2$ gives $\sigma_B^2/\sigma_A^2 = 1/3$; $C = 5$ gives $\sigma_B^2/\sigma_A^2 = 1/9$.

Confirm this by simulating an input series of 1,000 batches having masses that are randomly and uniformly distributed over the range of 10–20 kg. Blend these in a mixer with a volume of two batches and a blender with a volume of five batches. For convenience, assume the batches have a constant volume of 1 unit. Calculate the average input and output quality (kg/batch), and the variances of the input and the output.

20 Designing for Safety and Reliability

People ask whether a practice, product, condition, or investment is reliable and safe, even while knowing that many activities entail some risk. Better questions are:

- What can go wrong?
- How likely is it that something will go wrong?
- What will be the consequences if it does go wrong?

Murphy's law says that anything that can go wrong will go wrong. Quality and productivity improvement are about listening when "Murphy speaks" in order to improve the system so that the same things don't go wrong again and again. The day-to-day operation of a system will tell us when something is wrong and usually suggests what is wrong. If we don't listen to the system, the same thing will go wrong again and again.

Designing for safety and reliability involves assessing the probability of an accident, equipment failure, and damage or injury resulting from the accident or failure. The goal is to prevent accidents (fires, explosions, leaks, etc.) or failures in performance (pump stoppages, electrical outages, etc.). Perhaps a utopian fail-safe system can be designed, but it probably cannot be constructed. If it could be constructed, perfect maintenance and operation would be required to obtain zero failures. Unfortunately, equipment and control devices wear out, operators change and they may become inattentive to routine safety procedures, and failures and errors occur.

20.1 FAILURE ANALYSIS

Failure analysis includes the following steps, but the order is not fixed.

1. Define the problem. Do not narrow the focus too early. Do not start with a conclusion and work backward to find evidence to support that conclusion.
2. Document what you see and what people tell you. There is always a risk of litigation.
3. Identify, collect, and preserve evidence, such as environmental samples, damaged equipment, data records, documents, operating logs, purchase requisitions, design drawings, photographs, personnel records, and so on. Assemble whatever materials are needed to provide a clear description of what happened.
4. Examine the evidence.
5. Reconstruct what may have happened. Let facts, evidence, and test results speak for themselves. The reconstruction must account for all established facts and be consistent with all hard evidence. Usually the most direct and simple explanation that is consistent with the evidence is the correct one.
6. Determine why the failure happened. Both physical and procedural causes must be considered. People's actions, responses to signals, or failure to act or respond require careful assessment.
7. Formulate conclusions and recommendations. Recommendations relate to preventing similar and related incidents and for improving the performance of the plant, process, equipment, or procedures.
8. Report the results. Unless the results and recommendations are clearly communicated to management, the investigative effort will be wasted.

TABLE 20.1 Causes of Failure

People	Environment	Equipment
Defective communication	Unforeseen conditions	Malfunction
Role confusion	Unexpected deterioration	Improper repair
Improper attitude	Unexpected response	Design deficiency
Perceptual flaws	Inadequate control of conditions	Inadequate maintenance
Misplaced priorities	Demands exceed system limits	Technical fault or defect
Ignorance and uncertainty		Wrong or substandard material
Conceptual misunderstanding		
Inattention or distraction		

The causes of major failures fall into three categories: people, environment, and equipment. Examples are listed in Table 20.1.

Example 20.1 Chlorine Leak Caused by Control Valve Failure

Chlorine is a toxic, corrosive, non-flammable liquefied gas that fumes on contact with moisture in the air. It is extremely irritating to muscles, membrane, eyes, and the respiratory tract. In extreme cases, damage to lung tissues may cause pulmonary edema. Prolonged exposure or high concentrations is fatal.

Pigment manufacturing is done in batches, using three 4,000 glass-lined reactors, as shown in Figure 20.1. About 2 tons of chlorine is used per batch. Three liquid chlorine cylinders (1,000 kg) are attached to the chlorine pipeline. The liquid is passed through a strainer to the vaporizer tank where controlled steam supply is provided at 100°C. The vaporized chlorine gas flows to a 5 m^3 cushion tank and from there to the chlorine reactor (Patil 2014).

The chlorination process was in progress at 15:30 h and about 2,300 kg chlorine had been added in the process. At 15:45 h, the operator manually closed the process control valve (PCV 3) upstream of the rotameter to take a quality control test sample. This test takes approximately 10 min. The pressure of the chlorine addition system is maintained between 0.4 and 0.6 kg/

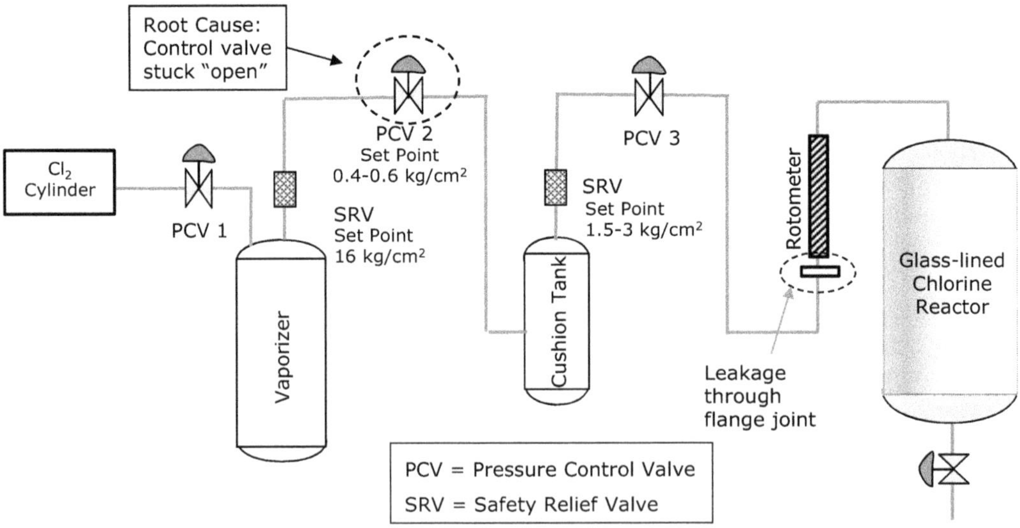

FIGURE 20.1 Chlorination process for pigment manufacturing.

cm^2 by PCV 1. During quality control test, the PCV 1 failed in an open position and the pressure in the system reached about 6 kg/cm^2. When the quality control test was completed, the operator opened the PCV 3, but the rise in pressure had caused the rotameter float to stick to the outlet of rotameter which stopped the chlorine addition to the reactor. The pressure rose to 6–10 kg/cm^2 and chlorine leakage was observed through a flange joint in the pipeline near the reactor. The operator immediately raised the alarm and shut off the valve manually.

It was discovered that the control valve was badly corroded. Eight people suffered minor injuries due to inhalation of chlorine gas and were admitted to a hospital. Everyone was discharged from the hospital after about 72 h.

The root cause was simple. The valve actuator was not properly lubricated. This caused it to stick in an open position where it could not control the chlorine pressure. Preventive maintenance of the control valve was not performed according to the preventive maintenance schedule.

The recommendations were:

- Regular preventive maintenance checks are needed of all safety relief valves, control valves, and equipment, including pipelines, flange joints, and tanks.
- Faulty equipment must be replaced immediately.
- Self-contained breathing apparatus should be provided and maintained in the factory. It should be periodically tested, and records should be maintained.
- The control room should be equipped with instruments for automatic detection of a small amount of chlorine gas.

20.2 PARETO DIAGRAMS FOR QUALITY IMPROVEMENT

Quality improvement programs define the dominant problems and solve them. The simple tools for recording data about quality and failures are run charts, scatter plots, histograms, and other visual displays (pie charts are not recommended). Plots of data are always helpful; tables of data usually are not except to construct graphical displays.

Check-sheets, such as Figure 20.2, are a quick and easy way of recording information on defects, failures, and customer complaints. Check-sheet entries record problems as they occur. The accumulated data can be displayed as a *Pareto diagram*, which is a tool for interpreting the information produced by the system.

The *Pareto principle*, or the law of the vital few, states that a large portion of all defects or complaints will be caused by a small number of conditions. A Pareto diagram shows which defects and problems are the most frequent. The quality improvement strategy is to work on the most frequently occurring problem and, when that is solved, to tackle the next most common problem.

Table 20.2 tallies and ranks by frequency of occurrence 8,711 pump failures due to mechanical causes. The data are plotted in Figure 20.3. Thirty percent of all failures were related to

Problem	Occurrences
High flow	✓✓✓✓✓ ✓✓✓✓
Calibration error	✓✓✓
Toxic spill	✓
Chemical shortage	✓✓✓✓✓ ✓✓
Power outage	✓✓
Truck breakdown	✓✓✓✓
Unknown	✓✓✓✓✓ ✓✓✓✓
Notes:	

FIGURE 20.2 An example check-sheet for a chemical treatment process.

TABLE 20.2 Data on 8,711 Instances of Pump Failures

Rank	Component	Number of Faults	Cumulative No. of Faults	Frequency (%)	Cumulative Frequency (%)
1	Mechanical seal	948	948	10.9	10.9
2	Gasket	752	1,700	8.6	19.5
3	Packings	750	2,450	8.6	28.1
4	Inboard bearing	710	3,160	8.2	36.3
5	Main bearing	706	3,866	8.1	44.4
6	Outboard bearing	704	4,570	8.1	52.5
7	Coupling	667	5,237	7.7	60.1
8	Gear and pinion	451	5,688	5.2	65.3
9	Impeller	439	6,127	5.0	70.3
10	Casing ring	360	6,487	4.1	74.5
11	Wearing ring	300	6,787	3.4	77.9
12	Casing	260	7,047	3.0	80.9
13	Fasteners	260	7,307	3.0	83.9
14	Inter cool	250	7,557	2.9	86.8
15	Suction pipe	244	7,801	2.8	89.6
16	Strainer	235	8,036	2.7	92.3
17	Inducer	225	8,261	2.6	94.8
18	Carbon ring	225	8,486	2.6	97.4
19	Flush pipe	225	8,711	2.6	100.0
Total		8,711			

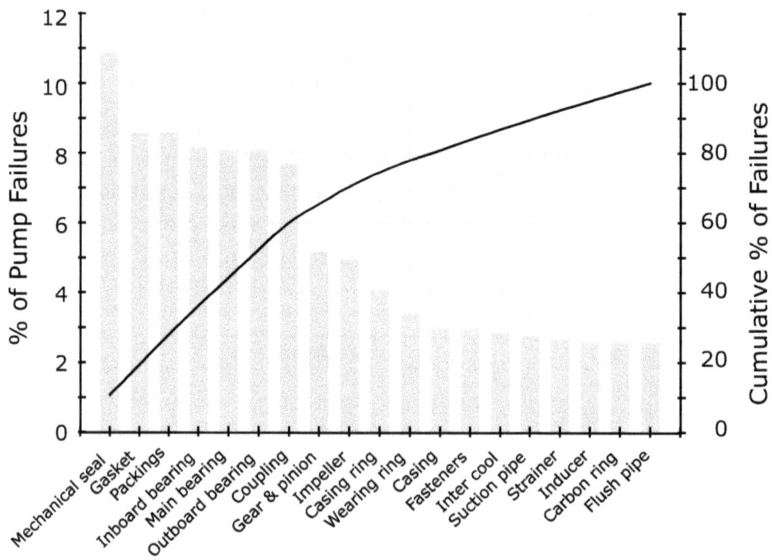

FIGURE 20.3 A Pareto diagram of cause of 8,711 mechanical pumps failures.

seals and gaskets; more than 50% were due to failures of packings and bearings. These failures could be subdivided into problems with lubrication, cooling, vibrations, misuse, or poor maintenance.

20.3 A CASE STUDY: ACTIVATED SLUDGE PROCESS FAILURE ANALYSIS

The activated sludge process consists of an aeration tank that is supplied with diffused air to provide dissolved oxygen for growing microorganisms coupled with a final clarifier (settling tank) that removes the microbial floc and a pump to return the microbial flow from the clarifier to the aeration basin. If the solids do not flocculate and settle properly, the process fails. This can happen when the dissolved oxygen concentration in the aeration tank is too low, the food to microorganism (F/M) loading rate is too high, or the system is shocked by an industrial spill. All of these conditions upset the balance of microorganisms needed for good settling. High flow rates reduce the detention time in the aeration tank and wash solids out of the final clarifier.

Figure 20.4 is the Pareto diagram of causes of upsets identified by examining five years of operating data from 15 well-operated activated sludge wastewater treatment plants (Berthouex and Fan 1986). Treatment technology today is more stable, and the effluent quality is much better than when these data were collected (1980–1985).

Upset was defined as an interval of time, which could be one day or a series of days when the effluent BOD or suspended solids were unusually high. It does not mean process failure or discharge permit violation. A variety of statistical methods were used to identify upset events in the data records. The causes of upsets were assigned during a plant visit to check the operator's log and interview the operators, who were remarkably good at this kind of diagnosis.

The data are given in Table 20.3. There were 1,624 days with upsets out of 27,385 days of operation at the 15 plants. This means that 5.9% of the operating days experienced some kind of an upset. Only 9% of the 1,624 upsets could not be assigned a cause; these were classified as unknown. Thirty-five of the upsets lasted only one day; 25% lasted for two days.

Low dissolved oxygen (DO) was the most common cause of upsets; 19% of upsets (315/1,624 = 0.194) were DO related. The 106 days of "aeration tank down" were mostly for cleaning air diffusers.

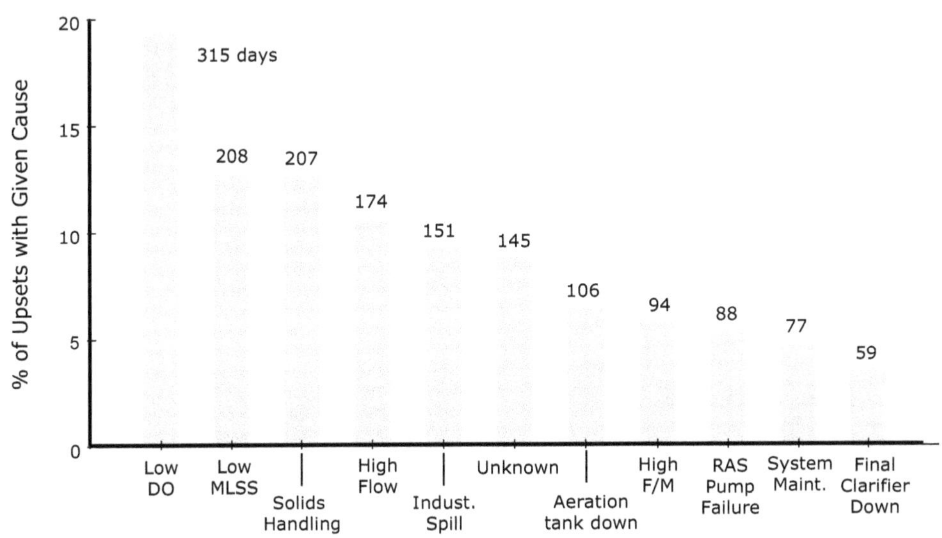

FIGURE 20.4 A Pareto diagram showing the causes of upsets in wastewater treatment plants.

TABLE 20.3 Number of Days of Plant Upsets and the Assigned Causes (Berthouex & Fan 1986)

Cause of Upset	Days of Upsets	Cumulative Days of Upsets	% of Upsets with Given Cause	Cumulative % of Upsets
Low dissolved oxygen	315	315	19.4	19.4
Low mixed liquor solids	208	523	12.8	32.2
Solids handling	207	730	12.7	45.0
High flow	174	904	10.7	55.7
Industrial spill	151	1,055	9.3	65.0
Unknown cause	145	1,200	8.9	73.9
Aeration tank down	106	1,306	6.5	80.4
High *F/M* ratio	94	1,400	5.8	86.2
RAS pump failure	88	1,488	5.4	91.6
System maintenance	77	1,565	4.7	96.4
Final clarifier down	59	1,624	3.6	100.0

As more was learned about the demand for oxygen in the activated sludge process, better diffusers were developed, and aeration systems were redesigned to deliver the proper volumes of air (Mueller & Boyle 1988, Boyle 1988, 2019). Solving the aeration problem made it easier to deal with some of the other difficulties. Low mixed liquor solids and high *F/M* ratio (which in a way are the same problem) can be related to low dissolved oxygen, as can certain kinds of solids-handling problems.

Another historical change was the improved design of the final clarifiers. Larger surface areas (lower overflow rates) were used for better retention of solids, and collection weirs, inlet structures, and sludge removal methods were improved.

20.4 HAZARDS AND RISKS

A *hazard* poses a theoretical risk of harm to life, health, property, or the environment. Trucks and rail cars that haul hazardous materials are clearly marked with the familiar sign, shown in Figure 20.5, that tells emergency responders exactly what hazard they face in case of an accident.

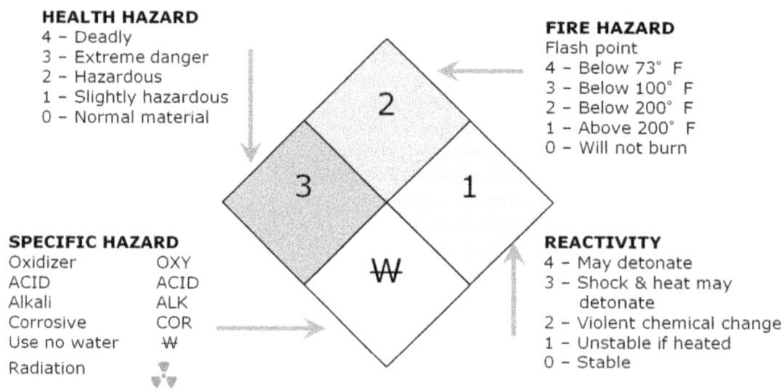

FIGURE 20.5 Hazardous material identification symbol showing the types of active hazard that could be created by an accidental release.

This sign shows that materials can be hazardous in many different ways: toxicity, flammability, and strong chemical reactions, including possible explosions.

Risk depends on the probability that a critical event (a *hazard*) will happen and the probability of exposure to that event.

$$\text{Risk} = (\text{Probability of a critical event})(\text{Probability of exposure to the event})$$

To have actual harm there must be both a hazardous event and an exposure. Toxicity is an intrinsic property of some chemicals, just as the heat of combustion is an intrinsic property of gasoline. The heat of combustion represents the potential for something to happen. Nothing happens until gasoline is ignited. What happens after ignition can be safe and useful, as in driving an automobile, or it could be a destructive explosion.

Likewise, toxicity represents a potential to do harm. Nothing will happen until an organism is exposed, and even then, nothing bad will happen unless the level of exposure is high enough for a sufficiently long time. *Toxicity risk assessment* is the business of learning what is "high enough" and "long enough." *Toxicity risk management* is the business of making the exposure low enough and short enough to guarantee safe living and working conditions.

Risk assessment models are simplifications of the real world that should rely heavily on experience and data, but at times must use assumptions and subjective judgments. This gives an element of uncertainty, but models that are wrong with respect to the small details can be correct with respect to the larger issues.

Risk is reduced by good engineering that reduces a hazard by limiting the dose or the exposure, or both. For example:

- Change raw materials used in manufacturing to eliminate the toxic substance.
- Provide waste treatment to reduce the discharge or emission of toxic substances.
- Encapsulate or contain the toxic substance to prevent exposure.
- Install barriers to keep people away from the substance.
- Control how the material is transported.

Example 20.2 Risk Of Exposure to a Carcinogenic Chemical

Figure 20.6 shows the lifetime risk of an individual getting cancer if exposed to a carcinogen at a level that causes cancer on average once in every 100,000 exposures. The risk to the individual is 1/100,000. The risk of added cancer deaths for an exposed population of 1,000,000 to a dose that causes cancer once in 100,000 exposures is (1/100,000)(1,000,000) = 10 cases of cancer.

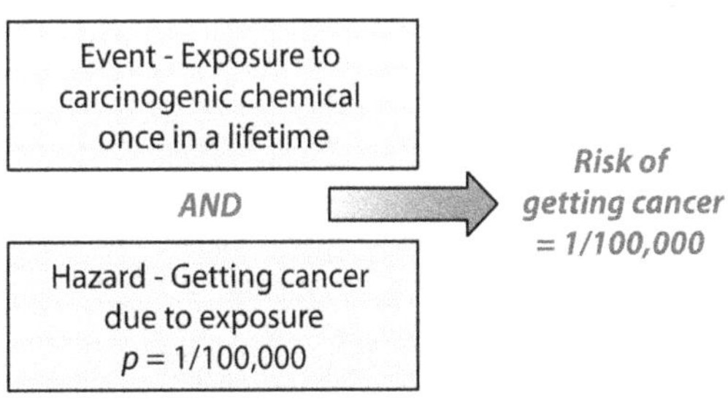

FIGURE 20.6 Illustration of risk as the combination of exposure and probability of exposure.

Example 20.3 Risk of Electrocution

Every year 200 people die accidentally in the United States by electrocution. Using the US population of 327.2 million people (in 2018), the annual risk of death by accidental electrocution is

$$\text{Annual risk} = (200\ \text{deaths/y})/(327.2 \times 10^6\ \text{people}) = 0.611 \times 10^{-6}/\text{y}$$

Taking 70 years as the average lifetime of an individual, we have a personal lifetime risk of electrocution of

$$\text{Personal lifetime risk} = (70\ \text{y})(0.611 \times 10^{-6}/\text{y}) = 42.8 \times 10^{-6}$$

20.5 RISK-MAPPING

Risk mapping is a semiquantitative risk management tool that combines subjective information (expert opinion) and objective information (measurements and calculations). Different approaches can be used depending on the amount and kind of information available. A hierarchy of decisions or actions can be prepared as part of the exercise.

Figure 20.7 is a risk profile matrix. The factors are the likelihood of an event occurring and the severity of the consequences if it does occur. The likelihood of an event ranges from being near zero (say less than 3%) to frequent (say more than 90%). The severity of the consequences, should an event happen, ranges from insignificant to catastrophic. Each factor is rated on a subjective scale of 1–5 and they are arbitrarily given weights of 1–5. The numbers in the cells are the products of the marginal weights. For example, a possible event with a major consequence has a risk score of (3)(4) = 12.

Proactive measures are needed to prevent the risk combinations in the lower right corner of the table. The goal is to reduce the likelihood of an event and to mitigate the damage if an event should occur.

		Consequence (Cost) of Failure				
Event Likelihood		Insignificant 1	Minor 2	Moderate 3	Major 4	Catastrophic 5
Rare	1	1	2	3	4	5
Unlikely	2	2	4	6	8	10
Possible	3	3	6	9	12	15
Likely	4	4	8	12	16	20
Almost Certain	5	5	10	15	20	25

FIGURE 20.7 A risk profile matrix for a semiquantitative risk assessment.

Event Likelihood		Consequence (Cost) of Failure: Insignificant 1	Minor 2	Moderate 3	Major 4	Catastrophic 5
Rare	1	1	2	3	4	5
Unlikely	2	2	4	6	8	10
Possible	3	3	6	9	12	15
Likely	4	4	8	12	16	20
Almost Certain	5	5	10	15	20	25

Cast-iron pipe

Chlorine pump

FIGURE 20.8 Risk profile matrix to rate asset for criticality.

Example 20.4 Criticality of Risk

Criticality depends on the probability of failure and the severity of the consequences of failure. Figure 20.8 shows the criticality rating for a cast-iron pipe and a chlorine delivery pump.

(a) Asset: A 56-y old, 10-inch cast iron pipe

Likelihood of failure = level 4. The pipe has had numerous breaks in the past five years. When it is repaired, it is still in reasonable condition.

Consequence of failure = level 2. The pipe serves three major subdivisions of only residential customers. A bypass is available for emergency service. The line is not in a major roadway, so repair is relatively easy.

(b) Asset: A chlorine pump delivers a solution of hypochlorite into the system for disinfection

Likelihood of failure = level 4. The pump has failed many times in the past ten years. The system has spare parts and a spare pump.

Consequence of failure = level 4. A failure may have major consequences because there is no continuous monitoring of the chlorine level in the water system and the chlorine level could be low for several days before the problem is discovered. The consequences are mitigated by the spare parts and spare pump.

20.6 THE RELIABILITY OF SYSTEMS

Reliability is the probability that a system will function when needed. The structure of a system affects the system reliability.

The strength of a chain is determined by the weakest link, not by the average strength of the links. Two chains hooked together end to end still have the strength of the weakest link in either chain. In contrast, if two chains equally share the load, the strength of the two chains is greater than one chain.

Figure 20.9 shows three systems. The series system will function only when components A and B both function. The parallel system will function if either A or B functions. The hybrid system is even more robust, with parallel redundancy in stages A and B.

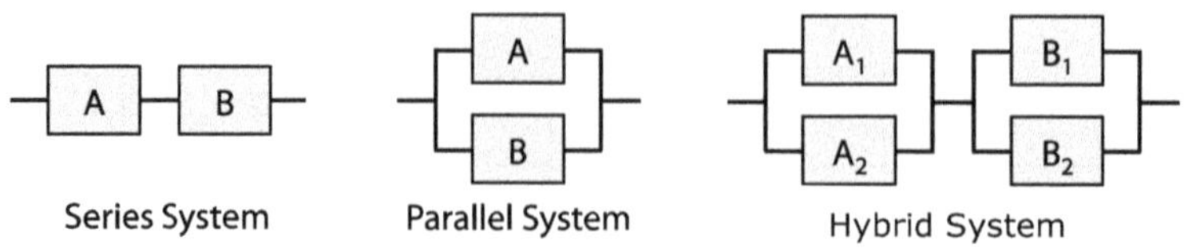

FIGURE 20.9 Structure of three systems.

The reliability of a serial system is less than the reliability of any single component. The reliability of a parallel system is greater than the reliability of a single component.

Reliability is calculated from the probabilities for the success (or failure) of elements of the system. This will be illustrated for the simplest situation, which is to assume that the failure events are random and independent. Events are independent if the occurrence of one event does not cause or prevent the occurrence of the other. Events are random if the probable state of one cannot be predicted by the probable state of the other. (*Note:* Assuming independent failures in complex systems can underestimate the probability of failure.)

Two independent events, A and B, have probabilities of occurrence of $P(\text{A})$ and $P(\text{B})$. The probability that both A *and* B will occur is

$$P(\text{A and B}) = P(\text{A})P(\text{B})$$

The probability that either A *or* B will occur is

$$P(\text{A or B}) = P(\text{A}) + P(\text{B})$$

The probability that one or both will occur is

$$P(\text{A and/or B}) = P(\text{A}) + P(\text{B}) - P(\text{A and B})$$

These three rules are sufficient for all problems that involve the probability of independent events.

Example 20.5 Independent Serial Units Subject to Random Failure

The system in Figure 20.10 with three units in series fails if any unit fails and succeeds if all units succeed. The three units have reliabilities of R_1 = 9/10 (90%), R_2 = 8/10 (80%), and R_3 = 98/100 (98%). A reliability of 9/10 means that there is a 90% probability that unit 1 will be functional when needed.

If the probabilities of success or failure are independent, the probability that all three units are simultaneously successful is

$$R_{\text{System}} = (R_1)(R_2)(R_3) = (0.9)(0.8)(0.95) = 0.684$$

There is a 68.4% probability the system will function and a (100% − 68.4%) = 31.6% probability that it will fail.

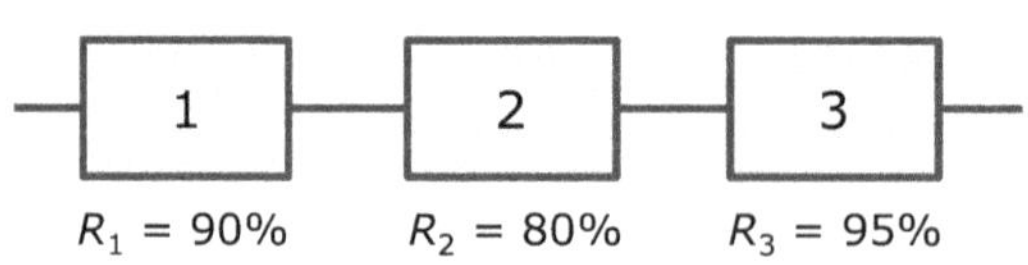

FIGURE 20.10 Three units in series.

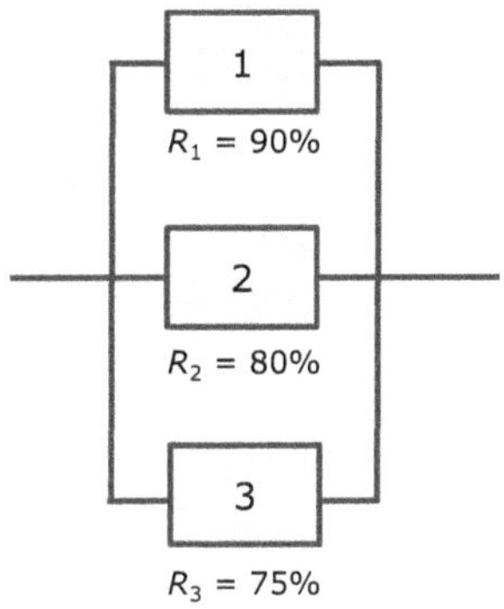

FIGURE 20.11 Three units in parallel.

Example 20.6 Three Units in Parallel

The parallel system shown in Figure 20.11 will serve its design function so long as any one component is operable. The three units have reliabilities of R_1 = 9/10 (90%), R_2 = 8/10 (80%), and R_3 = 75/100 (75%). The system will fail if all three components fail. Assume independence of the components with respect to failure.

The probability of system failure, F_S, is the probability that all components fail is

$$F_S = (1-R_1)(1-R_2)(1-R_3) = (1-0.90)(1-0.80)(1-0.75) = 0.005$$

or 0.5% probability of system failure.

The probability that at least one unit is functional is the system reliability, R_S.

$$R_S = 1.0 - F_S = 1.000 - 0.005 = 0.995 \text{ or } 99.5\%$$

Example 20.7 Reliability of a Three-Component Parallel System

Suppose that the parallel system in Figure 20.11 succeeds when any two of three units will operate as needed. System failure is when any two of three components, or all three components fail simultaneously.

An easy way to find the system reliability is by complete enumeration of the eight possible conditions, as shown in Table 20.4. The sum of the probabilities for all eight possible states must

TABLE 20.4 Eight Possible States for a System of Three Independent Components in Parallel

State	Unit 1	Unit 2	Unit 3	System State	Probability	
1	S	S	S	S	(9/10)(8/10)(3/4) =	216/400 = 0.540
2	S	S	F	S	(9/10)(8/10)(1–3/4) =	72/400 = 0.180
3	S	F	F	F	(9/10)(1–8/10)(1–3/4) =	18/400 = 0.045
4	S	F	S	S	(9/10)(1–8/10)(3/4) =	54/400 = 0.135
5	F	S	S	S	(1–9/10)(8/10)(3/4) =	24/400 = 0.060
6	F	F	S	F	(1–9/10)(1–8/10)(3/4) =	6/400 = 0.015
7	F	S	F	F	(1–9/10)(8/10)(1–3/4) =	8/400 = 0.020
8	F	F	F	F	(1–9/10)(1–8/10)(1–3/4) =	2/400 = 0.005
					Total =	400/400 = 1.000

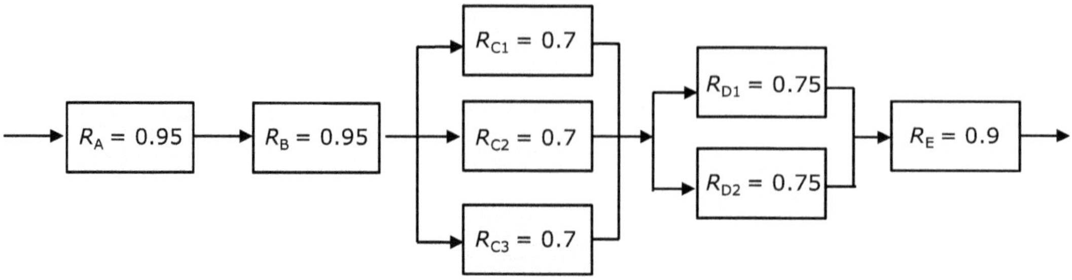

FIGURE 20.12 Compound system with five stages.

add up to 1. Four conditions (1, 2, 4, and 5) are successful. The probability of system success is the sum of the four probabilities.

$$216/400 + 72/400 + 54/400 + 24/400 = 366/400 = 0.915\ (91.5\%)$$

The probability of system failure is 100% – 91.5% = 8.5%.

Example 20.8 Compound System

Compound systems have series units and modules of parallel units. Figure 20.12 shows a compound system with five stages, two of which have parallel units.

Assuming the component reliabilities are independent, the stage reliabilities are

$$R_A = 0.95 \qquad R_B = 0.95 \qquad R_E = 0.9$$

$$R_C = 1 - (1 - 0.7)^3 = 1 - 0.027 = 0.973$$

$$R_D = 1 - (1 - 0.75)^2 = 1 - 0.0625 = 0.9375$$

The system reliability is

$$R_S = (0.95)(0.95)(0.973)(0.9375)(0.9) = 0.7409$$

20.7 CONDITIONAL PROBABILITY

Assuming that events or failures are independent allows a simple introduction to probability theory. Natural and engineered systems are not so simple. For example, the probability of a rear end collision is changed if the lead driver brakes suddenly, if the following driver is inattentive, if the street is snow covered or icy, or if visibility is poor.

Conditional takes into account that the probability that the occurrence of event A may depend on the occurrence of another event, B.

The probability of A occurring given that B occurs is

$$P(\text{A given B}) = \frac{P(\text{A and B})}{P(\text{B})}$$

Conditional probability is used to reevaluate the probability of an event in light of additional information. P(A given B) is an "updating" of P(A) based on the knowledge that B occurred.

Example 20.9 Parallel Components

A system consists of components A and B in parallel. Successful operation requires that both A and B function when needed. The probability that A functions when needed is 0.8, and the probability that B functions as needed is 0.9. However, if B performs satisfactorily during its design life, the probability of A also doing so increases to 0.95.

The performance of A and B are not independent, because the reliability of A increases from 0.8 to 0.95 when B performs satisfactorily. This is stated as the conditional probability of A functioning, given that B functions, or

$$P(\text{A given B}) = 0.95$$

The probability that both components will perform when needed is $P(\text{A and B})$, which is calculated by rearranging the conditional probability law:

$$P(\text{A and B}) = P(\text{A given B})P(\text{B}) = (0.95)(0.9) = 0.855$$

Note this is different from the probability if the two components acted independently:

$$P(\text{A and B}) = P(\text{A})P(\text{B}) = (0.8)(0.9) = 0.72$$

Example 20.10 Pump Failure

A system consists of two identical pumps A and B, and the system will operate if one pump fails. The added load on the operating pump makes it more likely to fail than when the load was shared by two pumps. Data show that at least one pump fails by the end of the pump design life in 7% of all systems and the probability that both pumps fail during that period in only 1%. Compare the probability of pump A failing, assuming independent events, with the conditional failure probability if pump A given the failure of pump B.

(a) The probability that either pump A or pump B will fail is found using the third rule in Section 20.6.

$$P(\text{at least one pump failing}) = P(\text{A and/or B}) = P(\text{A}) + P(\text{B}) - P(\text{A and B}) = 0.07$$

$$P(\text{both pumps failing}) = P(\text{A and B}) = 0.01$$

Since the pumps are identical, $P(\text{A}) = P(\text{B})$

Substituting into $P(\text{A and/or B})$ gives

$$P(\text{A and/or B}) = 0.07 = P(\text{A}) + P(\text{A}) - 0.01$$

and

$$P(\text{A}) = 0.04 \quad \text{and} \quad P(\text{B}) = 0.04$$

The probability that both pumps fail when needed is

$$P(\text{A and B}) = P(\text{A})P(\text{B}) = (0.04)(0.04) = 0.0016$$

(b) The conditional probability of pump A failing when pump B fails is

$$P(\text{A given B}) = \frac{P(\text{A and B})}{P(\text{B})} = \frac{0.01}{0.04} = 0.25$$

This is also the risk that both pumps are out of service and it is much higher than the risk that is estimated for independent failures. Beware of assuming independent events without first considering how the system structure may create conditional failures.

Example 20.11 Conditional Pump Operation

A pump has been working satisfactorily for five years, what is the probability that it will continue working for at least another three years, for a total of at least eight years. The failure rate for this type of pump is 0.2/y.

Event A = pump works at least eight years
Event B = pump works at least five years
Define the probabilities of event A as *P*(A) and event B as *P*(B).

The individual probabilities are calculated using the exponential time to failure model.

$$P(\mathrm{A}) = \exp[-(0.2/\mathrm{y})(8\mathrm{y})] = 0.202$$

$$P(\mathrm{B}) = \exp[-(0.2/\mathrm{y})(5\mathrm{y})] = 0.368$$

Noting that the order of the "and" events is interchangeable:

$$P(\mathrm{A\ and\ B}) = P(\mathrm{B\ and\ A})$$

and

$$P(\mathrm{B\ and\ A}) = P(\mathrm{B\ given\ A})P(\mathrm{A})$$

so that

$$P(\mathrm{A\ and\ B}) = P(\mathrm{B\ given\ A})P(\mathrm{A})$$

Substituting gives

$$P(\mathrm{A\ given\ B}) = \frac{P(\mathrm{A\ and\ B})}{P(\mathrm{B})} = \frac{P(\mathrm{B\ given\ A})\ P(\mathrm{A})}{P(\mathrm{B})}$$

Here *P*(B given A) = 1.0 because a pump that worked for at least eight years also worked for five years. Thus,

$$P(\mathrm{A\ given\ B}) = \frac{P(\mathrm{B\ given\ A})\ P(\mathrm{A})}{P(\mathrm{B})} = \frac{(1.0)(0.202)}{0.368} = 0.549$$

Because we know the pump has worked for five years, the probability that it will work another three years has been "updated" to 0.549 from a value of 0.368, if we know nothing of the pump's working history.

20.8 THE BINOMIAL THEOREM

Suppose that three units are installed and two must operate at all times; failure of one unit is allowed. These are the combinations for system success:

1. All 3 units operate
2. Unit 1 fails and units 2 and 3 operate
3. Unit 2 fails and units 1 and 3 operate
4. Unit 3 fails and units 1 and 2 operate

The probability of system success is

$$R_S = R_1R_2R_3 + (1-R_1)R_2R_3 + R_1(1-R_2)R_3 + R_1R_2(1-R_3)$$

If the reliabilities are equal for all units,

$$R_S = 3R^2 - 2R^3$$

This result can be obtained using the binomial theorem. This can only be used when the probability of success is the same for all units and the units act independently.

A system has n parallel units and k must operate for the system to succeed. From the binomial distribution, the probability that exactly k out of n units will successfully operate is $P(k|n)$:

$$P(k \mid n) = \binom{n}{k} p^n q^{n-k} = \frac{n!}{k!(n-k)!} p^n q^{n-k} = \frac{n!}{k!(n-k)!} p^n (1-p)^{n-k}$$

where

p = probability of success and $q = (1 - p)$ is the probability of failure.

The probability of obtaining at least k successes is

$$P(\text{at least } k) = \sum_{k}^{n} \binom{n}{k} p^k (1-p)^{n-k} = \sum_{k}^{n} \frac{n!}{k!(n-k)!} p^k (1-p)^{n-k}$$

The expanded function for a system with $n = 3$ units that operates when $k = 2$ units succeed is

$$P(\text{at least } k=2) = \sum_{k=2}^{3} \binom{n}{k} p^n (1-p)^{n-k} = \frac{3(2)(1)}{(2)(1)(1)} p^2 (1-p)^1 + \frac{3(2)(1)}{(3)(2)(1)(1)} p^3 (1-p)^0$$

$$= 3p^2(1-p) + p^3 = 3p^2 - 2p^3$$

20.9 RELIABILITY AND REDUNDANCY

Series systems are less reliable than the weakest element. Increasing the number of components in series reduces the system reliability. Making each component more reliable, or improving preventive maintenance, or scheduled replacement of failure-prone parts also will strengthen the system.

In contrast, parallel systems are more reliable than the weakest component. Reliability is further increased by adding active or passive redundant elements to parallel systems.

Again assuming independent events, the reliability for an n component parallel system in which system failure is the simultaneous failure of all components,

$$R_S = 1 - \prod_{i=1}^{n} (1 - R_i)$$

For $n = 2$,

$$R_S = 1-(1-R_1)(1-R_2) = 1-(1-R_1-R_2+R_1R_2) = R_1+R_2-R_1R_2$$

If $R_1 = R_2 = R$, then $R_S = 2R - R^2$.

TABLE 20.5 Diminishing Returns for Redundancy in Parallel Systems

Number of Components	Reliability		
	Component	System	Marginal % Increase
1	0.9	$1 - 0.1^1 = 0.9$	
2		$1 - 0.1^2 = 0.99$	10.0
3		$1 - 0.1^3 = 0.999$	0.91
4		$1 - 0.1^4 = 0.9999$	0.09
1	0.95	$1 - 0.05^1 = 0.95$	
2		$1 - 0.05^2 = 0.9975$	5.0
3		$1 - 0.05^3 = 0.99987$	0.24
4		$1 - 0.05^4 = 0.99999$	0.01
1	0.97	$1 - 0.03^1 = 0.97$	
2		$1 - 0.03^2 = 0.9991$	3.0
3		$1 - 0.03^3 = 0.999973$	0.09
4		$1 - 0.03^4 = 0.999999$	0.003
1	0.99	$1 - 0.01^1 = 0.99$	
2		$1 - 0.01^2 = 0.9999$	0.01
3		$1 - 0.01^3 = 0.999999$	0.0001

For $n = 3$ parallel units, all having the same reliability R, the system reliability is

$$R_S = 3R - 3R^2 + R^3$$

Parallel systems follow the law of diminishing returns – the increased reliability decreases for each added unit. Table 20.5 shows that increasing the number of parallel units is more interesting when the individual components have lower reliability.

Two pumps with a reliability of 95% give a system reliability equal to four pumps with reliability of 0.9. This allows the designer to balance savings (or lower losses), reliability, and system cost.

20.10 EVENT TREES

An *event* tree, or *decision tree*, has a branching structure in which each branching point (node) represents the test of an attribute that has two chance event outcomes. They show the opportunities for failure of physical components and human operational actions.

A *fault tree* is a related tool that visually connects events and causes that lead to failure and helps to organize and complete a reliability analysis.

Figure 20.13 is an event tree for determining the risk of damage if a leak develops in a chlorine storage tank. The risk of a leak is 0.01 per year. If the chlorine leak is detected by the alarm, the alarm will alert the operator. The operator may or may not respond to the alarm, and a response may be followed by the correct action or not.

There are four possible outcomes. Three lead to the possibility of damage. One outcome has no risk of damage. The probabilities of the four possible outcomes sum to 100% (there is either risk or no risk).

The probability that the alarm functions is 98%. The probability that the operator responds is 90%. The probability that the proper corrective action is taken is 90%. When each of these four

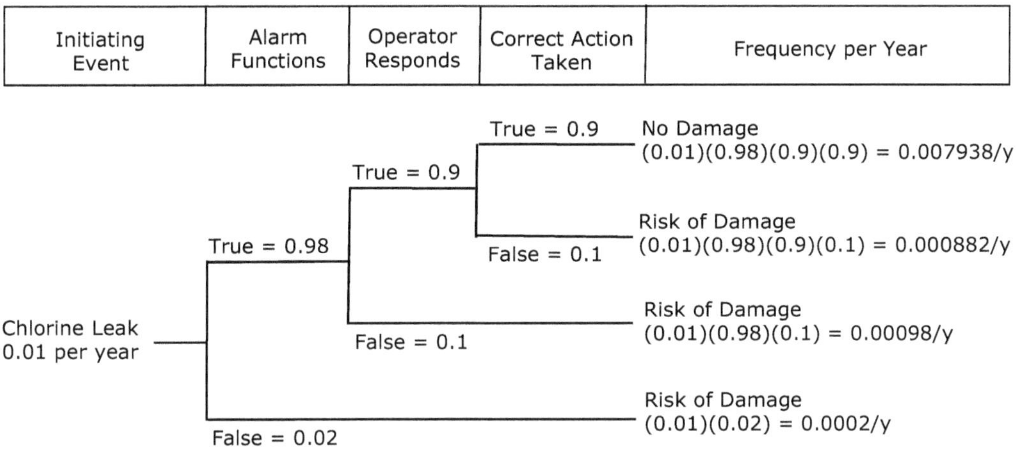

FIGURE 20.13 Event tree for response to a chlorine leak.

events is handled correctly, there will be no damage. This is the sequence of events at the top of the event tree. The probability of no damage is (0.98)(0.90)(0.90) = 0.7938 each time there is a chlorine leak. Given a rate of 0.01 chlorine leaks per year, this is (0.01)(0.7938) = 0.007938 events per year that are handled correctly.

The event tree shows that the process for responding to a chlorine leak fails with probability 0.0882 + 0.098 + 0.02 = 0.2062, due to failures in the alarm, operator response, or appropriate action stages. Given a rate of 0.01 chlorine leaks per year, this is (0.01)(0.2062) = 0.002062 events per year that might cause damage.

The event tree shows the weak steps in the protection scheme. In this case, they are the human elements. The operator may not respond to the alarm because it is not loud enough or it is in the wrong location. If so, this is a hardware problem that can be corrected. Failure to take the correct action is a human problem that should be solved by training and practice. Or, perhaps the response and action can be automated.

20.11 RELIABILITY AND HUMAN ERRORS

Human reliability is the probability of successful performance of only those human activities that are necessary to make a system work. Human error is some human output which is outside the tolerances established by the system requirements in which the person operates. Routine human actions are explicitly included in documents, procedures, specifications, etc. Non-routine actions require a cognitive process of understanding and decision-making prior to taking action.

The key events in Figure 20.13 were operator actions. A *human action representation tree* is a logic structure similar to the event tree that explains the different activities (decision and/or actuation) that operators could do when they have to respond to an actual situation.

Failure to respond to the alarm is an act of omission. Taking the wrong action is an act of commission.

The target failure rate for catastrophic events is 10^{-9} per year. This can never be achieved when humans are involved because human failure rates are in the ranges of

$< 0.001\%$ for routine repetitive acts
5×10^{-4} to 5×10^{-2} for rule-based acts
5×10^{-3} to 5×10^{-1} for knowledge-based acts

TABLE 20.6 Human Error Rate Table (Smith, D 2005)

Activity	Error Rate (per Task)
Simplest possible task	0.0001
Read a single alphanumeric wrongly	0.0002
Read a five-letter word wrongly	0.0003
Select wrong switch	0.0005
Repetitive simple task	0.001
Read a checklist or digital display wrongly	0.001
Set switch (multiposition) wrongly	0.002
Check wrong indicator in a display	0.003
Wrongly carry out visual inspection (e.g., leak)	0.003
Read analog indicator wrongly	0.005
Routine task with care needed	0.01
Mate a connector wrongly	0.01
Fail to reset valve after some related task	0.02
Record information or read graph wrongly	0.01
Type or punch character wrongly	0.01–0.02
Do simple arithmetic wrongly	0.02
Do simple algebra wrongly	0.02
Complex, infrequently performed task	0.1
Fail to notice incorrect status in roving inspection	0.1
Fail to notice wrong position of valves	0.5
Fail to act correctly after 1 min of emergency situation	0.9

Most human reliability data, such as given in Table 20.6, is collected in the laboratory, a simulator, or the field. Data are rarely available for the same system and the same circumstances, especially for a new design.

The human error rate will increase when acting under extreme stress. An action that has reliability of 99.9% under ideal conditions could be only 70% under great stress.

20.12 AVAILABILITY

Availability is a term that is used for repairable components. Reliability generally refers to non-repairable components, that is, components that are replaced when they fail. Pumps, motors, blowers, boilers, and generators are repairable. Valves may be repaired or replaced. Electrical components are replaced.

A standard design is to provide at least two pumps, each capable of handling the peak demand. Or the design might include more than two pumps with identical or different capacities. That will depend on the range of flows that must be handled, how much storage is available, and other local conditions. The time between a critical loss of pumping capacity and the time to bypass or pump station flooding is from 15 minutes to two hours.

Example 20.12 Pumping Station Availability

The hydraulic availability of a pumping station is defined as the fraction of time that the pumping station can meet the required flow capacity. Typically, several pumps are installed to improve reliability and adjust to changes in flow requirements.

TABLE 20.7 Analysis of Pump Station Capacity (Based on Operating 8,760 h/y)

State of System	Pumping Capacity (m^3/min)	Required Pumps Available	Probability Capacity Is met	Time Required (h/y)	Fraction of Time	Weighted Probability of Success
State 1	>44–56	A + B + C	0.9638	560	0.0639	0.0616
State 2	>25–44	State 1 plus A + B, A + C, B + C	0.9995	6,200	0.7078	0.7076
State 3	>15–25	State 2 plus A, B, or C	0.99994	2,000	0.2283	0.2283
	Probability the pumping station supplies the required capacity =					0.9975

A pumping station has three pumping units of equal capacity each single pump can discharge 25 m^3/min. Two pumps operating in parallel can discharge 44 m^3/min, and three pumps in parallel can discharge 56 m^3/min.

The fraction of time that the pumping station will be at a specific condition, based on 8,760 operating hours per year, is given in Table 20.7. The probability that the specific condition will be satisfied is the system reliability.

Each pump has a probability of failure of

$$F = 0.0122$$

giving a reliability, which is the same for all pumps, of

$$R_A = R_B = R_C = 0.9878$$

State 1: All three pumps operating
Probability of the system being in state 1 is

$$R_1 = R_A R_B R_C = (0.9878)(0.9878)(0.9878) = 0.9638$$

State 2: At least two pumps are available
The probability of the system being in state 2 is the probability of state 1 plus the probability of any two pumps available

Probability that pump A is not available, and pumps B and C are available is

$$(1 - R_A) R_B R_C = (0.0122)(0.9878)(0.9878) = 0.0119$$

Two of three pumps can be available in three ways; A + B, B + C, and A + C
Total probability of at least two of three pumps available = 3(0.0119) = 0.0357

$$R_2 = R_1 + 0.0357 = 0.9638 + 0.0357 = 0.9995$$

State 3: At least one pump is available
The probability of the system being in state 3 is the probability of state 2 plus the probability of any one pump available.

Probability that pump A is available, and pumps B and C are not available is

$$(R_A)(1 - R_B)(1 - R_C) = (0.9878)(0.0122)(0.0122) = 0.000147$$

Total probability of at least one of three pumps available = 3(0.000147) = 0.000441

$$R_3 = R_2 + 0.000441 = 0.9995 + 0.000441 = 0.999941$$

State 4: No pumps are available.
Probability is negligible (less than 2/100,000).

The state probabilities should be rounded to about 6%, 70%, and 23%, and the total probability should be stated as greater than 99.5%.

The probability of a pump failure, given as 0.0122, is an uncertain value, even though it may be based on failure data from a large number of pumps. The working environment of the pump and its maintenance will alter the failure rate. Under specific conditions, it could be 0.01–0.015.

Some comments and cautions are needed. The number of decimal places shown in the above calculation (the mathematical solution) gives a misleading impression about how well we can evaluate reliability (the engineering solution). The values put into the calculations are not known with a high degree of precision. It would be more correct to say that the reliability of a pumping unit is about 98% or 99%, and the probability of having only two pumps in operation is "99% or less" and that the probability of having only one pump operational is so small that it has no practical importance.

Under the assumption of equal pumping unit capacities and reliabilities, the hydraulic reliability and the mechanical reliability are the same. They are not if the pumping units have different capacities, or if the pumping units have different mechanical reliability. It will be true, nevertheless, that a pumping station will have a finite number of pump combinations, each having a known discharge and reliability, that can be enumerated and evaluated.

What are the consequences of not being able to meet the demand of, say 50 mgd, because one pump is not available? A demand of this level occurs about 6% of the time, but it occurs at intervals, perhaps once a week. Water distribution systems have storage tanks to increase reliability. Interceptor sewers, which is where pumping stations are usually found, do not flow full, so there is some available in-pipe storage that might provide a short interval for repairs to be made. Treatment plants often have pumps at the influent and sometimes at the effluent. Multiple pumps are installed, usually with sufficient capacity to meet the peak demand with largest pump out of service.

20.13 FAILURE RATES

All components and equipment are subject to failure during their normal lifetime of operation. These can be classified as *early* failures (infant mortality), *random failures* during a long stage of normal operation, and *wear-out* failures (old age). Figure 20.14, the so-called *bathtub curve*, shows three life stages for equipment.

Early failures are avoided by preoperational testing, or *burn-in*. This can be stress under constant operating conditions, power cycling (stress under turn-on and turn-off cycles), temperature cycling, and other methods. Mechanical equipment tends not to have infant mortality. All pollution control equipment is tested after installation before it is accepted by the owner.

Wear-out is failure due to the end of life that comes with a long useful life.

The normal useful lifetime starts when early failures have been eliminated and ends when wear-out failures begin. Things will go wrong and parts will break during normal operation, but the failures are *random*. Failures are not a function of the component's age. *Random* means that even though the user knows that breakdowns and repairs will be needed, there is no way to predict when they will happen except on a statistical basis.

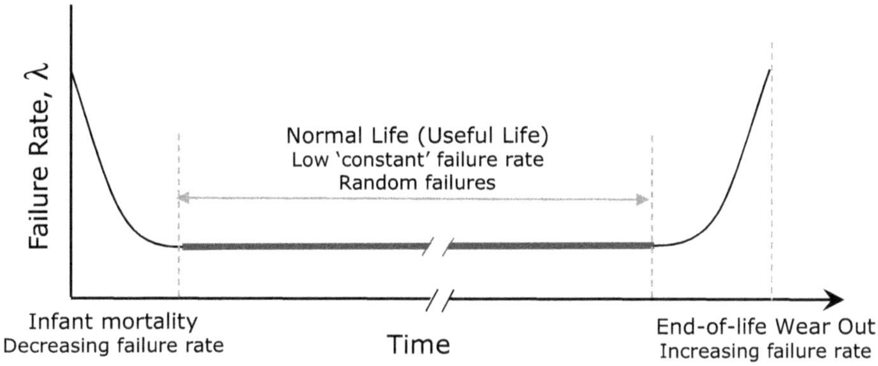

FIGURE 20.14 Bathtub curve showing early failures (infant mortality), wear-out failures, and in intermediate period of normal operation, which is characterized by random failures.

The failure rate is estimated as

$$\text{Failure rate} = \frac{\text{Observed number of failures}}{\text{Accumulated time in service}}$$

Table 20.8 gives the failure rate for come process control devices. The failure rate has the unit of 1/time. These failure rates are equivalent.

TABLE 20.8 Failure Rates for Some Process Control Devices (Ahmad, www.utm.my)

Instrument	Faults/year
Controller	0.20
Control valve	0.15
Flow measurements (fluids)	1.14
Flow measurements (solids)	3.75
Flow switch	1.12
Gas–liquid chromatograph	30.6
Hand valve	0.13
Indicator lamp	0.044
Level measurements (liquids)	1.70
Level measurements (solids)	6.86
Oxygen analyzer	5.65
pH meter	5.88
Pressure measurement	1.41
Pressure relief valve	0.022
Pressure switch	0.14
Solenoid valve	0.42
Stepper motor	0.044
Strip chart recorder	0.22
Thermocouple temperature measurement	0.52
Thermometer temperature measurement	0.027
Valve positioner	0.44

- 8.5 per 10^{-6} h
- 8.5×10^{-6}/h
- 0.85% per 1,000 h
- 0.074 per year

A widely used model of *normal operation* assumes that the failure rate λ is constant and random. Constant does not mean that failures occur at regular fixed intervals. They don't. They occur at random. Constant failure rate means that the probability of a failure in any given interval, Δt, is equal to the probability of a failure in any other interval of equal duration. The probability that a device will operate for time t without a failure is

$$P(t) = e^{-\lambda t}$$

where λ is the failure rate (failures/time).

The probability that a device is still working at $t = 1/\lambda$ is

$$P(t) = e^{-\lambda(1/\lambda)} = e^{-1} = 0.368$$

This means that 63.2% of the devices will have failed by time $t = 1/\lambda$.

The number of failures in a given time is $N(t)$. The expected number of failures during an accumulated time in service t is

$$E\left[N(t)\right] = \lambda t$$

The mean number of failures per unit time in operation is the failure rate:

$$\lambda = \frac{N(t)}{t}$$

The failures rates of machine parts are additive. Put another way, the overall failure rate can be decomposed. For example, the failure rate of a centrifugal pump is

$$\lambda_{\text{pump}} = \lambda_{\text{seal}} + \lambda_{\text{shaft}} + \lambda_{\text{bearing}} + \lambda_{\text{casing}} + \lambda_{\text{impellor}}$$

The shaft seal has the greatest failure rate and the pump casing has the least. The order of magnitude of these failure rates is

$$\lambda_{\text{casing}} < \lambda_{\text{impellor}} < \lambda_{\text{shaft}} < \lambda_{\text{bearing}} < \lambda_{\text{seal}}$$

The main source for failure rate data is either the equipment manufacturer or field data (see Figure 20.3).

Example 20.13 Probability of Failure

A device has a random failure rate of 0.000004/h. What is the probability that it will operate for 43,800 h (5 years) without failing?

$$P(t) = \exp[-(0.000{,}004/\text{h})(43{,}800\ \text{h})] = 0.839$$

This says that there is an 83.9% probability that the device will operate for five years without a failure, or that 83.9% of installed identical units will still be working at the five-year point.

TABLE 20.9 Component Failure Data

Component	1	2	3	4	5	6	7	8	9	10
Hours	1,000	1,000	467	1,000	630	590	1,000	285	648	882
Results	S	S	F	S	F	F	S	F	F	F

Example 20.14 Estimated Failure Rate

Ten identical components were tested until they failed or reached 1,000 h of successful operation, at which time the test was terminated for that component. The results are given in Table 20.9.

The total operating time of the ten components was 7,502 h.

$$\text{Estimated failure rate} = \frac{6 \text{ failures}}{7{,}502 \text{ h}} = 0.0008/\text{h} = 800\text{x}10^{-6}/\text{h}$$

Example 20.15 Failure of Control Valve

The flow control loop shown in Figure 20.15 has four components and all must function for the flow to be regulated as required. Failure rates for the loop components are given in Table 20.10.

Each component has several failure modes. A control valve, for example, can fail shut, fail open, or be slow to move.

The calculation used earlier to calculate the probability that all four devices function is

$$R_{\text{Loop}} == (0.861)(0.961)(0.942)(0.932) = 0.726$$

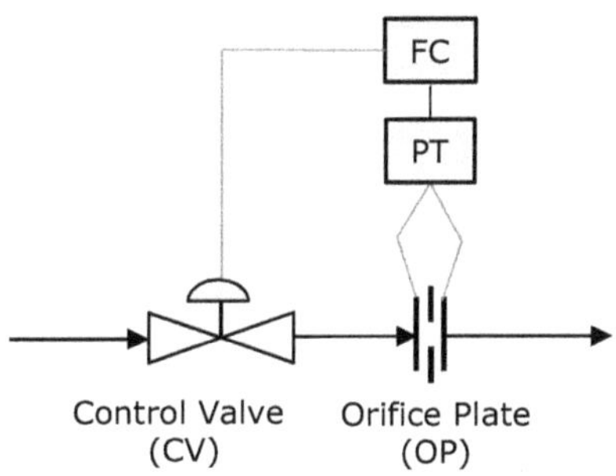

FIGURE 20.15 Process control loop for flow control.

TABLE 20.10 Failure Rates and Reliability for Flow Control Loop Components

Component	Failure Rate (λ) (Faults/Year)	Annual Reliability $R = \exp(-\lambda t)$	Annual Failure Probability $F = 1 - R$
Control valve (CV)	0.15	0.861	0.139
Orifice plate (OP)	0.04	0.961	0.039
Flow controller (FC)	0.06	0.942	0.058
Pressure transmitter (PT)	0.07	0.932	0.068

An alternative calculation using the failure rates is
Total failure rate

$$= 0.15/y + 0.04/y + 0.06/y + 0.07/y = 0.32/y$$

Probability all devices function for one year

$$= \exp[(-0.32/y)(1\ y)] = 0.726$$

Probability the control loop fails within the year

$$= 1 - 0.726 = 0.274$$

20.14 MEAN TIME TO FAILURE

Figure 20.16 illustrates some statistical measures of reliability: mean time to failure (MTTF), mean time between failure (MTBF), and mean time to recovery (MTTR).

Repairable components are characterized by the *mean time to failure* (MTTF), the *mean time between failures* (MTBF), and the *mean time to recovery* (MTTR). These are measured in units of time.

Mean time between failures (MTBF) is the expected (average) time between two successive failures. MTBF is the mean time to failure (MTTF) plus the mean time to recovery (MTTR).

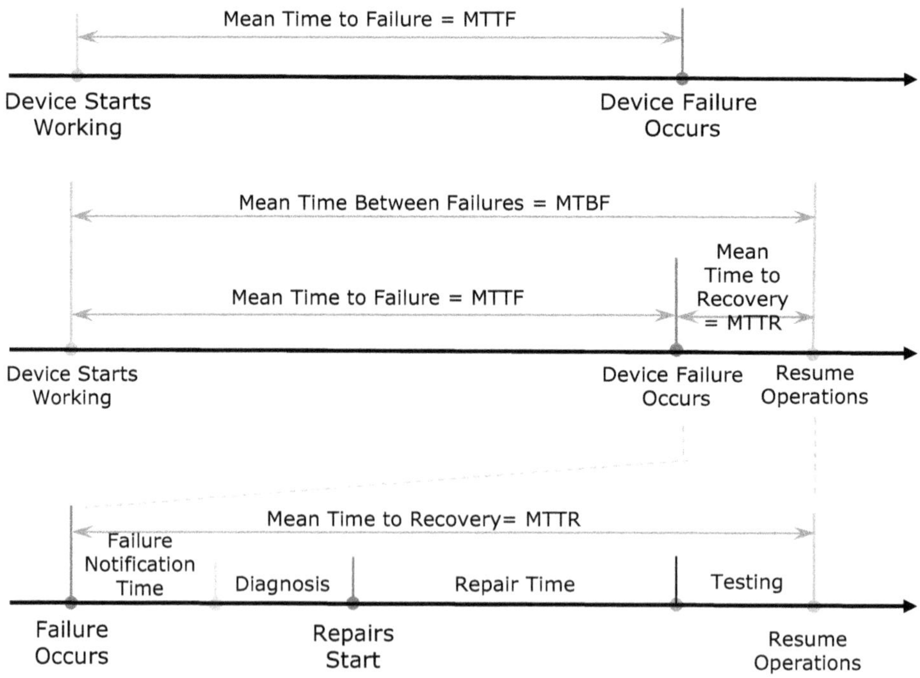

FIGURE 20.16 Mean time to failure (MTTF), mean time between failure (MTBF), and mean time to recovery (MTTR).

$$\text{MTBF} = \text{MTTF} + \text{MTTR}$$

Mean time to recovery (MTTR) is the expected length of time to detect the failure, bring a repair crew onsite, make the physical repair, test the repair, and return the device to normal service. Non-repairable components have a mean time to replacement. The component fails just once and is replaced, and the system can be restarted. This time depends on the availability of the replacement part.

If a device works for a long time without breakdown, it can be said to be highly reliable.

If it does not fail very often and can be quickly returned to service after it does fail, it is said to be highly available.

MTTR affects availability. If it takes a long time to repair a failed device and recover normal service, the system availability will be decreased. High availability is achieved by having MTBF large compared with MTTR. Scheduled maintenance and downtime are not included in MTTR, but they do reduce availability. Knowing the MTBF allows manufacturers to recommend how often components should be inspected, maintained, and replaced. Failure rates increase as components wear out. A product with a MTBF of ten years can wear out in two years if it is poorly maintained or subjected to harsh operating conditions. Operation and maintenance are the responsibility of the owner who wants to avoid wear-out before end of the design life.

MTBF is a measure of reliability, but it is not the expected life, the useful life, or the average life. It is used to characterize a large collection of devices. It cannot be used to predict the time of failure of any particular device. Nor does it specify the time (on average) when the probability of failure equals the probability of not having a failure (i.e., an availability of 50%). It is an aggregate statistic and does not predict a specific time to failure.

MTBF and MTTF are related to reliability and availability:

$$\text{Reliability} = \exp(\text{-Time/MTBF})$$

$$\text{Availability} = \text{MTTF/MTBF}$$

Example 20.16 Annualized Failure Rate

A component that runs continuously has a MTBF of 1,200,000 h. The failure rate is

$$\text{MTBF} = \frac{1{,}200{,}000\ \text{h}}{8{,}760\ \text{h/y}} = 137\ \text{y} \Rightarrow \frac{1\,\text{failure}}{137\ \text{y}} \times 100\% = 0.73\%/\text{y}$$

Expect about 0.73% of the population of this component to fail in the average year.

Example 20.17 Repairing Pumps

Centrifugal pumps in a sewerage district have a useful life of 15 years and a MTTF of 760 days. The MTTR once a pump has failed is five days. The mean time between failures is MTTF + MTTR = 760 + 5 = 765 days. The average repair cost is $12,500.

$$\text{Expected number of repairs per pump} = \frac{365\ \text{d/y}}{765\ \text{d}} = 0.477/\text{y}$$

Expected cost of repairs = (0.477 repairs/y) ($12,500/repair) = $5,964/y

Present value factor for 15 years at 6.5% interest = 9.403

Expected present value of repair costs = (9.043)($5,964) = $53,933

Example 20.18 Mean Time Between Failures

A device was operated for two years (17,520 h). During this time, the device failed four times; it was quickly repaired and returned to service. The mean time between failures is

$$\text{MTBF} = T/n = (17{,}520\ \text{h})/(4\ \text{failures}) = 4{,}380\ \text{h/failure}$$

The times of failure might have been, for example, at 8,600, 12,000, 14,500, and 17,000 h. The average of these times to failure is not the MTBF, which is defined so to give the same MTBF for *n* failures during a running time of *T* regardless of when the devices failed.

20.15 CONCLUSION

Several ways to improve the reliability are:

- Reduce complexity
- Increase component and subsystem reliabilities
- Parallel redundancy
- Stand-by redundancy
- Preventive maintenance
- Rapid detection and repair of failed components

Saying that component reliabilities are independent means that a downstream valve failure does not increase the probability of having an upstream valve failure, and that it does not alter the probability that the pipe will rupture. Assuming independent results makes it easy to understand the basic principles of failure analysis.

Real systems are more complicated. Failures of complex systems tend to involve conditional failures. Failure of one component causes another failure. System components may fail sequentially. A block of components may fail due to a common cause. This is how catastrophic failures occur.

If the failure probabilities are not independent, the system reliability will be lower than what is predicted when assuming independent events. These more complicated situations are analyzed using conditional probabilities.

20.16 PROBLEMS

20.1 PARETO DIAGRAM: LOST TIME

Construct the Pareto diagram for time lost due to the six kinds of failures in Table P20.1.

TABLE P20.1 Lost Time

Cause of Lost Time	Time Lost (min)
Generator failure	346
Conveyor failure	71
Training	34
Safety	26
Flooding	20
Pump failure	9

20.2 PARETO DIAGRAM: WIND TURBINE FAILURES

Construct Pareto diagrams for the wind turbine failure and time lost data in Table P20.2.

TABLE P20.2 Wind Turbine Failures

Type of Failure	Failures per Year	Days Lost per Failure
Electrical system	0.6	1.5
Electronic control	0.45	1.6
Sensors	0.25	1.4
Hydraulic system	0.245	1.2
Yaw system	0.20	2.4
Rotor hub	0.20	3.8
Mechanical brake	0.14	2.8
Rotor blades	0.12	3.0
Gearbox	0.11	6.2
Generator	0.10	5.8
Support and housing	0.14	4.0
Drive train	0.08	6.0

20.3 CHLORINATION SYSTEM

Figure P20.3 shows a chlorination system that is designed to prevent biological growth in a cooling water system. Chlorine is provided on the vapor side of a 1-ton cylinder. Pressure is reduced from the cylinder (normally 5.4 atm at 20°C) to 1.05 atm at the rotameter. The rotameter is adjusted manually to provide an average flow of 1.1–1.5 kg/h of chlorine to the pressure check valve. The pressure check valve reduces the pressure to zero so that the vacuum from a venturi draws a controlled amount of chlorine into the water stream. A small pump recirculates water in the cooling basin. The pressure reducing valve and the pressure control valve should not open unless there is a vacuum on the downstream side. If the first valve passes gas when a vacuum is not present, the second valve remains closed and contains the gas pressure in the unit. If the second valve also passes gas, the built-in pressure relief valve permits gas to pass out of the vent. Explain the ways the system can fail.

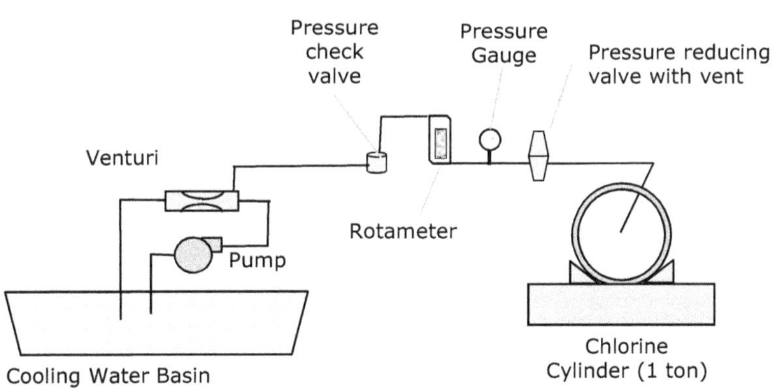

FIGURE P20.3 Chlorination system.

20.4 ASSET CRITICALITY

Rate the criticality of 56-year-old sewer that has had numerous breaks in the past five years. It is likely to continue to fail at this frequency. The sewer serves 400 residential households The line is not in a major roadway, so repair is relatively easy. It can be repaired to a reasonable condition. An outage of several hours does not cause damage to any houses.

20.5 SPILL PREVENTION PRIORITY NUMBER (SPPN)

The Spill Prevention Priority Number (SPPN) was suggested (Tansel (2001) as a way of scoring the risk of an accidental release (spill) when transporting oil by pipeline. The frequency of pipeline oil spills is 17×10^6 per mile, with an average size spill of 24,000 L. Table P20.5a lists the three rating criteria and the weights assigned to possible events. SPPN is the product of the failure mode frequency rating (FMFR), the spill detectability rating (DR), and the spill consequence rating (CR). The rating scheme is given in Table P20.5a.

$$\text{SPPN} = (\text{FMFR})(\text{DR})(\text{CR})$$

(a) Calculate the SPPN for the spill risk factors in Table P20.5b. Ratings are given for an old and a newer facility. Note that the environmental factors and "acts of God" are different because the facilities are in different locations.
(b) Display the results as a Pareto diagram.
(c) Suggest ways in which the SPPN for the old pipeline might be reduced.

TABLE P20.5a Spill Rating Criteria and Weights Assigned to Possible Events (Tansel 2001)

Failure Frequency	**Failure Mode Factor Rating (FMFR)**
Remote (spill is unlikely)	1
Low (relatively low possibility of a spill)	2–3
Moderate (occasional spills are likely)	4–6
High (spills would occur)	7–8
Very high (spill is inevitable)	9–10
Spill Mode Detectability	**Detectability Rating (DR)**
Very high detectability	1–2
High detectability	3–4
Moderate detectability	5–6
Low detectability	7–8
Very low detectability	9
Absolute certainty of non-detection	10
Spill Consequence	**Consequence Rating (CR)**
Minor effects	1–2
Moderate effects	3–4
Severe effects	5–6
Very severe effects	7–8
Catastrophic effects	9–10

TABLE P20.5b Risk Factors for Pipeline Spills (Tansel 2001)

Origin		Spill Risk Factor	Old Pipeline FMFR	Old Pipeline DR	Old Pipeline CR	New Pipeline FMFR	New Pipeline DR	New Pipeline CR
Design	D1	No. of components	8	5	8	2	2	5
	D2	Age	9	4	9	1	1	5
	D3	Pipe material	8	4	9	1	1	6
	D4	Pipe length	7	4	8	1	1	6
	D5	Pipe capacity	8	4	8	2	2	6
	D6	System redundancy	6	4	6	2	2	6
	D7	Degree of automation	8	7	8	1	2	5
	D8	Pipe pressure	7	6	7	2	2	4
	D9	Material being piped	7	3	6	2	2	4
Operational	O1	No. of people employed	5	2	5	2	1	4
	O2	Quantification	4	2	5	2	1	4
	O3	Periodic training program	5	1	5	1	1	4
	O4	Frequency of inspection	5	1	6	1	1	4
	O5	Work hours	5	3	6	1	1	3
Environment	E1	Geology	4	3	6	2	3	4
	E2	Geography	4	3	6	2	3	4
	E3	Weather	4	3	6	3	3	4
	E4	Vibration	5	3	6	3	5	4
	E5	Nearby activities	5	3	6	3	5	4
Acts of God	G1	Earthquake	2	9	9	2	9	9
	G2	Flood	1	8	9	2	8	9
	G3	Hurricane	2	3	9	2	3	9

20.6 UNITS IN SERIES

Calculate the probability of success for three units operating in series using unit reliabilities of $R_A = 9/10$, $R_B = 8/10$, and $R_C = 3/4$. If all three units must function to have system success, the probabilities of success and failure are as follows:

$$P(\text{system success}) = (9/10)(8/10)(3/4) = 216/400 = 0.54$$

$$P(\text{system failure}) = 1 - P(\text{system success}) = 1 - 0.54 = 0.46$$

Compute these probabilities by examining all possible states of the system and summing those that result in system failure.

20.7 UNITS IN PARALLEL 1

A system consists of five components in parallel. System success requires that at least three of these components must function. What is the probability of system success, if each component reliability is 0.9?

20.8 UNITS IN PARALLEL 2

Calculate the reliability of a system with three units operating in parallel when the unit reliabilities are $R_A = 9/10$, $R_B = 8/10$, and $R_C = 3/4$. System success requires that component A is operating and that either component B or component C is operating.

20.9 COMPOUND SYSTEM 1

Calculate the reliability of the three-stage compound system shown in Figure P20.9. The system operates if each stage operates. Each stage operates if at least one component is working.

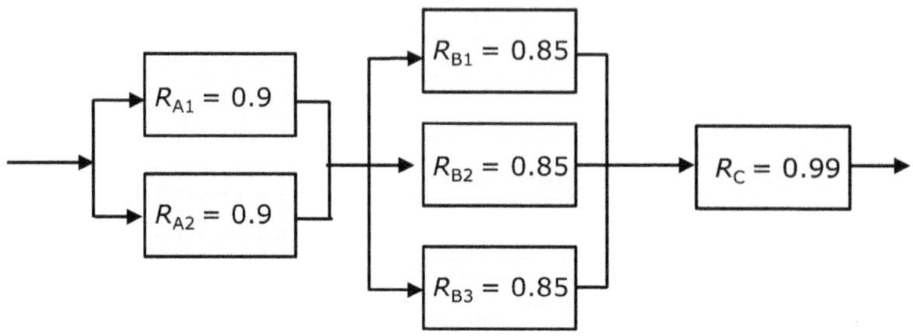

FIGURE P20.9 Three stage compound system.

20.10 COMPOUND SYSTEM 2

Figure P20.10 shows a two-stage system with each stage having two components in parallel. The system operates if at least one subsystem operates. What is the system reliability?

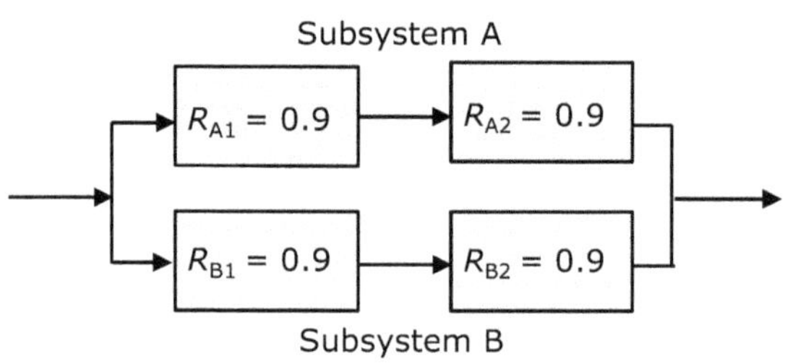

FIGURE P20.10 Two stage compound system.

20.11 REACTOR RELIABILITY

A waste processing plant has three separate processing lines and each line consists of two reactors in series with a capacity of 5,000 lb/h. The reactors are physically identical but can be operated in different ways. A seventh reactor is available to replace any of the six in use. Each reactor is available 95% of the time. When the heat transfer surfaces become fouled, a reactor must be shut down and washed with hot solvent three times. This, plus random mechanical failures, account for the 5% unavailability. Two suggestions have been made for improving the availability of the reactors, and both cost the same: (1) Increase the size of the hot wash facility so that a fouled reactor is out of service only 4% of the time (96% availability); (2) Purchase another reactor to serve as an additional spare. Which solution will probably lead to a higher production rate from the three production lines?

20.12 FIRE RISK DECISION TREE

There is a 0.01/y chance of an explosion that will start a fire. The fire will be suppressed with no damage if the sprinkler system activates as designed. There is a fire alarm as a backup to the sprinkler system, which if activated will give the operators a high probability of suppressing the fire. The decision tree and relevant probabilities at each stage are shown in Figure P20.12. Calculate the annual frequencies of the five potential outcomes.

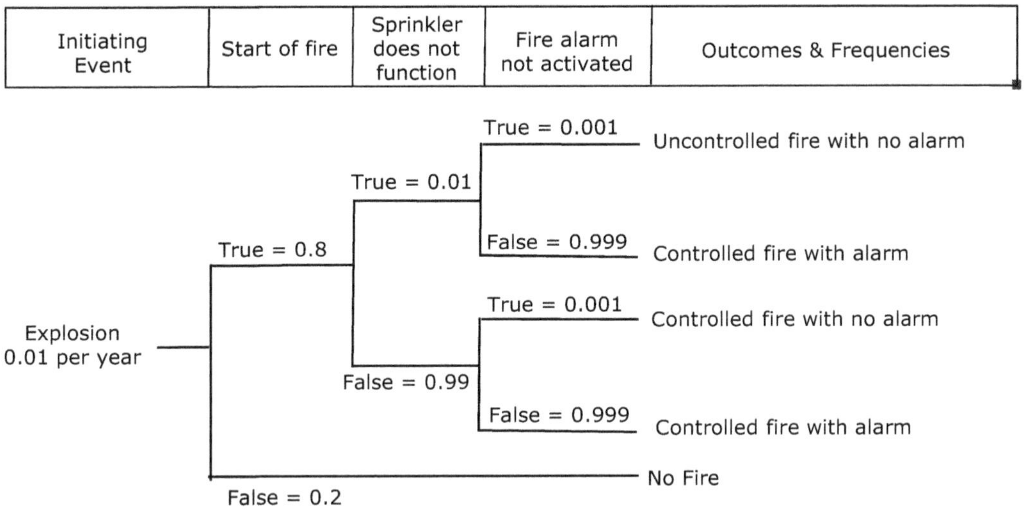

FIGURE P20.12 Decision tree for fire control system.

20.13 ELECTRIC LOAD

A system has a continuous load requirement of 80 MW. Find the expected load loss and the probability of load loss (loss of load probability) if the generation system has nine 10 MW units each with a probability of failure (unavailability) of 1.5%.

20.14 PUMPING STATION

A pumping station has two 20,000 gal/h pumps and is to have an additional 40,000 gal/h pump installed. Develop a system capacity outage probability table given that the unavailability (probability of failure) for the 20,000 and 40,000 gal/h pumps are 0.2 and 0.1, respectively. If the system requires at least 50,000 gal/h for successful operation, what is the probability of system success?

20.15 WATER SUPPLY PUMPING

The water supply system in Figure P20.15 consists of a pump system in series with a delivery system (consisting of double pipes). The pump system has four possible operating levels, ranging from 0% to 100% of the system requirement as shown below. Similarly, the water delivery system has three possible operating levels, as shown in Table P20.15.

Develop an operating level probability table for the water supply system. What is the probability of meeting 50% of the system requirement?

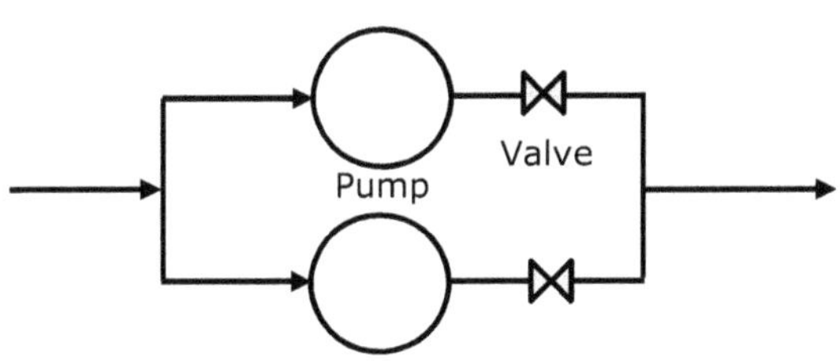

FIGURE P20.15 Water supply system.

TABLE P20.15 Output and Throughput of Water Supply System

Pump System		Delivery System	
Output	**Probability**	**Throughput**	**Probability**
100%	0.8	100%	0.7
75%	0.1	50%	0.1
25%	0.05	0%	0.2
0%	0.05		

20.16 CONDITIONAL PUMP OPERATION

A pump has been working satisfactorily for four years, what is the probability that it will continue working for at least another two years, or a total of at least six years. The failure rate for this type of pump is 0.25/y.

20.17 TURBINE SYSTEM RELIABILITY

Consider a system, Figure P20.17, composed of a heater ($R1$), two pumps ($R2$ and $R3$), and five turbines ($R4$ through $R8$), where R is the reliability of the component. The two pumps work in parallel, meaning that the pump subsystem operates if at least one of the pumps operates. The turbine subsystem operates if at least three turbines operate. The heater, pump, and turbine subsystems are connected in series, meaning that they must all work properly for system to succeed. Component failure probabilities ($F = 1 - R$) are given in Table P20.17. Assume failures are independent.

Calculate the reliability of the system.

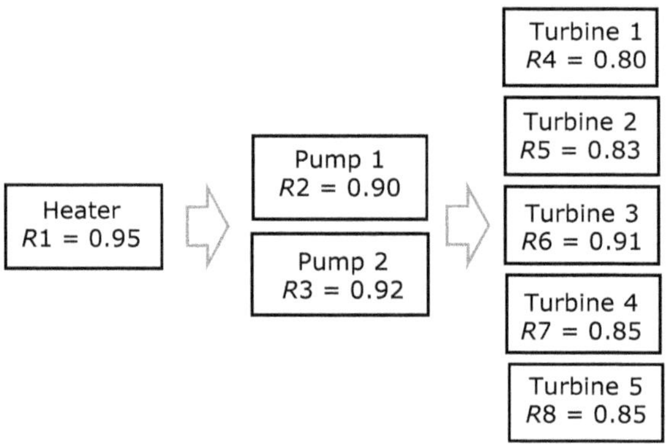

FIGURE P20.17 Turbine system.

TABLE P20.17 Independent Failure Rates For Turbine System

Heater	Pump		Turbine				
F1	*F2*	*F3*	*F4*	*F5*	*F6*	*F7*	*F8*
0.05	0.10	0.08	0.20	0.17	0.09	0.15	0.15

20.18 EXPONENTIAL DISTRIBUTION OF FAILURE DATA

Tests on a device during the constant failure rate phase yielded the data given in Table P20.18. Fit the data to an exponential model, $F = \exp(-\lambda t)$, to estimate the failure rate λ.

TABLE P20.18 Device Failure Data

Time to Failure (h)	Fraction of Total Failures
0–50	0.26
50–100	0.18
100–150	0.16
150–200	0.12
200–250	0.07
250–300	0.07
300–350	0.05
350–400	0.04
400–450	0.03
450–500	0.02

20.19 FITTING EXPONENTIAL FAILURE DATA

Tests on eight devices during the constant failure rate phase yielded the data in Table P20.19. Fit the data to an exponential distribution and estimate the failure rate λ. What is the mean time to failure? What is the probability that a device will fail before 500 h?

TABLE P20.19 Time to Failure for Eight Devices

Device tested	1	2	3	4	5	6	7	8
Time to failure (h)	80	134	148	186	238	450	581	890

20.20 FAILURE RATE AND MTTF

The average failure rate of a certain type of pressure relief valve is 0.10 failures per 10^6 h.

(a) Calculate the mean time to failure (MTTF).
(b) What is the probability that a valve of this type will operate for 43,800 h (5 years) without failing?
(c) If an industry has 2,000 of this type of valve, what is the expected number of failures over five years?

20.21 CHECK VALVE FAILURES

A study of failure rates for of check valves showed that the failure rate was 0.01 failures per valve year.

(a) Calculate the mean time between failures (MTBF).
(b) What is the probability that a single valve will operate for two years without failing?
(c) If a plant has 250 of these valves, what is the expected number of failures over five years?

20.22 CATALYTIC CONVERTERS

Twelve cars having a certain type of catalytic converter were driven until the converter failed or they reached 60,000 miles, at which time the test was terminated. The data are in Table P20.22.

Estimate the mean drive distance to failure.

What is the probability that a catalytic converter will still be functioning at 60,000 miles?

TABLE P20.22 Catalytic Converter Failures for Twelve Cars

Device	1	2	3	4	5	6	7	8	9	10	11	12
Miles driven (1,000s)	60	44.4	49.2	60	47.6	60	37.4	55.8	57.5	58.3	56.2	54.3
Results	S	F	F	S	F	S	F	F	F	F	F	F

20.23 FAILURE RATE AND MTTF

Table P20.23 gives the time to failure (TTF) in hours for 40 devices that were tested to failure. Estimate the MTTF, the constant failure rate λ, and the probability that a device will be operating at 40,000 h.

TABLE P20.23 Time to Failure (Hours)

Device	TTF	Device	TTF	Device	TTF	Device	TTF
1	2,480	11	6,760	21	5,540	31	17,920
2	2,480	12	6,760	22	5,540	32	17,920
3	15,780	13	23,340	23	62,780	33	9,340
4	29,880	14	30,000	24	8,000	34	20,580
5	34,720	15	18,640	25	19,060	35	27,420
6	4,300	16	14,480	26	54,400	36	1,240
7	4,300	17	14,480	27	54,400	37	1,240
8	21,720	18	12,060	28	32,560	38	3,260
9	46,640	19	16,460	29	10,700	39	8,100
10	15,420	20	23,620	30	39,900	40	44,080

20.24 PUMP RELIABILITY

The life cycle cost (LCC) of a pump is

$$\text{LCC} = C_{IC} + C_{IN} + C_E + C_O + C_M + C_S + C_{ENV} + C_D$$

where

C_{IC} = initial investment costs
C_{IN} = installation and start-up costs
C_E = energy costs
C_O = operation costs
C_M = maintenance and repair costs
C_S = downtime and loss of production costs
C_{ENV} = environmental costs, including disposal of parts and contaminated liquid
C_D = decommissioning/disposal costs, including restoration of the local environment

Maintenance and repair costs (C_M) are a significant cost factor, which is controlled by the repair interval (MTBF) and the cost per repair. The relevant data for two types of pump are given in Table P20.24. For each pump, calculate the expected number of repairs during the pump lifetime, the repair cost per year, and present value of the lifetime repair cost.

TABLE P20.24 Repair Costs (Australian $) of Canned versus Mechanical Seal Pumps (Inflation Rate = 5%)

Option	Dual Seal	Canned Motor
Service life	25 y	25 y
Cost per repair	$7,500	$10,000
Repair interval (MTBR)	3 y/repair	8 y/repair

References and Recommended Reading

AACE International 2019, *Cost Estimate Classification System –As Applied in Engineering, Procurement, and Construction for the Process Industries* 18R–97, Morgantown, WV.

AIChE 2001, *Practical Solutions for Reducing and Controlling Volatile Organic Compounds and Hazardous Air Pollutants*, American Institute of Chemical Engineers, Center for Waste Reduction Technologies, New York.

AIPP 1992, *A Primer for Financial Analysis of Pollution Prevention Projects*, American Institute for Pollution Prevention (AIPP), Cincinnati, OH.

Akesrisakul, K & Jiraprayuklert, A 2007. 'The Application of Statistical Techniques to Reduce Ethylene Glycol in Wastewater Produced by Esterification Reaction in Polyester Manufacturing Process', *Journal of Science & Technology*, vol. 29, pp 181–190.

Al-Bazadi, GH & Amin, M 2016, 'Comparison between Reverse Osmosis Desalination Cost Estimation Trends', *The Journal of Scientific and Engineering Research*, vol. 3, pp 56–62.

Aldrich, JR 1996, *Pollution Prevention Economics: Financial Impacts on Business and Industry*, McGraw-Hill, Inc., New York.

Aldrich, JR 1999a, 'P2 and the Bottom Line: Payback Period and Benefit-Cost Ratio', *Pollution Prevention Review*, vol. 99, pp 93–99.

Aldrich, JR 1999b, 'P2 and the Bottom Line: Internal Rate of Return', *Pollution Prevention Review*, vol. 9, pp 105–110.

Aldrich, JR 1999c, 'P2 and the Bottom Line: Net Present Value', *Pollution Prevention Review*, vol. 9, pp 103–107.

Alexander, H 1993, 'Manufacturing Firm Profits from Waste Minimization', *Water Environmental & Technology*, vol. 10, p. 47.

Amos, SJ 2004, *Skills & Knowledge of Cost Engineering*, 5th ed. AACE International, Morgantown, WV.

AMSA 2002, *Managing Public Infrastructure Assets*, Association of Metropolitan Sewerage Agencies, Washington, DC.

Atallah, S. 1980, 'Assessing and Managing Industrial Risk', *Chemical Engineer*, vol. 87, no. 18, pp 94–103.

AWWA 2008, 'Microfiltration and Ultrafiltration Membranes for Drinking Water', *Journal of American Water Works Association*, vol. 100, pp 84–97.

Bakkar, M 2014, 'Improving the Performance of Water Demand Forecasting Models by Using Weather Input', *Procedia Engineering*, vol. 70, pp 93–102.

Bannerman, RR 1980, 'Water and Waste Engineering in Africa', *Proceedings of the 6th WEDC International Conference*, Zaria, Nigeria, 24–28 March, pp. 12–19.

Berthouex, PM & Brown, LC 2002, *Statistics for Environmental Engineers*, 2nd ed., CRC Press, Boca Rotan.

Berthouex, PM & Brown, LC 2013, *Pollution Prevention and Control: Environmental Quality and Human Health*, Bookboon, Copenhagen.

Berthouex, PM & Brown, LC 2014, *Pollution Prevention and Control: Material and Energy Balances*, Bookboon, Copenhagen.

Berthouex, PM & Brown, LC 2016, *Chemical Processes for Pollution Prevention and Control*, CRC Press, Boca Rotan.

Berthouex, PM & Brown, LC 2018, *Energy Management for Pollution Prevention and Control*, Bookboon .com, Denmark.

Berthouex, PM & Brown, LC 2019, *Pollution Prevention and Control: Material Balances*, Bookboon.com, Denmark.

Berthouex, PM & Fan, R 1986, 'Evaluation of Treatment Plant Performance: Causes, Frequency, and Duration of Upsets', *Journal of Water Pollution Control Federation*, vol. 58, pp 368–375.

Berthouex, PM & Polkowski, LB 1970, 'Design Capacities to Accommodate Forecast Uncertainties', *Journal of the Sanitary Engineering Division*, ASCE, pp 1183–1210.

Berthouex, PM & Rudd, DF 1977, *Strategy of Pollution Control*, John Wiley, New York.

Berthouex, PM & Stevens, DK 1982, 'Computer Analysis of Settling Data', *Journal of the Sanitary Engineering Division*, ASCE, vol. 108, pp 1065–1069.

Besselievre, EB 1959, 'Industries Recover Valuable Water and By-Products from Their Wastes', *Wastes Engineering*, vol. 30, p 760.

Biesadda, M 2001, 'Simulations in Health Risk Assessment', *International Journal of Occupational Medicine and Environmental Health*, vol. 14, pp 397–402.

Blank, L & Tarquin, A 2018, *Engineering Economy*, 8th ed., McGraw-Hill, New York.

Box, GEP 1954, 'The Exploration and Exploitation of Response Surfaces: Some General Considerations and Examples', *Biometrics*, 10(1), 16–60.

Box, GEP 1999, 'The Invention of the Composite Design', *Quality Engineering*, 12(1), 119–122.

Box, GEP, Hunter, JS & Hunter WG 2005, *Statistics for Experimenters: Design, Innovation, and Discovery*, John Wiley & Sons, New York.

Box, GEP & Wilson KB 1951, 'On the Experimental Attainment of Optimum Conditions', *Journal of the Royal Statistical Society: Series B*, 13(1), 1–45.

Boyle, WC 1988, Fine Pore Aeration Design Manual, EPA/625/1-89/023, U. S Environmental Protection Agancy, Cincinnati, OH.

Boyle, WC 2002, 'A Brief History of Aeration of Wastewater', *Environmental and Water Resources History Sessions*, ASCE Conference and Exposition, Washington, DC, pp 13–21. (Published on-line in 2012).

Breen, JJ & Dellarco, MJ (eds.) 1992. 'Pollution Prevention in Industrial Processes: The Role of Process Analytical Chemistry', *American Chemical Society Symposium Series 508*, Washington, DC.

Briner, RP 1984. 'Making Pollution Prevention Pay', *EPA Journal*, pp 28–29.

Brown, M, et al. 2011, *Revenue Sources to Fund Recycling, Reuse, and Waste Reduction Programs*, Washington State Department of Ecology, Bellevue, WA.

Chen, WH, et al. 2008, *Delta Drinking Water Quality and Treatment Costs: Technical Appendix H*, Public Policy Institute of California, San Francisco, CA.

Ciric, AR & Huchett, S 1993. 'Economic Sensitivity Analysis of Waste Treatment Costs in Source Reduction Projects: Discrete Optimization Problems', *Industrial & Engineering Chemistry Research*, 32, 2636–2646.

Committee on Reducing Hazardous Waste Generation 1985. *Reducing Hazardous Waste Generation*, National Academy Press, Washington, DC.

Cristóvão, RO 2015, 'Fish Canning Wastewater Treatment by Activated Sludge: Application of Factorial Design Optimization', *Water Resources and Industry.*, vol. 10, pp 29–38.

Cullinane, JM 1989, *Methodologies for the Evaluation of Water Distribution System Reliability/Availability*, Ph.D. Thesis, University of Texas.

Donkor, EA et al. 2014, 'Urban Water Demand Forecasting: Review of Methods and Models', *Journal of Water Resource Planning and Management*, vol. 140, pp 146–159.

DSM Envir. Services n.d., *Solid Waste Management and Municipal Finance*, Connecticut Governor's Recycling Working Group.

Eaton, G & Lutras, J 2008, 'Turning Methane into Money: Cost-Effective Methane Co-Generation Using Microturbines at a Small Waste Water Plant', in *Ensuring a Sustainable Future: An Energy Management Guidebook for Wastewater and Water Facilities*, US EPA, EPA 832-RA-08-002.

Enouy, RW et al. 2015, 'An Implicit Model for Water Rate Setting Within Municipal Utilities', *Journal of American Water Works Association*, vol. 107, pp E445–E453.

Faller and Ryder, 1991. 'Clarification and Filtration to Meet Low Turbidity Reclaimed Water Standards', *WET*, pp 68–74.

Feiedel, D & Arroyo, J 2005, *Industrial vs. Utility Emissions Control Equipment – Analysis of EPA Methods, Assumptions, References, and Costs*, Black & Vetch.

Gere, AW 1997, 'Microfiltration Operating Costs', *Journal of American Water Works Association*, vol. 89, pp 40–49.

Ghafari, S et al. 2009, 'Application of Response Surface Methodology (RSM) to Optimize Coagulation–Flocculation Treatment of Leachate Using Poly-Aluminum Chloride (PAC) and Alum, *Journal of Hazardous Materials*, vol. 163, pp 650–656.

Gillot, S 1999, 'Optimization of Wastewater Treatment Plant Design and Operation Using Simulation and Cost Analysis', *Proceedings of the 72nd Annual WEF Conference*, New Orleans, USA.

GIZ 2015, *Economic Instruments in Solid Waste Management Applying Economic Instruments for Sustainable Solid Waste Management in Low and Middle-Income Countries*, Deutsche Gesellschaft fur Internationale Zusammenarbeit (GIZ), Federal Ministry for Economic Cooperation and Development, Bonn.

Goldstein, M 2002, *Electricity Requirements for Wastewater Treatment by POTWs and Private Facilities*, Electrical Power Research Institute, Washington, DC.
Goldstein, M & Ritterling, J 2001, *A Practical Guide to Estimating Cleanup Costs*, U. S. Environmental Agency, Washington, DC.
Gorsek, A & Tramsek, M 2007, 'Kefir Grains Production—An Approach for Volume Optimization of Two-Stage Bioreactor System', *Biochemical Engineering Journal*, vol. 42, pp 153–158.
Graff, G et al. 1998, *Snapshots of Environmental Cost Accounting*, U.S. Environmental Protection Agency. EPA 742-R-98-006.
Greeley & Hansen 2015, *CSS Long Term Control Plan Update: Basis for Cost Opinions*, City of Alexandria, VA.
Hahn, GJ & Shapiro, SS 1967, *Statistical Models in Engineering*, Wiley, New York.
Hartmen, P & Cleland, J 2007, *Wastewater Treatment Performance and Cost Data to Support an Affordability Analysis for Water Quality Standards*, Montana Department of Quality, Helena.
Hazen & Sawyer 2013, *Tampa Bay Water Demand Management Plan-Final Report*, Tampa Bay Water.
Henze, M (ed.) 2008, *Biological Wastewater Treatment: Principles, Modelling and Design*, IWA Publishing, London.
Hillier, FS & Lieberman, GJ 2001, *Introduction to Operations Research*, 7th ed., McGraw-Hill, New York.
Hobson, MJ & Mills, NF 1990, 'Chemostat Studies of Mixed Culture Growing of Phenolics', *Journal of the Water Pollution Control Federation*, 62, 684–691.
Hodgson, J & Walters T 2002, 'Optimizing Pumping Systems to Minimize First or Life-Cycle Costs, (presented at the 19th Int'l Pump Users Symposium)', *Pumps & Systems Magazine*, October.
Holland, M 2012, *Current Practice in Cost Benefit Analysis of Air Pollution Policies*, Royal Society, London.
Humphreys, KK 1991, *Jelen's Cost and Optimization Engineering*, 3rd ed., McGraw-Hill, New York.
Humphreys, KK 2005, *Project and Cost Engineer's Handbook*, 4th ed., Marcel Dekker, New York, NY.
Hunter Water Corporation 2013, *Operating and Maintenance Cost Estimating Guidelines*, Newcastle, New S. Wales, Australia.
Jelen, FC & Black, JH 1983, *Cost and Optimization Engineering*, McGraw Hill, New York, NY.
Keifer, J 2015, 'The Causes of Water Demand Uncertainty and Their Importance for Water Supply Risk Assessment and Management', *2015 AWWA Spring Specialty Conference*, Los Angeles, CA.
Keoleian, G & Menerey, D 1994. 'Sustainable Development by Design: Review of Life Cycle Design and Related Approaches', *Journal of Air and Waste Management Association*, 44(5), 645–668.
Kinneman, TC 2010, 'The Costs of Municipal Curbside Recycling and Waste Collection', *Bucknell Digital Commons*, Bucknell University, http://digitalcommons.bucknell.edu/fac_journ.
Kitto, JB Jr, et al. 2016, 'World-Class Technology for the Newest Waste-to-Energy Plant in the United States — Palm Beach Renewable Energy Facility No. 2', Paper BR-1935, Presented to: Renewable Energy World International, Orlando, Florida, U.S.A.
Kittrel, JR & Watson, CC 1966, 'Don't Overdesign Process Equipment', *Chemical Engineer Progress*, vol. 64, p 79.
Knopp, PV 1970, 'Wastewater Treatment with Powdered Activated Carbon Regenerated by Wet Air Oxidation', *25th Purdue Indusrial Waste Conference*, 1970.
Kopp, R & Hazilla, M 1990, 'Social Cost of Environmental Quality Regulations', *Journal of Political Economy*, vol. 98, pp 853–873.
Malcolm Pirney 2009, *Prevention of Significant Deterioration (PSD) Permit Application, Solid Waste Authority of Palm Beach County, Appendix D – Cost Estimates*, Solid Waste Authority of Palm Beach, County, FL.
Male, JM, Walski, TM & Slutsky, AH 1990. 'Analyzing Water Main Replacement Policies,' *Journal of Water Research Planning and Management*, vol. 116, pp. 362–74.
Manne, A 1967, *Investments for Capacity Expansion: Size, Location and Time Phasing*, MIT Press, Cambridge, MA.
Margolin, BH 1985, 'Experimental Design and Response Surface Methodology – Introduction', *The Collected Works of George E. P. Box*, Volume 1, edited by George Tiao, pp 271–276, Wadsworth Books, Belmont, CA.
Marlow, D, et al. 2007, *Condition Assessment Strategies and Protocols for Water and Wastewater Utility Assets, Water Environmental*, Research Foundation, AWWA Research Foundation.
McGivney, WT & Kawarmura, S 2008, *Cost Estimating Manual for Water Treatment Facilities*, Wiley, New York, NY.
McMasters, RL 2002, 'Estimating Unit Cost in a Cogeneration Plant Using Least Squares', *IEEE Power Engineering Review*, vol. 22, pp 78–78.

Monod, J 1949, 'The Growth of Bacterial Culture', *Annual Review of Microbiology*, vol. 3, pp 371–394.
Mueller, JA & Boyle, WC 1988, 'Oxygen Transfer under Process Conditions', *Journal of the Water Pollution Control Federation*, vol. 60, pp. 332–341
Mueller, JA, Boyle, WC & Popel, HJ 2002. *Aeration: Principles and Practice*, CRC Press, Boca Raton. FL.
Nadh, T 2016, *The Buried Pipeline Replacement Era: A Cost Effectiveness Analysis of Pipeline Replacement Strategies for the Santa Clara Valley Water District*, M.S Thesis, San Jose State University.
NASA 2015, *NASA Cost Estimating Handbook*, version 4.0, Houston, TX.
Nevins, T & Weber, C 2009. 'Reduce Pumping Energy Costs by Strategic Management of Distribution, Storage and Pumping', presented at the AWWA DSS 2009, September 1.
New Mexico Environmental Finance Center 2007, *Cost Estimating Guide for Water, Wastewater, Roads, and Buildings*, Albuquerque, NM.
Newman, DG et al. 2011, *Engineering Economic Analysis*, Oxford University Press, USA.
Newman, DG, Lavelle, JP & Eschebach, TG 2013, *Engineering Economic Analysis*, 8th ed., Oxford University Press, Oxford, England.
Orhan, AH, et al. 2006, 'Cost of Leakage', *Journal of Applied Science Research*, vol. 2, 276–278.
Pabi, S et al. 2013, *Electricity Use and Management in the Municipal Water Supply and Wastewater Industries*, Electric Power Research Institute, Palo Alto, CA.
Pacific Institute n. d., *Water Rates: Water Demand Forecasting*, www.pacinst.org.
Parasivam, R, et al. 1981, 'Slow Sand Filter Design and Construction in Developing Countries', *Journal of American Water Works Association*, vol. 73, pp 178–185.
Patil, R et al. 2014, 'Chlorine Leakage from Control Valve in Charging Station Pipeline: A Cas Study', *Journal of Industrial Safety Engineering*, vol. 1, pp 1–5.
Pennsylvania Dept. of Envir. Protection n.d. *Applicant Guidance for Wastewater Facility Cost-Effectiveness Analysis*, 10 pp.
Peters, M, et al. 2002. *Plant Design and Economics for Chemical Engineers*, 3rd ed., McGraw Hill, New York, NY.
Puchajda, B & Oleszkiewicz, J 2008, 'Impact of Sludge Thickening on Energy Recovery from Anaerobic Digestion', *Water Science & Technology*, vol. 57, pp 395–401.
Qasim, SR & Motely, EM 2000, *Water Works Engineering: Planning, Design and Operations*, Prentice-Hall, Upper Saddle River, NJ.
Razif, M, et al. 2015, 'Implementation of Regression Linear Method to Predict WWTP Cost for EIA: Case Study of Ten Malls in Surabaya City', *Procedia Environmental Sciences*, vol. 28, pp 158–165.
Rinaudo, JD 2015, 'Long-Term Water Demand Forecasting. Understanding and Managing Urban Water in Transition', *HAL Archives*, pp 239–268.
Rogers, CS 2008, *Economic Costs of Conventional Surface-water Treatment: A Case Study of the McAllen Northwest Facility*, M.S. Thesis, Texas A & M University.
Rudd, DF & Watson, CC 1966, *Strategy of Process Design*, Wiley, New York, NY.
Sarkar, RS & Newton, CS 2008, *Optimization Modeling: A Practical Approach*, CRC Press, Boca Rotan, FL.
Schauer, R 2016 *Solid Waste Authority of Palm Beach Count, Director, Presentation at the International Solid Waste Association*, May 26.
Schaur, R 2016, Paper presented to the International Solid Waste Association.
Schroepfer, GJ 1951, 'Determination of Fair Sewage Service Chargers for Industrial Wastes', *Sewage and Industrial Wastes*, vol. 23, p 1493.
Schultz, KL & Lipe, RS 1964, 'Relationship Between Substrate Concentration, Growth Rate, and Respiration Rate of Escherichia Coli in Continuous Culture', *Archiv für Mikrobiologie*, vol. 48, pp 1–20.
Schultz, O et al. 2013, 'Forecasting Water Demand for the City of Hamburg', *Colloquium of the EAU&3E Project at ASTEE Congress in Nantes*, 6 June.
Sedlak, R (ed.) 1991. *Phosphorus and Nitrogen Removal from Municipal Wastewater: Principles and Practice*, 2nd ed., The Soap and Detergent Association, New York.
Smith, D 2005, *Reliability and Maintainability and Risk*, 7th ed., Elsevier, Amsterdam, Netherland.
Soderquist, MR 1971, 'Activated Carbon Renovation of Spent Cherry Brine', *Journal Water Pollution Control Federation*, vol. 43, no. 8, pp 1600–1608.
Steiner, HM 1992. *Engineering Economic Principles*, McGraw Hill, New York, NY.
Stumm, W & Lee, GF 1961, 'Oxygenation of Ferrous Iron', *Industrial & Engineering Chemistry*, vol. 53, pp 143–146.
Symister, J 2016, *An Analysis of Capital Cost Estimation Techniques for Chemical Processing*, M.S. Thesis, Florida Institute of Technology, Melbourne, FL.
Tansel, B 2001, 'Spill Prevention Priority Analysis for Reducing Accidental Release Risks during Pipeline Transport', *Journal of Environmental Systems*, vol. 28, pp 319–335.

Tiessier, G 1936, 'Les Lois Quantitatives De La Croissance', *Annales de Physiologie et de Physiochimie Biologique*, vol. 12, pp 527–573.

Thackston, EL 1966, *Longitudinal Mixing and Reaeration in Natural Streams*, Ph.D. Thesis, Vanderbilt Univ., Nashville, TN.

Treadwell & Rollo, 2009. *Probabilistic Cost Estimate for Remediation of Environmental Impacts Pier 70*, San Francisco, CA, August.

Turton, R et al. 2009, *Analysis, Synthesis and Design of Chemical Processes*, Prentice Hall, Upper Saddle River, NJ.

Ulrich, GD & Vasudevan, PT 2006, 'How to Estimate Utility Costs', *Chemical Engineering*, April 26, pp 66–69.

U.S. DOE 2002, *Appleton Papers Plant-wide Energy Assessment Saves Energy and Reduces Waste*, Office of Industrial Technologies, U.S. Dept. of Energy, Washington, DC.

U.S. EPA 1980, *Construction Costs for Municipal Wastewater Treatment Plants 1973–1978*, EPA/430/9-80-003, U.S Environmental Protection Agency, Washington, DC.

U.S. EPA 1981, *Construction Costs for Municipal Wastewater Conveyance 1973–1978*, EPA/430/9-81-003, U.S Environmental Protection Agency, Washington, DC.

U.S. EPA 1989, *Design Manual: Fine Pore Diffuser Systems*, Office of Research and Development, EPA/625/1-89/023, U.S Environmental Protection Agency, Cincinnati, OH.

U.S. EPA 1992, *Total Cost Assessment: Accelerating Industrial Pollution Prevention Through Innovative Project Financial Analysis*, 168 pp., U.S Environmental Protection Agency, Washington, DC.

U.S. EPA 1997, *Guiding Principles for Monte Carlo Analysis*, EPA/630/R-97/001, Risk Assessment Forum, U.S. Environmental Protection Agency, Washington, DC 20460.

U.S. EPA 1998, *Cost Accounting and Budgeting for Improved Wastewater Treatment*, Office of Policy, Planning and Evaluation and Office of Water, Washington, DC.

U.S. EPA 2002a, *EPA Air Pollution Control Cost Manual*, 6th ed., EPA/452/B-02-001, Office of Air Quality Planning and Standards, Research Triangle Park, NC.

U.S. EPA 2002b, *Waste Transfer Stations: A Manual for Decision-Making*, EPA530-R-02-002, Office of Solid Wastes, U.S Environmental Protection Agency, Washington, DC.

U.S. EPA 2009, *Landfill Gas Energy Project Development Handbook*, U.S Environmental Protection Agency, Washington, DC.

U.S. EPA 2010. *Evaluation of Energy Conservation Measures*, EPA 832-R-10-005, U.S Environmental Protection Agency, Washington, DC.

U. S EPA 2016, *Control Strategy Tool (Cost): Cost Equations Documentation*, Office of Air Quality Planning and Standards, U.S Environmental Protection Agency, Washington, Research Triangle Park, NC 27711.

U.S. EPA & U.S Army 2000, *A Guide to Developing and Documenting Cost Estimates During the Feasibility Study*, EPA 540-R-00-002, U.S Environmental Protection Agency, Washington, DC.

Van Haandel, AC & van der Lubbe, JGM 2012, *Handbook of Biological Wastewater Treatment: Design and Optimisation of Activated Sludge Systems*, 2nd ed., IWA Publishing, London.

Vatavuk, WM 2002 'Updating the CE Plant Cost Index', *Chemical Engineering*, January, pp 62–70.

Vik, TE 2003, *Anaerobic Digester Methane to Energy: A Statewide Assessment*, Wisconsin Focus on Energy, Wisconsin Public Service Commission, Madison, WI.

von Rooy, PT et al. 1993, 'Weighing the Risks of Accidental Releases', *Water Environmental and Technology*, vol. 3, no. 4, pp 37–40.

Walski, TM 2012, 'Planning-Level Capital Cost Estimates for Pumping', *Journal of Water Research Planning and Management, ASCE*, vo. 138, pp 307–309.

Walski, TM & Pelliccia, A 1982, 'Economic Analysis of Water Main Breaks', *Journal of American Water Works Association*, 74, pp 149–147.

Walski, TM & Wade, R 1986, *New York City Water Supply Infrastructure Study: Vol. III – The Bronx and Queens, Tech., Report EL 87–9*, U.S. Army Engineer Waterways Experiment Station, Vicksburg, MI.

Weber, WJ & Norton, JW 2008, *Financial and Technological Analysis of Water Treatment Technology Implementation Using Distributed Optimal Technology Network (DOT Net) Concepts*, National Water Research Institute, Fountain Valley, CA.

WEF 2018, *Financing and Charges for Wastewater Systems*, MOP 27, Water Environment Federation, Alexandra, VA.

WEF 2009, *Energy Conservation in Water and Wastewater Facilities, Water Environment Federation Manual of Practice No. 32*, McGraw-Hill, New York.

Weiss, P et al. 2005, *The Cost and Effectiveness of Stormwater Management Practices*, Minnesota Department of Transportation, St. Paul, MN.

22 Appendix A *Staged Construction for Linear Growth*

The cost of each stage will be

$$C = KQ^M$$

The present value of the initial stage, built at time $t = 0$, is

$$PV_0 = KQ^M$$

Discounting from future to present values is done using the exponential terms, exp(ti). This is the equivalent of $1/(1 + i)^n$ where time is continuous instead of discrete.

The second stage will have capacity Q and will be built at $t = Q/r$. The present value of this stage is

$$PV_1 = KQ^M e^{-ti} = KQ^M e^{-iQ/r}$$

The second expansion, or third stage, has a present value

$$PV_2 = KQ^M e^{-2ti} = KQ^M e^{-2iQ/r}$$

The present value for expansions into the future is

$$PV_{\text{Total}} = PV_0 + PV_1 + PV_2 + \cdots$$

$$PV_{\text{Total}} = KQ^M + KQ^M e^{-iQ/r} + KQ^M e^{-2iQ/r} + KQ^M e^{-3iQ/r} + \cdots$$

This can be reduced to (Rudd & Watson 1966)

$$PV_{\text{Total}} = \frac{KQ^M}{1 - e^{-iQ/r}}$$

Define the optimal capacity Q^* as the value that minimizes the total present value. This balances the savings possible from building large plants against the excess interest charges, which result from having large plants idle over part of the design life.

The value of Q^* is determined from the derivative, which must vanish at the minimum. Thus,

$$\frac{d(PV_{\text{Total}})}{dQ} = M\left(1 - e^{-iQ^*/r}\right) - \frac{iQ^*}{r} e^{-iQ^*/r} = 0$$

The graphical solution to this equation was given in Figure 3.10.

23 Appendix B
Economy of Scale Factors

TABLE 23.1 Economy of Scale of Waste Treatment Processes

Process or Equipment	**Size Range**	**Exponent *M***
Refuse disposal processes		
Composting	≤100 ton/day	0.65
Composting	>500 ton/day	0.95
Incineration	<10 ton/day	0.78
Incineration	>10 ton/day	0.90–0.95
Landfill		0.93
Wastewater treatment processes		
Aerated lagoon basin	Volume	0.67–0.71
Activated sludge process	Flow	0.85
Activated sludge, aeration basin	Volume	0.50–0.70
Biofilter	Flow	0.80
Biofilter	Area	0.74
Clarifier, final	Area	0.57–0.76
Clarifier, primary	Area	0.60–0.76
Clarifier, circular tanks	Area	0.66
Clarifier, rectangular tanks	Area	0.56
Detritor	Flow	0.40
Equalization	Flow	0.52
Hydrocyclone degritter	Flow	0.35
Neutralization	Flow	0.70
Oil separation	Flow	0.84
Preliminary treatment	Flow	0.63
Wastewater process equipment		
Aerators, surface (small units)	Total power	0.96
Aerators, surface (large units))	Total power	0.72
Air compressor (125 psig)	Airflow	0.28
Agitated mixing tank reactor	Volume	0.5
Agitators, propeller	Power	0.5
Agitators, turbine	Power	0.3
Aqueous chrome reduction	Volume	0.38
Aqueous cyanide oxidation	Volume	0.38
Centrifuges, solid bowl	Flow	0.73
Clarifier equipment, primary	Area	0.32
Clarifier equipment, final	Area	0.32
Dissolved air flotation	>3 m^3 cell capacity	0.74
Electric generators	kV-amp	0.71
Heat exchangers	Area	0.58
Incinerators	Mass dry solids/h	0.58
Ion exchangers	Resin volume	0.85

(Continued)

TABLE 23.1 Economy of Scale of Waste Treatment Processes

Process or Equipment	Size Range	Exponent *M*
Motors, electric (10–100 hp)	hp	0.80
Pump, centrifugal (with motor)	gpm × psi	0.51
Pump, metering, plunger	Flow	0.59
Pressure filter, plate, and frame	Area	0.58
Screen, vibrating	Area	0.58
Vacuum filter, rotary drum	Area	0.68
Air pollution control equipment		
Cyclones	0.1–1 ton/h	0.75
Scrubbers	Airflow	0.9
Incinerators for combustible gases	Airflow	0.40
Dust collector, cyclone	Airflow	0.8
Dust collector, cloth filter	Airflow	0.68
Dust collector, electrostatic precipitator	Airflow	0.75

24 Appendix C
Factors for Economic Calculations

24.1 CONVERT A UNIFORM ANNUAL COST (*A*) TO A PRESENT VALUE (*PV*)

Convert a uniform annual cost *A* to a present value *PV* (Table 24.1):

$$PV_{n,i} = \frac{A}{(1+i)^1} + \frac{A}{(1+i)^2} + \cdots + \frac{A}{(1+i)^n} = A\left(\frac{1}{(1+i)^1} + \frac{1}{(1+i)^2} + \cdots + \frac{1}{(1+i)^n}\right)$$

TABLE 24.1 Factors to Convert a Series of Uniform Annual Costs *A* to a Present Value *PV*

	Interest Rate							
Year	**0.03**	**0.04**	**0.05**	**0.06**	**0.07**	**0.08**	**0.09**	**0.1**
1	0.97087	0.96154	0.95238	0.94340	0.93458	0.92593	0.91743	0.90909
2	1.91347	1.88609	1.85941	1.83339	1.80802	1.78326	1.75911	1.73554
3	2.82861	2.77509	2.72325	2.67301	2.62432	2.57710	2.53129	2.48685
4	3.71710	3.62990	3.54595	3.46511	3.38721	3.31213	3.23972	3.16987
5	4.57971	4.45182	4.32948	4.21236	4.10020	3.99271	3.88965	3.79079
6	5.41719	5.24214	5.07569	4.91732	4.76654	4.62288	4.48592	4.35526
7	6.23028	6.00205	5.78637	5.58238	5.38929	5.20637	5.03295	4.86842
8	7.01969	6.73274	6.46321	6.20979	5.97130	5.74664	5.53482	5.33493
9	7.78611	7.43533	7.10782	6.80169	6.51523	6.24689	5.99525	5.75902
10	8.53020	8.11090	7.72173	7.36009	7.02358	6.71008	6.41766	6.14457
11	9.25262	8.76048	8.30641	7.88687	7.49867	7.13896	6.80519	6.49506
12	9.95400	9.38507	8.86325	8.38384	7.94269	7.53608	7.16073	6.81369
13	10.63496	9.98565	9.39357	8.85268	8.35765	7.90378	7.48690	7.10336
14	11.29607	10.56312	9.89864	9.29498	8.74547	8.24424	7.78615	7.36669
15	11.93794	11.11839	10.37966	9.71225	9.10791	8.55948	8.06069	7.60608
16	12.56110	11.65230	10.83777	10.10590	9.44665	8.85137	8.31256	7.82371
17	13.16612	12.16567	11.27407	10.47726	9.76322	9.12164	8.54363	8.02155
18	13.75351	12.65930	11.68959	10.82760	10.05909	9.37189	8.75563	8.20141
19	14.32380	13.13394	12.08532	11.15812	10.33560	9.60360	8.95011	8.36492
20	14.87747	13.59033	12.46221	11.46992	10.59401	9.81815	9.12855	8.51356
21	15.41502	14.02916	12.82115	11.76408	10.83553	10.01680	9.29224	8.64869
22	15.93692	14.45112	13.16300	12.04158	11.06124	10.20074	9.44243	8.77154
23	16.44361	14.85684	13.48857	12.30338	11.27219	10.37106	9.58021	8.88322
24	16.93554	15.24696	13.79864	12.55036	11.46933	10.52876	9.70661	8.98474
25	17.41315	15.62208	14.09394	12.78336	11.65358	10.67478	9.82258	9.07704
26	17.87684	15.98277	14.37519	13.00317	11.82578	10.80998	9.92897	9.16095
27	18.32703	16.32959	14.64303	13.21053	11.98671	10.93516	10.02658	9.23722
28	18.76411	16.66306	14.89813	13.40616	12.13711	11.05108	10.11613	9.30657
29	19.18845	16.98371	15.14107	13.59072	12.27767	11.15841	10.19828	9.36961
30	19.60044	17.29203	15.37245	13.76483	12.40904	11.25778	10.27365	9.42691

The summation of the $1/(1 + i)^j$ – terms is the present value factor:

$$\text{Present value factor} = F_{AP,\, n,\, i} = \frac{(1+i)^n - 1}{i(1+i)^n}$$

24.2 CONVERT A PRESENT VALUE (*PV*) TO A UNIFORM ANNUAL COST (*A*)

The factor to convert a present value (e.g., a capital cost or the amount of a loan) to a series of uniform annual costs is called the capital recovery factor. It is the inverse of the factors in Table 24.1.

$$\text{Capital recovery factor} = \text{CRF}_{n,\, i} = F_{PA,\, n,\, i} = \frac{i(1+i)^n}{(1+i)^n - 1}$$

$$A = (\text{CRF}_{n,\, i})(PV) = (F_{RP,\, n,\, i})(PV)$$

TABLE 24.2 Factors to Convert a Present Value *PV* to a Series of Uniform Annual Costs *A*

	Interest Rate							
Year	**0.03**	**0.04**	**0.05**	**0.06**	**0.07**	**0.08**	**0.09**	**0.1**
1	1.03000	1.04000	1.05000	1.06000	1.07000	1.08000	1.09000	1.10000
2	0.52261	0.53020	0.53780	0.54544	0.55309	0.56077	0.56847	0.57619
3	0.35353	0.36035	0.36721	0.37411	0.38105	0.38803	0.39505	0.40211
4	0.26903	0.27549	0.28201	0.28859	0.29523	0.30192	0.30867	0.31547
5	0.21835	0.22463	0.23097	0.23740	0.24389	0.25046	0.25709	0.26380
6	0.18460	0.19076	0.19702	0.20336	0.20980	0.21632	0.22292	0.22961
7	0.16051	0.16661	0.17282	0.17914	0.18555	0.19207	0.19869	0.20541
8	0.14246	0.14853	0.15472	0.16104	0.16747	0.17401	0.18067	0.18744
9	0.12843	0.13449	0.14069	0.14702	0.15349	0.16008	0.16680	0.17364
10	0.11723	0.12329	0.12950	0.13587	0.14238	0.14903	0.15582	0.16275
11	0.10808	0.11415	0.12039	0.12679	0.13336	0.14008	0.14695	0.15396
12	0.10046	0.10655	0.11283	0.11928	0.12590	0.13270	0.13965	0.14676
13	0.09403	0.10014	0.10646	0.11296	0.11965	0.12652	0.13357	0.14078
14	0.08853	0.09467	0.10102	0.10758	0.11434	0.12130	0.12843	0.13575
15	0.08377	0.08994	0.09634	0.10296	0.10979	0.11683	0.12406	0.13147
16	0.07961	0.08582	0.09227	0.09895	0.10586	0.11298	0.12030	0.12782
17	0.07595	0.08220	0.08870	0.09544	0.10243	0.10963	0.11705	0.12466
18	0.07271	0.07899	0.08555	0.09236	0.09941	0.10670	0.11421	0.12193
19	0.06981	0.07614	0.08275	0.08962	0.09675	0.10413	0.11173	0.11955
20	0.06722	0.07358	0.08024	0.08718	0.09439	0.10185	0.10955	0.11746
21	0.06487	0.07128	0.07800	0.08500	0.09229	0.09983	0.10762	0.11562
22	0.06275	0.06920	0.07597	0.08305	0.09041	0.09803	0.10590	0.11401
23	0.06081	0.06731	0.07414	0.08128	0.08871	0.09642	0.10438	0.11257
24	0.05905	0.06559	0.07247	0.07968	0.08719	0.09498	0.10302	0.11130
25	0.05743	0.06401	0.07095	0.07823	0.08581	0.09368	0.10181	0.11017
26	0.05594	0.06257	0.06956	0.07690	0.08456	0.09251	0.10072	0.10916
27	0.05456	0.06124	0.06829	0.07570	0.08343	0.09145	0.09973	0.10826
28	0.05329	0.06001	0.06712	0.07459	0.08239	0.09049	0.09885	0.10745
29	0.05211	0.05888	0.06605	0.07358	0.08145	0.08962	0.09806	0.10673
30	0.05102	0.05783	0.06505	0.07265	0.08059	0.08883	0.09734	0.10608

25 Appendix D *Interpreting the LINGO Solution Report*

The LINGO output provides more than the value of the objective function and variables. This additional information is not essential for the purpose of learning how to formulate and solve linear programming problems. It is useful for users who deal with larger problems.

25.1 THE LINGO REPORT FOR EXAMPLE 14.1

LINGO gives each equation a *row* number. Comment lines are not counted as rows. The program with comment lines omitted is

```
1 MAX = 1*A + 0.8*B + 1.3*C;
2 A >= 5000;
3 B >= 3000;
4 2*A + 1.5*B + 2.6*C <= 50000;
5 1*A + 1.2*B + 0.9*C <=9000;
6 END
```

The LINGO solution report for Example 14.1 is reproduced below. Table 25.1 is the raw LINGO solution report for Example 14.1. The raw report is rearranged in Table 25.2 to show the equations. The *Reduced Cost*, *Slack* or *Surplus*, and *Dual Price* provide some sensitivity analysis. A brief explanation follows.

Reduced Cost. There are two valid, equivalent interpretations of a reduced cost. First, a variable's reduced cost is the amount by which the objective coefficient of the variable would have to improve before it would become profitable to give the variable in question a positive value in the optimal solution. For example, if a variable had a reduced cost of 10, the objective coefficient of that variable would have to increase by 10 units in a maximization problem and/or decrease by 10 units in a minimization problem for the variable to become an attractive alternative to enter into the solution. A variable in the optimal solution automatically has a reduced cost of zero.

The second interpretation is that the reduced cost of a variable is the amount of penalty you would have to pay to introduce one unit of that variable into the solution. Again, if you have a variable with a reduced cost of 10, you would have to pay a penalty of 10 units to introduce the variable into the solution. In other words, the objective value would fall by 10 units in a maximization model or increase by 10 units in a minimization model.

Slack or Surplus. The Slack or Surplus column tells you how close you are to satisfying a constraint as an equality. This quantity, on less-than-or-equal-to (≤) constraints, is generally referred to as slack. On greater-than-or-equal-to (≥) constraints, this quantity is called a surplus.

If a constraint is exactly satisfied as an equality, the slack or surplus value will be zero. If a constraint is violated, as in an infeasible solution, the slack or surplus value will be negative. Knowing this can help you find the violated constraints in an infeasible model – a model for which there doesn't exist a set of variable values that simultaneously satisfies all constraints. Non-binding constraints will have positive, non-zero values in this column.

TABLE 25.1 Raw Report from LINGO For Example 14.1

Variable	Value	Reduced Cost
A	5000.00	0.000
B	3000.00	0.000
C	444.44	0.000
Row	**Slack/Surplus**	**Dual Price**
1	7977.78	1.000
2	0.00	–0.444
3	0.00	–0.933
4	34,344.44	0.000
5	0.00	1.444

TABLE 25.2 Summary of the LINGO Output Report for Example 14.1

Solution			Sensitivity Analysis Information			
Variable	Value	Reduced Cost	Row	Equation	Slack or Surplus	Dual Price
A	5,000	0	1	Max = 1*A + 0.8*B + 1.3*C	7977.778	1.000
B	3,000	0	2	A ≥ 5000	0	–0.444
C	444.44	0	3	B ≥ 3000	0	–0.933
			4	2*A + 1.5*B + 2.6*C ≤ 50,000	34,344.44	0
			5	1*A + 1.2*B + 0.9*C ≤ 9,000	0	1.444

Dual Price. The dual price is the amount that the objective would improve as the right-hand side, or constant term, of the constraint is increased by one unit. In a maximization problem, improve means the objective value would increase. In a minimization problem, the objective value would decrease if you were to increase the right-hand side of a constraint with a positive dual price.

Dual prices are sometimes called shadow prices because they tell you how much you should be willing to pay for additional units of a resource.

Table 25.3 is the LINGO report (slightly rearranged) for Example 14.2 solid waste reclamation problem.

TABLE 25.3 LINGO Solution Report for Example 14.2

Solution			Sensitivity Analysis		
Variable	Value	Reduced Cost	Row	Slack/Surplus	Dual Price
REVENUE	46500	0	1	4500	1.0
PRETREAT	42000	0	2	0	1.0
A	3667	0	3	0	−1.0
B	2000	0	4	0	5.5
C	4333	0	5	0	5.0
M1	3000	0	6	0	4.0
M2	2000	0	7	0	−4.0
M3	4000	0	8	0	−6.5
M4	1000	0	9	0	−4.0
M1A	0	0	10	0	−6.5
M2A	1467	0	11	0	1.0
M3A	1467	0	12	0	0.5
M4A	733	0	13	0	0.0
M1B	1000	0	14	0	1.5
M2B	533	0	15	1667	0.0
M3B	200	0	16	0	−0.5
M4B	267	0	17	667	0.0
M1C	2000	0	18	1100	0.0
M2C	0	2.5	19	0	−2.5
M3C	2333	0	20	367	0.0
M4C	0	2.5	21	0	−2.5
			22	0	2.5
			23	333	0.0
			24	0	2.5
			25	1033	0.0

26 Appendix E
Generating Random Numbers in Excel

This appendix explains how to use Excel to generate random numbers for Monte Carlo simulations.

26.1 UNIFORM DISTRIBUTION

$R_U(0, 1)$ is a random number from a standard uniform distribution having a lower bound of 0 and upper bound of 1.

$R_U(0, 1)$ can be generated using the RAND() function.

A uniformly distributed random value $Y_U(a, b)$, having lower bound a and upper bound b can be obtained from

$$Y_U(a,b) = (b-a)\mathrm{R_U}(0,1) + a$$

For example,

$$Y_U(a,b) = \mathrm{RAND()*(10 - 5) + 5}$$

will generate uniform random numbers between 5 and 10.

26.2 NORMAL DISTRIBUTION

$R_N(0, 1)$ is a random number from a standard normal distribution having a mean of 0 and a standard deviation of 1 ($\mu = 0, \sigma = 1$).

$R_N(0, 1)$ can be generated using the NORMINV(RAND(), 0, 1) function.

$R_N(0, 1)$ can be scaled to give normally distributed random numbers with mean (μ) and standard deviation (σ) using:

$$Y_N(\mu,\sigma) = \sigma R_N(0,1) + \mu$$

$R_N(\mu, \sigma)$ can be generated using NORMINV(RAND(), μ, σ)

For example,

NORMINV(RAND(), 5, 2)

generates normally distributed random numbers with mean 5 and standard deviation 2.

26.3 LOG NORMAL DISTRIBUTION

A random variable x is log-normally distributed if ln(x) has a normal distribution. (Usually natural logarithms are used, but $\log_{10}(x)$ also will give a normal distribution.)

$R_{LN}(\mu, \sigma)$ is a random number having a log normal distribution where μ and σ are the mean and standard deviations of the log-transformed values.

$R_{LN}(\mu, \sigma)$ can be generated using exp[NORMINV(RAND(), μ, σ)], The "exp" converts the normally distributed random variable to the metric of the random variable x.

For example, log-normally distributed random numbers are generated with

exp[NORMINV(RAND(), 2, 1)]

where the mean of the logarithms = 2 and standard deviation of the logarithms = 1.

Note: This method requires knowledge or estimates of the mean and standard deviation of the log-transformed variable. In many situations, we often know (or have estimates of) the values of the mean and standard deviation of the log normal random variable x (for simplicity, let's call them α and β), and need to estimate μ and σ, the mean and standard deviation of the log-transformed variable, in order to generate random numbers as described above. The following transformation equations accomplish this.

$$\sigma = \sqrt{\ln(1+(\beta/\alpha)^2)} \qquad \mu = \ln(\alpha) - \frac{\sigma^2}{2}$$

where

α = mean of x
β = standard deviation of x
μ = mean of $\ln(x)$
σ = standard deviation of $\ln(x)$

For example, a log normal variable x, having mean $\alpha = 5$ and standard deviation $\beta = 4$, the mean μ, and standard deviation σ of the normally distributed logarithms are

$$\sigma = \sqrt{\ln(1+(4/5)^2)} = \sqrt{\ln(1.64)} = \sqrt{0.495} = 0.703$$

$$\mu = \ln(5) - \frac{0.703^2}{2} = 1.609 - 0.247 = 1.362$$

26.4 TRIANGULAR DISTRIBUTION

Random numbers having a triangular distribution $R_T(a, b, c)$ with lower bound a, upper bound b, and peak value c can be generated from standard uniformly distributed random numbers $R_U(0, 1)$ using

$$R_T(a,b,c) = \begin{cases} a + \sqrt{R_U(0,1)(b-a)(c-a)} & \text{for} \quad 0 < R_U(0,1) < F(c) \\ b - \sqrt{(1-R_U(0,1))(b-a)(b-c)} & \text{for} \quad F(c) < R_U(0,1) < 1 \end{cases}$$

where $F(c)$ is the value of the cumulative distribution function at the peak value c:

$$F(c) = \frac{c-a}{b-a}$$

Using the RAND() function in Excel gives

$$R_T(a,b,c) = \begin{cases} a + \sqrt{\text{RAND()}(b-a)(c-a)} & \text{for} \quad 0 < \text{RAND()} < F(c) \\ b - \sqrt{(1-\text{RAND()})(b-a)(b-c)} & \text{for} \quad F(c) < \text{RAND()} < 1 \end{cases}$$

For example, a triangular distribution with $a = 2$, $b = 12$, and $c = 9$, with RAND() = 0.39

$$F(c) = (9-2)/(12-2) = 7/10 = 0.7$$

Since 0 < RAND() < $F(c)$

$$R_T(2,12,9) = 2 + \sqrt{(0.39)(12-2)(9-2)} = \sqrt{27.3} = 5.22$$

26.5 EXPONENTIAL DISTRIBUTION

Random numbers having an exponential distribution $R_E(\lambda)$ with parameter λ can be generated from standard uniformly distributed random numbers, $R_U(0, 1)$ using

$$R_E(\lambda) = -\frac{1}{\lambda}\ln(1 - R_U(0,1))$$

where λ = the average rate at which events (arrivals, failures, etc.) occur.

The RAND() function is used to generate

$$R_E(\lambda) = -\frac{1}{\lambda}\ln(1 - \text{RAND()})$$

For example, for λ = 4 arrivals/h and RAND() = 0.68, the time to the next arrival is

$$R_E(4) = -\frac{1}{4/\text{h}}\ln(1-0.68) = 0.285\text{ h}$$

Note: If the parameter λ is defined as the mean time between events, then

$$R_E(\lambda) = -\lambda\ln(1 - \text{RAND()})$$

For example, for λ = 0.25 h/arrival and RAND() = 0.68, the time to the next arrival is

$$R_E(4) = -(0.25\text{h})\ln(1-0.68) = 0.285\text{h}$$

Index

For Product Safety Concerns and Information please contact our EU
representative GPSR@taylorandfrancis.com
Taylor & Francis Verlag GmbH, Kaufingerstraße 24, 80331 München, Germany

www.ingramcontent.com/pod-product-compliance
Ingram Content Group UK Ltd.
Pitfield, Milton Keynes, MK11 3LW, UK
UKHW050057200926
475702UK00019B/2878
* 9 7 8 0 3 6 7 7 2 4 1 5 3 *